MATERIALS PROCESSING IN SPACE

Theory, Experiments, and Technology

VOLUME 1

MATERIALS PROCESSING IN SPACE

Theory, Experiments, and Technology

VOLUME 1

L. L. Regel'

Institute of Space Research
Academy of Sciences of the USSR
Moscow, USSR

Edited by
Academician R. Z. Sagdeev

Translated from Russian by

J. E. S. Bradley

CONSULTANTS BUREAU
NEW YORK AND LONDON

Library of Congress Cataloging in Publication Data

Regel', L. L.
  [Kosmicheskoe materialovedenie. English]
  Materials processing in space: theory, experiments, and technology / L. L.
Regel'; edited by R. Z. Sagdeev; translated from Russian by J.E.S. Bradley.
      p.   cm.
  Translation of: Kosmicheskoe materialovedenie.
  Bibliography: p.
  ISBN 0-306-11026-1
  1. Materials—Effect of space environment on. 2. Space industrialization. I.
Sagdeev, R. Z. II. Title.
TA418.59.R4413   1989   89-7222
620'.419—dc20   CIP

ISBN-13: 978-1-4684-1685-5   e-ISBN-13: 978-1-4684-1683-1
DOI: 10.1007/978-1-4684-1683-1

This translation of Kosmicheskoe Materialovedenie, Chast' II, published in
Moscow in 1987 by VINITI as Volume 29 in Itogi Nauki i Tekhnika, Seriya
Issledovanie Kosmicheskogo Prostranstva, is published under an agreement
with VAAP, the Copyright Agency of the USSR.

© 1990 Consultants Bureau, New York
Softcover reprint of the hardcover 1st edition 1990
A Division of Plenum Publishing Corporation
233 Spring Street, New York, N.Y. 10013

PREFACE

There has been considerable interest recently in microgravity physics and the effects of gravitation on crystal growth, alloy solidification, and other processes in space manufacturing. Regel' [1] has provided an extensive but not exhaustive bibliography on microgravity physics and materials science in space, in which the major aspects are discussed along with the state of the art and future research prospects. The literature survey in [1] covered a period of about 10 years, including some publications appearing in 1983 that reflected not only theoretical and experimental studies completed by 1983 but also a list of experiments to be carried out in the next few years. In particular, the closing part of the survey [1] enumerated experiments planned under the Intercosmos program and by the European Space Agency (ESA) for the flight of Spacelab-1 and D-1 in 1985 and under the Eureka programs. Some of the space experiments planned in 1983 have now been completed, and the results have been published. It is therefore desirable to survey again research on materials science in space for the last few years and extend the literature survey begun in [1].

The literature listing on materials science in space begun in [1] is supplemented (there were 1061 citations in [1]) by recent publications (beginning with 1982). The tasks are to deal with the current state of research in microgravity physics and materials science in space – as well as the suggestions made by specialists in this area as regards future research and the scope for organizing space manufacturing. As in [1], the purpose has been to list the publications, including the titles, which enables one to judge the main topics covered by this research. However, the text also emphasizes the most important results obtained in experiments and in certain theoretical studies.

The bibliography includes almost the same parts as in [1]:

1. Part I lists books and collections of papers on materials science in space, including the proceedings of international conferences and certain national conferences held after 1982 [1–23].

2. Part II contains reviews, as well as publications that summarize research previously carried out under international or national programs, and plans for future research [24–113].

3. Part III contains theoretical papers on materials science in space, viz., hydromechanics and heat and mass transfer in microgravity, the theory of gravitational convection and thermocapillary convection (Marangoni convection), impurity segregation theory, gas-bubble formation, and bubble and particle migration in liquids. As in [1], this part includes certain experimental work on the physics of liquids closely related to the theories. Here it should be noted that it is sometimes difficult to distinguish between theoretical and experimental research on the physics of liquids and various other topics, which is reflected in the papers cited in Part III [114–316].

4. Part IV consists of papers on experiments on semiconductor crystal growth from liquids [317–395], while Part V deals with crystal growth from the vapor [396–411].

5. Part VI deals with research on solidification of metals, composites, and eutectics under microgravity [412–513].

6. Part VII deals with glasses [514–524], and Part VIII with crystal growth from aqueous solution [525–533].

7. Part IX deals with new equipment used under microgravity in research on the physics of liquids and the growth of crystals, as well as with methods used under microgravity, such as levitation and skin technology [534–661].

8. Part X covers some experiments (far from all) performed recently with brief weightlessness [662–688].

9. Part XI lists new publications not considered in [1] on crystal growth and metal solidification under elevated gravity in centrifuges [689–693].

10. Part XII deals with microgravity electrophoresis [694–713].

Finally, Part XIII lists papers not fitting into these parts, including microgravity combustion, biological processes, polymer separation and synthesis, etc. [714–737].

It is worth repeating [1] that the distribution between parts is not necessarily clearcut. For example, it is difficult to distinguish experiments under brief weightlessness conditions from general experiments under microgravity.

All the same, the classification characterizes the distribution to some extent, and one can compare the total number of papers on microgravity physics and materials science in space for the last 3–4 years with the number for the 10 years in [1], which shows that the topics have continued to be of considerable interest.

Each chapter in this book deals with the conclusions on the papers in the listing, especially the novel features distinguishing them from the papers considered in [1], where it has to be recognized that many authors in recent years have made reference to previous experiments and in part reproduce conclusions from earlier papers.

The chapter titles and section headings show that the new investigations have been surveyed in somewhat the same order as the bibliography, viz., in the same topic parts, although there are some deviations: for example, papers from Parts IV and V are dealt with in Chapter 2, with Part V being covered in Section 2.4.

The analysis here differs somewhat from that of [1], where the contents of many papers were surveyed in tabular form. As the number of papers cited is only 737 instead of 1061 in [1], that form has not been used, and the individual papers are characterized as far as possible, even though briefly, without resort to tables.

The conclusions at the end of each chapter deal not only with the review and other publications not discussed in the chapter, but also with research results and development prospects.

Chapter 10, *Research in Materials Science in Space*, characterizes the state and development prospects in this area on the basis of recent forecasts and from my own surveys.

I am indebted to Academician R. Z. Sagdeev, the Director of the Space Research Institute, for editing this book, to Academicians A. P. Aleksandrov, Ya. B. Zel'dovich, and Yu. A. Osip'yan for their support in this undertaking, and to Professor V. R. Regel' for advice on the preparation of the manuscript.

CONTENTS

Chapter 1

THEORY OF LIQUID BEHAVIOR AND CRYSTAL GROWTH UNDER REDUCED GRAVITY

There continue to be many papers on crystal growth and metal and glass solidification theory as well as on liquid behavior under microgravity, which have been included mainly in Part III of the bibliography [114–316]. Theoretical aspects of materials science in space are also covered in two books [1, 14], several reviews [35, 46, 51, 52, 84, 109], and in various papers in the other sections of the bibliography. As in [1], Part III includes both experimental and purely theoretical studies on the physics of liquids . In the sections below, the theoretical studies have been divided among several topics, each of which give brief characteristics of the state of research.

1.1. FLUID MECHANICS AND HEAT AND MASS TRANSFER IN MICROGRAVITY [14, 35, 46, 51, 52, 84, 109, 121, 123, 154, 160, 179, 190, 208, 210, 300–302, 307, 312]

The book by Avduyevsky et al. [14] deals, on the whole, with the scientific principles of space manufacturing, where Chapter 4 is devoted to fluid mechanics, which deals with microgravity heat and mass transfer models. It is impossible to present the content of that book in detail here, and I merely enumerate briefly the topics covered there. A classification and general characterization of liquid motion under microgravity is followed by a discussion of the basic equations of motion and heat and mass transfer, the Boussinesq approximation, equations in terms of vortices and stream functions, dimensionless groups, and the orders of magnitude of these. There is then a discussion of simulation methods and programs, followed by theoretical results on gravitational and nongravitational convection and on the interactions between these forms of motion. Some aspects of heat transfer under rotation are also considered. Finally, a model is considered for heat and mass transfer during directional solidification, which is directly related to microgravity crystal growth, viz., heat transfer through tube walls and solidification-front movement.

TABLE 1. Dimensionless Parameters and Similarity Criteria

Name	Symbol	Expression in dimensional quantities	Physical significance
Reynolds number	Re	$\Omega L^2/\nu$	Ratio of inertial forces to viscous forces (Ω is angular velocity)
Grashof number	Gr	$(g\beta^T L^3/\nu^2)\Delta T$	Ratio of buoyancy forces in a nonisothermal homogeneous liquid to frictional forces, $\beta_T = 1/\rho \cdot \partial S/\partial T$ bulk thermal-expansion coefficient
Modified Grashof number	Gr*	$(g\beta_T L^4/\nu^2\lambda)q$	Analogous, but temperature scale $\Delta T = qL/\lambda$
Prandtl number	Pr	ν/a	Ratio of viscosity to thermal diffusivity a
Rayleigh number	Ra	$(g\beta_T L^3/\nu a)\Delta T$	Product GrPr
Diffusion Grashof number	Gr_D	$(g\beta_c L^3/\nu^2)\Delta C$	Ratio of buoyancy forces due to composition change in isothermal medium to frictional forces
Schmidt number	Sc	ν/D	Ratio between viscosity ν and diffusion coefficient D
Marangoni number	Mg	$(\sigma\beta_\sigma T L/\rho\nu^2)\Delta T$	Ratio of gradient in surface-tension forces σ due to temperature difference ΔT to viscous forces ($\beta_{\sigma T} = \partial\sigma/\partial T$)
Modified Marangoni number	Mg*	$(\sigma\beta_\sigma T L^2/\rho\nu^2\lambda)q$	Similar, but temperature scale $\Delta T = qL/\lambda$
Diffusion Marangoni number	Mg_Δ	$(\sigma\beta_{\sigma c} L/\rho\nu^2)\Delta C$	Ratio of gradient in surface-tension forces σ due to concentration difference ΔC to viscous forces $\beta_{\sigma c} = \partial\sigma/\partial C$
Biot number	Bi	$\alpha L/\lambda$	Dimensionless heat-transfer coefficient at outer surface
Nusselt number	Nu	$qL/\lambda\Delta T$	Dimensionless heat-transfer coefficient

Table 1 (cont.)

Name	Symbol	Expression in dimensional quantities	Physical significance
Fourier number	Fu	$\alpha t/L^2$	Dimensionless time
Equilibrium distribution coefficient	K_0	C_s/C_L	Ratio between impurity concentration in solid C_s and that in liquid C_L at crystallization front
Diffusion number	Bi_D	$(K_0 - 1/D)V_f R$	Dimensionless mass-transfer coefficient at crystallization front: D, diffusion-layer thickness; V_f, crystallization rate
Bond number	Bo	$\rho g L^2/\sigma$	Ratio between gravitational force and capillary forces
Weber number	We	$\rho\Omega^2 L^3/2\sigma$	Ratio of centrifugal force to capillary force (Ω is angular velocity)

An extensive review deals with fluid mechanics and with heat and mass transfer in crystal growth [51], and there is also a detailed survey on microgravity convection and heat-transfer research [52]. These two reviews have many features in common with what has been said above [14]. As in [14], there are treatments [51, 52] of general topics in fluid mechanics designed mainly to provide rational control for crystal growth, since only a proper understanding of hydrodynamic phenomena and of heat and mass transfer enables one to relate growth conditions to properties and structure and allows one to judge the elementary steps in growth as well as other aspects of crystal growth physics.

An attempt has been made to divide the publications in Parts II and III into particular topics, especially on the basis of the classification used for hydrodynamic processes and heat and mass transfer in crystal growth.

I used only the general classification of [51] in schematic form (Fig. 1) together with the classification of thermal-concentration conditions given in [51] as well as in [14] and [52] (Fig. 2). Most theoretical papers contain the nonstationary Navier-Stokes equations (the Boussinesq approximation) in dimensionless form [14, 51, 52], which are not given here. Instead, I merely give a list of the dimensionless parameters used in fluid mechanics and in physical simulation. This list is given in [51] as a table (Table 1), which is particularly useful for the reader wishing to become acquainted with the area and is convenient for reference in the subsequent text. In [1] I have already given information on some dimensionless parameters used in fluid mechanics, but Table 1 characterizes them more fully.

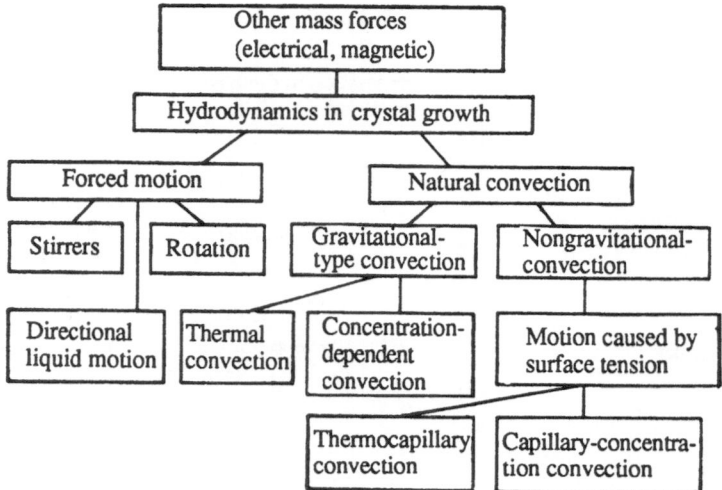

Fig. 1. Classification of hydromechanical and heat- and mass-transfer processes in crystal growth [51].

1.2. EFFECTS OF GRAVITATIONAL CONVECTION ON CRYSTAL GROWTH IN MICROGRAVITY [14, 51, 52, 114, 119, 130, 147–150, 155, 190, 244, 254, 272, 292, 310, 316]

Gravitational convection is much reduced in microgravity, but there are still substantial contributions to liquid mixing during crystallization and to impurity segregation under the reduced gravity found in space vehicles, where g/g_0 is 10^{-5}–10^{-2}. One therefore needs a theoretical discussion of the interactions of gravitational and nongravitational types of motion [14, 51, 52, 254, 272]. Some treatments are concerned with gravitational convection under ground conditions with various mutual orientations between the gravitational forces and the temperature gradient, or with estimating the effects of accelerations on microscopic inhomogeneities during crystallization [14, 51, 52, 155].

Gravitational convection has been considered for crystal growth in model experiments [114, 119] for growth from liquids [244] and in epitaxial film deposition [147]. Effects from tube walls and cell shapes can be eliminated [119, 292] from the crystallization of spheres out of contact with tube walls and the simulation of convection under microgravity [130, 149, 150, 316].

Figure 2 illustrates the importance of thermal-concentration convection arising from gravitational forces, as well as the complexities [52].

Here I give only a brief list of papers on this because the effects of g perturbations are discussed in Section 1.10.

Although it is necessary to consider gravitational convection for microgravity conditions, as is evident from the papers cited above (as well as those in Section 1.10 below), most of the papers from Parts II and III in the bibliography relate to nongravitational convection caused by surface tension.

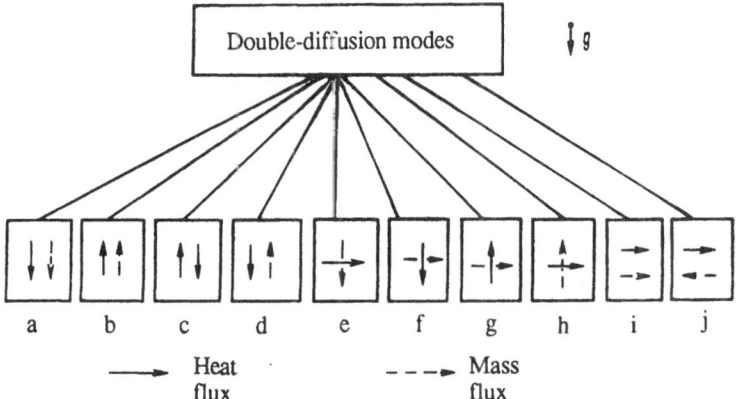

Fig. 2. Classification of thermal-concentration state (double diffusion): a) absolutely stable stratification; b) unstable equilibrium as regards temperature and impurity concentration; c) equilibrium unstable with respect to temperature but stable with respect to concentration; d) equilibrium stable with respect to temperature but unstable with respect to concentration; e) concentration-stable equilibrium with a horizontal temperature difference; f) temperature-stable equilibrium with a horizontal concentration difference; g) temperature-unstable equilibrium with a horizontal concentration difference; h) concentration-unstable equilibrium with a horizontal temperature difference; i) horizontal heat and concentration fluxes (in the same direction); j) horizontal heat and concentration fluxes (in opposite directions) [51].

1.3. THERMOCAPILLARY AND CAPILLARY-CONCENTRATION CONVECTION (MARANGONI CONVECTION) [14, 51, 52, 118, 120, 127, 131–133, 143, 170, 171, 176, 177, 185, 191, 192, 200, 202, 206, 207, 211, 221, 224, 234–238, 243, 245–248, 252, 254, 258, 261–272, 275–286, 291, 294–299, 306]

Marangoni convection in microgravity is of interest because it is particularly convenient to examine any nongravitational convection when the gravitational contribution is reduced, including convection caused by surface-tension gradients, especially since Marangoni convection becomes very prominent in crystal growth under microgravity, where it is sometimes decisive. A list of papers on this topic is to be found in [1].

Research in this area continues to develop rapidly. Various theoretical problems have been solved, and model and other experiments have been performed.

Without going into details, one can list the main topics here.

Marangoni convection has been examined theoretically for various conditions: layers heated from below [132] or from the side [170, 172], including thin films with temperature distributions [176, 207, 248, 261], two-layer systems [191, 192], the planar stationary case with a free boundary [120, 143], axially symmetric boundary Marangoni layers, square or rectangular cavities [131, 170, 202, 211], during crystallization by drawing from the melt [133], in crucible-free zone melting with liquid bridges and columns [206, 212, 234, 236, 238, 245, 275, 277, 280, 281, 283, 284], and in gas–liquid mixtures [177, 203, 206]. Studies have also been made on thermocapillary stability, stability effects related to free-boundary deformation at external pressure [118], and stationary shapes taken by nonisothermal liquid layers. The dynamic contact angle problem has also been solved [120], and conditions have been defined for instability onset and oscillating Marangoni convection, viz., the transition from laminar to turbulent motion [246, 261, 266, 282, 291, 294, 296–298, 306], which has included the determination of critical Marangoni numbers for such transitions [296–298]. Some papers deal with the interaction between the fluxes due to gravitational and thermocapillary forms of convection [159, 176, 200, 279, 282] and fluxes caused by rotation [185]. Some of these papers and [221] deal with the effects of convection on impurity distributions and layered structures due to Marangoni convection. Some specific cases of thermocapillary and capillary-concentration convection have been examined, such as in sinusoidal heat supply [171], in systems with temperature-dependent surface tension, or near the surface tension minimum [267, 268, 270, 289, 306], as well as thermocapillary convection in liquid metals [243]. There have also been studies on bubble or droplet drift, the latter of one liquid in another, on account of Marangoni convection [127, 141, 142, 247, 250] and the resulting phase separation [262–264]. There have also been many experiments on microgravity Marangoni convection, and sometimes also for ground conditions. Examples are [127, 131, 170–172, 207, 224, 275, 280, 283, 286, 296–298], and certain others of those cited above. Convection currents can be displayed in model experiments by the use of aluminum tracing particles as described by Schwabe et al. in papers cited in [1].

1.4. EXTERNAL-FORCE CONTROL OF LIQUID-SURFACE STABILITY   [14, 35, 51, 52, 129, 134–140, 144, 151–153, 161, 163, 166–168, 209, 210, 217, 220, 230, 259, 260]

Many papers in Part III in the bibliography deal with the effects on hydrodynamics from mass forces other than gravitational (electrical and magnetic), which can be used, along with induced motion arising from vibration of electroconvection, for microgravity stability control.

If the liquid has an inhomogeneous distribution for the electrical or magnetic parameters, an electrical or magnetic field will give rise to mass forces (Fig. 1) that resemble gravitational or Marangoni forces in moving or stabilizing the liquid. They have therefore been considered as means of controlling liquids on the ground and under microgravity [52]. Practical requirements imply that one has to research microgravity stability with time-varying inputs, which may have destabilizing or stabilizing effects. Destabilization is useful in mixing and in making emulsions and suspensions, while stabilization is useful for preventing surface changes due to instabilities caused by space-vehicle accelerations. In [52] there is a survey of suppressing convection in liquids during crystal growth. The papers cited above also indicate interest in this area. For example, in [35] and [129] one finds the physical principles of control without feedback applied to a free-surface stability or a boundary between liquids, where alternating fields may be applied in any orientations to liquids with any properties. Equations have been discussed for the perturbations and stability conditions, which have been tested with alternating electric fields, while there have been studies on suppressing Rayleigh–Taylor instability by vibration. Electroconvective heat transfer and equilibrium stability have been examined in [136–139, 161, 168, 217, 220]. These topics have been discussed in relation to electrohydrodynamic (EHD) effects in practical applications to microgravity heat and mass transfer, where it is difficult or impossible to obtain solutions by other means. There have been interesting studies on EHD pumps, including ones for microgravity. Measurements have been made [138] on the effects of electric fields on boiling, condensation, and liquid dispersal under microgravity, where gravitational forces have less effect than electrical forces. A spraying model has been proposed, and the prospects for realizing it under microgravity have been discussed. Stability conditions for a planar horizontal surface have been considered [161] in the presence of electric fields, as has the equilibrium of a spherical drop suspended in a liquid insulator [168]. Studies have also been made on the electroconvection in a closed cavity under microgravity, where one gets vortex convection and current oscillations. Such solutions have been used in discussing [220] electroconvective flows applied to heat-transfer acceleration under microgravity.

Centrosymmetric convection can be simulated [230] by means of ponderomotive electrical forces in spherical layers of insulating liquid.

Korovin's paper [259] concerns a spherical liquid-metal drop in a high-frequency electromagnetic field produced by two current-carrying turns; there is a discussion of the ultrasonic field produced by the electromagnetic forces and the bulk density of the forces acting on the drop.

The acoustic flow pattern has been derived. In [260], estimates were made of the rotational (Lorentz) forces acting on a drop in a medium having a different conductivity and the resulting drop motion.

There are also effects of an electric field on heat transport in microgravity, such as the use of electrodynamic fluidization (EF) for microgravity heat-transfer control [134]. Charged particles in a capacitor gap are entrained in the motion by the field; when the particles collide with the electrodes, the charge of the powder material reverses. Heat-transfer and thermostatic devices for microgravity can be based on EF layers and have some advantages [134].

Other forms of regular flow in microgravity are of interest, including a vibrational mechanism for thermal convection arising when the vessel containing the liquid oscillates harmonically. Estimates [152] show that such convection effects may be important and can produce heat transfer comparable with or greater than that due to thermal conduction. In [153] stability in such flows in a planar layer is studied, in [135] that type of convection is discussed for cylindrical layers, and in [140] for some other new configurations. In [166] microgravity convection was examined for a binary mixture between walls in an oscillating gravitational field, and it was found that convection can arise with any vibration direction. Some recommendations were made on vibrational control of convection.

In [163, 167] experiments were made on vibrational–thermal convection and heat transfer under these conditions.

1.5. LABORATORY SIMULATION OF MICROGRAVITY PROCESSES [14, 47, 52, 145, 146, 155–159, 183, 184, 186, 189, 191, 195, 198, 199, 205, 206, 221, 224, 239, 242, 245, 251, 311]

Progress in materials science in space requires computer calculations [1], as well as simulation and model experiments under ground conditions in order to resolve microgravity problems that can be handled but only by means of expensive experiments in space laboratories. Some theoreticians consider that many problems in this area should be solved primarily theoretically, i.e., by computer methods, and only when it is demonstrated that a detailed theoretical solution is impossible, should space experiments be performed.

Some theoretical problems mentioned above [155, 191, 206, 221, 224, 245] have been solved by simulation.

Simulation methods and programs have been covered in reviews [14, 51, 52], particularly a software package written at the Institute of Mechanics Problems, Academy of Sciences of the USSR.

This package [51] enables one to solve equation systems for heat and mass transfer in binary mixtures in planar and cylindrical geometries by finite-difference methods for all the g given in Fig. 2 and for all cases of relative orientation between the concentration and temperature gradients with various types of boundary condition and various modes of gravitational-force variation.

Recently, the package has been supplemented with facilities for b/w and graphics display, which output data to a TV screen and to film. A special vector processor speeds up the calculations considerably, and some of them can be performed in real time. Another software package [51], which is based on finite-element methods, handles convection equations written in terms of velocity and pressure. It can handle complicated boundaries and can provide a finite-element net with any variable spacing. In [51] there is also mention of other special models and software packages, including ones for fluid mechanics and for heat and mass transfer in directional crystal growth, where allowance is made for front motion [194, 195], as well as for Czochralski's method with a rotating crystal or seed and crucible [205], for liquid epitaxy [147, 201, 323], and for electrophoretic separation in a free flow [701].

Simulation methods are well developed and have already solved many current topics in this area, as is evident from the above publication list. A recent book by Paskonov et al also deals with this topic.*

The theoretical studies mentioned in Sections 1.1–1.5 above on fluid mechanics have a bearing on research on microgravity crystal growth and the physics of liquids. This includes certain papers considered in the sections below.

1.6. THEORY OF IMPURITY DISTRIBUTION, SEGREGATION, AND STRATIFICATION [119, 155, 165, 180, 194–196, 205, 222, 223, 225, 232, 663]

It has previously been emphasized [1] that it is particularly important to study impurity distributions in microgravity-grown crystals because it is considered that

---

*V. M. Paskonov, V. I. Polezhaev, and L. A. Chudov, *Simulation in Heat and Mass Transfer* [in Russian], Nauka, Moscow (1984).

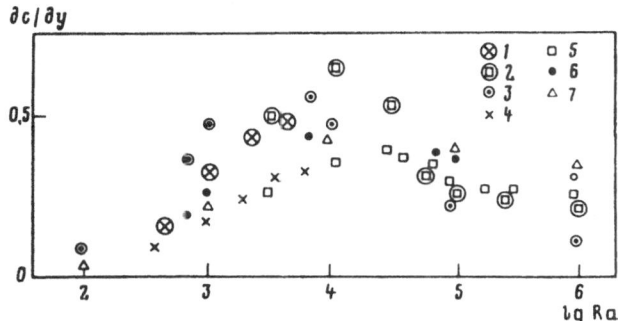

Fig. 3. The impurity-distribution inhomogeneity-maximum effect, where the concentration gradient $\partial c/\partial y$ at the center (1–3) is accompanied by concentration differences (4–7) transverse to the region (1) in accordance with the value of the Rayleigh number; $Ra_D = 0$ for a model of 19×19 mm (1, 4) and for a model 51×51 mm (2, 5); $Ra_D = 0$, $Pr_D = 0.75$, $Pr = 0.65$ (3, 6) [52].

these conditions enable one to avoid macroscopic and microscopic inhomogeneities, particularly in semiconductor materials. There have been many experimental and theoretical papers on this, as cited in [1]. Many theoretical studies, including recent ones, have been dealt with in reviews [14, 51, 52].

In [52] it is stated that weak thermal gravitational convection can arise in a closed region in the presence of a horizontal heat flux. This convection can give rise to inhomogeneous temperature and impurity distributions perpendicular to the heat-flux direction, which may become maximal at a certain Rayleigh number (Fig. 3) because the weak ascending currents do not mix the impurity thoroughly, but merely redistribute it. This effect extends to the concentration patterns that can be caused by weak thermal or concentration-dependent convection, or by interaction between these two, in accordance with case (h) in Fig. 2. It also occurs in the impurity distribution along a crystallization front, and may be very substantial if there is uneven impurity uptake there [194, 195]. It may also be responsible for greater inhomogeneity under microgravity conditions if the ground process corresponds to a Rayleigh number to the right of the maximum, while the microgravity process corresponds to a point in Fig. 3 in the maximum zone.

In [52] the layered and microinhomogeneity formation conditions for binary systems in case (e) in Fig. 2, where there is a lateral heat flux and stable vertical concentration stratification, are discussed. Here the ascending vertical flows are influenced by the stable concentration gradient and split up into horizontal structures. Calculations [155] show that convection under certain conditions can lead to stratification with stepped vertical impurity distribution.

In [51] there is a discussion more detailed than that of [52] on practical cases where there are temperature and concentration gradients in the liquid, where all the cases stated in Fig. 2 are examined in terms of the classification there. There are also discussions of impurity distributions for various growth techniques: directional crystallization, Czochralski's method, and others, including crystals grown under microgravity. In experiment MA-150 (1977), the Si and Sb distributions in the Ge–Si–Sb system showed longitudinal and transverse inhomogeneity substantially exceeding that under surface conditions, which has been explained [51] (see also [194]) from a theoretical model for directional crystallization. The calculations [194] imply that thermal-gravitation and thermocapillary forms of convection have substantial effects on impurity distributions under microgravity and can cause, in particular, transverse inhomogeneity. The two mechanisms must be considered together in interpreting the measurements. It is suggested [194] that this should explain the observed macroscopic inhomogeneity under microgravity.*

The reviews [51, 52] and the papers already cited [155, 194] are accompanied by theoretical studies on impurity distributions in crystal growth as in [119, 180, 195, 196, 205, 222, 223, 225, 663]; some aspects of the theory are also considered in the corresponding experimental studies, such as for impurities in semiconductors [320, 321, 325, 239–335, 351, 354–356].

It has been shown [119] that the macroscopic impurity segregation along the tube decreases as the radius increases in directional crystallization, but at the same time the radial inhomogeneities increase. A numerical dependence has been derived for the distribution as a function of the distance to the center [180].

A two-dimensional model for directional crystallization with a planar front has been used [195] to derive the effective distribution as a function of the thermal convection rate for the ranges in Gr, Pr, and Sc (Table 1) characteristic of microgravity semiconductor crystallization. Numerical calculations have been performed [196] on impurity distributions in the Ge–Ga system for periodic pulling-rate perturbations.

Calculations have been performed for Czochralski's method involving crystal rotation [205], which have shown that microgravity gives a distribution at the front that is much more uniform, with much less macroscopic segregation (reduced to 3–5%) by comparison with surface conditions (20–26%).

A model has been proposed for microgravity directional crystallization [225] on the basis of experiment MA-060 under the Soyuz–Apollo program (1977). In [225] theoretical calculations on Te in PbTe and measurements on zone melting with a moving heater are compared.

Considerable attention has recently been given to theoretical studies of impurity segregation in microgravity-grown crystals; considerable progress has been made.

1.7. NUCLEATION THEORY AND GAS-BUBBLE MOTION IN MICROGRAVITY: LIQUID BOILING [121, 141, 142, 173–175, 193, 226–228, 235, 237, 247, 250, 253, 257, 304, 314]†

Bubble nucleation theory is based on fundamental theoretical studies published over 50 years ago (see the citations in [421]). The principles from these old papers are presented in [421], where it was considered that the radius of a bubble nucleus is governed by equality between the pressures and chemical potentials for the liquid and gas phases. The nucleation rate I_{gb} is

$$I_{gb} = Ce^{-16\pi\sigma^3/3kTP'^2} \cdot e^{-4\pi r^2\sigma/3kT}, \tag{1.1}$$

---

*Chapter 2 (see also [351, 352]) discusses other possible reasons for macroscopic inhomogeneity in microgravity crystal growth.

†See also [345, 350, 387, 415, 421, 444, 445] and other experimental studies on this topic.

where σ is the surface tension, r the radius, p' the pressure in the liquid, and C the preexponential factor. An expression for the nucleation rate for crystals or droplets applies by analogy with (1.1):

$$I_k = N B e^{-A/kT},\qquad(1.2)$$

where N is the number of molecules per unit volume of metastable phase, B is the preexponential factor, and $A = 4\pi r^2 \sigma / 3$ is the nucleation energy.

In [421] it is stated that the pressure in the liquid is equal to the sum of the external pressure P_0 and the hydrostatic pressure $\rho g h_0$ $(P' = \rho g h + P_0)$, and reduced gravitation not only reduces $\rho g h$ but also suppresses free convection, so the conditions should favor more rapid bubble nucleation, i.e., they reduce the critical size and, correspondingly, the formation energy, while favoring more uniform bubble distribution. With sealed tubes (incompletely filled with liquid), the external pressure is equal to the sum of the gas pressure P'' and the capillary pressure at the interface, with principal radii of curvature R_1 and R_2:

$$P_0 = P''_0 + \sigma\left(\frac{1}{R_1} + \frac{1}{R_2}\right).\qquad(1.3)$$

The metastable-equilibrium region at a given temperature T is governed by

$$P' < P_s(T),\qquad(1.4)$$

where P_s is the saturation vapor pressure above a flat surface. In [421] it is stated that the capillary pressure is directed into the liquid if it does not wet the wall, so (1.3) implies that $P_0 > P_s$, i.e., $P_0'' \geq P_s$, so condition (1.4) is violated, and bubble nucleation is impossible under equilibrium conditions. In such a system, bubbles can form while the gas pressure is substantially less than the saturation vapor pressure.

If the liquid wets the wall, compliance with (1.4) is dependent mainly on the hydrostatic pressure, so microgravity favors nucleation.

These theoretical arguments have been utilized in experiments on bubble nucleation under microgravity, as in [421] and in other such studies [228, 257, 345, 350, 387, 415], and the results have been utilized in the theory of foam metal production under microgravity [226].

Theoretical and practical interest also attaches to microgravity bubble migration caused by various forces, particularly Marangoni effects, i.e., thermocapillary bubble drift caused by surface-tension gradients due to temperature or concentration differences. Calculations have been performed on the Stokes forces acting on bubbles on account of capillary effects [304] in general temperature or concentration patterns. There have also been discussions on droplet–droplet interaction in temperature gradients. Measurements have been made [141] on air bubble movement in a thin horizontal liquid layer heated from the side and bounded above and below by solid surfaces. The bubble drift speed v has been measured as a function of temperature gradient, liquid parameters, and bubble shape and size. Theoretical models have been constructed for various ratios of the bubble radius to liquid layer thickness h. A friction correction has been applied on the assumption that there is a liquid film with thickness δ separating the bubble from the wall. This δ is an adjustable parameter in comparing theory with experiment. As there are difficulties in making measurements on thermocapillary drift in space, it has been suggested [142] that the phenomenon should be simulated on the ground.

Theoretical calculations have also been performed on liquids containing bubbles [121], which have given the flow structure and temperature distribution in a cylinder for various bubble positions. The calculations were compared with measurements on the Pion apparatus. The apparatus and the experiments performed with it have provided the basis for various theoretical studies and special experiments. This link between theory and experiment has been particularly fruitful.

In some papers not considered here, attention is given to microgravity bubble boiling [173–175, 227, 314], as well as to gas release in electrolysis [193], droplet evaporation [253], and drop oscillations under microgravity [237].

1.8. THE PHYSICS OF LIQUIDS, LIQUID BRIDGES UNDER MICROGRAVITY, FLOATING ZONES IN ZONE MELTING [126, 182, 212–214, 234, 236, 238, 240, 241, 246, 254, 255, 263, 265, 273–275, 293], AND LIQUID FREE-SURFACE SHAPES [123, 124, 181, 197, 204]

Of the liquid-physics experiments recently performed, we consider only ones in the FPM module for the physics of liquids on Spacelab-1 (SL-1), whose results have been surveyed in [263], together with two experiments on the Texus rockets, while experiments with the Pion apparatus [14, 534–537] will be considered in Chapter 6.

The FPM module was intended for studying large liquid bridges between disks of diameter up to 100 mm, which could be separated by up to 130 mm and could vibrate, rotate, be displaced laterally, and be heated. An electric field could be applied between the disks. The illumination and the photographic recording gave the bridge shapes and oscillation frequencies. Even minor deviations from the theoretical predictions could be detected.

The FPM module on SL-1 had provision for some liquid-physics experiments enumerated in [1] (see Table 12 in [1], pp. 158–160). These experiments have been characterized in [263] as follows.

We consider a large liquid bridge satisfying the equilibrium conditions, i.e., the Gauss–Laplace equation relating the surface tension σ, the radii of curvature r_1 and r_2, and the pressure P: $\sigma(1/r_1 + 1/r_2) = P$. The largest pressure change occurs near the edge because of the solid surface (Fig. 4), which deforms the meniscus considerably. The large menisci greatly accentuate the local curvature changes. These considerations were used in experiment ES-329 [287] in order to estimate long-range van der Waals forces. Two experiments were performed on SL-1 with different stability parameters and different capillary pressures in the bridges (less by factors of 10^3 than under ground conditions). The equilibrium shapes were attained very rapidly. A study of zone formation and disruption revealed a new microgravity effect: internal wetting. It is possible to measure the van der Waals forces, but the film thicknesses near the edge (from 10 to 30 μm) exceeds the usual range for such forces.

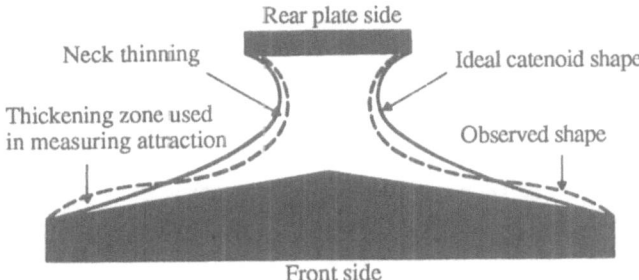

Fig. 4. Surface tension in thin liquid films [263].

Fig. 5. Oscillations in a drop retained by a disk [263].

The large liquid volume means that the oscillation periods are high, as are the bridge failure times. The circular frequencies ω_n for droplet oscillation are defined by

$$\omega_n^2 \rho r^3 / \sigma = n\,(n-1)\,(n+2). \tag{1.5}$$

The resonant frequency for a mode having n nodal lines decreases as the radius r increases as $r^{-3/2}$. A long liquid column thus gives good prospects for determining the resonant frequencies and shape changes, and such measurements were made on SL-1 in experiment ES-326 (Fig. 5). Such studies are particularly important because there is at present no theory on the oscillations of a drop supported by a disk. There are approximate models applicable only to inviscid and very viscid liquids. In [293] qualitative relationships were derived for the amplitude A, resonant frequency F, and damping time t for the bulk oscillations of a liquid (silicone oil) as a function of diameter D; A decreases as D increases, while t increases. It was impossible to test the damping system because there were unexpected tendencies for the liquid to spread over the disk, on which more detailed research is needed. Experiments in preparation for [293] were performed in [240, 241]. Here we may note the results from another study [182], which also dealt with phenomena similar to those in experiment ES-326 but were based on a microzone and neutral buoyancy. In [182] there was a discussion of effects near the resonant frequency, viz., amplitude breakaway and frequency hysteresis. At comparatively large contact angles ($\psi \sim 100°$), there was a rotation effect because unsymmetrical oscillations were involved in rotational motion. At large contact angles ($\psi > 100°$), it was found that the drop oscillations were excited.

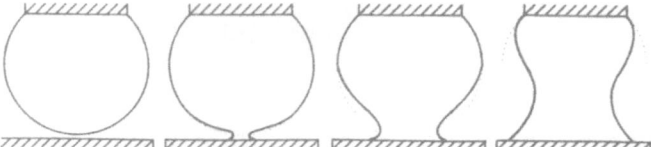

Fig. 6. Drop propagation between coaxial disks [263].

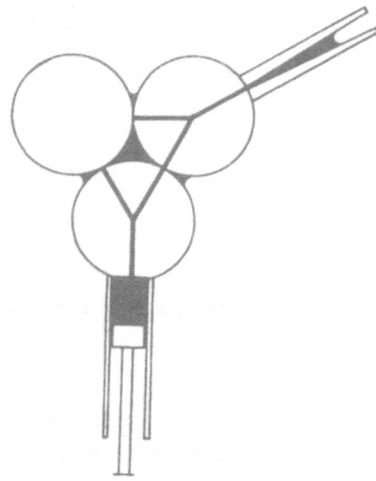

Fig. 7. Liquid propagation between spheres [263].

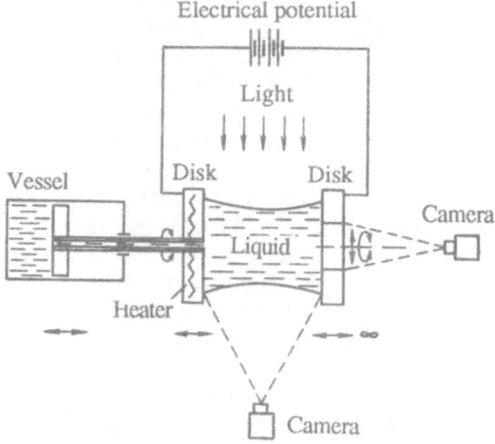

Fig. 8. Scheme for liquid-physics module [280].

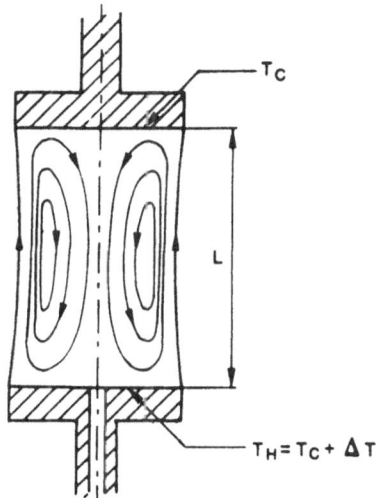

Fig. 9. Marangoni convection in a liquid
bridge [263].

This pumping effect occurs for large low-viscosity drops even at quite small
amplitudes near the linear range, which represents an intermediate phase in the
transition from linear (small) oscillations to droplet detachment. On the whole, the
excitation can be represented as: 1) droplet mass redistribution in prominent zones,
with the surface shape remaining similar to that for resonant linear oscillation, 2)
transformation to a shape close to cylindrical, 3) cylinder stretching and thinning,
and 4) motion in the thin regions and mass redistribution into prominent zones,
which was accompanied by excitation and frequency hysteresis, where the loop
width was determined by the forcing frequency and droplet size.

Experiments ES-327 and ES-339 [255, 263] were designed to examine droplet
spread, contact-angle dynamics, and capillary hysteresis under microgravity. In
spite of equipment failure (film damage), some useful and interesting qualitative
observations were made, in particular on the behavior between coaxial disks (Fig.
6) and spheres (Fig. 7). Liquid-bridge rotation and breakage were also examined.
as well as satellite-drop formation and electrostatic effects.

In experiment ES-328 [263, 275, 277, 280], Marangoni convection was examined in a large liquid bridge in microgravity; the bridge volume was two orders
of magnitude greater than that under ground conditions. The liquid-phase
configuration was analogous to that in the floating-zone method used in crystal
growth. Silicone oil was the model liquid. Figure 8 shows the FPM apparatus,
while Fig. 9 shows the Marangoni convective fluxes. The experiment showed that
Marangoni convection occurs in microgravity and confirms the prediction that
Marangoni boundary layers may then be obtained (see theoretical studies [234, 236,
278, 279]). No effect was found from an electric field in this case (with this
geometry and liquid).

We have seen in Section 1.3 that there are recent papers dealing with
Marangoni convection, while Section 1.7 deals with gas-bubble migration due to it.
Here we can add, as regards experiment ES-328 on SL-1, that various experiments

have recently been performed on Marangoni convection on the Texus rockets. In [265] the Texus-7 experiment was described, which dealt with mixture separation and Marangoni convection in drops. The model system was methanol containing 0.35 mass % cyclohexane. The chamber was heated above the critical point (up to 50°C above it) before the flight and cooled monotonically to 10°C before the start of microgravity. A film showed the cooling-front propagation, the cyclohexane droplet growth behind the front, and the marked Marangoni convection toward the front caused by thermal and concentration surface-tension gradients. These experiments confirmed the theoretical predictions on mixture separation related to droplet migration from Marangoni convection.

A further microgravity experiment with the Texus-7 [246] was designed to examine whether Marangoni convection occurs in a liquid column at previously unattained high Marangoni numbers such as $M_g = 9.4 \cdot 10^4$ and $M_g = 1.24 \cdot 10^5$, which greatly exceed the critical values for the first oscillatory instability, the object being to examine whether the motion remains oscillatory or goes over to the turbulent state, as well as the effects of the ratio of length to diameter on the instability. In [246] turbulent and oscillatory states were observed, and the criteria for higher oscillatory modes were discussed, including the conversion to turbulent nonperiodic structures, which are influenced by the relative dimensions.

We now turn to the analysis in [263] of SL-1 experiments on this.

In experiment ES-331 [263, 273], which was concerned with floating-zone stability, a cylindrical liquid bridge about 10 cm long was formed between two coaxial disks, where the surface deformation and internal motion were examined as affected by rotation, vibration, stretching, and so on (Fig. 10). On the SL-1 flight, there were difficulties in controlling the necessary surfaces (the liquid, which was silicone oil, resisted installation in certain forms, and showed poor adhesion to the solids). Physicochemical barriers were found to be unimportant for large liquid masses, i.e., large bridges. However, the experiment revealed many interesting details, and the information is useful for future experiments. The topics covered by experiment ES-331 have been considered theoretically in [274], where analytic expressions were given for the stability of a long axisymmetric bridge attached to coaxial disks of radii R_1 and R_2, separated by a distance L. The discussion concerned the effects of L, axial gravitational forces, and the disk-radius ratio on the limiting stable volume; a comparison was made with the corresponding numerical results.

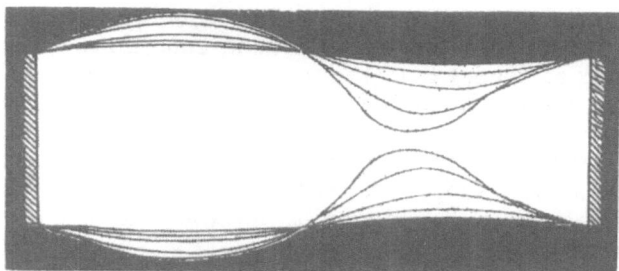

Fig. 10. Liquid-bridge vibration and stability [263].

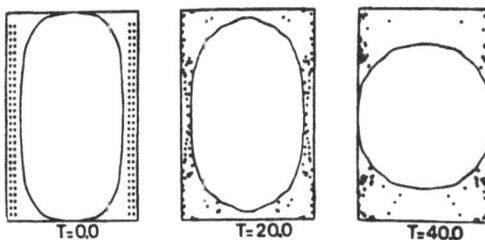

Fig. 11. Liquid shapes adopted on rotating a partially filled cylinder: computer simulation of the tracks (points) from reflected and refracted rays; T) time from start of retardation [263].

Unequal disks ($R_1 \neq R_2$) were also examined, and the expression was extended to noncylindrical bulk configurations via linear analysis of the equilibrium surface forms. As the Bond number Bo increases, the neck position on the column shifts, and at the critical value B_{cr}, the neck occurs at the middle.

The last of seven liquid-physics experiments, ES-330 [263, 307, 308], on SL-1 was prepared to examine the behavior of an axisymmetric free surface of a moving liquid in a rotating partially filled transparent Plexiglas container (outside diameter 6 cm) having various forms of internal cavity (cylindrical, spherical, annular, rectangular box, and combined). The cavities were half-filled with silicone oil containing aluminum particles. The container in each case rotated at 35 rpm, and a film was recorded during acceleration and breaking. Although there were difficulties with the exposure and there was damage to the film, some interesting, unexpected results were obtained. The silicone-oil wetting led to air bubbles and a spherical meniscus having a sharp contact line. This meant that capillary surface resonance excitation was difficult in the cylindrical and spherical containers. An unexpected effect was that there were bright reflected and refracted rays, whose line and point images enabled one to examine the behavior. In [307, 308] a study was made of using this in subsequent experiments. A simulation program has been written for a cylindrical container [307, 308]. In general, computer simulation is required for all liquid-physics experiments [263]. Computers enable one to identify results that, in principle, are known or are essentially novel. Figure 11 shows computer simulations for experiment ES-330 for a cylindrical container.

It is intended that SL-D1 shall carry out four new experiments on liquid physics:

1. D1-FPM-01, the effects of convection at a boundary on mass transfer during evaporation of acetone from water into air;
2. D1-FPM-02, bubble and droplet migration under constant temperature gradient;
3. D1-FPM-03, the contribution from boundary convection to immiscible-liquid separation on cooling; and
4. D1-FPM-04, retarded boundary convection in the temperature range near minimum surface tension.

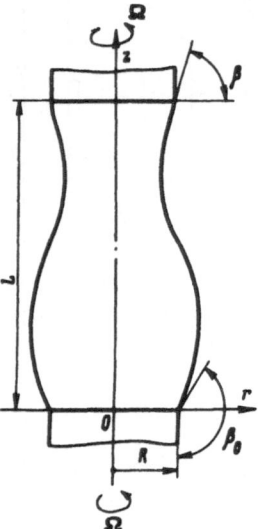

Fig. 12. Scheme and co-ordinate system [126].

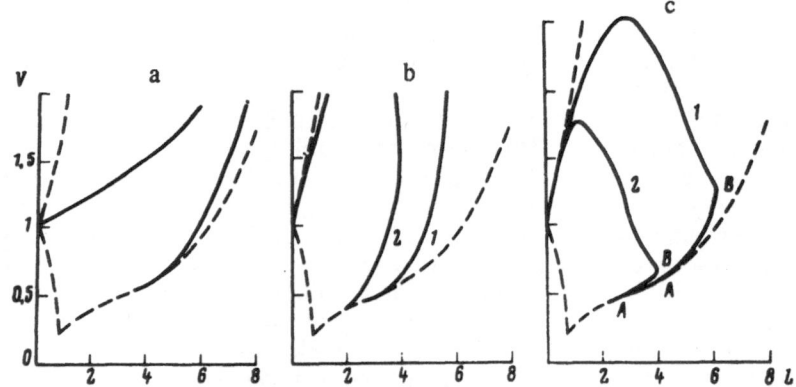

Fig. 13. Effects of wetting angle (a), axial mode (b), and rotation (c) on the stability region [126].

New containers and an improved FPM system will be used in the new series. The new container provides safety measures for the use of liquids other than water and silicone oil previously permitted in FPM. The cylindrical FPM system is promising for liquid-physics experiments.

To conclude, I give briefly the contents of theoretical studies [126, 214, 274] on liquid-bridge behavior under microgravity.

In [126] it was stated that bridge dimensions could be increased under microgravity conditions because the Bond number is reduced by several orders of magnitude. There was also a discussion of calculating the shape for a liquid zone of any volume on rotation and in the presence of an axial acceleration, together with stability tests on the shape. Figure 12 shows the scheme and the coordinate system. Barmin and Senchenkov's solution [125] to the linear problem was used in [126] to obtain an expression for the shape:

$$U(Z) = \frac{(V-1)\sin(xz/2)\sin\left[\frac{x}{2}(L-Z)\right]}{\frac{2}{kL}\sin(kL/2) - \cos(kL/2)} + \frac{BoL}{k^2}\left[\frac{Z}{L} - \frac{\sin(kZ/2)}{\sin(kL/2)}\cos\left(\frac{k}{2}L - Z\right)\right],$$

$$(1.6)$$

where $U = r - 1$, V is the dimensionless volume of the zone, $k = 1 + 2We$, and We and Bo are the Weber and Bond numbers (Table 1).

The shape described by (1.6) enables one to decide crystal shapes and to determine the conditions (the values of Bo and We) for growing constant-diameter crystals. Even a short zone does not allow one to neglect the small forces occurring on orbiting satellites in growing large-diameter crystals. In [126] numerical values of V are given, as occurring in growing germanium crystals with given Bo and We.

The shape calculations in [126] were accompanied by stability-limit definition, viz., the effects of wetting angle β, axial acceleration, and rotation on the limiting V. Figure 13 shows the results. Figure 13a shows the effect of wetting angle on the stability region. The solid lines correspond to β = 17° and Bo = 0.01, while the dashed lines correspond to complete weightlessness (Bo = 0) and β = 0 or 180°. It is evident that the stability limits for minimal L (zone lengths) are displaced appreciably to the right for real materials. Figure 13b shows the effects of Bo. Curve 1 is for Bo = 0.1, curve 2 for Bo = 0.5, and the dashed line for complete weightlessness (Bo = 0). Figure 13 also illustrates the effects of rotation. A new boundary appears, which defines the limiting volume. With an immobile zone (dashed line), one can find a length for which the zone is stable at any volume, whereas the picture is essentially different with rotation. At the new limit, planar boundaries are hazardous, whereas axisymmetric ones are so on part AB. Curve 1 corresponds to We = 0.1 and curve 2 to We = 0.5.

Hydrostatic topics were considered in [214] to determine zone shape and stability during zone melting not only under conditions of complete weightlessness but also under gravity or centrifugal forces on rotation around the axis. The solutions in [214] are illustrated by graphs, which are not given here.

Many studies have been made on equilibrium shapes and stability limits for long liquid bridges. Citations are given in [126, 214] and in [274]; most of these papers deal with numerical treatments. In [274] there is an analytic treatment, where analytic and numerical results are compared. For example, the following expression is given for the minimal stable volume [274]:

$$V = 2\Lambda(2\Lambda - \pi) + 6(3/2)^{1.3}\pi\Lambda[Bo - H/(2\Lambda - 2\sin\Lambda]^{2.3}, \qquad (1.7)$$

where $\Lambda = L/2R_0$; $R_0 = (R_1 + R_2)^2$; $H = 1 - k^2/1 + k^2$; $k = R_1/R_2$, R_1 and R_2 being the lower and upper radii of the disks supporting the bridge. Formula (1.7) applies for small Bond numbers (Bo ≈ 0) and for almost-equal disks, $k ≈ 1$.

1.9. CAPILLARY FORCES AND LIQUID SHAPING IN MICRO-GRAVITY, INCLUDING CONTACT PHENOMENA AT LIQUID–TUBE BOUNDARIES [46, 115, 119, 122–126, 164, 204, 214–216, 218, 219, 229, 237, 273, 274, 287, 288, 300–302, 307, 315]

A few recent reviews [14, 51] deal with hydrodynamics, while one review [46] deals with hydrostatics and methods of solving for capillary-liquid statics. There are 55 papers on this topic cited in [46], which include some of those cited in [1].

The term "capillary liquid" [46] is applied to a liquid under conditions such that surface-tension effects must be considered, which arise, in particular, if the liquid has a free surface and is on board a freely moving space vehicle.

Simple problems in capillary-liquid mechanics were considered during the last century, but extensive and detailed studies have been made only quite recently, in part in relation to materials science in space. In the previous section, we have examined and cited various treatments of liquid physics under microgravity, particularly free-surface shapes and stability under various perturbations [123, 126, 197, 204, 214–216, 273, 274, 287, 307], and especially at the end of Section 1.8, where we cited [126, 214, 274], which have a direct bearing on hydrostatic aspects. In [46] existing methods of treating static topics are enumerated, and it is stated that these are partly qualitative or approximate and therefore need experimental confirmation because of inexact correspondence with the assumptions.

In [46] a differential equation is given for the equilibrium surface as derived from the Laplace formula in general form and for the axisymmetric case. Closed and unclosed equilibrium surfaces are considered for a liquid at rest, as well as an annular surface and two-dimensional and three-dimensional cases.

The stability conditions are also examined. The principle of minimum potential energy is applied, and if the potential energy has a minimum at the equilibrium position, the equilibrium is stable, otherwise it is unstable. Stability is usually considered in relation to infinitely small perturbations, first in general form and then for the simpler, common case of an axisymmetric surface. General methods are considered for determining stability in axisymmetric equilibrium, as well as for deriving equilibrium stability in a slot. The stability margin in the equilibrium state is considered not only as the depth of the minimum in the potential energy, in particular for a state close to critical, but also as the difference between the potential-energy levels corresponding to the unstable or critical states for those parameter values and the stable one. Some particular applications are listed for which the stability margins have been estimated: plane-parallel equilibrium in a short rectangular channel and a rotating cylindrical column of weightless liquid between two parallel plates (i.e., a microgravity liquid bridge).

In conclusion, [46] deals with surface simulation of microgravity hydrostatics, where all the methods are based on geometrical similarity between the actual and model systems, where Bo and We have the same values as does the wetting angle φ if there is a wetting line.

To maintain Bo, the reduced gravity can be balanced by reducing the linear dimensions, but this often leads to very small models. Another approach is to replace the liquid–gas system by a liquid–liquid system, but this can alter φ. One can simulate complete weightlessness (Bo = 0) if the two liquids have the same density (Plateau's method), but experiments on rotation here give results different from those for weightlessness, i.e., the phenomena are to be explained in terms cf the relation between the surface-tension and viscous-stress forces at the interfaces.

The liquid is placed in a narrow slot between transparent plates to simulate two-dimensional treatments, but no systematic study has been made on the effects of the plates themselves on the equilibrium shapes.

This brief survey of [46] may be completed by noting that the review enumerates major topics in microgravity hydrostatics, and the list can be used in analyzing other such papers cited above. Unfortunately, there are no illustrations in [46], and the literature cited contains papers only up to 1982, inclusive. It is impossible here to consider individually all the papers cited in the heading for this section, and it is necessary to characterize briefly only papers on capillary wetting [229, 300–302] and then to consider crystallization with microgravity capillary shaping [218, 219]. The contents of the other papers can be judged from their titles.

Microgravity capillary wetting is considered in [300–302] from experiments with the Texus rockets and in a ground laboratory, where weightlessness was simulated by Plateau's method, i.e., by suspending the test liquid (a cyclohexane–carbon tetrachloride mixture) in a supporting liquid (fresh water), which has the same density. No supporting liquid is required in space, and the test liquid was water. The one difference between the ground and space experiments was that in the former there was a liquid–liquid interface, while in the latter there was a gas (air)–liquid interface. In [300–302] measurements were made on the liquid rise in cylindrical, conical, and sinusoidal capillaries. In [300] theoretical studies were made on hydrodynamic forces (capillary, inertial, and so on) arising in capillary rise on simulating zero gravity or microgravity.

In [300] the hydrodynamic forces were considered for a homogeneous incompressible wetting liquid in a circular vertical tube having a constant or variable circular cross section, where a theorem on linear momentum was applied in integral form (Fig. 14):

$$\int_V \frac{\partial}{\partial t} (\rho \bar{u}) \, dv + \rho \, (\bar{u}_2^2 A_2 - \bar{u}_1^2 A_1) = F_c - F_v. \tag{1.8}$$

The first term is the local rate of change in momentum and the second is the convective momentum flux from the region dv (Fig. 14). On the right, F_c is the capillary force (driving force), F_v the resistance (losses at inlet and outlet, friction in laminar flow, and so on), ρ the density, $\bar{u} = (\dot{v}/\pi R^2)$ the mean velocity along the capillary axis z ($\dot{v}$ is the bulk velocity through the tube), A the tube cross section, R the radius, and t the time. In [300] detailed studies are made on F_c and F_v together with the inertial forces, and curves are given relating the hydrodynamic forces to time for cylindrical and sinusoidal tubes. We cannot consider the calculation details and merely give curves from [300] for cylindrical and sinusoidal tubes (Figs. 15 and 16). A theoretical study has also been made on liquid flow through a capillary under microgravity [229], whereas in [300] theorems were used on mass and momentum conservation. Then the projection of the momentum on the capillary (tube) axis is

$$\frac{d}{dt} (mv) = SP + f\sigma + f_m + f_g. \tag{1.9}$$

We restrict consideration to a cylindrical capillary with radius R. Then $m = \rho \pi R^2 l$ in (1.9) is the liquid mass in the capillary, v the mean velocity, ρ the density,

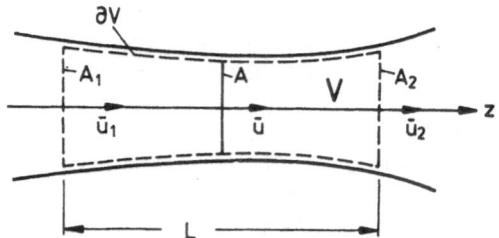

Fig. 14. Hydrodynamic forces for a liquid rising in a vertical tube [300].

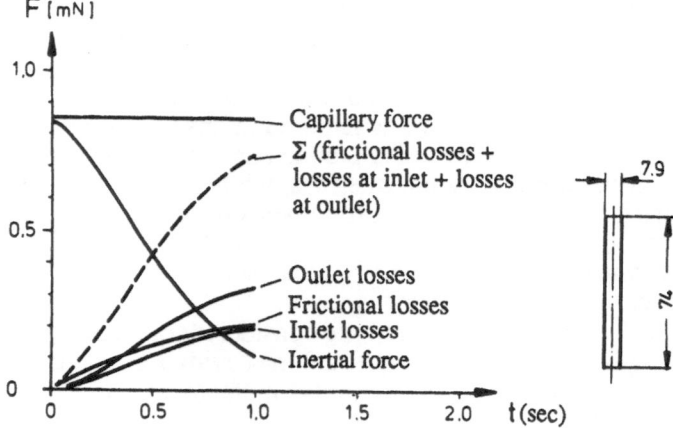

Fig. 15. Time dependence of hydrodynamic forces for a cylindrical tube [300].

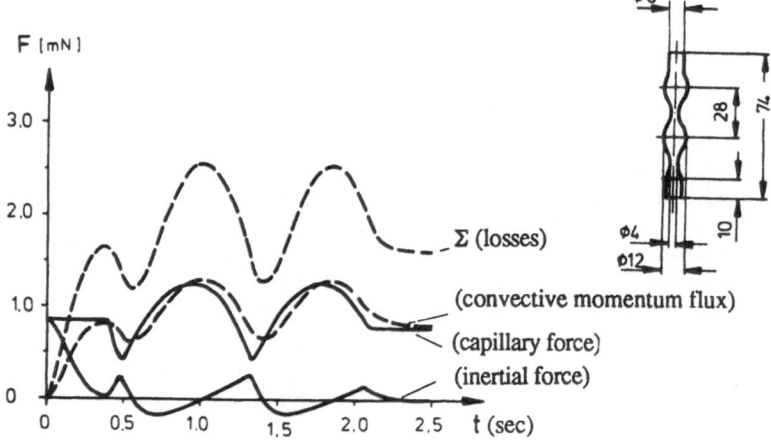

Fig. 16. Time dependence of hydrodynamic forces for a sinusoidal tube [300].

l the column length, P the total pressure in the liquid, σ the surface tension, ψ the wetting angle, $f_M = 8\pi\mu lv$ the viscous friction in laminar flow (μ the dynamic viscosity), $f_g = -\rho g \pi R^2 (l - h) \cos \psi$ the external force (gravity, inertia, etc.), $\vec{g} = ng_0$, n the loading factor, g_0 the acceleration due to gravity on the ground, ψ the inclination of the capillary to the $\vec{g}$ vector, and h the liquid layer height outside the capillary. A difference of [229] from [300] was that no allowance was made for phenomena at the inlet and outlet (they were neglected).

The pressure P, in general, is the sum of the external P and the capillary pressure $P_\sigma = \sigma(1/R_1 + 1/R_2)$, where R_1 and R_2 are the principal radii of curvature in the free surface.

In weightlessness ($g = 0$) or with the capillary perpendicular to $\vec{g}$ (i.e., with $\psi = \pi/2$), the solution to (1.8) is

$$l_v = \frac{R^2}{8\mu}\left(P\left(t\right) + \frac{8\sigma}{R}\cos\psi\right)\left[1 - \exp\left(-\frac{8\mu}{\rho R^2}t\right)\right].\qquad(1.10)$$

Furthermore, [229] deals with a long capillary and cases where $g \neq 0$ and $\psi \neq \pi/2$, where there is a detailed study of the conditions in the Rumanian–USSR "Kapillyar" ("Capillary") experiment. In [229] planar and cylindrical capillaries are considered. This ends our brief characterization of [229, 300–302], and we now consider [218, 219, 433], which deal with capillary shaping in microgravity crystallization, particularly profiled-crystal growth (Stepanov's method) in microgravity.

In [218, 219] it is emphasized that the main interest attaches to modes of crystallization from the liquid under microgravity in which the side surface does not contact the container walls: Czochralski's, Stepanov's, Kyropoulos's, and Verneuil's methods, as well as floating-zone techniques. In all these cases, the crystal shape and size are determined by capillary forces governing the meniscus in the phase-boundary zone as well as by the heat- and mass-transfer conditions in the solid–liquid system.

Capillary forces are also involved in tube crystallization under microgravity, since usually the liquid is not in contact with the tube, and the same applies in directional crystallization in freely suspended spherical droplets.

The following equation system has to be solved to determine crystal–liquid behavior in general: the equation of motion for the liquid (the Navier–Stokes equation) with boundary conditions at the free surface (the capillary Laplace equation), the continuity equation (mass conservation), the thermal-conduction equation (energy conservation), and the diffusion equation (impurity-mass conservation), with the condition for constant growth angle specific for crystallization from the liquid. In [218] the general formulation is followed by an analysis of microgravity crystallization conditions for Czochralski's, Stepanov's, and Verneuil's methods. In [218] inertial forces affecting crystallization are neglected by comparison with the capillary forces, and therefore solution of the Navier–Stokes equation is replaced by treatment of the Laplace capillary equation for the equilibrium surface shape for the liquid $z(u, v)$:

$$\sigma\left(\frac{1}{R_1} + \frac{1}{R_2}\right) + \rho g z = \text{const.}\qquad(1.11)$$

This equation can be written as

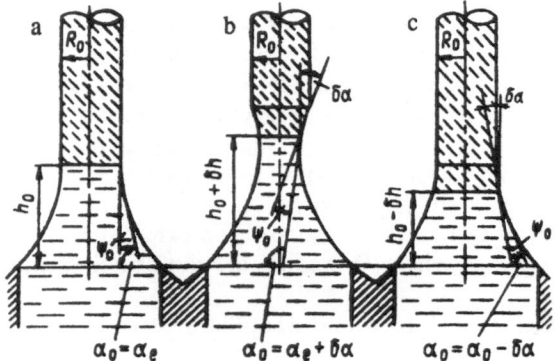

Fig. 17. Scheme for growing a circular rod by Stepanov's method [218]: a) $\alpha_0 = \alpha_e$, stable growth; b) $\alpha_0 > \alpha_e$, rod contracts; c) $\alpha_0 < \alpha_e$, rod expands.

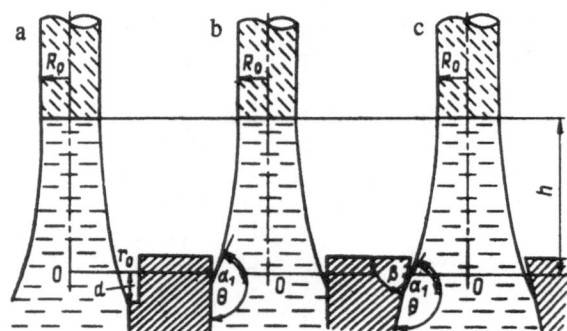

Fig. 18. Scheme for boundary conditions in Stepanov's method related to shaping [218]: a) attachment; b) wetting (straight walls); c) wetting (inclined walls).

$$\text{div}\,(H\,\text{grad}\,z) - \rho g z / \sigma = \text{const} = -n, \qquad (1.12)$$

where $H = [1 + (\partial z/\partial u)^2 + (\partial z/\partial v)^2]^{-1/2}$, and the parameter n is dependent on the choice of origin for the z coordinate.

On replacing the variables for a circular crystal, we get from (1.12) that

$$\frac{z''}{(1+z'^2)^{3/2}} + \frac{z'}{r\,(1+z'^2)^{1/2}} + \frac{1}{\sigma}\,(\pm n - \rho g z) = 0. \qquad (1.13)$$

We introduce the dimensionless coordinates $\sqrt{2\sigma/\rho g} = a$, $z/a = y$, $r/a = x$, $na/2\gamma = d$, i.e., the unit for the linear dimensions is the capillary constant a, while the pressure unit is the weight of a liquid column of height one capillary constant. For microgravity conditions we neglect the weight of the liquid column by comparison with the capillary forces to get

$$y''x + y'(1+y'^2)^2 \pm 2d(1+y'^2)^{3/2}x = 0. \tag{1.14}$$

Two boundary conditions have to be specified for this differential equation for the fixed liquid pressure: one at the crystallization front and the other at the bottom of the liquid column.

The condition at the crystal–liquid boundary is common to all methods of crystallization from the melt and records the fact that the angle φ (Fig. 17) between the tangents to the crystal and liquid at the liquid–solid–gas intersection remains unchanged. Here ψ_0 is the wetting angle in a state of equilibrium on the crystal, which is to be distinguished from the angle θ of liquid crystal wetting under equilibrium conditions.

In [218] the condition at the crystal–liquid boundary is written as

$$y'|_{x=R_0} = -\tan\alpha_e, \tag{1.15}$$

where R_0 is the radius of the growing crystal and $\alpha_e = \delta/2 - \varphi_0$. If the angle α_0 formed by the tangent to the liquid at the triple point (Fig. 17) is equal to α_e, a crystal of constant cross section grows; if α_0 differs from α_e, the size does not remain constant. For $\alpha_0 = \alpha_e + \delta\alpha_0$, the crystal narrows (Fig. 17b), and for $\alpha_0 = \alpha_e - \delta\alpha_0$, it expands (Fig. 17c).

Condition at Liquid Column Base. Figure 18 illustrates three forms of such conditions for Stepanov's method:

a. The attachment condition, where the liquid is attached to a sharp edge (Fig. 18a):

$$y'_{x=r_0} = -d. \tag{1.16a}$$

b. The wetting condition; for a vertical shaper wall (Fig. 18b)

$$y'|_{x=r_e} = -\tan\left(\theta - \frac{\pi}{2}\right). \tag{1.16b}$$

c. For an inclined shaper wall (Fig. 18c)

$$y'|_{x=r_0 - y\tan\beta} = -\tan(\theta + \beta - 180°). \tag{1.16c}$$

One can solve the boundary-value problem for (1.14) with boundary conditions (1.15) and (1.16) to relate the crystal size to the front position, boundary conditions, angles α, and pressure.

We consider only the capillary problem in crystal growth here and note that in [218] that formulation was used in considering capillary shaping during pulling under microgravity for a circular rod and for a plate. In [218] there is also a brief discussion of the thermal pattern and the effects of Marangoni convection on the meniscus and on the heat transport in the liquid. There was also a study of the stability in Czochralski's, Stepanov's, and Verneuil's methods. We cannot go into the details of this part of [218]* and merely state, in conclusion, that the theoretical analysis there shows that microgravity improves the capillary shaping conditions in Stepanov's and Verneuil's method. It is considered [218] that this should stimulate the use of these methods in liquid crystallization in space.

1.10. GRAVITATIONAL CONDITIONS ON SPACE VEHICLES AND THE DESTABILIZING EFFECTS OF RESIDUAL ACCELERATION ON SOLIDIFICATION AND THERMAL OSCILLATIONS [166, 290, 584, 585]

The actual acceleration patterns on orbiting satellites have received considerable attention, since g perturbations sometimes affect experiments in materials science in space considerably. Most such studies have been experimental [554, 569, 574, 575, 577] (see also the list of earlier papers in [1]). However, there have also been theoretical studies, an example being [290] (see also [166, 584, 585]), of which a summary is given below. In [290] a theoretical model is discussed for the actual force pattern in a natural orbit for manned and unmanned satellites.

The perturbing forces are [290] classified as follows:

   I. gravitational perturbations (inhomogeneity in the earth's gravitational field and gravity due to the space vehicle);
  II. perturbations from the cosmic environment (effects of the environment on the satellite components);
 III. perturbations associated with satellite system operation (engine operations, personnel actions, life support systems, etc.).

The perturbing forces are calculated by integrating the equations of motion around the center of mass, which for a planar orbit are put [290] as

$$\frac{d\theta}{dt} = \frac{P}{B},$$

$$\frac{dP}{dt} = -\left[3\frac{\mu}{R^3}(A-C)\sin\theta\cos\theta + B\frac{d\omega_0}{dt}\right] + M_a,$$

(1.17)

where θ is the angle by which the longitudinal axis deviates from the vertical, P is the generalized momentum, A, B, and C are the principal internal moments of inertia, ω_0 is the angular orbital velocity, M_a is the aerodynamic moment from the retarding forces, μ is the gravitational constant, and R is the orbit radius. The additional acceleration in the center of mass is given by

$$a = F_a/m,$$

(1.18)

where m is the satellite mass.

---

*All the topics considered in [218] are examined in detail for growing crystals on the ground in *Making Profiled Single Crystals and Components by Stepanov's Method* [in Russian], Nauka, Leningrad (1981).

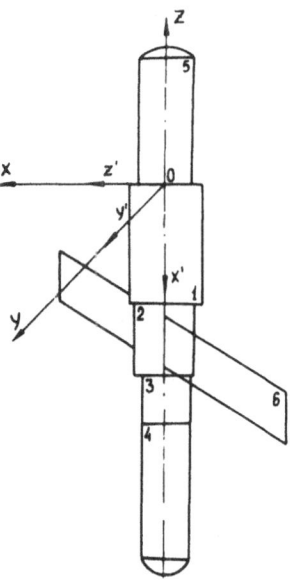

Fig. 19. Scheme for the Salyut–Soyuz space vehicle: x, y, z) orbital coordinate system; x', y', z') satellite coordinate system; 1, 2, 3) components in the Salyut station model; 4, 5) components in the Soyuz space vehicle model; 6) solar battery model [290].

Perturbations of types I and II are usually [290] sign-varying and close to harmonic, the frequency being comparable with that of the satellite's rotation around the earth. Inhomogeneity in the earth's gravitational field means that g perturbations arise for points in the satellite not coincident with the center of mass, which are given by

$$a_x = 0; \quad a_y = -\mu_i' R^2 \cdot y / R, \quad a_z = 3\mu / R^2 \circ z / R,$$

$$y = y', \quad z = z' \cos \theta - x' \sin \theta,$$

$$(1.19)$$

where x, y, z and x', y', z' are the distances from the center of mass correspondingly in the orbital coordinate system and in the system related to the satellite (Fig. 19).

Systems (1.17), (1.18), and (1.19) may be integrated to give the mass-force variations (Grashof number, Gr) during experiments on the satellite. Figure 20 gives calculations for Salyut–Soyuz and the Shuttle (dashed lines) for parameters close to the actual ones. Figure 20 illustrates the change in K defined by

$$K = \frac{\sqrt{a_0^2 + a_1^2 + a_2^2}}{\sqrt{a_y^2 + a_z^2}},$$

$$(1.20)$$

where a is the acceleration of the center of mass, a_1 and a_2 are the centrifugal and rotational accelerations, and a_y and a_z are the gravitational-acceleration components [see formula (1.19)].

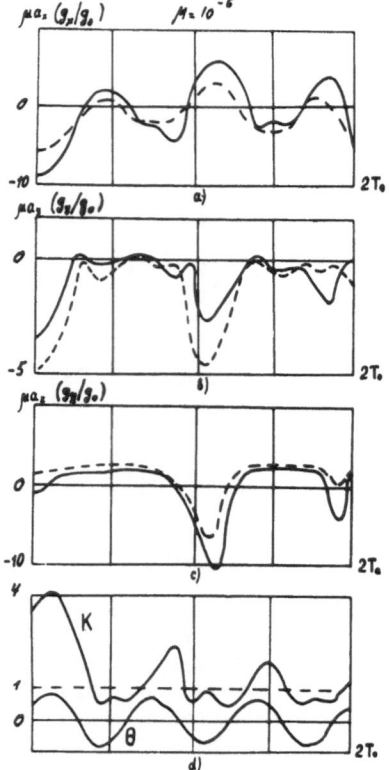

Fig. 20. Variations in the projection of microgravity on the x', y', z' axes (a, b, c). Functions $K(t)$ and angle θ for Salyut (solid lines) and the Shuttle (dashed line) for two periods ($T_0 = 90$ min) ($H = 200$ km, $l = 0.01$, $\theta_0 = 0.5$, $\theta = 10^{-3}$ sec^{-1}) [290].

The function $K(f)$ shows that one can approximate the complicated model for orbital microgravity by means of the comparatively simple (1.19).

Figure 21 shows the general behavior of the maximal, minimal, and average microgravity oscillation amplitudes at orbit heights for Salyut–Soyuz and the Shuttle; sign-varying behavior is typical for Salyut with a mean value close to zero, while the Shuttle (at a height $H = 250$–300 km) had sign-constant microgravity with nonzero mean.

The initial satellite parameters (θ_0 and $\dot{\theta}_0$) have the most effects on the microgravity oscillation amplitudes. As the inaccuracy in satellite orientation increases, the amplitudes in the microgravity variations increase sharply, and the satellite may lose gravitational stability and begin to rotate slowly. Then the microgravity-variation component due to centrifugal forces becomes predominant. The behavior of the average mass forces changes as the mass configuration changes and as the flight mode varies, but oscillatory (sign-varying) microgravity is typical of any satellite oriented by gravitational forces.

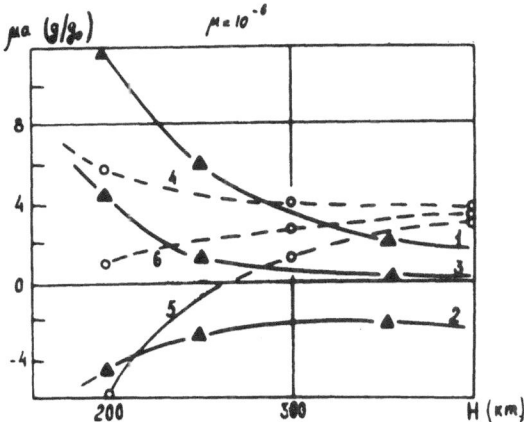

Fig. 21. Oscillation amplitude (1, 2, 4, 5) and mean microgravity levels (3, 6) in relation to height of the Salyut orbit (solid lines) and the Shuttle (dashed lines) [290].

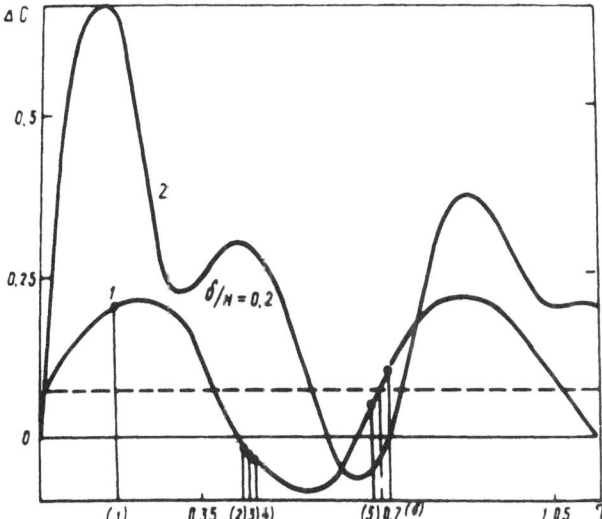

Fig. 22. Impurity concentrations (curve 2) on low-frequency harmonic variation in the mass forces, amplitude of Ra = 500, mean Ra = 375, period $T_0 = 0.7$, $\delta = 0.2$, Pr = 0.01 [290].

On the whole, the calculations in [290] show that the mass forces on an actual flight vary considerably in time and space, and the levels shown in Fig. 20 represent the limit to the state of weightlessness that can be realized on an orbiting satellite without feedback from the vehicle command and orientation system. These

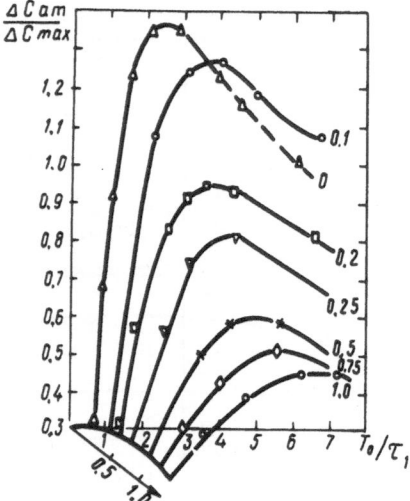

Fig. 23. Oscillation amplitude in impurity-stratification concentration in a steady state in relation to force pulsation period and initial concentration gradient [290].

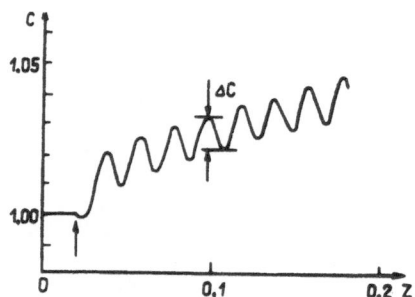

Fig. 24. Impurity distribution along crystal ($f = 10$, $Gr_t = 3 \cdot 10^4$) [290].

microgravity features must be considered in relation to convective and hydrostatic phenomena. This has been done in [290], where curves were given for impurity stratification on low-frequency harmonic mass-force variation. The effects may be very considerable, as is evident from Figs. 22 and 23, which, however, we do not examine in detail. In [290] there is a discussion not only of the effects from low-frequency g perturbations (types I and II) on the impurity distribution in a growing crystal but also those from high-frequency g perturbations (type III) caused by engineering systems and spacemen's actions. A solution has been obtained for a constant-speed front from the data such as [360]. We restrict ourselves to presenting Fig. 24 from [290], which shows the impurity distribution in the part transitional from diffusion-limited to vibration-influenced growth (the transition point is indicated by an arrow). The impurity concentration oscillates with an amplitude of about 1%, which qualitatively resembles the banding typical of crystals grown under ground conditions. Quantitatively, the effect, however, is not as large as that from low-frequency g perturbations, i.e., high-frequency perturbations do not produce appreciable microscopic inhomogeneity.

1.11. CONCLUSIONS

The theoretical developments on space manufacturing considered in this section from recent publications indicate that considerable progress has been made in this area of microgravity research. One, therefore, has to consider whether these published theoretical studies define the trends of microgravity experiments or whether they merely explain the observations and do not provide a stimulus to new experiments. In other words, we have to determine whether the current theory is the decisive factor in formulating new space-manufacturing experiments, as it should be, or whether the selection is based on other considerations such as intuitive ones or others with even less basis. As I am not a theoretician, I consider that the theory has made a considerable contribution to the solution of some manufacturing problems in space, but far from all of the theoretical studies give clear formulations of the tasks to be dealt with in microgravity experiments.

Theoreticians should be involved more in discussing proposed experiments in order to provide better selection, or, further, theoreticians should suggest experiments that would advance the theoretical principles of space manufacturing and that would solve fundamental problems in liquid and solid-state physics (in particular, crystal growth theory). I have come to this conclusion not only from familiarity with various theoretical publications, including those cited in [1] and in this chapter, but also from discussions with certain leading theoreticians (who, admittedly, do not work on space manufacturing). Some of them consider that the results from most of the expensive space experiments could have been predicted theoretically. The same view is found in papers on simulating microgravity processes.

To complete this discussion of the theory, I deal with theoretical evaluations of microgravity crystallization features. I cite particularly the conclusions of [58], where experiments on crystal growth in space were examined.

In [58], it is stated that microgravity crystallization and the actual structures of the crystals produced have some features related to changes in heat and mass transfer, which involve the following:

1. *Gas-phase growth*: $A^{II}B^{VI}$ compounds grown under microgravity result in a stoichiometry different from that for surface crystals evidently because of more pronounced diffusion effects.
2. *Growth from solution*: under microgravity, more bubbles are trapped than on the ground. The same applies to growth from liquids. There are signs of attraction and intergrowth. A possible reason is that the momentum transfer in crystallization is proportional to the growth rate, and the latter is less for parts in two crystals facing one another leaving a gap narrow by comparison with the crystal size.
3. *The liquid* does not wet the tube under microgravity (Ge or InSb) if the tube is silica, although the reason has not been established. It may be that contaminants such as carbon and roughness play an important part. It is

necessary to perform experiments with clean smooth surfaces and with drops smaller in size than the capillary constant.

4. *Impurity distribution in melt growth*: the zoning amplitude is less, but it does not vanish completely. The general impurity distribution along a cylindrical crystal made by directional crystallization was different in the Soviet experiments than that in the American ones, the two types being virtually purely diffusion-controlled and of the Pfann type. These phenomena may arise from instability in the microgravity or in the temperature, or from Marangoni convection.

5. *Dislocation densities*: these are 2–4 orders of magnitude lower in crystals grown from the liquid in microgravity than those made under analogous surface conditions. The accepted view is that the liquid does not wet the tube, but this is not a sufficient reason, since the most important and most general cause of growth dislocations is thermal stress.

In conclusion, in [58] it is emphasized that the level of experiments on crystal growth in microgravity for various objective reasons is lower than that under ground conditions because of poor thermal stabilization and inadequacies in the measured data and observations. This makes many of the statements on the causes rather hypothetical. Experiments designed merely to see what happens, or those on growing many different substances without any particular physical basis, should be replaced by strictly formulated experiments on particular phenomena, with apparatus providing good photographs and quantitative results.

The general conclusion from Chapter 1 is that theoretical research on microgravity physics and space manufacturing is fruitful, and it is desirable to extend it considerably, as it is essential to involve highly qualified theoreticians in discussing plans for future experiments in these areas.

Chapter 2

SEMICONDUCTOR GROWTH FROM MELTS
AND VAPORS UNDER MICROGRAVITY

In [1] (Chapter 2, pp. 37–78) and in [15] there are surveys on semiconductor growth under microgravity taking the state of the subject up to 1983, where about 140 papers are considered. Tables 1–3 and Chapter 2 in [1] indicate that the conclusions from those papers are mainly optimistic. Improved structure and properties occur in space crystals on account of the elimination of harmful gravitational convection and the prospects for growing the crystals without contact with the tube walls. Some physical phenomena important to semiconductor crystal growth under microgravity indicate that the heat and mass transferred there, as well as the growth generally, are determined mainly by diffusion because natural gravitational convection is suppressed. Under these conditions, there is accentuated nucleation, while the supercooling in phase transitions is increased, and nonequilibrium processes become more important, which can lead to various effects, which are not always useful in making materials, e.g., increased macroscopic impurity segregation. On the other hand, most experiments have shown reduced dislocation densities and less microstriation (microsegregation) along with other useful effects.

However, some experiments show that there are more gas bubbles in the liquid, while in certain cases the crystals are more defective, not less, and so on. Sometimes, the effects are more complicated than theoretical arguments suggest, i.e., existing theories do not explain all the phenomena. Furthermore, there is a lack of statistically sound measurements that could be used in theoretical developments on technological processes in space.

Examples have been given in [1] of current structure and electrophysical-property research methods applied to semiconductors made on the ground and in space, and there was a discussion of the most promising experiments on microgravity growth for the near future (Chapter 2.5 and Tables 12 and 13). The forecasts for new experimental processes in making semiconductors were accompanied by the conclusion that there is a pronounced need for microgravity experiments to solve fundamental problems in crystal growth, although great care is needed over the number of such experiments and their correct formulation.

Recent publications [317–395] on semiconductor crystal growth in microgravity may be considered in that light. Before we examined some of these papers, we noted that many of them dealt with more detailed studies on crystals made under microgravity, sometimes quite a long while ago, e.g., even during the Apollo–Soyuz flight of 1975. The new papers are naturally discussed on the basis of comparing the new results with those from earlier studies, as presented, in particular, in condensed form in [1] (Tables 2 and 3 in Chapter 2). The comparison is given in subsequent sections in relation to structure and properties of particular crystals grown under microgravity.

2.1. MICROGRAVITY GROWTH OF Ge AND Si CRYSTALS FROM MELTS [322, 344, 353, 354, 356, 360, 362, 366, 371, 376, 384, 392]

Germanium crystals doped with Ga and other substances have previously been tested for the segregation on the ground and in space, as have doped InSb crystals (experiments M 559, M 560, and MA 060 on pp. 45 and 48–49 in [1]), and the same applies to experiments with the Kristall apparatus (p. 55 in [1]). The experiments showed that microgravity gives a more uniform dopant distribution. Section 2.6, pp. 72–76, of [1] discusses these studies and states that diffusion-controlled segregation predominates under microgravity without appreciable signs of convection due to the residual accelerations, g perturbations, or Marangoni convection, and that further experiments are required to identify these effects. Space-grown Ge crystals show radial composition variations, which is ascribed to growth surface morphology: the surface deviated from planar, and it was concluded that thermal conditions should be selected for space growth giving flat fronts.

The new papers include [353, 354, 356, 392], which deal with dopant segregation in Ge under microgravity.

In [353] there is a theoretical analysis of Marangoni convection as affecting dopant segregation in Ge; it was concluded that Marangoni convection and residual gravitational convection must be suppressed simultaneously to attain the diffusion-limited condition under microgravity.

In [354] chemical etching was applied to the Ga distribution in Ge; an initial Ga concentration of 10^{18} cm^{-3} or more gave banded inhomogeneity under ground and space conditions. However, the band contrast was more prominent in the ground specimens, with the band spacing 30–50 μm, while in space specimens it was from 50 to 70 μm. It was concluded that the bands are not due to faults in the crystal drive mechanism but are due, instead, to temperature fluctuations arising from convection or the mode of melt temperature control.

In [356] estimates are made of the effects from melt mixing in directional crystallization of Ga-doped Ge on the ground and under microgravity; the mixing rate varies during growth, and there are related changes in diffusion-layer thickness at the front and in the effective partition coefficient. The segregation was found to be close to equilibrium at 1 g_0 and 10^{-3} g_0. The segregation was estimated also for 10^{-7} g_0, and it was shown that one could make crystals having improved structural and electrophysical homogeneity if nongravitational convection types can be suppressed.

In [392] solidification measurements were made on concentrated Ge–Si binary mixtures and dilute Ge–Ga mixtures on the ground with various orientations for the tube axes in relation to gravity. Bridgman's method was used in a furnace where the radial temperature gradients were minimized. The boundary-layer thicknesses δ differed considerably (δ_{Ge-Si} = 3 cm; δ_{Ge-Ga} < 2.5 mm) under identical thermal conditions, which could be explained from simple hydrodynamic models. It was found that: 1) convective Ga transport in the Ge–Ga system occurs on account of residual horizontal temperature gradients associated with the continuity in the thermal-parameter changes at the liquid–crystal boundary, and 2) the large boundary-layer thickness for Ge–Si and the correspondingly pure diffusion transport are due to the stabilization from the longitudinal concentration gradient. Convective effects at the gravitation level can thus be eliminated in dilute systems. On the other hand, pure diffusion transport can be attained in the vertical form of Bridgman's method at 1 g_0 with concentrated systems, where the solution concentration effect is stabilizing.

In [371] two rocket experiments are described to examine convection conditions as affecting dopant distribution in Ge growth with the addition of Ga at about $1.5 \cdot 10^{18}$ cm^{-3} in zone melting with 1-Hz pulses to provide labels at the front. The crystals were 5 mm in diameter; the growth rates were 2.5 and 5 mm/min, and the grown crystal lengths were 5 and 7 mm, respectively. Although both of the specimens were polycrystalline, interesting segregation profiles were observed. The markers and leakage-resistance measurement indicated the macroscopic segregation. The space crystals showed segregation profiles corresponding to conditions close to diffusion-controlled growth. Ground specimens showed marked effects from gravitational convection, much more so than the space specimens, which showed none of the striation typical of convective instabilities, whereas such striation occurred in the surface ones.

Germanium has been used [344, 362, 366, 371] as a model material for research on microgravity capillary shaping and floating-zone shape stability. These studies might be assigned to the capillary-phenomena section and research on wetting considered in Section 1.9 of Chapter 1, although a semiconductor material was used. In [344] it was confirmed that forms of Stepanov's method can be used in space to make shaped components, particularly from semiconductors. It is effective to supply the material through a capillary under microgravity. Theoretical estimates were derived from model calculations. The space specimen showed a more uniform dopant distribution along the capillary than did the control ground specimen. In [362] laminar flow through planar and cylindrical capillaries under microgravity was discussed via models for the conditions used in [344], where numerical estimates were made.

In [366] shape stability was considered for molten semiconductor and other materials in floating-zone growth. Constant-diameter crystals, i.e., those without deviations from strictly cylindrical shape, require a plot of volume against thinning, which is available only for complete weightlessness and for experiments in the absence of rotation. No general solution is available, and there was no discussion of the contact angles between the crystal and melt. In [366] a zone stability diagram was derived for a rotating liquid of any volume subject to axial acceleration, where allowance was made for these contact angles. Germanium zone-melting results illustrated the treatment.

We now consider microgravity solidification for spherical Ge and Si droplets [322, 384].

In [322] rocket experiments were described on crystallization of Ge doped with As, Cu, and other substances, with a comparison being with Ag. Disks, 10 mm in diameter and several mm high, were placed in evacuated tubes heated to 1200–1300°C (an outside package) for about 80 sec, with the material remaining in the molten state for 200–300 sec, which was followed by passive cooling at 50–100 K/min.

Most of the Ag droplets crystallized under microgravity were spherical, while the Ge drops had a shrunken form. The tip in the shrinkage pattern corresponded to the final crystallization point. Crystallographic markers occurred on the surfaces: ⟨111⟩ bases, dentate edges, and large ridges corresponding to ⟨100⟩ zones, as well as zones with a wave structure, which might be due to capillary waves in the liquid. The Ge + As droplets had an internal morphology showing that the crystallization began at the center, where there were signs of internal stress. The crystallization may have been initiated at a foreign particle or gas bubble. Surface nucleation was not observed even in the oxide film.

The most perfect crystal derived from a partially melted rod grew in the ⟨100⟩ direction, as the growth rate along it was higher than that along ⟨111⟩ and ⟨110⟩.

In the Ag specimens, nucleation occurred at the surface, in contrast to Ge, possibly because of microscopic inclusions. The heat transfer in the Ge was through the liquid, but in the Ag through the crystal.

The interstitial dopant Cu and the substitutional As used with the Ge showed that the As does not have time to enter the lattice, which differs from the situation in the normal Czochralski method. On the whole, fast microgravity crystallization gave substantial differences in shape and structure, which may be due to features of the heat transfer through the solid and liquid phases under microgravity.

The above are new papers dealing with germanium crystallization under microgravity. Experiments have been performed with Ge crystals in space even in the early 1970s, for the simple reason that the melting point is only 937.2°C. Silicon has a higher melting point, 1415°C, and research on it began only recently. In [1] it has been stated that such experiments were planned for SL-1 (ES-321 and ES-324, Table 12 in [1]). These experiments have now been performed and have been described in [376, 384].

These papers are considered in somewhat more detail than for those on Ge because it is new to crystallize Si under microgravity.

In [384] the purpose was to determine the effects of Marangoni convection on the impurity distribution in Si by melting and crystallizing the end of a silicon rod under microgravity. The MHF ellipsoidal mirror oven was used. The Si rod was grown by Czochralski's method on the ground, diameter 11 mm and length 50 mm, and was melted at the free end to produce a drop, about 1 cm^3. In experiment ES-324a, the rod rotated at 10 rpm, and after the drop had formed, it was withdrawn from the hot zone at 1 mm/min. Directional crystallization started from the Si rod, which acted as a (100) single-crystal seed. In experiment ES-324a, the rotation continued throughout the crystallization. In a second experiment (ES-324b), it was planned to crystallize the rod without rotation, but this was not done because of a furnace fault.

Although the drop under microgravity was spherical (ES-324a), the crystal deviated markedly from spherical because of the wetting angle between the liquid and solid Si; Fig. 25 shows the Si, which appears clean, with a lustrous surface. However, the picture was taken after the specimen had been etched, which removed a film containing a carbon compound. The reason for this contaminated layer was uncertain, although it was known that the subsequent experiment (ES-321) did not

Fig. 25. Silicon rod with sphere after solidification, magnification × 3 [384].

give such contamination. It may be that the molten Si in experiment ES-324a collected all the contaminant in the furnace.

This crystal was sectioned lengthwise and etched, which revealed marked rotational striation.

The distances between bands should be constant at about 50 μm, but in fact they varied: there was initially an increase from 40 μm to 300 μm along the axis and then a rapid fall to 80 μm. This was presumably due to growth-rate changes associated initially with the carbon dissolving in Si and thus affecting the melting point and then the deposition of SiC particles.

Parts a and b of Fig. 26 are two of the photographs illustrating the striation in [384], which differ in magnification, the obvious rotation striation is accompanied by weaker bands between the latter, which were ascribed to Marangoni convection. It was considered necessary [384] to perform a further experiment without rotation to confirm this. Clearly, in the next experiment it would be necessary to eliminate the surface contamination observed in experiment ES-324a.

Experiment ES-321 on SL-1 involved growing Si under microgravity by zone melting [376]; dislocation-free Si single crystals were prepared, each having a neck of diameter 3.5 mm at the start and passing gradually into a cylinder of diameter 10 mm. The total length was 110 mm. Figure 27A shows the initial form.

Table 13 on pp. 167–168 in [1] deals with the automated production of such specimens. The molten zone was produced by an MHF two-ellipsoid mirror furnace, which contained two 400-W lamps. The radiative heating eliminated electromagnetic forces, and the microgravity experiments eliminated gravitational convection.

Fig. 26. Striation at 5 mm from surface, magnification ×
90 (a), with many nonrotational bands as well as rotational
bands, magnification × 180 (b) [384].

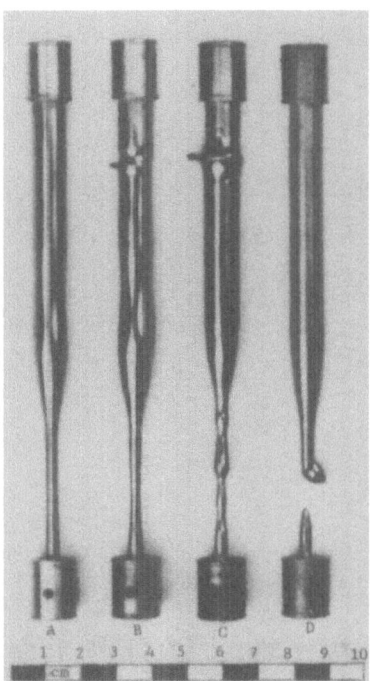

Fig. 27. Si crystals: A) initial specimen;
B) specimen made under ground conditions;
C) specimen obtained successfully under
microgravity; D) specimen obtained under
microgravity after zone breakage in neck
[376].

The purpose was to examine the striation in microgravity-grown Si crystals for comparison with specimens treated in the same MHF furnace but on the ground.

Two experiments were performed under microgravity. In one of them, the zone was passed along the entire specimen, while in the other, the specimen broke at the start (at the neck). Figure 27 shows four specimens: the initial specimen A, the surface (synchronous) specimen B, a specimen successfully recrystallized under microgravity C, and the specimen where the zone broke in the neck at the start, D.

The experiment with microgravity recrystallization throughout the length lasted 210 min. The accelerations during that time were $(0.5-2) \cdot 10^{-3} \, g_0$. The zone melting in the cylindrical part was partly with the furnace rotating around the fixed specimen and partly without rotation, in order to record the rotational and nonrotational striation on the ground and in space.

It is difficult to choose the lamp power for the MHF furnace [376] to maintain the required zone length under microgravity.

Under microgravity, there is no convection in the surrounding gas, so the heat transfer is rather different from that on the ground. On the other hand, if the lamp power is set too low, the zone length may contract to 0, i.e., to contact between the

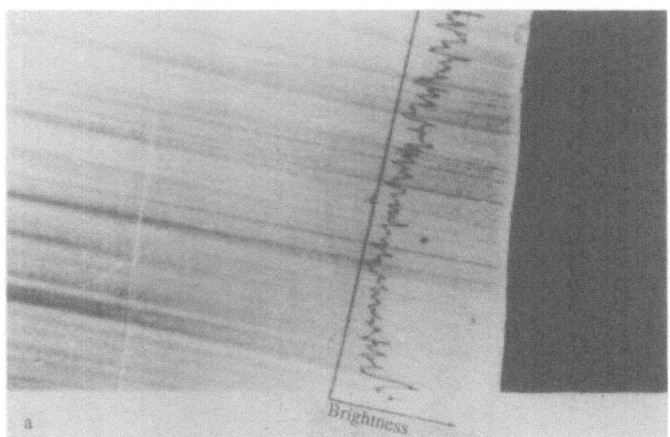

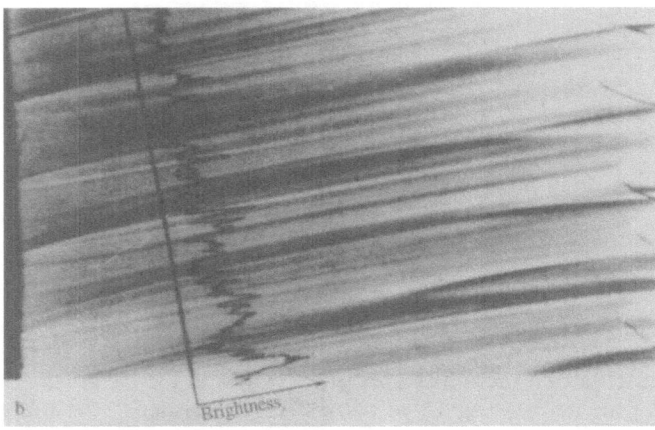

Fig. 28. Photomicrographs from the space specimen (a) and the ground specimen (b) at corresponding stages. The band brightness is indicated by the photometric curves [376].

upper and lower solid parts, which gives rise to several crystallization centers there and, thus, to the growth of several grains, i.e., the initial crystal is converted to a polycrystalline product. If, on the other hand, the power is higher than a certain limit, the zone length increases so much that it breaks. Both of these cases occurred in the SL-1 experiments.

The microgravity-recrystallized specimen showed the zone length contracting to such a contact; a polycrystalline material then grew (although the grains were larger).

The other specimen broke in the liquid zone because the lamp power was adjusted manually in space. In spite of these difficulties, it was possible to compare the two forms of striation. The recrystallized specimens were cut lengthwise with a diamond saw and etched. Photomicrographs were taken (350 in all), which provided information on the topography. The striation was also characterized by photometric scanning, which gave the contrast and the band separations, which indicated the periodicity. Figure 28 shows examples for ground and space specimens with the photometric curves. These pictures were taken from characteristic rotational parts. In [376] there are also pictures from nonrotational parts and examples of topograms from the part corresponding to the end of the experiment, with the zone halting in the upper part and solidifying when the heating was switched off. There are also schemes for major details, in particular the shape and the form of the solid–liquid boundary. The latter are different for the surface specimen.

The reasons for this are related mainly [376] to the heat-transfer conditions. Here we consider only the striation results for the two specimens.

The main conclusion is that the microstriation for the microgravity-grown Si characterizes the inhomogeneity due to thermocapillary convection; the similarities between the two specimens show that the microstriation on the ground is due to the same process, which is confirmed by Si crystals grown by zone melting on the ground with surface film protection, which did not show striation.

Finally, the microstriation separations are not periodic, which is due to turbulence in the molten Si going over to oscillating convection as theoretically predicted and as observed on other materials. The transition occurs if the Marangoni number exceeds a critical value. In [376] there is also a discussion of the reasons for the less contrasty microstriation in microgravity-grown crystals and the reasons for the less-prominent rotational bands in them, as is evident from Fig. 28.

The microgravity experiments with Si were fully justified, as they provided additional data on the growth of semiconductors more refractory than Ge in space, as well as information on the microstriation, which reflects the impurity distribution. Subsequent experiments in space should deal with Si to elucidate topics indicated in [376, 384].

2.2. MICROGRAVITY GROWTH OF $A^{III}B^V$ SEMICONDUCTOR COMPOUND CRYSTALS

$A^{III}B^V$ semiconductor compounds are widely used, and techniques for growing them on the ground are highly developed. Nevertheless, there are prospects for producing even better crystals under microgravity, and such studies have been

made since the 1970s and still continue. Microgravity growth for them is interesting also to provide further information governing fundamental concepts on gravitational and nongravitational convection affecting crystal growth, perfection, and
properties.

In [1] information was given on publications up to 1982–1983 relating to these
microgravity-grown crystals; in what follows, we deal with certain recent papers
relating to compounds in this class such as InSb, GaSb, GaAs, and GaP. We
consider separately papers not cited in [1] on each of these.

2.2.1. Indium Antimonide [320, 321, 329, 331, 333, 335, 338, 355, 364]

In [1] mention was made of experiments EK-6 (Table 1, p. 38), M 562 (Table
2, p. 45), and M 563 (Table 2, p. 45), as well as experiments under the USSR national program (Table 3, p. 56) dealing with InSb crystal growth under microgravity. In [1] there is also a discussion of Gatos's review (Section 2.6, pp. 72–76),
which surveys impurity segregation in microgravity-grown InSb. The conclusions
on InSb are as for Ge (see Section 2.1 above); viz., microgravity gives better crystals, since the structure and the impurity segregation are controlled mainly by diffusion without appreciable effects from convection due to residual accelerations, g
perturbations, or Marangoni convection. On the other hand, there are radial variations in impurity content in InSb and Ge, which are due to the crystallization front
not being flat. This effect can be eliminated by ensuring a flat front.

Gatos's conclusions in [1] (Section 2.6) enable one to consider the new papers, some of which relate to 1982 [329, 331], while the rest are later.

In [329, 331] there are discussions on mass transfer in the liquid and the Te
dopant distribution in InSb grown by directional crystallization under microgravity
and on the ground. The space specimens showed a contactless crystallization region, which was the most perfect part. On the whole, the space specimens were
much more perfect than the ground specimens, as they did not have the complicated
inhomogeneity of the latter and had low dislocation densities and few inclusions
decorating the $p–n–p$ junctions, while the junctions were close to flat. This shows
that microgravity reduces thermogravitational convection considerably, i.e., the
diffusion mechanism is predominant. The radial Te distribution in InSb was also
examined. The results in [329, 331] agree with Gatos's conclusions given in [1].
On the other hand, it is emphasized in [329, 331] that there was a thermocapillary
convection zone where the crystallization occurred without contact with the walls,
i.e., such convection affected the dopant distribution appreciably. It was also suggested that residual microaccelerations may have affected the transverse distributions in liquid and crystal.

In [320] there is a discussion of a theoretical treatment of the InSb impurity
distribution from microgravity, which incorporated Marangoni convection, but the
agreement between theory and experiment was not complete.

In [321] impurity segregation in InSb was discussed in relation to growth conditions.

In [333] and [335], as well as in [364], there are studies on impurity distributions and structure in InSb crystals containing $p–n$ junctions grown under microgravity. A distinctive point in these experiments was that two specimens were

placed in a tube instead of one: the first was seed p-type InSb doped with Zn acceptor and the second n-type InSb containing donor Te. The crystallization front shape could be judged from the p–n junction form. A neck appeared at the boundary in the space specimen, which was ascribed to Marangoni convection, which tended to equalize the dopant concentrations in that zone.

In [364] it was shown that the best part of a space crystal is near the neck, where the growth occurs without contact with the walls. An almost flat p–n junction was obtained under microgravity. It was concluded [364] that highly perfect InSb crystals can be made by directional crystallization under microgravity.

In [355] some other topics were discussed in the crystallization of InSb and InAs under microgravity, viz., shaping during drop crystallization, with the drop retained on the seed by surface tension. Directional crystallization in space causes the spherical drop to be distorted and a pit to be formed. The growth angle for small InSb droplets on the ground was 30°, while in space it was 15° (19.5° in the initial stage). The variation was probably due to a change in front curvature during growth and the corresponding change in the direction of the surface tension at the liquid–solid boundary with respect to the tangent to the liquid (a similar conclusion has been drawn [384] on the crystallization of Si under microgravity; see above in Section 2.2).

These results on growth of InSb under microgravity indicate that the conclusions drawn there do not always agree with those drawn by Gatos and given in [1] (Section 2.6). The new papers indicate that the microgravity front is almost planar and emphasize the importance of Marangoni convection, which is advantageous for impurity concentration equalization, while it is suggested that the residual accelerations may have an effect, whereas Gatos considered that front deviation from planar was the main reason for the radial inhomogeneity, while Marangoni convection and residual accelerations had only slight effects on the InSb crystals, although they required further study.

2.2.2. Gallium Antimonide [329–331, 341–344, 346, 363, 367, 368, 382, 385, 386]

In [1] there is a discussion of experiments M 563 (Table 2 on p. 46) and EK-7 (Table 1 on p. 38 and Table 3 on p. 56), in which microgravity conditions were used in an attempt to make GaSb crystals more perfect than those obtained on the ground, i.e., free from defects and having uniform dopant distributions.

In experiment EK-7 under the Soviet Hungarian Eötvös program, the Kristall apparatus was used with Bridgman's method without the seed to make GaSb under ground and space conditions. In [1] (pp. 62 and 63), some results are given on the morphology, and it is stated that the space specimen was a bicrystal of better quality than the ground specimen, which consisted of a set of elongated crystals approximately parallel to the growth direction. In subsequent publications [341–343, 346, 363, 382, 385], there are detailed studies on the morphology, structure, and electrophysical parameters for space crystals, which are compared with ground crystals by electron and optical microscopy, x-ray diffraction, electron channeling, and Rutherford ion backscattering.

In [341–343] comparative studies were made on ground and space specimens by these methods; we cannot go into the details and do not give the figures (they are given in part in [1]), and merely repeat that the results show that the space-grown GaSb specimens had better structures.

In [382, 385] results were given on GaSb crystals grown by Bridgman's method under microgravity. We give the closing discussion from [385], which concerns the morphology of the products obtained under microgravity.

In [385] it is stated that GaSb crystals resemble others grown previously under microgravity in contacting the tube surfaces only at isolated points, which substantially altered the heat-transfer conditions by comparison with the ground (reduced the radial temperature gradients) and provided higher perfection (pure crystallization centers at the surfaces). At the start, molten GaSb tends to acquire the most favorable form: a cylinder with hemispheres at the ends. The approximately 8% volume reduction on melting allows the liquid to leave the pointed end of the tube and form a curved surface. Nucleation begins on cooling not at the spherical end but along a ring at the upper end of the molten hemisphere, which is closest to the wall. The spherical end is free from ridges, and the surface above the ring bearing the nucleation centers is thickly covered with them. These regions were carefully examined by ECP, but no boundaries were observed. This confirms that the ridges are formed near the perfect volume, but all attempts at explanation were unsuccessful. It proved impossible to explain the details in terms of wetting, convection caused by surface-tension gradients, or other effects that cause liquid flow.

The electrophysical parameters have been characterized [363] from Hall-effect measurements by van der Pauw's method, along with conductivity-type determination and resistivity, major-carrier concentration, and mobility measurements at 300 and 77°K. It was found that the conductivity was p-type, and the carrier concentration $(1–2)\cdot10^{17}$ cm^{-3} at 300°K. There was no marked difference in electrophysical parameters between the space and ground specimens [363].

In [346] GaSb crystals made on the ground and under microgravity were compared for the kinetic effects: resistivity ρ and Hall coefficient R as functions of magnetic field (up to 31 kOe) and temperature (between 1.9 and 300°K). The following formula applies for the T dependence of the carrier concentration at 20–100°K:

$$\rho = \frac{1}{\beta}\, N_V T^{3/2}\frac{1-K}{K}\exp\left(\frac{E_L}{kT}\right), \tag{2.1}$$

where $\beta = 4$ is the degree of valence-band degeneracy, $K = N_d/N_a$ is the degree of acceptor compensation, and ε_L is the acceptor-dopant activation energy. The $\rho(H)$ and $R(H)$ curves have inflections at 5–7 kOe, which is due to there being two types of holes, which differ in effective mass and mobility. At lower temperatures, the formula becomes $\rho = \rho_0 \exp(E_3/kT)$, where E_3 is the conduction activation energy. Values of ρ_0 and E_3 are given in [346]. Below 3°K there is a transition to conduction having a variable activation energy. Below 16°K, ρ for the space specimen was an order of magnitude higher than that for the surface specimen because it is exponentially dependent on the distances between impurity centers, which differed for the two specimens.

These EK-7 microgravity-grown GaSb crystals made by Bridgman's method showed some distinctive features in morphology, structure, and parameters, but there were various aspects requiring further research.

We now consider [367, 368, 386], which deal with experiment ES-323 on GaSb crystals on SL-1, which had been planned long before (Table 12 on p. 158 in

[1]) and was designed to attain stationary transport conditions in making gallium antimonide crystals under microgravity with a moving heater with the object of producing crystals more perfect and homogeneous than those made on the ground.

The crystals were grown in an MHF two-ellipsoid mirror furnace; the Ga + 3% Sb solution zone, height 6 mm, was located between the GaSb seed and the source material (a polycrystalline substance), which were located in a vertical silica tube. The crystal diameter was 10 mm, the growth rate 4.5 mm/day, and the crystal rotation speed 8.4 rpm.

It was planned to grow a crystal of height 3.5 mm, but premature termination of heating produced only 150–200 μm. Nevertheless, the dopant distribution was examined by sectioning the crystal along the growth axis and etching the surfaces, which provided a comparison of the striation with ground specimens.

A space specimen was less inhomogeneous than the ground specimen; there were merely a few rotational bands in the initial 40 μm, principally in the growth facets, whereas the ground specimen showed prominent rotational striation throughout the length.

Two crystals also showed nonrotational striation at the middle (from 70 to 100 μm); between the bands due to growth-kinetic fluctuations there were other banded structures of unknown nature. In the space crystal, this previously unknown striation appeared only close to defects such as inclusions and at the end of the growth facets. In contrast, the ground crystal showed this striation throughout the section. The end of the space crystal was completely free from striation.

The space crystal showed only a few rotation bands, so it is concluded that those bands are formed on the ground mainly as a result of convection.

The ES-323 results on the whole indicate that growth in space can give GaSb crystals with improved homogeneity. However, the results require confirmation, and it is planned to repeat the experiment on Spacelab D-1 and, at the same time, to perform a new experiment on growing InP crystals by the THM method with the use of a new single-ellipsoid furnace. In other words, it is considered promising to develop the moving-heater method for microgravity conditions to make perfect homogeneous semiconductor crystals of $A^{III}B^V$ type.

2.2.3. Gallium Arsenide [323, 324, 326, 343, 359, 361, 379, 380]

As we have considered InSb and GaSb crystal growth under microgravity, we can deal more briefly with other $A^{III}B^V$ crystals, the more so since certain papers such as [323, 324, 326, 343] deal simultaneously with GaSb and GaAs crystals and with epitaxial structures based on them.

In [323, 324] two experiments are described: "Diffusion" (a) and "Antimonide" (b) (Fig. 29). The purpose of the first was to examine As transport when GaAs dissolves in its liquid, with a view to making GaAs films by liquid epitaxy on GaAs substrates under microgravity. The purpose of the second was to examine transport during GaSb crystallization under microgravity. The initial Te-doped GaSb ($n \sim 10^{18}$ cm^{-3}) had prominent inhomogeneity striation. It was assumed that the striation should be reduced under microgravity. In experiment (a) it was found possible to perform liquid epitaxy for GaAs under microgravity. A model was proposed for the effects of residual accelerations on the transfer in (a) and (b), although it was concluded that precise evaluations require more knowledge on the physicochemical properties of these liquid materials, the temperature distribution, and the residual accelerations.

There were certain deviations from the specified thermal conditions under microgravity in (b), so specimens with less striation were not obtained.

In [326] ground developments on zone melting were used in examining factors influencing transport in making Ge, Si, GaSb, and GaAs crystals by crucible-free zone melting under microgravity. It was found impossible to produce a planar front in the ground experiments. There was also a discussion of the effects from the oxide film on the heat and mass transfer in the liquid, as well as the conditions giving rise to radial inhomogeneity with diffusion-limited transfer under microgravity.

In [359] new experiments were formulated on growing GaAs under microgravity; in [361] ground and space conditions were used with a moving solvent to grow GaAs crystals in silica tubes, diameter 12 mm, in a gradient zone ($dT/dX = 80°C/cm$); under microgravity ($g < 10^{-4}$ m·sec^{-2}), mass transfer is controlled by diffusion and the crystals are more perfect. In [379] it is emphasized that certain firms in the USA are interested in growing GaAs in space and that such experiments have been performed under a NASA contract on Space Shuttle flights.

An interesting innovation has been proposed in [380] for growing GaAs from the liquid under microgravity; instead of the ordinary cylindrical tube, a three-faced prism would be used, where full advantage can be taken of space conditions. This new configuration enables one to monitor the stoichiometry. Exact maintenance of stoichiometry ensures good GaAs crystals.

The liquid takes the shape corresponding to minimum surface energy under conditions of weightlessness; the liquid in a three-faced prism becomes cylindrical, with a circular cross section, if it does not wet the surfaces, as shown schematically in Fig. 30a. The GaAs fills the three-faced section completely (Fig. 30b) only if there is good wetting. The empty spaces at the corners constitute space for adaptation to expansion on solidification. They also form three channels through which the vapor with a controlled pressure is in contact with the liquid. Figure 30c shows GaAs grown by Bridgman's method in such a tube. The As vapor pressure required to control the stoichiometry is adjusted via the temperature of the As source at the end. The tube may be made of silica or of boron nitride, as molten GaAs does not wet these.

Preliminary experiments on the ground have confirmed that the approach is useful. Three-faced containers were made from pyrolytic BN having sides of 3 mm, where it was shown that the liquid does not fill the cavity completely and leaves channels as predicted. In [380] emphasis is placed on the unique advantages of microgravity for growing GaAs from the liquid, as one cannot only control segregation and concentration-dependent supercooling but also monitor the stoichiometry exactly for undoped crystals, which is necessary in order to provide semiinsulating GaAs.

The favorable effects from suppressing convection·have been demonstrated [380] on the ground by employing a magnetic field, which increases the kinematic viscosity; undoped GaAs was found to show pronounced striation on selective etching and from cathodoluminescence in the part grown without the field. The inhomogeneity decreased as the field H increased, and at 1500 Oe the microscopic inhomogeneity was reduced by an order of magnitude (Fig. 31). It was assumed that the inhomogeneity was due to stoichiometry deviations, not ordinary impurity segregation. Consequently, suppressing convection favors reduced inhomogeneity in stoichiometry, not merely in the impurities.

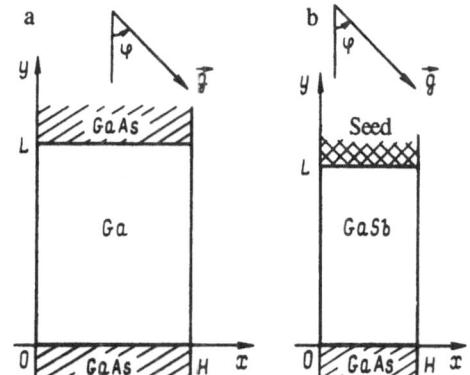

Fig. 29. Model for the "Diffusion" experiment [323].

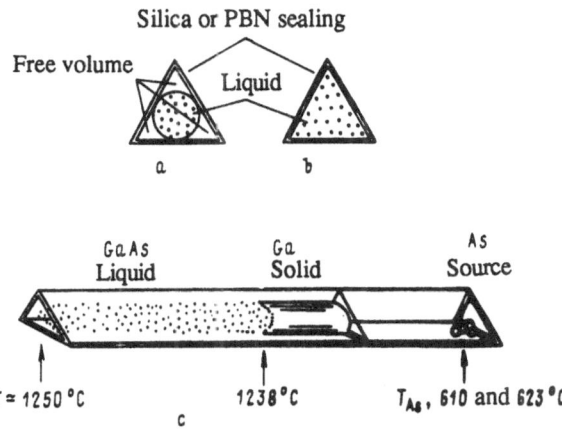

Fig. 30. See the text [380].

2.2.4. Gallium Phosphide [190, 349, 350, 387–389]

In [1] mention has been made of papers on GaP grown under microgravity with a moving solvent (Table 1 on p. 39 and Table 3 on p. 57). The papers cited in [1] and recent ones [190, 349, 350, 387–389] give results that largely reproduce or confirm the above conclusions on crystal growth under microgravity: the pure diffusion transfer mechanism in the liquid and the complicated interaction between the liquid and the wall, which improved the structure and homogeneity.

However, some new results have been obtained, particularly since the GaP system has the largest concentration dependence for the density in the $A^{III}B^V$ group [190], which should accentuate disequilibrium effects under microgravity [349, 389]. A model has been proposed for periodic growth from a supercooled

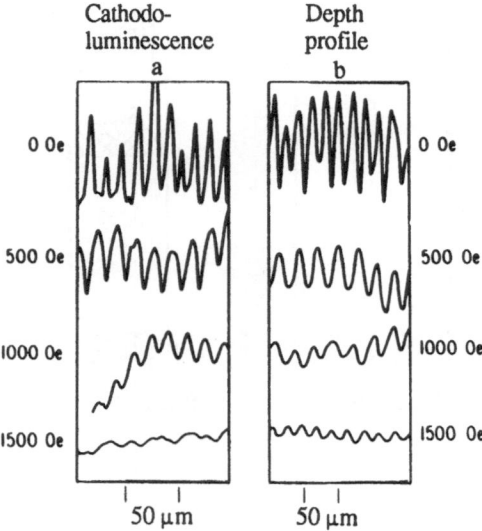

Fig. 31. See the text.

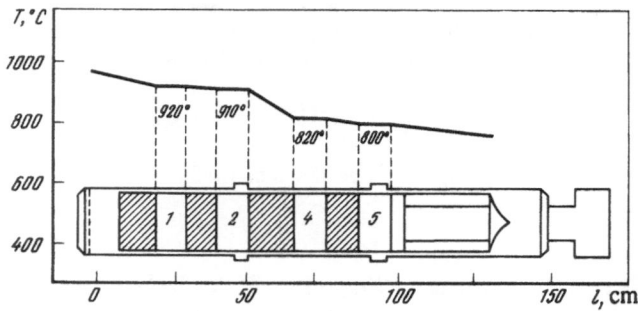

Fig. 32. Capsule used for GaAlAs homoepitaxy and heteroepitaxy under microgravity: 1, 2, 4, 5) graphite containers holding GaAs specimens and Ga solution; 3) silica tube; 6) metal cylinder; the temperature distribution at the start of growth is shown [328].

liquid, which has been used in estimating the impurity-concentration oscillation period. The period and amplitude are substantially dependent on the liquid transport mechanism. Under pure diffusion conditions, concentration-dependent supercooling can lead to macroscopic growth periodicity, whereas convective transport gives rise to the same mainly from microscopic segregation, with much smaller period and amplitude. Further checks are required on the suggestion that nonequilibrium phenomena have pronounced effects on binary-alloy growth under microgravity.

In [349, 350, 387, 388] there are results on gas bubbles in GaP [and also in $Bi_2(Te, Se)_3$] formed under various growth conditions. Heterogeneous formation at surfaces has been suggested arising from gallium oxide films interacting with the liquid gallium, with the release of a volatile lower gallium oxide. The bubble data indicate that this harmful phenomenon in GaP can be suppressed by removing the oxide layer by first wetting the crystal with gallium as well as by reducing the layer growth temperature. One can thus anticipate a considerable improvement in GaP crystals grown from fluxes under microgravity.

2.2.5. GaAlAs–GaAs Epitaxial Systems [328, 375]

In [328, 375] there are results on growing epitaxial GaAs homostructures and GaAlAs–GaAs heterostructures under microgravity with the Splav apparatus. The experiment tested a liquid-epitaxy technique under microgravity involving a thin layer of Ga liquid as flux.

Figure 32 shows the capsule used for the space experiment in [328, 375]; each of the zones 1, 2, 4, and 5 has a graphite container, each containing four circular GaAs:Te plates previously wetted on the ground with gallium at 650°C in hydrogen. The plates are assembled into two sandwiches by means of two separating graphite rings. Between the specimens in zones 1 and 4 there was Ga–Al–Zn solution, and GaZn in zones 2 and 5. Figure 32 shows the initial longitudinal temperature distribution. The epitaxial films were produced on controlled temperature reduction at 0.75 deg/min over a range of 200°C, which was followed by passive cooling.

Here we merely state that the films showed an irregular morphology due to merging in the film zones, which grew independently evidently because the substrate was poorly wetted.

Diodes were made from the specimens, whose voltage–current curves confirmed that there were $p–n$ junctions, in which the donor and acceptor concentrations were of the same order.

It was concluded that such experiments should be continued, with particular attention to producing stable GaAlAs growth centers on GaAs, as well as the diffusion of Al and Ga and the Zn contents in the solid phase.

We complete this discussion for $A^{III}B^V$ crystals grown under microgravity before proceeding to discuss other compounds such as $A^{II}B^{VI}$ and $A^{IV}B^{IV}$ by indicating two basic trends in semiconductor materials science. The first is associated with improving techniques for making known compounds, particularly Si and $A^{III}B^V$ considered above. The second is related to novel or improved technologies for making and doping new materials such as ternary solid solutions containing $A^{II}B^{IV}$ and $A^{IV}B^{VI}$. Section 2.3 deals with experiments under microgravity on making these important compounds.

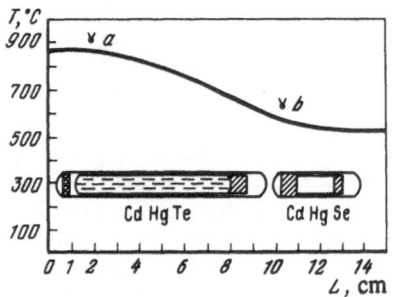

Fig. 33. Temperature curve for the Splav-01 furnace at the start [327].

A

B

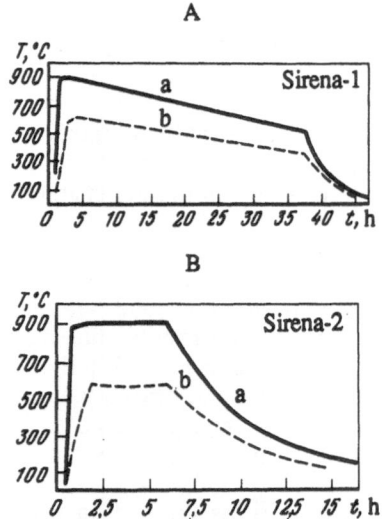

Fig. 34. Temperature as a function of time in the hot region (a) and cool region (b) in the first experiment (A) and the second experiment (B) [327].

2.3. MICROGRAVITY GROWTH OF A^{II}B^{VI}, A^{IV}B^{VI}, AND OTHER SEMICONDUCTOR SOLID SOLUTIONS

In [1] (Section 2.5.1, pp. 64–72) mention has been made of the technical significance and advantages provided by A^{II}B^{VI} compounds and their solid solutions grown under microgravity. It is difficult to make perfect specimens of solid solutions for these compounds, but they are extremely valuable for two reasons. First, the basic parameters can be varied continuously over wide ranges in the solid solutions; second, solid solutions of these compounds, in principle, extend the materials for which conduction-type inversion is possible, which is necessary in making homojunctions. Also, these solid solutions are important to semiconductor physics, since they demonstrate new effects, one of which is the transition from the semiconducting state to the gap-free state, and others are band-spectrum rearrangement on crystal structure change, and so on. In what follows, I examine papers not cited in [1] on some of these important materials.

2.3.1. Cadmium Telluride CdTe [393]

In [393] experiment ES-322 is reported (see [1], Table 12, p. 158, for the basis), which involved growing a CdTe crystal on SL-1 with a moving heater and Te as solvent. A silica tube ($l = 110$ mm, $d = 10$ mm) contained a cylindrical CdTe seed (length 18 mm), a solvent Te zone (length 4 mm), and a source CdTe rod (length 37 mm). The tube was heated in an MHF mirror furnace. The speed was 6 mm/day. Unfortunately, after growth for 6 h (instead of the planned 17), the furnace was switched off by the control computer because of a cooling fault. At that time, a layer of only about 1.5 mm had grown on the seed (instead of the planned 4.5 mm). The heating was switched off suddenly, i.e., there was no slow controlled cooling, so the system was quenched and the specimen was subject to large thermal stresses. The solidified zone and the adjacent part of the crystal therefore contained many microcracks, which, in conjunction with the small volume, prevented one from obtaining conclusions on the microgravity characteristics. All the same, these ES-322 data from detailed comparison with a ground specimen showed that the moving-heater method can be used under microgravity to grow CdTe crystals better than ground crystals.

2.3.2. Ternary Compounds: CdHgTe (CMT) and CdHgSe (CMS) [327]

In [1] (Table 1, p. 40 and Table 3, p. 57) mention has been made of CMT and CMS crystal growth from the liquid under microgravity in the Splav apparatus.

In [327] structure and property data are given, so we quote some of the results.

The purpose was to determine the effects of microgravity on the chemical homogeneity in these three-component semiconductors for various growth rates.

The initial $Cd_xHg_{1-x}Te$ was a polycrystalline specimen containing 22 mole % CdTe and 78 mole % HgTe made from the liquid phase by rapid cooling. The initial $Cd_xHg_{1-x}Se$ was a polycrystalline material containing 10 mole % CdSe and 90 mole % HgSe. The two specimens in separate tubes were placed one behind the other in a single capsule, as shown in Fig. 33, which shows the initial temperature curve, while Fig. 34 shows the temperature variation over time in the two different experiments. The crystallization rates differed substantially (by a factor six). The space experiments were successful, and the specimens were examined in detail initially by nondestructive methods (x-ray diffraction and electron microprobe), and then the crystals were sectioned and the Cd distributions were examined by microprobe analysis. The electrical parameters were also measured.

A new, unexpected effect was that there was a thin surface layer (about 30 μm) having an elevated Cd content, no matter what the growth rate, probably formed on account of surface tension. The sharply defined surface layer ruled out an explanation, for example, from mercury diffusing outside the material. The effect was not observed in ground specimens.

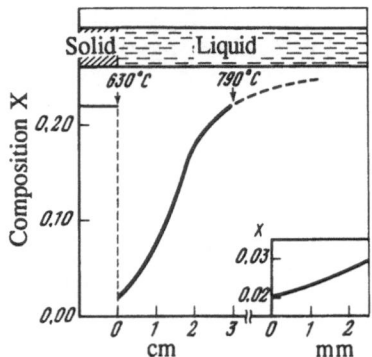

Fig. 35. Computed diffusion profile
for Cd in the liquid near the boundary
[327].

We do not consider the results of [327] on these crystals in any more detail and
merely give a brief discussion of them. The directional crystallization in [327] differed in general from that of crystallization from the liquid. Only part of the crystal
melted, and the melting ceased when the boundary temperature reached the solidus
value T_S for the given composition. That temperature was 680°C for $x = 0.22$. The
phase diagram shows that the liquid at the boundary should have $x = 0.02$. Far
from the boundary, the liquid had the composition of the load, $x = 0.22$. The composition gradient caused diffusion to the boundary, which may be affected by convection. A liquid having $x = 0.22$ can exist below 790°C only in supercooled form,
as has been observed on the ground, which leads to composition inhomogeneity
between the seed and the first recrystallized part, range from $x = 0.2$ to $x = 0.4$.
This was not observed in space at growth rates less than 3 mm/h; at 12 mm/h, the
space specimen showed considerable inhomogeneity, but less than in the synchronous ground experiment.

Figure 35 shows the Cd profile before the start calculated approximately in
[327]; if this pattern is realistic (although the calculation did not incorporate some
relevant factors), then supercooling effects do not perturb the diffusion profile, and
the crystal should be homogeneous. When the growth rate becomes too high, the
boundary lies in a region with a high composition gradient, and stationary conditions do not apply. Liquid composition instability at the boundary leads to simultaneous crystallization at different points and to temperature fluctuations, which give
rise to spontaneous inclusions and pores. The main difference between the two experiments with CdHgTe is that one cannot get the full profile of Fig. 35 on the
ground, so supercooling effects become very important, which complicate obtaining homogeneous material. The following are some of the conclusions in [327]:

1. At growth rates less than 3 mm/h under microgravity, the process is controlled by diffusion, and one obtains a homogeneous material of improved
   structure, but these conditions are not attained in reference specimens
   grown at the same speed on the ground.
2. Microgravity growth at a low enough rate can eliminate the cavitation associated with the high mercury vapor pressure (up to 20 atm).

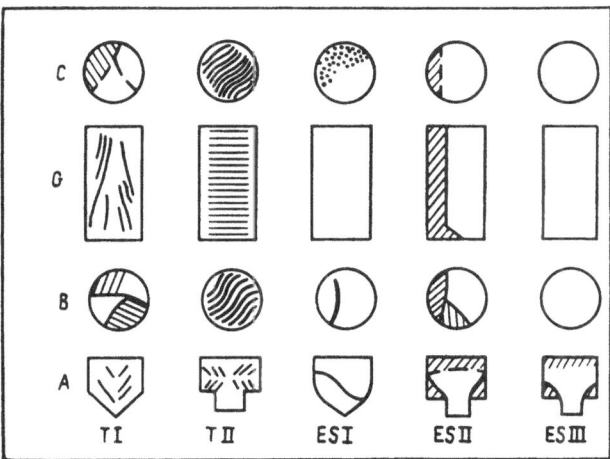

Fig. 36. Schematic representation of PbTe and striation seen at the surface after polishing and electrolytic etching [391].

3. It is desirable to continue this research on ternary compounds (particularly mercury and lead chalcogenides) in space to provide good material for use in infrared detectors and tunable lasers.

2.3.3. Lead Telluride PbTe [347, 348, 391]

In [1] (Table 1, pp. 39 and 40 and Table 3, p. 56) papers are cited dealing with microgravity growth of $A^{IV}B^{VI}$ compounds, and it is also stated (Table 12, p. 156) that experiment ES-318 was planned for SL-1. Here I consider new papers on this topic.

In [391] ES-318 results on microgravity-grown PbTe are compared with ground crystals. The GHF gradient furnace was used with Bridgman's method. One crystal was grown without a seed and two with seeds. The thermal cycle was operated normally. The two crystals with seeds grew well. The third suffered from marked g perturbations during the flight (unexpected accelerations during growth). The surfaces showed signs of changes in microgravity. In [391] the space and ground crystals were compared for morphology, imperfections and banded Ag distribution (10^{18} cm^{-3}). Figure 36 gives the main results on a comparison of the structure of the ground specimens TI and TII with the space specimens (ES-I, ES-II, and ES-III). Specimens TI and ES-I were grown without seeds, and the others with them. The structures were examined by metallography (including selective etching) and x-ray diffraction, while the dopant distribution uniformity was examined by microprobe (IMMA).

The ES-318 results indicated [391] that large and perfect crystals can be grown in space if substantial g perturbations can be excluded.

In [347, 348] detailed measurements were reported on the electrophysical and thermal parameters of p-PbTe single crystals at 2–100°K grown by directional crystallization on the ground and under microgravity on Salyut-6 with the Splav-01

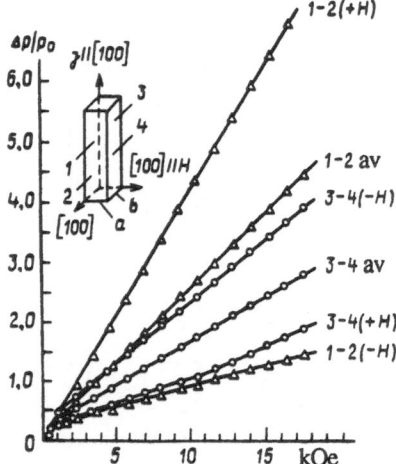

Fig. 37. Magnetoresistance M_c ($\Delta\rho/\rho_0$) at 4.2°K as a function of field strength and direction for a space specimen in measurements on face a. The left-hand side shows the disposition of the measurement contacts and the field direction [348].

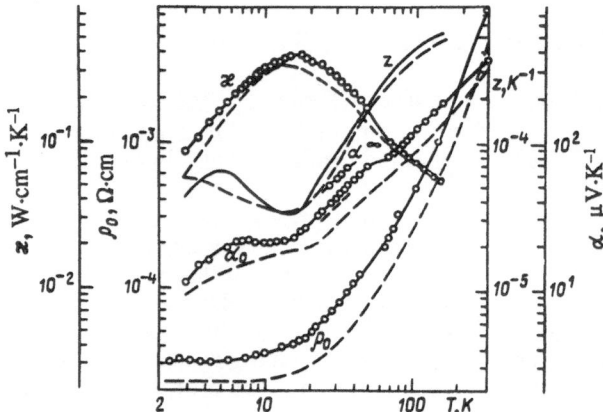

Fig. 38. Temperature dependence of the resistivity ρ_0, thermo-emf a_0, thermal conductivity $\varkappa$, and thermoelectric quality factor z for space specimens (solid lines) and ground specimens (dashed lines) [348].

apparatus. The low-temperature transport effects were compared for the two. The galvanomagnetic and thermomagnetic effects were examined in fields up to 18 kOe. The parameters and temperature curves are given for the two specimens. The results are:

1. The magnetoresistance in a transverse field in both cases showed a switching effect, with the maximum value $[\Delta\rho(+H)/\Delta\rho(-H)] = 6$ for 18 kOe at 4.2°K (Fig. 37). Hall-effect measurements showed hole-concentration inhomogeneity; $\Delta\rho/\rho$ is a linear function of H for strong fields (Fig. 37a).

2. The thermo-emf is of diffusion type above 20°K, but below that point there is a considerable effect from phonon-hole entrainment. The diffusion emf at $20 < T < 40$°K in strong magnetic fields gives a calculated mass at the Fermi level the same for the two specimens: $m^*d(\varepsilon_F) \approx 0.19m_0$ at 25°K. The mass found for holes at the band bottom in the Kane approximation agrees with published values.

3. The temperature dependence of the thermo-emf α, thermal conductivity $\varkappa$, and resistivity gives the thermoelectric quality factor $z = \alpha_2(\varkappa\rho)$ (Fig. 38), which was much the same for the two specimens.

In [348] it is pointed out that the space and ground specimens differed in hole concentration by about a factor 1.5, but there was no substantial difference in the integral transport parameters. However, the space specimen had a carrier distribution more stable along the crystal and transverse to it.

2.3.4. PbSnTe (LTT) Semiconductor Compounds [365, 370, 372–374, 377, 378]

All papers cited so far deal with the need for growing LTT crystals on Spacelab and ground development experiments, mainly with Bridgman's method. Space preparations involve theoretical calculations and developing methods of examining microscopic composition variations.

In [365] a new electrochemical etching method is given, and thermodynamic calculations are used to propose three etchants that give good indications of composition inhomogeneity in LTT crystals. An electrochemical method used there revealed microscopic inhomogeneities in LTT crystals grown in a thermally unstable Bridgman system (on the ground), and it was shown that the etch patterns are reproducible in the sense that grinding and polishing followed by fresh etching will reproduce the original picture. The lines in it show that there was oscillating growth. The solid–liquid boundary is concave to the solid. As the growth advanced, the oscillations began to extend along the interface and ultimately completely disrupt it.

This method enables one to examine growth front kinetics.

In [372] a series of LTT growth studies was performed by Bridgman's method; theoretical calculations showed that convection does not influence the temperature pattern in the liquid at small Prandtl numbers but does affect the composition distribution substantially, as well as that convection exists in Bridgman's method even in a thermally stable configuration. A method was developed for labeling the front by inserting electrodes in the vacuum tube. This showed that an anomalous initial stage was followed by convection-controlled growth as predicted by Pfann. The thermophysical parameters are required for more precise theoretical growth evaluation. In [372] measurements are reported on the temperature dependence of the specific heat up to 722°C, thermal diffusion up to 1010°C, and PbTe–SnTe mutual diffusion coefficient.

In [370] theoretical calculations and ground experiments by Bridgman's method with a vertical furnace showed that convection occurs on the ground even for very narrow tubes (2 mm), while the growth conditions in space differ for alloys differing in solubility.

In [372] there is a discussion of an experiment to be performed under the NASA Shuttle program in November 1984, but it is evident that it was not performed in 1984 because in [377, 378] the same authors discuss ground developments and the basis for growing LTT crystals on a Shuttle flight during the fourth quarter of 1985. We give here merely a few details from [378], where there is a discussion of directional solidification in semiconductor crystals, and it is shown that the corresponding effects influence the device quality. Liquid flow dynamic factors influence the solidification front and the composition. Reduced gravity reduces the convection, but does not eliminate it completely. Theoretical and experimental results are given on the liquid dynamics. Ground experiments showed that there is considerable gravitational convection in molten LTT.

Convection affects device quality not only due to composition inhomogeneity but also because of defects. Infrared detector sensitivity and laser emission are adversely affected as a direct result of convection and can be derived as functions of the inhomogeneity.

Stationary and moving furnaces have been used on the Shuttle in directional solidification of $Pb_{1-x}Sn_xTe$; the experiments enable one to determine how far convection is suppressed and the result from reducing convection on the properties. The main attention has been given to measuring and monitoring the temperature pattern.

The main facilities used in determining the effects of growth technology on crystal properties used in [378] included optical microscopy following electrochemical etching, x-ray analysis, and electron-probe methods.

These studies have led to a sound formulation for growing LTT crystals under microgravity, and an experiment is planned for the end of 1985 under the NASA program.

2.3.5. Solid Solutions of Various Semiconductor Compounds [317, 330, 336, 339, 345, 351, 352, 358, 388, 389]

Solid solutions of other semiconducting compounds have also been grown under microgravity; here I list some of the new papers dealing with such research.

a) **In–Sb–Bi crystals** [330, 336]. In [330, 336] results are given on the crystallization of In–Sb–Bi alloys for two compositions: 1) 7 mass % InSb + 93% BiSb; 2) 48% InSb + 52% BiSb; bulk and directional crystallization were used as well as moving-flux methods. Structure comparisons for space and ground specimens indicated some microgravity features: one can make In–Sb–Bi alloys with uniform distributions and having compositions $InSb_{1-x}Bi_x$ in the bulk. The number of such crystals in the product is much smaller than in ground specimens, and the sizes are larger, the crystallographic forms also being more regular; furthermore, there is an unsymmetrical dissolution boundary for the InSb in contact with the InSb–InBi liquid, which indicates that there is a residual liquation effect under the microaccelerations on Salyut-6. The composition inhomogeneity in the InSb–InBi liquid along the boundary is related to the gravity level.

These results on ternary systems are of interest for the general laws governing the production of other such compounds with uniform phase distributions under microgravity.

b) **$Pb_{1-x}In_xTe$ crystals** [317]. In [1] (Table 3, p. 52) some information was given on growing these crystals under microgravity. In [317] $Pb_{1-x}In_xTe$ crystals ($x = 0.0041$) grown on the ground and under microgravity were compared, where the Kristall apparatus was used on Salyut-6. Scanning electron microscopy and local x-ray structure analysis were applied to determine the In distributions and morphology in the two specimens.

These results confirmed the limited liquid contact with the tube known from other crystals and also the better space homogeneity. In addition, it is pointed out [317] that there is growth from the vapor as well as the liquid because of the mass transfer to the colder part of the tube and some parts of the crystallized material.

There was pronounced radial composition inhomogeneity, particularly for In, in the space specimen obtained from the liquid: the In was displaced to one of the sides.

c) **$(Bi_{1-x}Sb_x)_2Te_3$ crystals** [339, 358]. In [339, 358] results are given on the thermoelectric parameters of Bi–Sb–Te crystals grown on the ground and in space by directional crystallization. Crystals of $(Bi_{0.5}Sb_{0.5})_2Te_3$ and $(Bi_{0.25}Sb_{0.75})_2Te_3$ have been made. The ground specimens were grown with various orientations of $\vec{g}_0$ relative to the temperature gradient ∇T. Space specimens and those made on the ground with $\vec{g}_0 \parallel \nabla T$ (the upper end hot) showed no differences in the thermoelectric and galvanomagnetic transport coefficients, or in the sizes and distributions for the single-crystal regions, or even in the concentration distributions. A diffusion-controlled concentration profile applied. Crystals grown on the ground with $g_0 \uparrow\downarrow \nabla T$ showed convection. There was a considerable improvement in thermoelectric performance for microscopically homogeneous crystals having the optimum charge-carrier concentration, which can be obtained under microgravity with appropriate thermal conditions.

d) **$Bi_2(Te, Se)_3$ crystals** [345, 349, 351, 352, 388, 389]. In [1] (Table 3. p. 57) results have been given on morphology, structure, and electrophysical parameters for $Bi_2(Te, Se)_3$ crystals grown on the ground and in space, where it was shown that nonequilibrium effects influence the growth more under microgravity. In [349] it is confirmed that some morphological features (extended grooves and facets) are closely related to changes in structure (reduced block size) and composition (periodic concentration variation), which reflect nonequilibrium growth conditions (changes in growth rate, supercooling, concentration, and temperature). In [352] the transport measurements at low temperatures showed that space and ground $Bi_2(Te, Se)_3$ specimens differ in electron-concentration distribution as well as in the longitudinal and transverse inhomogeneity. On the other hand, there were no substantial differences between the two types as regards integral transport parameters such as the temperature dependence of the resistivity, thermo-emf, and thermal conductivity at $2-300°K$. In [345] an attempt was made to incorporate the dynamic effects from bubble formation in the $Bi_2Te_3-Bi_2Se_3$ liquid; in particular, there was a discussion of the time dependence, the vapor pressure, and the related critical bubble radius, where estimates were made of the time course of the liquid and gas compositions. This paper is thus mainly one dealing with bubbles forming under microgravity in any liquid and does not characterize crystallization of binary semiconductor alloys, so we restrict ourselves here to what has been said (see Section 1.7 for more details).

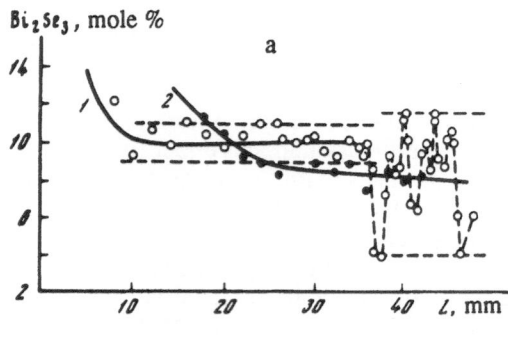

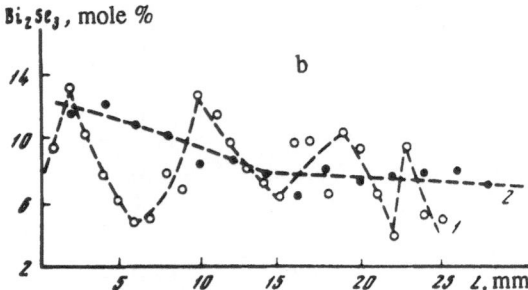

Fig. 39. Bi$_2$Se$_3$ distributions along a space crystal (1) and a ground crystal (2): a) KhK-1; b) KhK-6 [351].

In [351] selenium segregation was examined for Bi$_2$Te$_{2.7}$Se$_{0.3}$ crystals made in two experiments. In experiment KhK-1, a crystal, diameter 0.9 cm and length 5.5 cm, was made by drawing an ampul from the Kristall furnace at 0.188 mm/min. In experiment KhK-6, a crystal, diameter 1.2 cm and length 4.2 cm, was made with the Splav apparatus with controlled cooling at 2.8 deg/h. The Bi$_2$Se$_3$ distributions in space and ground crystals were examined by local x-ray spectral analysis (Fig. 39). After a short initial period, the space specimen (KhK-1) showed steady-state segregation and a stable Bi$_2$Se$_3$ composition (10 ± 1 mole %). In the corresponding part of the ground specimen, the Bi$_2$Se$_3$ content decreased monotonically in accordance with normal solidification. At the end of the space specimen, there was a sudden change in setting rate when the Kristall furnace was switched off, and there were considerable fluctuations in Bi$_2$Se$_3$ content (from 4 to 12 mole %) with a period of about 2 mm.

Figure 39b shows that the Bi$_2$Se$_3$ distribution along the two specimens in experiment KhK-6 shows major fluctuations, with diminishing period and amplitude, which are accompanied by changes in structure and morphology as mentioned in [349]. The distribution along the entire ground specimen shows a monotonic fall without appreciable oscillations, as in experiment KhK-1.

Figure 39 shows that Bi$_2$Se$_3$ segregation in Bi$_2$(Te, Se)$_3$ solid solutions is very much dependent on the growth conditions. There are considerable differences in selenium distribution between the two specimens due, in the main, to the mass-transfer mechanism. Under microgravity, the transfer rates are low, and high growth rates or fluctuations may readily result in nonequilibrium conditions and thus produce a macroscopically periodic distribution. On the ground, on the other hand, the increased transfer rates from convection may produce growth more stable under external fluctuations. We cannot deal here with the more detailed analysis of [351] concerned with major composition fluctuations in space binary-alloy crystals and merely state that the theoretical explanation has been considered in [1] (Section 1.3.3, pp. 27–29). Then [351] provides a good illustration of the theoretical arguments in [1].

2.3.6. Te–Se Solid Solutions [318]

In [318] studies were continued on gravitational effects for Te–Se solid solutions containing 10 at. % Se, as mentioned in [1] (Table 3, p. 55). Gravitational effects were demonstrated, since specimens made at $0\,g$ (on an aircraft and with brief weightlessness), on the ground at $1\,g_0$, and in a centrifuge at $10\,g_0$ were compared. Gravitation affected the electrophysical parameters, and it was stated that dendritic crystallization is suppressed in space and one gets a homogeneous cellular structure. Local x-ray spectral analysis was used [318] to examine the same Te–Se specimens made with various gravity levels. The x-ray pictures and the Se profiles showed the typical distribution characteristic of cellular growth in normal crystallization with an impurity that reduces the melting point. There were elevated Se contents at the cell boundaries and reduced contents at the centers, particularly in specimens made in space, which indicates less convection there.

2.4. MICROGRAVITY GROWTH OF SEMICONDUCTOR CRYSTALS FROM THE VAPOR STATE [334, 396–411]

In [1] 30 papers were cited on growing crystals from vapor under microgravity, but the contents of the papers were briefly reflected only in tables: Table 1, pp. 39, 40, and 42; Table 2, p. 44 (M 566) and p. 49 (MA-085); and Table 3, pp. 58–59. The main results from the papers cited amount to confirming the predicted advantage from microgravity on crystal quality: improved structure and homogeneity, better surface, and bulk perfection. Mass transfer under microgravity is mainly by diffusion, although there is a contribution from other mechanisms, as is evident from an increase in the crystallization rate as the transport-agent pressure increases. As regards the M-566 results, it was pointed out that the observed transport rates are larger than theory predicts for microgravity, and it was suggested that this is due to thermochemical convection in the gas–solid system. Several theoretical papers cited in [1] deal with Stefan flow effects on mass transfer in a sealed tube. The discrepancies between theory and experiment in some cases have been ascribed to lack of appropriate checks on source and substrate temperatures and errors in the diffusion coefficients, and also because the assumption of pure diffusion-limited transport does not incorporate vapor-phase reactions.

In light of those papers cited in [1], we consider new papers, where the experiments are classified by materials as above for crystallization from melts.

2.4.1. Chemical Transport of Ge by Iodine [402, 403, 406]

These papers confirm the conclusions in [1], although further studies have refined some quantitative transport aspects under microgravity and have indicated

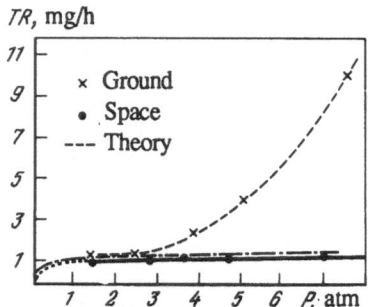

Fig. 40. Dependence of transport fraction on total pressure on the ground and in space [402].

how gravity affects ground experiments, where the gravitational vector has been varied in orientation relative to the temperature gradient.

In [402] it was found that the microgravity transport rate is only slightly dependent on pressure, so there is virtually no thermal convection, and a pure diffusion mechanism applies. Figure 40 illustrates this, as it differs from the conclusions in experiment M-566 and corrects them, as emphasized in [402]. Furthermore, in [402] there was a detailed thermodynamic description of the Ge–I system, with a state barogram and an analysis of the gas composition in chemical transport.

In [403] the discrepancies between the calculated and observed mass-transport rates in the Ge–GeI$_4$ system were explained by a model that incorporates the interaction between the forward and reverse flows (GeI$_2$, GeI$_4$, and I$_2$) with the walls, which gives rise to ordered flows and reduces the transfer rate by comparison with the one-dimensional model prediction.

The gas concentration gradients under microgravity produce concentration-dependent convection, which causes deviation from the diffusion-limited state.

In [406] the effects of total pressure on the transport rate in the Ge–GeI$_4$ system under microgravity were examined (the transport rate increased with pressure), and it was asserted that the chosen tube geometry does not produce pure diffusion-limited transport, which differs from the conclusion in [402], possibly because of the difference in geometry between [406] and [402].

2.4.2. Vapor-Phase Crystallization of Ge–Si Solid Solutions [334, 394, 395, 397, 398]

These papers deal with Ge–Si solid-solution whiskers, particularly the mass transport, structure, and Si distribution when the whiskers are made under microgravity and on the ground by chemical transport reaction in a closed Ge–Si–Br system. The temperature was kept at 1230°K in the source region for 10 h, and 930°K in the crystallization zone, and then the tube was allowed to cool passively along with the Splav furnace. The transport rate under microgravity was 12 mg/h, as against 110 mg/h on the ground, while the theoretical estimate for pure diffusion was 13.4 mg/h. It was therefore concluded [402] that a microgravity diffusion mechanism applies.

The whisker structures showed that the space specimens were actually worse than the ground ones, which was not explained; the internal stresses in the space specimens were so high that the Si distribution could not be determined by luminescence methods, so x-ray topography was used instead. In the ground specimens, the Si contents varied from crystal to crystal from tenths of an atomic percent to 1.5%, although any given crystal was homogeneous. The Si distribution was very uneven in the space specimens (see Figs. 2 and 3 in [394]).

2.4.3. Germanium Selenide: Chemical and Physical Gas Transport [410, 411]

In [410, 411] new experiments were described on the GeSe–Xe system in order to elucidate the reasons for the anomalies found in experiment M-566 on Skylab in 1975 for chemical transport under microgravity in the GeSe–GeI$_4$ system (the transport rate was much higher than expected for pure diffusion).

The reacting vapor is here replaced by inert Xe; i.e., physical transport replaces chemical. Microgravity then gave pure diffusion, which confirmed that the anomalies in GeSe–GeI$_4$ are due to thermochemical convection.

The GeSe–Xe experiments were performed on the ground with various angles between the temperature gradient and g_0 and under microgravity on two flights: one of them (STS-7) on the Shuttle and the other on the recent D-1 mission.

The new data clarify the interpretation of the apparently conflicting results on crystal growth from a vapor–gas mixture.

2.4.4. Mass Transfer in the ZnO–H$_2$–(Ar) System [404, 405]

In [404] gas-kinetic theory was used on the basis of diffusion, thermal diffusion, and convection to consider the ZnO–H$_2$–(Ar) gas-transport system in the presence and absence of the inert gas Ar; the calculations showed that the ZnO–H$_2$ system has an appreciable thermal-diffusion ratio ($K_T \sim 0.12$) for the H$_2$–Zn pair at low H$_2$ pressures. The effect becomes less as the pressure increases. Adding Ar increases the separation of Ar and H$_2$ by thermal diffusion ($K_T \sim 0.16$ for approximately 0.2 mole fraction of Ar). Also, the Ar produces convection destabilization. The ZnO–H$_2$–(Ar) system has a range where there is anomalous thermal diffusion and convection, and it represents a convenient model for studying transfer, including under microgravity. The system is also promising for growing ZnO crystals.

In [405] there is a survey of all the available data on ZnO–H$_2$ mass transfer with the Kristall apparatus on the ground and under microgravity on Salyut-6. The measurements were compared with the theoretical model of [404]. Estimates were made of the contributions from the different mass flows. The program included 40 ground and 13 space experiments with various initial conditions: hydrogen pressure P, ampul dimensions ($L/d = 11.0$, 17.5, and 26.5), and furnace position in the gravitational field. Where there are kinetic constraints ($P_0 \leq 0.3$ atm), the two sets of experiments correspond, but more complicated mass transport patterns occur as P increases. Under microgravity, the diffusion model applies closely, with a 15% contribution of laminar flow. The contributions from other nongravitational convection types could not be determined, since they were within the measurement errors. The contributions from thermal concentration-dependent convection were estimated for $P = 0.8$ atm on the ground with the tube axis at $0°$ and $180°$ relative to $\vec{g}_0$ (i.e., relative to the temperature gradient).

The general conclusion from [405] is that comparative experiments on the ground and in space can be used to determine the contributions from gravitational and other types of convection in gas-transport systems important in growing single crystals and epitaxial films.

Theoretical conclusions [404] on the effects of Ar on $ZnO–H_2$ behavior have so far [405] not been tested by experiment.

2.4.5. Gas Effects on Vapor-Phase Growth of α-HgI_2 Crystals [408]

Optimizing space growth for large α-HgI_2 single crystals (over 500 g) for use in γ- and x-ray detectors requires a better understanding of the physical gas transport for this material. Although Hg_2I_2 evaporates without dissociating, suitable crystals could not be grown on account of nonstoichiometry and the lack of pure initial material. Mass spectrometry showed that the growth atmosphere may contain surplus components (I_2 or Hg), residual gases (N_2, O_2, CO_2, and H_2O), and gas impurities (hydrocarbons and possibly also H_2). Calculations were made on the HgI_2 flux as affected by supercooling and gas partial pressures, which showed that the nonstoichiometric components and high-molecular-weight hydrocarbons reduce the transport rates considerably, while the partial pressures of those gases may be comparable with that of HgI_2. A large-diameter growth vessel with quasipoint crystal cooling is expected to produce convection independent of time on the ground because of the combination of radial temperature gradients with the gas composition, although the Rayleigh numbers are fairly low for pure HgI_2 vapor. It is therefore expected that microgravity growth would improve the homogeneity in large HgI_2 crystals. This agrees with an SL-3 experiment, where fairly small HgI_2 crystals (7.2 g) were grown.

2.4.6. Vapor-Phase Growth of CdSe Crystals [399, 400]

In [399, 400] theoretical and experimental studies were made on the ground and in space on mass transfer in vapor-phase growth of CdSe crystals, group $A^{II}B^{VI}$. Theoretical estimates were made on the diffusion constraints by means of a one-dimensional model. The following conditions applied: seed temperature 1030°C, source temperature 30–40° higher, working medium H_2 or He, and initial pressure varying from 15 to 150 mm Hg. The internal tube diameter was 8 to 25 mm, the ratio of source-seed distance L to tube diameter D from 1 to 5.

Calculations show that H_2 pressures over 50 mm Hg and not more than 58 mm Hg caused the rate-limiting stage to be vapor diffusion; below these pressures, surface kinetics were important. The component diffusion coefficients in He are 20% higher than in H_2, so there was faster mass transfer in the CdSe–He system. There was additional transport in the CdSe–H_2 system because Se is transported by H_2, but this is estimated at most at 6%. The microgravity calculations agreed within 5% with the measurements, whereas the deviations on the ground for certain L/D ratios were very substantial. The additional mass-transfer increase was ascribed [399] to concentration-dependent convection.

In the range $1.5 < L/D < 2.0$, circulation due to concentration-dependent convection produced a particularly large increase in the transfer by comparison with diffusion.

It has thus been shown [399, 400] that the one-dimensional diffusion model gives good agreement with experiment under microgravity in the $CdSe–H_2$ and $CdSe–He$ systems, whereas concentration-dependent convection must be incorporated for ground conditions.

2.4.7. Vapor-Phase Growth of Lead and Tin Telluride Crystals [407]

Lead telluride and solid solutions based on it can be grown as crystals on the ground and under microgravity in sealed tubes and in demountable reactors under vacuum or in inert gas.

The working conditions govern the growth rate, which is 0.2–0.4 mm/h; single crystals have been made, diameter up to 25 mm, height up to 45 mm, and dislocation densities $5 \cdot 10^4 – 1 \cdot 10^5$ cm^{-2}.

The growth rate is dependent on the inert-gas pressure, the temperature gradient in the growth zone, and the source–crystal distance.

The longitudinal dislocation distribution indicates considerable stresses arising at the contact between the seed and holder during cooling at the end. Studies have been made on how the holder material affects the perfection. There are also discussions of effects from source composition and growth rate on the solid-solution component distributions under microgravity and in synchronous ground experiments.

2.4.8. Other Studies on Vapor-Phase Growth of Crystals [396, 401, 409]

We have mentioned above some theoretical studies and comparisons with experiment for vapor-phase growth, as for ZnO and CdSe; in [401] there is a general discussion of optimizing growth and doping under microgravity without specifying particular materials.

In [409] there is also a discussion of gravitational effects on crystal growth from vapor. There is a discussion of the common suggestion that ground convection can be eliminated by removing horizontal temperature gradients and heating from the upper end, while ensuring that the interface is horizontal. Model calculations [409] show that this is correct only for strictly congruent vapor–solid transitions. Segregating components diffused with viscous interaction with the tube walls, which leads to convectively destabilizing (horizontal) density gradients. This result can be extended to any liquid–solid phase transition. As impurity segregation is always important, it is concluded that convective stability is virtually unattainable in making materials on the ground.

In [396] there is a study of effects from the gravity-vector direction on gas–solid transition with reference to cryocondensates of CO_2, N_2O_4, and Xe. Without going into details, we can say that such experiments simulate some general aspects of gas–solid transitions important in vapor-phase growth of crystals.

2.5. CONCLUSIONS [319, 332, 337, 338, 340, 357, 369, 383]

The papers discussed above on microgravity growth for particular crystals are accompanied by certain papers, each dealing with several different materials, from which general conclusions can be drawn. We consider some of these briefly.

In [319] there is a study of the general scope for using crucible-free zone melting under microgravity, and it is concluded that it is particularly promising to make large-diameter semiconductor crystals in this way, as well as crystals doped with heavy elements that tend to segregate on density, in addition to compounds that decompose, such as $A^{III}B^{V}$ and $A^{II}B^{VI}$. Preliminary results have been given on the facilities in a special apparatus for such zone melting in space, along with tests on the ground.

In [332] there is a survey of growing various semiconductor crystals under microgravity on Skylab, Soyuz–Apollo, and Salyut-6 plus Soyuz, which emphasizes that in all cases (apart from Ge + Ga on the Kristall apparatus), the transport mechanism is diffusion-limited. It is considered [332] unsound to explain the complicated impurity distributions in terms of Marangoni convection alone, and it is stated that it may be accompanied by g perturbations and vibrational convection, whose effects must be considered.

In [357] there is a survey of growth results in space with the Kristall and Magma equipment applied to Ge, InSb, InAs, and GaAs grown from the liquid, where there is a summary of conclusions drawn in papers on each of these individually.

In [340] there is also a study of general topics relating to technological processes on space vehicles, where it is concluded that a further wide range of research is required on heat and mass transfer, microgravity crystallization, and the design of automated equipment.

In [383] there is a reporter paper dealing with growing semiconductor crystals under microgravity on Spacelab-1. Two experiments were considered with Si crystals grown by zone melting and by solidification of a Si drop retained on a Si rod (see Section 2.1 for a description of these experiments [376, 384]), which showed that Marangoni convection gives rise to impurity layering under microgravity. Protective films suppress this convection and improve the homogeneity considerably on the ground. This is a good demonstration that microgravity experiments can be used to improve ground technology. Three other experiments on growing PbTe, GaSb, and CdTe did not attain the targets because of apparatus deficiencies and external perturbations, but they still showed that these crystals can be grown in space.

On the whole, the SL-1 experiments provided some valuable indications on better apparatus use and the scope for experiments with unmanned space vehicles. It is clear that research on microgravity growth will be desirable in the future, although at present it is mainly an academic topic. The author considers that it is hardly feasible in the foreseeable future to make such crystals for commercial purposes.

Many others disagree with that conclusion and consider that industrial use of the method may be promising for certain semiconductors made under microgravity in the near future. For example, in [369] it is asserted that highly perfect GaAs crystals made in space and currently costing $250,000 per kilogram are suitable for special applications where one needs high speed, reliability, and radiation resistance. The NRA Company projects the production of the first crystal batch in 1991. To make the operation competitive, the company considers that it should produce 20 kg of GaAs every year. To support these results, the firm has drawn up a contract with NASA for certain preparatory experiments on seven Shuttle flights.

The prospects for commercializing space will also be discussed in the general conclusions at the end of this book, where aspects of using space for making high-grade semiconductors constitute only a small fraction of the possible commercial applications. Here we may summarize research on semiconductor growth under microgravity considered in this section, viz., from liquids and vapors.

Without repeating the general conclusions on the features of microgravity growth, we would point out that there are unsolved questions that need to be discussed and resolved in future fundamental research on semiconductor materials science in space.

As regards liquid growth, there are unelucidated topics concerning nongravitational convection (Marangoni convection, g perturbations, and vibrational convection) as affecting dopant distributions for doped crystals, as well as composition homogeneity and stoichiometry deviations for semiconductor compounds. Aspects of nonequilibrium phenomena (in particular, supercooling) under microgravity remain unelucidated, and these under some conditions can produce structure deterioration and adversely affect the composition homogeneity under microgravity, even by comparison with ground conditions. Finally, the effects from partial liquid contact with the tube remain not entirely elucidated, although they have been confirmed repeatedly by experiment under microgravity, especially for the case where the liquid does not wet the tube. A complete elucidation of all details here should include quantitative description.

Vapor-phase growth under microgravity requires additional fundamental research (or "academic" research, the term used in [383]) in order to elucidate the details, particularly when physical gas transport occurs in combination with chemical, where one has to incorporate thermochemical convection effects. Section 2.4 on gas transport under microgravity shows that there are some unresolved topics to be discussed here.

In conclusion, it seems clear that solutions to some fundamental problems in semiconductor materials science in space can provide a basis for commercial production of certain semiconductors as well as for improving the technology for making them on the ground.

Chapter 3

MICROGRAVITY SOLIDIFICATION OF METALS, EUTECTICS, AND COMPOSITES

In [1] it has already been stated that the problems on the preparation of semi-conductors under microgravity are similar to those for metals as regards solidifica-tion and crystallization physics. Therefore, the topics considered in [412–513], which form Part VI in the bibliography, are also dealt with in reviews [40, 41, 48, 71, 74, 76, 77, 84, 96, 108] and are largely analogous to the topics discussed for the appropriate research on semiconductors [317–395]. All the same, as in [1], one can distinguish certain topics having specific experimental features for metals, eu-tectics, and composites containing metal matrices. A list of these specific topics has been given in [1], and it is also reflected in the section headings in this chapter.

3.1. IMMISCIBLE-ALLOY SEPARATION [74, 84, 417–419, 428, 441, 442, 446, 451, 452, 459–462, 465, 467, 468, 471, 473–478, 481, 487, 489, 490, 494, 497, 512, 513]

In research on immiscible-alloy solidification under microgravity, it is assumed that the lack of gravitational sedimentation will enable one to obtain a homogeneous droplet distribution in the metal matrix, i.e., to obtain a composite of frozen-emul-sion type. However, even the first space experiments revealed unexpected macro-scopic segregation (see Section 3.10 in [1]). Many experiments with prolonged or brief weightlessness have been performed to elucidate this. A special seminar [18] dealt with the analysis of such experiments [84, 441, 452–454, 459–461, 471, 481, 490, 494, 512]. Much attention has been given to this theme also on Spacelab in experiments published in collections [18–20], as well as in individual papers [442, 451, 464, 467, 475, 478, 489, 513], in addition to some other studies [417–419, 428, 446, 462, 465, 468, 471, 474, 477, 487] not appearing in [18–20]. The state of the topic can be characterized from a condensed exposition of a review [84] with brief supplements from certain other papers.

 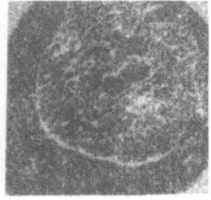

Expected   Obtained

Fig. 41. Microgravity separation [84].

TABLE 2

No.	Separation mechanism
1	Free convection ⎫ Dependent on
2	Sedimentation ⎰ gravitation
3	Difference in local nucleation rates
4	Wettability difference
5	Diffusion-limited growth
6	Differences in surface energy (tension)
7	Capillary forces
8	Surface convection (Marangoni convection)
9	Impulse of force (momentum) due to directional growth
10	Impulse of force (momentum) due to difference in diffusion rate
11	Ostwald ripening (coalescence)
12	Internal-energy minimization
13	Solidification-front repulsion
14	Solidification shrinkage

The purposes of [84] were to discuss the mechanisms contributing to immiscible-alloy segregation, to demonstrate segregation on the transparent model (methanol–cyclohexane), and to consider the scope for avoiding macroscopic separation.

Table 2 gives a list of the mechanisms that may contribute to immiscible-component separation; we draw on the conclusions of [84] to consider the possible effects from the mechanisms given in Table 2. Mechanisms 1 and 2 (gravitational convection and sedimentation) are decisive in the earth's strong gravitational field, whereas they can be neglected to a first approximation under microgravity.

Nucleation-rate differences (mechanism 3) are closely related to differences in tube wetting (mechanism 4). A metal that wets the container well has good scope for forming clusters at the walls and thus nucleating droplets, i.e., better wetting implies better adhesion. Droplet formation near the tube surface, in turn, favors wetting. The dark metal in Fig. 41 evidently wets the tube better than the light metal. For example, In wets an Al_2O_3 tube better than does Al.

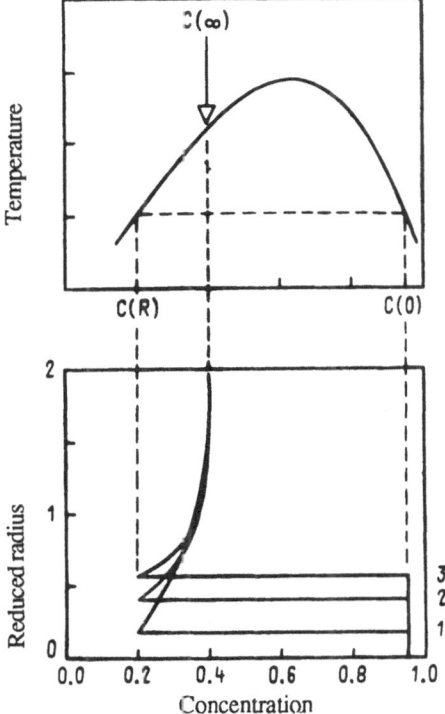

Fig. 42. Growth due to diffusion [84].

If a cluster is formed whose bulk (positive) enthalpy exceeds the (negative) surface enthalpy, it grows by diffusion (mechanism 5).

The particle radius R increases in proportion to the square root of the time t, so the radial coordinate r_D outside the particle and the particle radius R_D can be written as

$$r_D = r/2\sqrt{D \cdot t}; \quad R_D = R/2\sqrt{D \cdot t}, \tag{3.1}$$

where D is the diffusion coefficient. The relation between R_D and the concentration $c(r)$ outside the particle is [84]

$$\frac{c(r) - c(\infty)}{c(R) - c(0)} = 2R_D^n \exp R_D^2 \int_{r_D}^{\infty} d\xi \exp(\xi)^2/\xi^{m-1}, \tag{3.2}$$

where c_∞ is the liquid concentration before separation, and $c(R)$ and $c(0)$ are the equilibrium concentrations of the two separating components as given by the mixing diagram (Fig. 42); $m = 3$ for spherical particles, while $m = 2$ and $m = 1$ for cylindrical and planar particles, respectively.

Surface-tension differences (mechanism 6) and capillary forces (mechanism 7) have to be considered in the separation and the free-surface behavior, as there is always a free surface in a closed tube, since the tube includes a free volume, which is

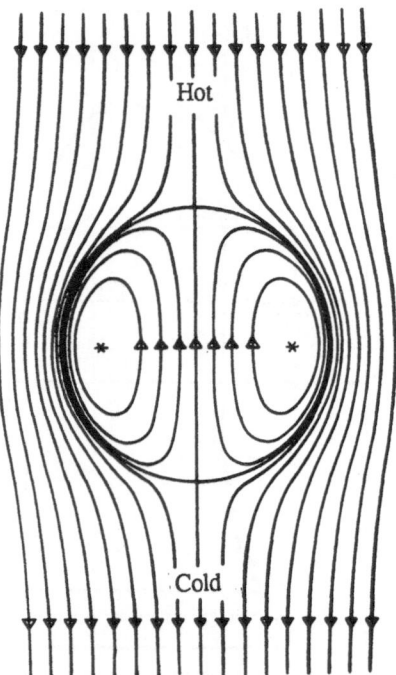

Fig. 43. Thermal Marangoni convection [84].

necessary to allow for expansion on heating. The component having the lower surface tension will spread along the free surfaces. Drops of the component having the lower surface energy also move to surface regions with lower average curvature.

Marangoni convection (mechanism 8) makes the most important contribution, as flows arise within and outside a drop as shown schematically in Fig. 43. The external force F acting on a drop of radius R, the velocity V, and the surface-tension gradient $\nabla\sigma$ are related by [84]

$$F = \frac{2\pi\eta_{ext} \cdot R}{\eta_{ext} + \eta_{int}} \left[(2\eta_{ext} + 3\eta_{int})\,v + \frac{2}{3}\nabla\sigma R \right], \tag{3.3}$$

where η_{ext} and η_{int} are the viscosities inside and outside the drop. Figure 44 shows an example illustrating the effects of Marangoni convection on drops in the transparent cyclohexane–methanol system, which was heated above the immiscibility range (to 50°C) before the rocket flew, and during the flight was cooled monotonically to 10°C. Cyclohexane droplets were formed, and marked Marangoni convection was observed for about 30 sec. Speeds up to 4 mm/sec were observed for 0.2-mm-diameter drops, which is more than on the ground, where Marangoni and gravitational forces act in opposite senses and balance partially.

Mechanism 9 is related to the impulse of the force arising if a drop accumulates atoms mainly from one side, where the concentration of the corresponding component is elevated. The drop moves to the side where it collects atoms.

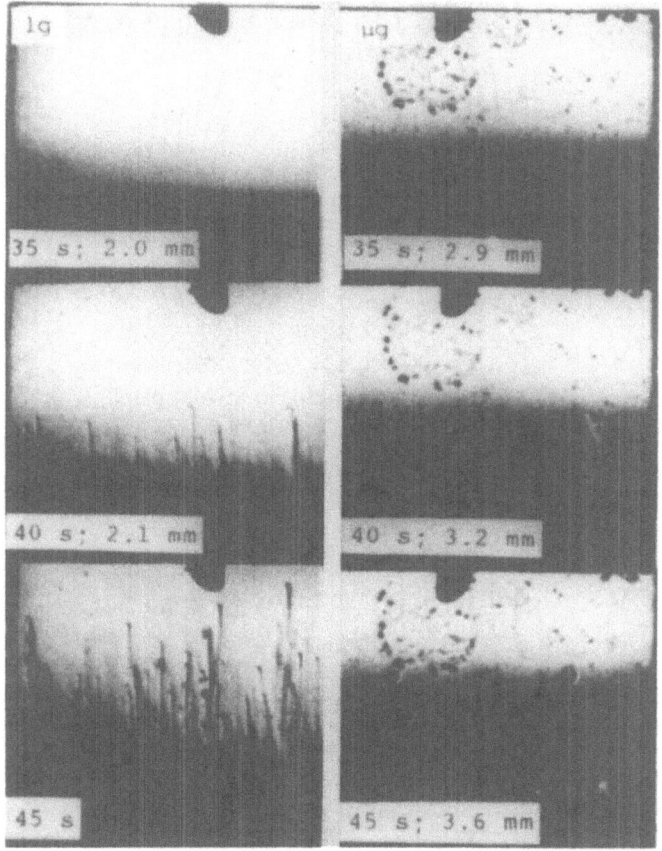

Fig. 44. Marangoni convection for a cyclohexane drop at 1 g and under microgravity on a Texus-7 rocket [84].

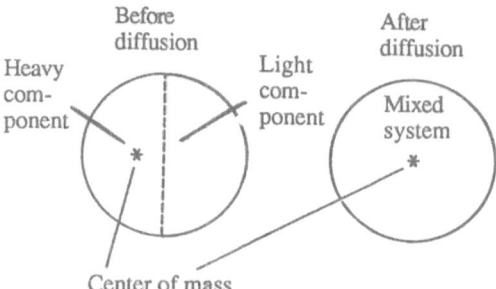

Fig. 45. Mutual diffusion for liquids differing in density [84].

A force can also arise from internal diffusion of atoms differing in mass (mechanism 10). We consider a drop consisting of equal volumes of a heavy liquid

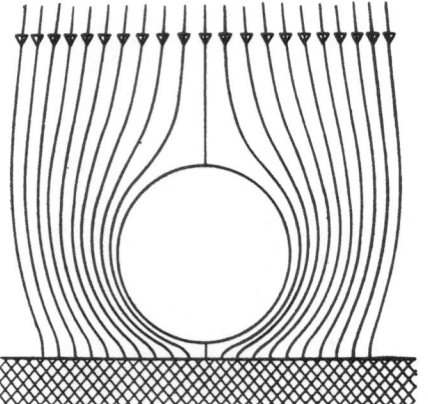

Fig. 46. Flow lines around a solid particle ahead of a solidification front [84].

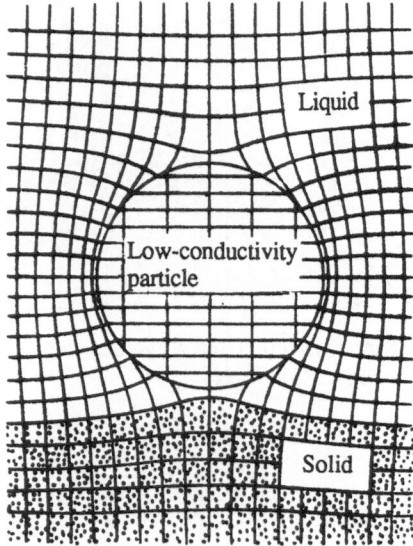

Fig. 47. Isotherms and solidification
front behind a low-conductivity particle [84].

on the left and a light one on the right (Fig. 45). The center of mass is displaced to
the left (from the center). The center of mass shifts toward the center on internal
diffusion.

The Ostwald ripening (coalescence) effect (mechanism 11) is dependent on the
solubility variation with size and is comparatively small by comparison with the
other effects in Table 2.

By internal-energy minimization (mechanism 12) one means that the liquids while not completely mixed by diffusion have a surface energy (interface energy). The cyclohexane–methanol model system has been used [84] to demonstrate curved surfaces arising from this effect (the illustrations are not given to save space).

Any region in the liquid whose temperature or concentration differs from that in the surroundings should be considered as another liquid having a tendency to become spherical.

Solidification-front repulsion (mechanism 13) also contributes to the separation. For a particle ahead of the front to be displaced, there must be a continuous liquid flow between it and the solid front (Fig. 46), and the repulsive force must be larger than the viscous resistance, which is proportional to the front speed V. The repulsion itself is not dependent on V, so the particle may be trapped at high V, while being repelled at low V. The stable particle–front distance is dependent on the radius R. The viscous force increases more rapidly with R than does the van der Waals repulsion, so the critical setting rate decreases as R increases.

Solidification shrinkage (mechanism 14) interferes with particle motion ahead of the front; more liquid needs to penetrate the gap. For a material such as Sb or Bi, which expands on setting, the particle motion is maintained, i.e., less liquid enters the gap.

In [84] there is a discussion of setting-front deformation ahead of a particle whose thermal conductivity is different from that of the liquid. The growing crystal is colder than the liquid; i.e., there is heat flux from the latter to the former, which is perturbed if the particle's conductivity is less than the liquid's. The heat flux passes around the particle, and the isotherms and the solidification front are curved toward it (Fig. 47). Conversely, a higher particle conductivity causes the crystal to be concaved by the particle. If the particle ahead of the front is liquid, e.g., molten metal, the viscous forces due to the liquid in the gap are substantially reduced. The liquid within the drop may track the external flow. The flow speeds inside and outside the drop at the interface are equal. In [84], schemes are given for flows in such a drop ahead of a front with and without Marangoni convection. The schemes are derived from computations, but to save space they are not given.

The conclusions in [84] list the scope for reducing or even eliminating macroscopic separation in immiscible alloys:

1) change in cooling method – rapid cooling suppresses diffusion-limited growth but increases Marangoni convection, while slow cooling retards the latter but gives it more time;

2) change in tube material and shape to alter the wetting, nucleation, and cooling time;

3) rotating the tube around a horizontal axis so that particles heavier than the liquid move in circles and do not sink;

4) attempts to equilibrate the Marangoni convection (acting upward) and gravitation (downward), as they act in opposite senses, which is complicated, since equilibrium can be attained only for one particle size;

5) suppressing capillary forces and Marangoni convection at free surfaces by volume compensation, an appropriate example being the use of unwetted capillaries;

6) acoustic stirring, which has the advantage that it can be started in the immiscibility range, although it works better under microgravity than on the ground.

To complete this exposition of [84], we note that certain papers cited above deal with topics not considered in [84]; as examples, we consider briefly the contents of papers published in [18].

In [512] the mechanisms enumerated in Table 2 as influencing the separation were accompanied also by particle coalescence on collision, and it was stated that the minor-component concentration in a mixture has an important effect on the separation. With Al–Pb, Al–In, and Al–Bi mixtures, dispersions were obtained only with volume contents of the heavy components less than 10%. A model considered in [512] was the Ag–silicate glass system as treated by powder-metallurgy methods. In [481] there is a discussion of alloy-separation thermodynamics, which demonstrated how the interfacial energy affects the process under microgravity, as well as effects from the volume phase ratio on the equilibrium distribution. In [481] there are theoretical and experimental studies on deposited-particle size distributions; a difference from [84] was that Zn–Bi and Cu–Pb alloys were used instead of model systems. In [459, 460] there is a discussion more detailed than that in [84] of the Ostwald swelling mechanism as involved in separation, and it was concluded that one can estimate diffusion coefficients from this under microgravity, although it has only a small effect on mixture separation, which agrees with what is said in [84] (the same topic is considered in [74]).

In [489] radiography was applied to Hg droplet dispersion in the Ga–Hg system in a Teflon container. In [452] measurements were made on acoustic stirring under microgravity to prevent coagulation in the Pb–Zn system. In [494] the Al–Zn system was considered as regards applying filtration theory to describing immiscible-alloy setting. In [471] data are given on Al–Pb alloy setting (up to 8% Pb) under microgravity. The main conclusions of all these papers appearing in [18] agree with the above conclusions in [84], while the contents can be judged from their titles.

These details and the list of other papers cited above show that immiscible-alloy separation under microgravity remains at the center of attention and requires further theoretical and experimental research.

3.2. METAL-MATRIX COMPOSITE MELTING AND CRYSTALLIZATION [453–455, 464, 465, 473–476]

As stated in [1], many previous experiments have been based on the assumption that microgravity should result in hardening particles or fibers being more evenly distributed in a molten metal matrix. In [1] experiments were listed that were planned to provide dispersion-hardened composites in flight and on rockets. Some of these experiments have recently been performed, and the results have been published.

In [454, 455] experiments are reported from the Texus-6 and Texus-7 rockets made with copper-matrix composites – Cu–SiC, CuSiO$_2$, and CuW – which were done in order to determine the forces displacing the particles in the liquid, with particular attention to boundary properties. Unexpected effects were observed, and it was concluded that an almost homogeneous dispersion can be formed from unwetted particles under microgravity if the matrix wets the crucible. No dispersion was obtained in the same experiments on the ground or under microgravity but with an unwetted crucible.

In [464, 465] SL-1 experiments are reported, which involved composites with Al matrices. The space specimens were more homogeneous than the ground specimens and had better properties: more uniform hardness and better resistance to abrasive wear. On the other hand, complete homogeneity was not guaranteed even under microgravity because there are other particle redistribution mechanisms than sedimentation and gravitational convection, some of which have been discussed in Section 3.1.

In all previous experiments with these composites hardened by particles under microgravity, the distributions were examined on the ground after the space experiments, whereas in [473–476] the model transparent CsCl system containing Pb particles and bubbles was used in direct observations on SL-1 to detect Marangoni convection as well as to obtain information on the interactions with the setting front. Displacement of Pb droplets by thermocapillary convection was not observed in [473] because the Pb drops were coated with oxide and thus acted as solid particles, i.e., Marangoni convection did not occur in them. At the start, the particles were repelled by the front, and there was a slight tendency for them to accumulate ahead of it, but at a front speed of 4 μm/sec, 40-μm particles were trapped by the setting CsCl, although the calculated critical setting rate was 10 mm/sec. The Pb droplets were also redistributed laterally, as they tended to accumulate in a layer of about 2 mm at the sides, but no final explanation for this is given in [473–476]. It is suggested that the reason for this lateral Pb displacement is thermal radiation. Interesting observations were made [474–476] on bubble displacement. The initial specimen contained some gas, which segregated as bubbles, size 150–500 μm, on melting. These bubbles showed collective motion: adjacent bubbles, no matter what their sizes, were displaced at the same speed by convection currents. There was virtually no independent motion due to Marangoni convection. The individual bubbles moved with speeds up to $U = 200$ μm/sec, although according to Young's formula

$$U = -\frac{\sigma_T}{\eta} \cdot \dot{T}, \tag{3.4}$$

where σ_T is the temperature coefficient of surface tension, $\dot{T}$ is the temperature gradient, and η is the matrix viscosity; so speeds of about 100 diameters per second are expected, i.e., two orders of magnitude larger than those observed. In [473] it is concluded that Young's model needs to be corrected, where allowance must be made for the effects of migration on the temperature distribution, as well as for the surface tension as a function of time due to impurities reaching the fresh surface. The behavior differences may be due to differences in origin: bubbles present in the initial specimen may be more contaminated than those at the phase boundary consisting of gas previously dissolved in the liquid.

Bubbles at the boundary grow quite large (diameter up to 1 mm) and move along with the boundary [473–476]; near them, there were short pointed Pb particles (Fig. 48a–c), which demonstrate very rapid Marangoni convection (some cm/sec). The Pb particles here acted as tracers.

This bubble-generated front Marangoni convection has a direct effect on the heat flux; it distorts the temperature pattern as well as the phase boundary (Fig. 48). These flows transport so much heat that there may even be local matrix remelting. Initially, the solidification is completely controlled by the external conditions, but the processes become very complicated and rather undefined when the bubbles and

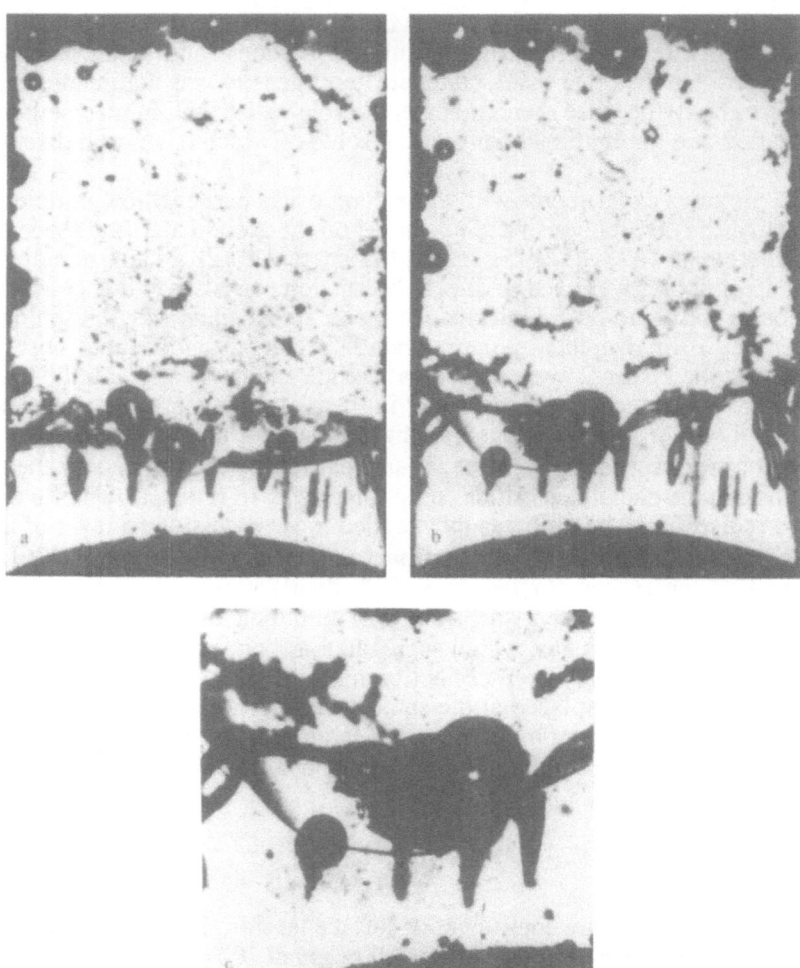

Fig. 48. a) 540-sec cooling; bubbles formed at front begin to produce deformation, with part of the gas entering the solid and forming elongated channels [473]. b) 756-sec cooling; bubbles grown large and considerable deformation [473]. c) Enlarged part of Fig. 48b, Pb particles near a bubble moving rapidly because of Marangoni convection and seen as streaks.

inclusions are formed: the local temperature gradient is no longer equal to the external one, but instead varies with position and time, while the boundary is no longer planar and the setting rate is spatially variable and can even be negative in places (remelting effects).

On the whole, these studies [473–476] show that valuable information can be obtained by direct observation on composite setting under microgravity.

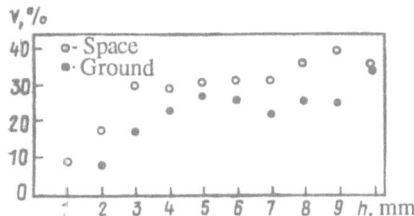

Fig. 49. Specific volume in foam aluminum as a function of distance from the edge [415].

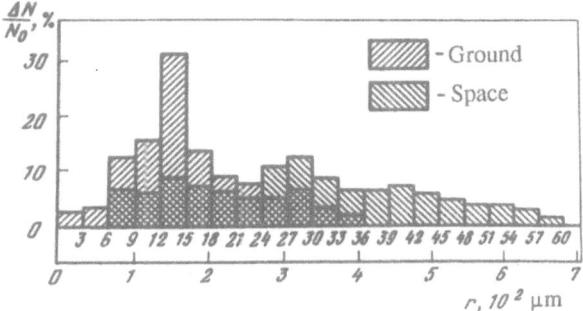

Fig. 50. Size distribution $\Delta N/N_0$ for bubbles in foam aluminum, viz., number of bubbles with sizes from r to $r + \Delta r$ refer to the total number of bubbles in a cross section remote from the edge [415].

3.3. FOAM METAL PRODUCTION
[345, 350, 387, 415, 421, 444, 445, 495]

The composites hardened by particles mentioned in Sections 3.1 and 3.2 to some extent include foam metals, which are not merely lightened by numerous bubbles but also strengthened by them. Section 1.7 has dealt with the theory of bubble nucleation, growth, and agglomeration (particularly for [421]). Section 3.2 has dealt with direct observations on bubbles during setting [473–475]. Here this is supplemented by a consideration of experiments on making foam aluminum under microgravity (under the Soviet–Bulgarian Pirin Program) [415], in which a theoretical analysis was presented for a space experiment on bubble formation in molten Te–Se, as well as of SL-1 directional-solidification results for an Al–Zn vapor emulsion under microgravity [495].

The foam aluminum was made [415] on the ground and in space by melting an Al alloy (silumin) in the presence of a pore generator (titanium hydride) and a bonding substance (silicon nitride). Titanium hydride is stable at the melting point of the Al alloy, but on superheating it dissociates and gives hydrogen bubbles.

We give the main results of [415] without going into the details. Figure 49 shows that the specific porosity in the space specimens was higher than in the

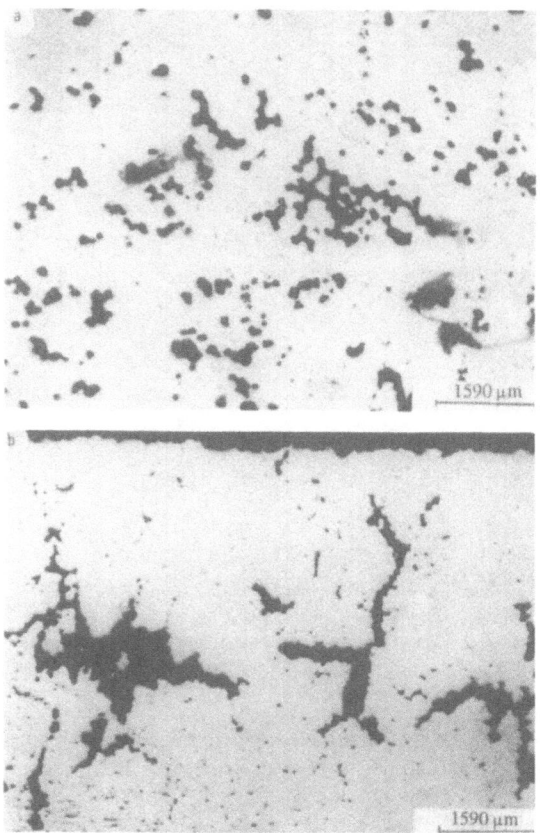

Fig. 51. a) Pores formed at the end of setting under microgravity for an Al + 5% Zn specimen; b) region solidified at
1 *g*, with shrinkage cavities and pores visible.

ground specimens, while Fig. 50 shows that the size distribution in the space
specimens was substantially different. The microhardnesses of the space specimens were much higher than the ground specimens (34.6 and 18.9 kgf/mm^2,
respectively). The difference between the ground and space results was due [415]
in the main to the bubbles floating up and out of the liquid, which on the ground
reduces the proportion of relatively large bubbles.

The general conclusion in [415] is that microgravity favors more rapid nucleation in the liquid, which reduces the bubble size, reduces the energy, and increases
the formation rate.

A difference from [415] was that in [495] another method was used for making
metals containing bubbles without using hydrogen. Microgravity (on SL-1) and
ground conditions were used in directional solidification of Al–Zn alloys at controlled partial pressures and controlled temperatures. The conditions were chosen
such that there was a nonequilibrium Zn pressure at the crystallization front, which
provided for bubble nucleation in the liquid there.

Six microgravity specimens were made with Zn contents (in at. %) of 5, 2.5, and 1. The space specimens did not wet the SiC tubes, whereas wetting occurred on the ground. Also, all six space specimens were displaced along the tubes by 3–5 mm. The lack of microgravity wetting was ascribed to gravity forcing the liquid into contact with the tube. Evaporation of Zn hinders this, and that factor predominates under microgravity. The displacement is due to differences in Zn vapor pressure at the ends, although there may have also been g perturbation effects on SL-1, which were not recorded in that experiment.

We cannot give all the microstructure data from [495], particularly since the structure varied between the parts (dendritic, cellular, and with gas cavities), and we merely give two figures characterizing the bubble distributions in certain parts of the space specimen (Fig. 51a) and the ground specimen (Fig. 51b). This part of the space specimen shows a bubble distribution in the Al more uniform than that in the ground specimen, but this was only part of the purpose of the experiment (to show that bubble dispersions in Al can be made in this way), as various problems were left unsolved.

Bubble formation in metals has been researched to produce the corresponding dispersions, i.e., to devise methods of making foam metals, whereas with semiconductors, the investigations have been designed to eliminate bubbles [345, 350, 387]. In spite of these opposite purposes, the fundamental aspects are analogous: it is necessary to elucidate the processes in nucleation, growth, and agglomeration, as well as in bubble loss.

3.4. EUTECTIC CRYSTALLIZATION [412, 426, 429, 432, 435, 438, 439, 443, 457, 458, 466, 482–484, 488, 507]

Some eutectic-crystallization experiments have been performed on SL-1 [443, 482–484, 488], which have provided new information, but which also have given rise to new problems for fundamental research. Only the experiment of [484] was unsuccessful.

In [457, 458] three eutectics differing in structure were used – Al–Al$_2$Cu platy, Al–Al$_3$Ni filamentary, and AgGe irregular – the purposes being to determine how microgravity affects the phase distribution, regularity, and crystallography in all three cases. Space and ground specimens were made with identical thermal conditions, which showed that reduced gravity did not alter the Al–Al$_2$Cu and AgGe structures substantially; i.e., going from 1 g to microgravity did not affect these platy and irregular eutectics, so convective fluctuations are not the essential defect sources there.

However, the filamentary Al–Al$_3$Ni structure showed that there were no appreciable differences in the Al$_3$Ni filament regularity in the Al matrix under microgravity, although the distances λ between them were increased by about 15%. The Jackson–Hunt eutectic-growth theory was modified with allowance for trans-eutectic effects in [457, 458] to explain this result, where it was shown that the parameter c in the relation between λ and the solidification front speed V,

$$\lambda^2 V = \text{const} = c, \tag{3.5}$$

is dependent on the volume fraction of the solidified part, which itself is dependent on convection and thus on gravitation. The following formula was derived for the dependence of λ on gravity:

$$\frac{\lambda_{1g} - \lambda_{og}}{\lambda_{og}} = \left[\frac{1 + AC_E}{1 + A\,(C_E - \Delta C_E)} - \frac{\Delta C_E}{K} \, \frac{A}{1 + A\,(C_E - \Delta C_E)} \right]^{1/2} - 1, \qquad (3.6)$$

where $k = 1 - e^{-\Delta}$, $\Delta = \delta/(D/V)$, $\Delta C_E = C_E - C_L$; A is a parameter characterizing the asymmetry in the phase diagram near the eutectic point, and Δ is the ratio of the boundary-layer thickness δ to the diffusion length D/V (D is diffusion coefficient). Parameter A can vary from 0 (for a symmetrical diagram) to 100 (for a very unsymmetrical diagram). If $A = 0$, there are no changes in λ even if $\Delta C_E \neq 0$ (the trans-eutectic case) or $k \neq 1$ (convection). Systems giving platy structures are symmetrical ($A = 0$). This explains why the platy Al–Al$_2$Cu structure showed no changes. With filamentary structures having eutectic compositions at very low volume fractions ($C_E \sim 1$–3%), convection may have a marked effect if there is considerable diagram asymmetry (A from about 10 to 100). For $\Delta = 1$, such a system may give $|\Delta\lambda/\lambda|$ from 30 to 50%, with $\Delta\lambda/\lambda$ positive for trans-eutectic compositions and negative for sub-eutectic ones; i.e., λ may increase or decrease on crystallization in space. An example of such a eutectic where λ decreases in space is InSb–NiSb, which was examined on SL-1 in [488] (see also [439]). Figure 52 illustrates the directional-solidification region for InSb–NiSb (length 33 mm), the scheme for sectioning the specimens, the choice of points for photography in the section, and one such photograph. That processing showed [488] that λ, the distance between NiSb filamentary crystals in the InSb, under microgravity was reduced by about 30%. Figure 53 shows the filament concentration N in a section as a function of the crystallization rate V [488]. This agrees with Eq. (3.5), $\lambda^2 V = $ const $= c$, since $\lambda = \sqrt{4/N\pi}$.

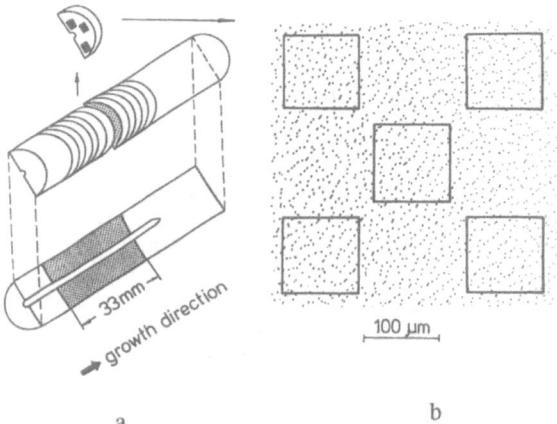

a   b

Fig. 52. a) Specimen sectioning scheme. b) Typical eutectic structure for a space specimen. The regions used to determine N are indicated [488].

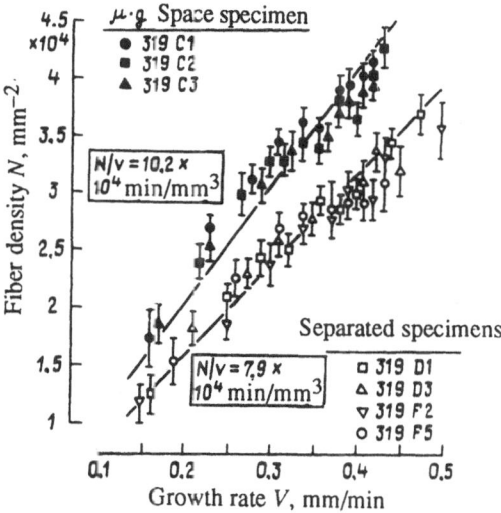

Fig. 53. Fiber density N as a function of growth rate V [488].

Another SL-1 experiment [443, 482, 483] dealt with a eutectic iron alloy, with composition $Fe + 3.9\ C + 15\%\ Si + 0.25\%\ S + 0.01\ P$, to examine the effect of microgravity on S transport in a liquid and graphite growth. The S concentration profile showed that convection occurred even under microgravity. Rapidly cooled parts had oscillatory S concentrations, which were due to segregation. The graphite flakes were no more prominent under microgravity than at $1\ g_0$, but the mean distances between flakes were smaller with microgravity.

There was an interruption in the power supply when the oven was first heated in this SL-1 experiment, and the cycle had to be repeated. The sudden interruption and fast cooling caused white iron to be formed from the initial gray. The gray iron contracts on melting, while the white expands, so cavities arose in the first heating, which persisted under microgravity during the second. Various other interesting effects occurred: a specimen broke into two parts, the material evaporated and condensed, and there were interactions with the tube walls and the skin layer, which will not be considered in more detail here because they are not specific to eutectics. We merely mention that studies have been made on cast iron containing flaky and spheroidal graphite [507], which have been examined on brief weightlessness in aircraft. Eutectic solidification experiments have also been performed on the Salyut–Soyuz combination [412, 426, 432, 435, 438, 439, 466], but we can indicate only briefly some of the results. In [412] it was suggested that there was more tendency to nonequilibrium solidification under microgravity because of differences in nucleation resulting from large supercooling. So gravity may affect the phase diagram, with the solidus curve displaced (because the liquidus is displaced), and a fine structure may arise, which does not occur on the ground. The Morava-2 experiment series will be designed to confirm this.

Measurements have been made [426, 432, 435, 438] on Al–Cu alloys, but the results were not entirely unambiguous. For example, in [435, 438] no new microstructural features were found for $Al + 4\%\ Cu$, or any effect from gravitation or the dendritic-cell size (such as occurred in [457] for Al–Al$_2$Cu). A difference from [426] is that it was stated that the eutectic formation mechanism in Al + 5% Cu takes one form on the ground and a different form under microgravity, as there are differences in Al matrix crystallization texture along the temperature gradient: in space there is preferred orientation along ⟨110⟩, but along ⟨100⟩ on the ground.

In [432] Al–5% Cu and Al–30% Cu were used. The Al–5% Cu casting structures consisted of α-solution grains and intergranular eutectic. The grains were cellular, and the morphology differed as between space and ground: the ground specimens showed mainly equiaxial cells, while the space specimens were elongated. The grains were more homogeneous in the space specimens. The Al–30% Cu specimens consisted of α-solution dendrites and eutectic, with the dendrites in space rather larger than on the ground. On the whole, in [432] and in [426] it is stated that microgravity has an appreciable effect on Al–Cu crystallization and on the structural elements. The gravitational level also affected the dendritic crystallization in Ag + 28% Cu [438].

These differing conclusions may be due, for example, to differences in impurity levels, but it is clear that unambiguous and statistically sound conclusions require new experiments.

3.5. SUPERCONDUCTOR, INTERMETALLIDE, AND MAGNETIC-MATERIAL CRYSTALLIZATION [424, 430, 431, 434, 486, 502–504, 509]

These papers deal with more detailed studies on space specimens and confirm the main conclusions in [1] (Sections 3.4 and 3.7), which are derived from the earlier papers and relate to structure features of superconducting and magnetic materials recrystallized under microgravity. The materials here are those mentioned in [1]. The new papers contain the following main results.

In [424, 431] ground developments are given for space experiments; in [424] an oven providing temperatures up to 1750°C was devised for use on rockets, while in [431] the Splav apparatus was used, which was intended for prolonged space experiments.

In [434, 502–504, 509], as previously, it has been confirmed that the structure is better and the magnetic characteristics are improved in the intermetallides Gd$_3$Co and (Tb$_{0.8}$Gd$_{0.2}$)$_3$Co, although the Gd$_3$Co space specimens contained numerous spherical macroscopic and microscopic pores, with sizes from 40 to 250 μm. The (Tb$_{0.8}$Gd$_{0.2}$)$_3$Co specimens had less porosity. We may quote the main conclusions from a recent publication in this series [486], although it repeats much of what has been said in [1]. In [486] results are presented on the structure and properties of superconducting MoCo, NbSn, VGa, and Pb–Sn alloys and magnetically ordered Gd$_3$Co and (Tb$_{0.8}$Gd$_{0.2}$)$_3$Co crystallized on the Salyut-6 + Soyuz station. Liquid-state diffusion produced considerable changes in reaction-layer formation (growth mechanism and thickness and phase composition) in Mo–Ga and Nb–Sn superconducting alloys. Ground specimens contained MoGa$_5$, NbSn$_2$, and NbSn$_6$, while space specimens had MoGa$_5$, Mo$_3$Ga, Nb$_3$Sn, and Nb$_6$Sn$_5$.

These diffusion-layer composition changes indicated superconducting transitions at 18.3 and 5.7°K in the space specimens, but only one transition at 6.9°K for the ground specimen.

The Pb–Sn space specimen (eutectic system) showed an appreciable increase in the critical current in a 500-Oe field on account of the microstructure change relative to the ground specimen (reduced number and greater size of the primary Pb crystals).

Microprobe and x-ray studies showed improved homogeneity and perfection in Gd_3Co and $(Tb_{0.8}Gd_{0.2})_3Co$ space specimens; new data were given on the behavior of the ground and space specimens during magnetization and in magnetic phase transitions (shifts in starting fields and Néel temperatures). On the whole, [486] summarizes that experiment series and confirms the previous conclusion that microgravity affects the structure and properties of superconducting and magnetic materials.

3.6. DENDRITIC AND CELLULAR CRYSTALLIZATION [413, 432, 438, 448–450, 456, 466, 469, 479, 492, 500, 503, 508, 510, 511]

In [1] (Section 3.11, p. 90) it is stated that it is possible to reduce gravitational-convection effects in dendritic growth by means of microgravity, which has provided a substantial contribution to theoretical and experimental research there. Microgravity results have provided the first exact measurements on convection as affecting dendritic growth for metals. In [469] these studies are surveyed, and the tasks of further research there are discussed.

Recently, considerable attention has been given to microgravity effects on dendritic growth for metals; as an example, we consider two recent papers on this [511, 479].

In [511] there was a theoretical study of convection affecting dendritic crystallization and an experimental test for eutectic Al + 40 wt. % Cu after solidification at two different rates R. The dendritic structure is quite simple: there are no secondary branches in the dendrites.

A simple theoretical analysis was applied to estimate the order of the effects; it was assumed that the liquidus has a constant slope m_L, as shown in Fig. 54. One distinguishes the three zones shown schematically in Fig. 55: 1) liquid alloy; 2) rough zone containing dendrites (component S) and liquid between them; and 3) solid formed by dendrites S and eutectic E. Figure 55 also shows the temperature and concentration patterns.

Here we cannot go into the details of this approximate treatment concerning solution transport in the bulk and in the gaps between dendrites or the speeds of the flow between them, or the descriptions of the metallographic pictures in [511], but the main conclusions can be stated.

In [511] a distinction is made between two types of effect from convection on solution transport here, viz., from bulk convection in the liquid (region 1 in Fig. 55) and interdendrite flow (regions 2 and 3 in Fig. 55).

The calculations showed that the bulk-convection contribution is on the order of $A\Delta$, where $A = [(D{\cdot}Gr)/Rm_L(C_E - C_S)]$ is a dendritic parameter and $\Delta = (R{\cdot}\delta)/D = \min [1{\cdot}Pe(Gr{\cdot}Sc)]^{-1/4}$ is a boundary-layer parameter. Here D is the solution diffusion coefficient, while Pe, Gr, and Sc are the Peclet, Grashof, and Schmidt numbers for the bulk liquid (see Table 1 in Section 1.1), and C is the temperature gradient.

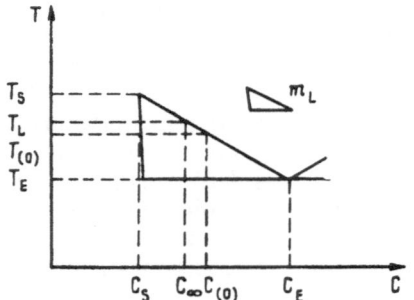

Fig. 54. Schematic part of alloy phase diagram [511].

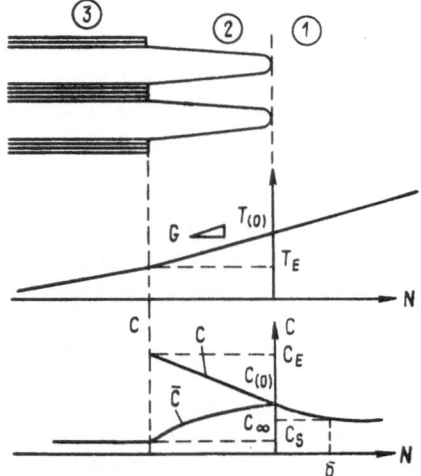

Fig. 55. Schematic representation of dendritic solidification [511].

As $A < 1$ and $\Delta \ll 1$, the bulk convection contribution is small, while the inter-dendrite flow contribution is more substantial. The difference in dimensionless liquid concentration at the end of the dendrite and the mean solid-phase concentration may be estimated from the ratio Γ of the flow speed U within the dendrites to the growth rate:

$$\Gamma = \mathrm{Re}^{-1} \cdot \mathrm{Gr} \quad (\text{if } \mathrm{Gr} < 1). \tag{3.7}$$

Here Re and Gr are the Reynolds number and Grashof number for the interdendrite channels, which were about 50 μm wide in this Al–Cu alloy, according to metallography [511]. In the dendritic-crystallization parameter range, U is simply proportional to Gr_1, i.e., to the horizontal density gradient and gravity level. As Γ is usually on the order of 1 in typical binary alloys at 1 g_0, one concludes that interdendrite flow effects can be eliminated in two ways: either under microgravity (simply by reducing the gravitation to 10^{-1}–10^{-2} g_0) or by solidification in the vertical position at 1 g_0, but with a low horizontal temperature gradient, which may require a special oven design. One can add to what is said in [511] that microgravity effects in dendrite growth in metals have been examined several times [413, 432, 438, 450, 469, 492, 500], including for Al–Cu. We cannot deal with the contents of those papers and merely state that dendrites grown in space [413, 432, 492] are substantially larger than ground dendrites. To supplement [511], we consider here briefly the results from a theoretical study [479] on growth stability conditions.

The stability in a planar phase boundary advancing in binary-alloy solidification is described either by means of a simple criterion of concentration-dependent supercooling for low speeds or from classical morphological stability analysis, which incorporates small-perturbation growth or damping on a planar surface. That analysis predicts a new stability state for a planar phase boundary at very high growth rates: the absolute stability limit. The dendritic-growth stability conditions lie between the latter and the concentration-dependent supercooling limit.

Dendritic growth in a pure liquid occurs only with a negative temperature gradient in the liquid ahead of the dendrite tips (i.e., in the supercooled liquid), while in a binary alloy it can occur with negative or positive gradients. In [479] a positive gradient is considered, the purpose being to show that the dimensionless parameter for dendrite-growth stability

$$\sigma_c = 2\gamma_c D_L / R r_t^2 \qquad (3.8)$$

can be derived by considering stationary dendrite behavior without invoking perturbation theory. In (3.8), γ_c is the capillary length, D_L the solute diffusion coefficient in the liquid, R the growth rate, and r_t the dendrite tip radius. The model [479] incorporated growing-dendrite interaction, i.e., the presence of adjacent dendrites, not merely the growth of an isolated dendrite in a supercooled bath, as is the case for other studies. We cannot consider the detailed expressions and arguments [479] and merely note that the study was successful and that (3.8) was confirmed for the dendrite-growth stability parameter from a stability-condition analysis without invoking perturbations.

Convection can result in dendritic or cellular growth by disrupting the stability conditions for planar-front growth.

Measurements have been made [448, 449] on cellular growth in Pb–Te systems with various orientations of the growth rate vector $\vec{v}$ relative to the gravity vector $\vec{g_0}$. Under ground conditions, both $\vec{v} \parallel \vec{g_0}$ and $\vec{v} \not\parallel \vec{g_0}$ orientations produced thermal gravitation or thermal concentration convection in concentrated alloys. For $\vec{v} \not\parallel \vec{g_0}$ the result was longitudinal macroscopic segregation, but the front remained macroscopically planar and the growth rate was constant. The cells are well formed and regular. For $\vec{v} \parallel \vec{g_0}$ there were modes lacking axial symmetry for specimens with various diameters and temperature gradients. A simple hydrodynamic-stability analysis similar to that in [511] or [479] is qualitatively suitable for explaining this. The front becomes corrugated and the growth is discontinuous. The cell periodicity on average is close to that for $\vec{v} \not\parallel \vec{g_0}$, but the structure is much less regular. It has thus been shown [448, 449] that there is always convection of some type on the ground. One can examine the effects of different convection types on cellular growth by varying the orientation of $\vec{v}$ relative to $\vec{g_0}$ (this is well illustrated in [449]).

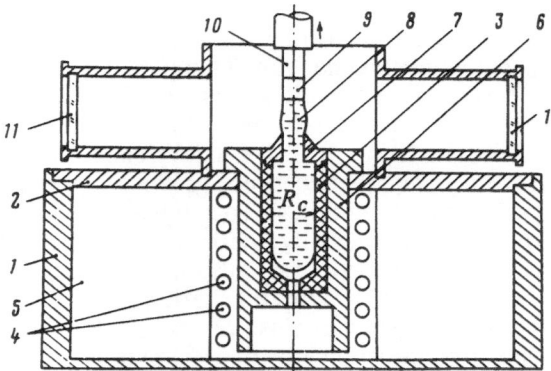

Fig. 56. Crystal growth device [433].

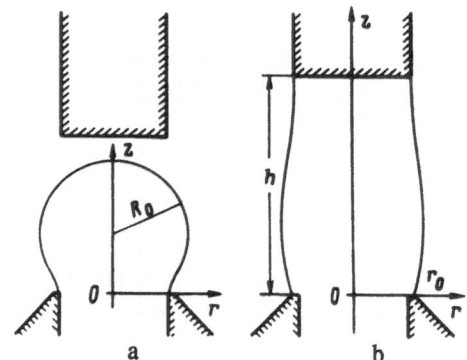

Fig. 57. Drop shape at shaper edge (a) and resulting meniscus (b) [433].

Only microgravity enables one to suppress convection due to radial temperature gradients and thus produce predominant heat and mass transfer by diffusion. The experiments of [448, 449] are considered preparatory for subsequent research (on SL-D1 [450]).

The above discussion, on the whole, shows that microgravity experiments remain both necessary and promising for fundamental research on dendritic and cellular crystallization of metal alloys.

3.7. CAPILLARY AND SURFACE FORCES AT CONTACTS WITH SOLIDS UNDER MICROGRAVITY [414, 433, 436, 437, 493, 499]

Section 1.9 has already dealt with theoretical aspects of capillary forces under zero gravity. Here I consider some experiments on this with metals, although similar tasks can be undertaken with semiconductors and glasses.

In [436, 437] the conclusions completely agreed with those of [1] (Section 3.1, pp. 80–81) on wetting angles in the Soviet–Bulgarian Pirin Program. The equilibrium wetting angle, which is determined by thermodynamic considerations alone, cannot alter under microgravity, as has been confirmed by measurements in [414].

In [436, 437] bundles of carbon fibers and carbon strips were examined with molten silumin (Al + 12% Si) with the object of making composites reinforced by carbon fibers, and it was found that there were no appreciable differences in wetting under microgravity and on the ground.

However, the special role of capillary forces is prominent in other microgravity phenomena, particularly in shaping with crystallization from a free surface, including crystal growth by Stepanov's method, where there is a liquid column between the crystallization front and the shaper edges. In [433] preliminary results were given on a technological experiment on Salyut-7, with indium crystallizing by Stepanov's method.

Figure 56 shows the growth device. Plastic body 1 is fitted with cover 2 and contains graphite container 3, which is filled with indium. Resistance heater 4 is insulated by porous material 5. The heat passes from the heater to the container through copper capsule 6, which also holds copper tip 7 (the shaper proper) and has a hole to equalize the inert-gas pressures inside and outside container 3. Meniscus 8 is formed initially between copper seed rod 9 mounted on shaft 10 and the edge of shaper 7. The meniscus is viewed by a movie camera, with illumination through window 11.

Figure 57 shows the drop shape on the edge of the shaper (a) and the resulting meniscus (b); the drop shape and the obvious relation $P = 1/R_0$ (R_0 is the drop radius) give $P \approx 1.3$. Then $Pr_0 \approx 0.72$, which is about 40% in excess of that necessary to produce a cylindrical column.

When the liquid has wetted the seed rod, the latter is pulled at about 3 mm·min^{-1}, although the pulling rate for a controlled specimen grown in an analogous device on the ground could not be raised above 0.2 mm·min^{-1}. Figure 57b shows a typical meniscus. The meniscus generator is a profiled curve having a multivalued projection on the abscissa, with $h/r_0 \approx 3.6 > \pi$, so the meniscus height attained by experiment exceeded the limiting value for the cylindrical column. The space specimen had an average diameter of 5.5–5.6 mm; the control specimen, 2.9–3.2 mm. The two specimens were largely single crystals, but the space specimen contained numerous bubbles of various shapes and sizes, whereas the ground specimen had no inclusions.

The research on liquid–solid interaction in wetting is supplemented by microgravity studies on interactions between solids and surface-energy estimates for solid boundaries. This topic is discussed in [439, 499], which are identical in formulation. It is clear that any change in gravity cannot have any appreciable effect on the forces between solids at contacts, but it has been shown [493, 499] that microgravity conditions favor measurements. The only difference between [493] and [499] is that in the former the results are those already attained on SL-1, while in [499] an analogous apparatus is described and the basis is given for a similar experiment on a rocket under the Texus program with brief weightlessness. The apparatus and basis for the experiment of [493] have already been mentioned in [1] (see pp. 113 and 161). A stainless-steel ball struck a planar metal target at a comparatively low speed and bounced repeatedly from it with gradually decreasing velocity until ultimately it adhered to the target because of the interaction forces. At each collision, the contact force was recorded as a function of time, along with the contact time and the recovery coefficient, i.e., the ratio of the rebound velocity to the incident velocity. These parameters allow one to examine the contact processes and provide information on the energy-dissipation channels and, thus, on the mechanisms controlling contact and adhesion. In that way, one can characterize not only the contact forces but also the surface energies of the contacting bodies. In [493, 499] it was shown that such an experiment can be successful only under microgravity, where the interfering gravitational forces are excluded. Other conditions are careful cleaning and the use of a high vacuum. The SL-1 experiments unfortunately were not very successful. In 80 experiments, only nine collisions gave information enabling one to estimate the surface energy at the contact as $\gamma = 1.15 \pm 0.02$ J·m^{-2}. The apparatus was tested before and after the flight, and was found to be fault-free. It was assumed that the failure was due to electromagnetic interference from adjacent apparatus, which made it impossible to record the data. In [499] there is a theoretical analysis of this experiment together with computer simulation results, which enable one to select the best parameters for a future rocket experiment.

3.8. DIFFUSION, THERMAL DIFFUSION, AND ELECTRICAL TRANSPORT [413, 422, 427, 447, 463, 485, 491, 496]

Gravitational convection is excluded under microgravity, so diffusion and thermal-diffusion coefficients can be determined more accurately, as is needed, in particular, for theoretical concepts on alloy setting. In [1] papers from past years on microgravity determination of these constants for liquid metals have been enumerated (Sections 3.1 and 3.2, pp. 80–82). Here such papers are only mentioned, and I merely consider the main results from experiment ES-335 [463] on SL-1, where the diffusion of ^{112}Sn and ^{124}Sn in liquid tin was examined. The preparations for this experiment have been indicated in [1] (p. 161). A special thermostat provided eight different temperatures in the range from 240 to 1400°C, which was used with 1-mm and 3-mm capillaries and tin specimens containing the ^{112}Sn and ^{124}Sn tracers. Most of the experiments were successful, although those at three temperatures (260, 312, and 1400°C) had to be repeated. The diffusion profiles have so far been examined with the 1-mm capillaries at 398, 810, and 1090°C, while they have not been examined for 3-mm capillaries. Therefore, at present the contribution from the capillary walls to the diffusion coefficient is uncertain. However, the preliminary data show that the typical convective contribution to the measured diffusion coefficient on the ground is from 30 to 50%, while it is almost excluded under space conditions. Therefore, the apparent diffusion coefficient on the ground is much larger than in space. The points on the diffusion profile in space have an unusually small spread (about 0.5%) by comparison with the ground (from 5 to 30%), as is evident from Fig. 58a, b. Table 3 gives the microgravity coefficients for two temperatures together with the isotope effect E calculated from the data:

$$E = -\frac{\Delta m_i^{-1}}{2\bar{m}^{-1}} \cdot \frac{\Delta D i}{\bar{D}}, \quad \text{where} \quad \Delta m = m_i - \bar{m}; \quad \Delta D = D_i - \bar{D},$$

where D is the average diffusion coefficient (for an average isotope mass $m = 119$).

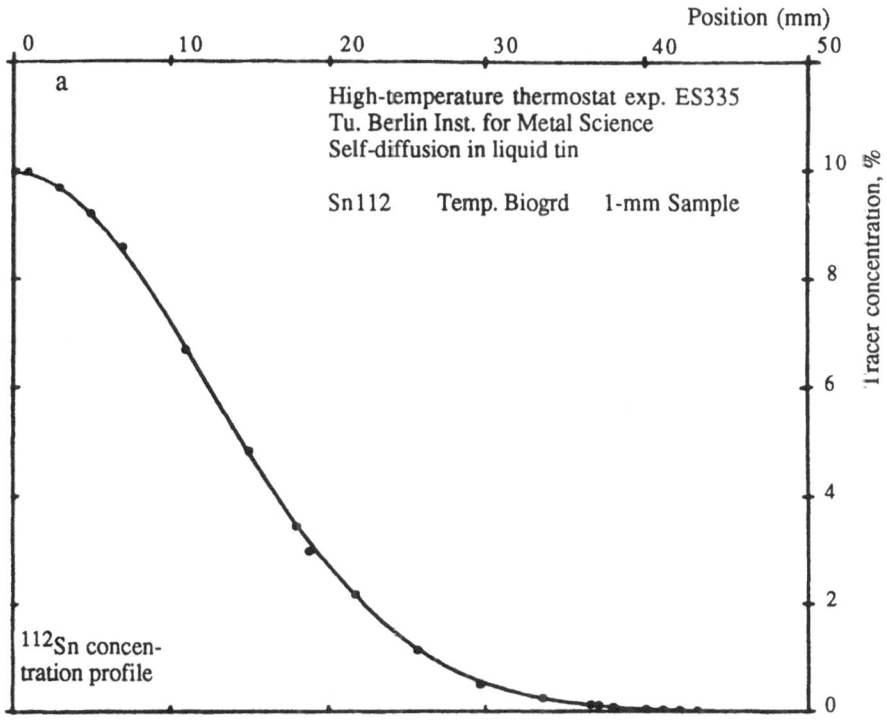

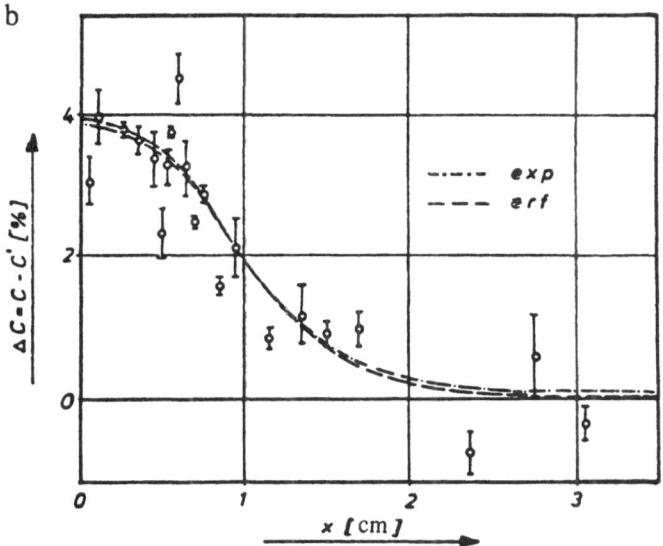

Fig. 58. ^{112}Sn concentration profiles in space specimen (a) and ground specimen (b) at 810°C, 1-mm capillary [463].

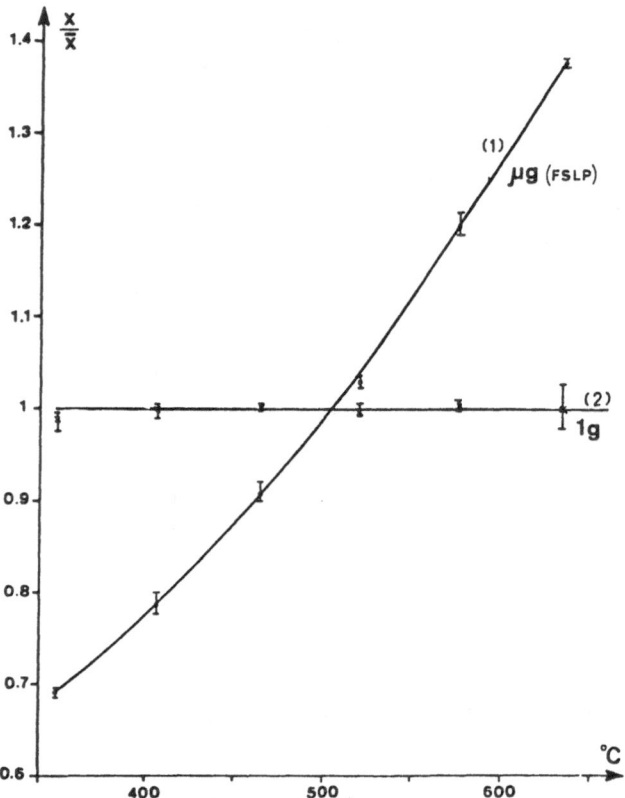

Fig. 59. Relative Co concentration as a function of temperature, with absolute Co and Sn concentrations in the range from 350 to 1400 ppm [485].

TABLE 3

T	D_{112} (10^{-5} cm/sec)	D_{124} (10^{-5} cm/sec)	E
398	3.65	3.61	0.10
810	9.07	8.67	0.89

The small isotope effect at 398°C indicated marked particle correlation, whereas the correlation factor tends to 1 at higher temperatures (800°C), so single particles are migrating. One can estimate the enthalpy H_i and preexponential factor D_{0i} in the Arrhenius formula for the diffusion coefficient $D = D_{0i} \exp(-H_i/kT)$, but it is considered [463] that these should be refined when all the specimens from SL-1 have been examined. The data from those studies will be published subsequently. On the whole, [463] showed the advantages of microgravity experiments for determining important constants such as the diffusion coefficient. The same may be said for microgravity measurements on the Soret coefficient S_T, which characterizes component thermal migration in response to temperature gradients and the separation arising from such thermal diffusion. Such studies have been performed in [447, 485].

In [485] results are described from experiment ES-320 on SL-1; this was concerned with the thermal migration of about 0.4% Co in liquid Sn in capillaries, diameter 2 mm and length 18 mm, with a temperature gradient of about 200°C/cm. A gradient oven was used. The system had been kept at a set temperature for 6 h (in [485] it had been shown that this is sufficient for equilibration and determining S_T), and then the specimen was removed and the Co distribution was determined by an activation method. The space experiments were successful, and 12 specimens were obtained, which showed Co distributions quite different from those on the ground (Fig. 59). The Co concentration was constant along the length in the ground specimen, whereas in the space specimen the Co concentration at the hot end was twice that at the cold end, which was due to convective mixing in the ground specimen. The space data gave the thermal migration of Co in liquid Sn together with the transport heat Q (cal/mole) $= -(3.72 + 0.08)T$ (K) for $593 < T < 933$ K and the Soret coefficient $S_T = 1.86/T$ ($S_T = 2.5\cdot10^{-3}$ K^{-1} for $T = 733$ K).

In [447], although the microgravity experiments had not been performed, there was a detailed discussion of the theoretical basis and the apparatus for thermal diffusion and determination of the Soret coefficient under microgravity for a $AgI_{0.75}KI_{0.25}$ mixture. The calculations showed that one could get separation corresponding to the pure Soret effect even with the residual accelerations under microgravity by using a gradient oven. A AgI–KI mixture is convenient in that the separation can be measured from the change in the electromotive force.

To complete this discussion, I emphasize that microgravity thermal diffusion is important in alloy solidification, since a temperature gradient arises above the solidification front, which leads to component separation by thermal diffusion, which must be incorporated in a full description of any binary or more complicated system. The advantages of microgravity research on diffusion and thermal diffusion are accompanied by the prospects for useful research on electrotransport in metals under microgravity.

If a direct current is passed through a homogeneous alloy, it causes partial separation, which is called electrotransport of electromigration. In [491] an apparatus is described for making observations *in situ* on the migration by continuous electrochemical analysis. Various types of convection interfere with such observations on the ground, although their origins are not always known. In [491] a method was described for establishing whether convection is due to gravitation or electrical phenomena. The cell was tested on a Hg–Na mixture, since a reliable electrochemical analysis method has so far been developed only for Na. However, it is intended that analyses should be performed for Ag, Cu, Ga, K, and Rb. It is considered [491] that the method should be useful in many microgravity operations, and it is proposed for use on the Eureka flights.

In conclusion, we can say that microgravity experiments give valuable information on physical constants in diffusion, thermal diffusion, and electrotransport, which it is difficult or impossible to obtain on the ground, and therefore microgravity should be promising for such research.

3.9. WELDING, SOLDERING, AND CUTTING
[40, 41, 420, 423, 425, 440, 480, 501, 506]

There is no need to repeat here what has been said in [1, 15] on the great practical importance for space research of welding and soldering methods in space and for research on fundamental trends governing such operations under weightlessness.

In [1] (Section 3.9, pp. 86–88) and in [15] (Section 1.5, pp. 37–41 and Section 3.6, pp. 142–153), it is emphasized that priority in such research belongs to the USSR, and a publication list is given, with descriptions of space soldering and welding apparatus. Here, therefore, I give only a brief characterization of recent publications on this topic in the USSR [40, 41, 420, 423, 425, 440] that are not reflected in [1, 15]; in addition, I describe research on soldering under microgravity performed on SL-1 [480, 501, 506].

In [425] there is a historical survey of space welding beginning with research on a "flying" laboratory (i.e., on an aircraft) on a TU-104 in 1965 and ending with the Isparitel' electron-beam apparatus, where the first experiments were performed in 1979. In that apparatus, the material is heated in a refractory crucible by an electron beam from a special gun and is deposited on various substrates. Over 180 films were made, and it was concluded that repairs can be made in orbit. Some of the coatings had unique properties not observed on the ground. Recently, the Electrical Welding Institute, Academy of Sciences of the Ukrainian SSR, has devised a method of making welded (thin-walled) convertible structures, which are assembled on the ground and made into a compact packet and disassembled or transformed in space.

In [440] there is a report on spacemen S. E. Savitskaya and V. A. Dzhanibekov conducting a unique experiment in open space on welding, cutting, and soldering metals, as well as depositing coatings by thermal evaporation.

In [420] there is a report on features of gas cutting under vacuum and, thus, in open space.

In [423] experiments are reported on welding thin sheet metal under conditions simulating space, viz., on aircraft laboratories. The reasons for burned areas occurring in thin sheets were considered. Eliminating this is a basic problem for space welding, particularly for transformable structures. These burns give rise to cavities in welds arising from flow out of the welding trough. When thin metal is welded, one sometimes gets specific defects similar to such burns but not related to efflux. The nature of these is not yet certain. An apparatus was built with movie recording for the welding and oscilloscope recording for the microplasma burner power. The melting time t_m and existence time t_e were related to the thickness under various gravitational conditions. The burning time is also related to t_m and t_e. Experiments and calculations on heat transfer showed that thin sheet can be welded by concentrated sources if there is stable tool displacement along the joint (without halts), along with controlled thermal pulses, whose frequency and spacing are adjusted to the task, and the use of additional metal and assemblies with minimum possible gaps.

As regards space welding methods, there are papers [40, 41] dealing with the prospects for solid-state metal joining by heating to temperatures below the melting point and applying pressure. This is called diffusion welding; in [40, 41] it is suggested that future space technology will use this method. However, microgravity does not appear to play any important part in weld processes in that method.

There are the ES-304/305 results on vacuum microgravity soldering on SL-1 [480, 501, 507]; technical Ni is soldered with a silver solder (71.81 Ag, 28.02 Cu, 0.148 Li) on a specimen consisting of three sections differing in shape and thickness: a) a multislot section, b) with an annular slot, and c) with a sickle-shaped slot. Such specimens provided a comparison of the slot filling mechanisms under varying gravitational conditions, the solder flow, and the microstructure in the solidified solder in relation to slot shape and width. All the parameters apart from gravity were the same in the ground and space experiments. The specimens were heated in a gradient oven to a temperature exceeding the melting point of the solder by 50°K to fill the slots and were then cooled to solidification. The cyclic heating patterns for the two experiments were identical.

The slot filling was determined by computer processing applied to radiographs (autoradiograms) after the end of soldering.

In [507] there is a detailed description of the solder filling in relation to slot shape and width, as well as certain details of the microstructure formed by the solder reacting with the Ni, which is dependent on the phase states and Ag–Cu–Ni diagram. On the whole, the structures in the ground and space specimens were identically dependent on the slot form and width and on the thermal conditions. However, the gravitational level had the following effects:

1. under microgravity, the porosity was increased; i.e., bubbles formed by gas release or due to instabilities in the liquid motion are not removed by buoyancy, as occurs on the ground.
2. At 1110°K, the density difference between the liquid and the CuNi dendrites (containing about 10 mass % Ni) is 0.95 g/cm^3, so there is appreciable segregation at 1 g_0, particularly for wide slots. This did not occur under microgravity.

3.10. CONCLUSIONS

I complete this survey of papers in bibliography Part VI by noting that certain specific topics in [416, 470, 498] and those in a survey [458] have not been considered above.

These topics include catalysts made under microgravity [416, 498]. In [416] there is a discussion of improving a silver catalyst by using microgravity at certain stages to make this composite and provide the optimum trace-component (promoter) distribution during formation. In [498] the structure and properties were examined for a nickel catalyst made on the Shuttle STS-7 in June 1983, where it was found that microgravity had a substantial effect on the reaction kinetics with this catalyst (in the decomposition of hydrogen peroxide, for the hydrogenation of N_2O and alkenes).

In [470] a patented method is described together with the corresponding apparatus for producing metal and components during space flight. The blank, a rod, is heated in a cavity by means of the sun's rays focused by a parabolic mirror. The metal vapor passes through nozzles to a mold, where it is deposited. That technique is particularly convenient for making weldless pressure vessels, flat surfaces for unbounded strips, and also more complicated components with smooth surfaces. The apparatus can also be used to make high-purity metals.

Metallurgical research in space can be summarized by supplementing the results of Sections 3.1–3.9 with the contents of a reporter paper [458], which deals with results on this obtained on SL-1. In [458] all studies here are classified under three headings:

1. Testing metallurgical techniques (soldering and welding) for direct use in space. Section 3.9 shows that welding and soldering are major operations in joining up large structures in orbit. Considerable progress has been made. It is obvious that there are prospects for developing and using these methods.

2. Microgravity measurements on thermophysical and physical parameters and theory checking.

In [458] good prospects are anticipated for research on diffusion and thermal diffusion under microgravity, where gravitational convection is eliminated (see Section 3.8), and also on adhesions (see Section 3.7), where interfering gravitational forces are excluded. In [458] it is pointed out that there is an infinite range of potential systems, from which a proper choice must be made for further research. One should select, first, systems of primary importance to the theory and, second, specific systems of interest to space metallurgy.

Microgravity provides a powerful tool for testing metallurgical theories, particularly concerning heat and mass transfer controlled primarily by diffusion. This is clear from the evidence given in this chapter. Results with immiscible alloys (Section 3.1) and eutectics (Section 3.4) under microgravity represent good data for comparison with theories. Such fundamental topics should be researched under space conditions. Space is a tool that resembles any other complicated and costly apparatus serving science. However, the following recommendations are made [458] for using this tool:

1. Only sound projects should be investigated, and it should be established that gravitation is a relevant parameter.
2. The experiments should be as simple and clear-cut as possible in order to obtain reproducible results.
3. Scientific societies should be set up concerned with experiments with the same objects and serving similar purposes.

3. Preparation of new or improved materials. Space activities in the future will be evaluated not merely as regards fundamental research, and it is now necessary to examine the scope for using space in applications. Unfortunately, no definite detailed suggestions have yet been made. However, each new method of handling materials must be tested carefully. Particular attention should be given to the fact that not all mechanisms dependent on gravitation and interacting with one another are involved in a given process. This means that the analysis can become very troublesome and require special methods. In [458] an effective scheme is proposed for making important materials in space.

The data in this chapter on the whole demonstrate considerable progress in fundamental research on solidification for metals, alloys, and composites under microgravity as well as in the technically important operations of welding, soldering, and coating. On the other hand, these studies have raised many problems for future solution.

CHAPTER 4

MICROGRAVITY SOLIDIFICATION
OF GLASS

Several papers cited in [1] (Chapter 4, pp. 93–100) deal with the prospects for using space conditions to make high-grade optical glasses. The advantages of microgravity in processing optical glasses are emphasized also in [15] (Section 3.7, pp. 153–158).

Here we consider briefly some new papers on microgravity optical-glass setting, glass reaction kinetics, and bubble formation and elimination by diffusion [514–516, 520–522, 524], as well as certain papers [518, 519, 523] dealing with making metallic and optical glasses. In that case, nucleation problems are examined in particular detail for supercooled alloys not only in the bulk (homogeneous nucleation) but also at foreign centers in the bulk or at surfaces (heterogeneous nucleation). This survey of current papers begins with this line of research.

4.1. NUCLEATION, METAL-GLASS FORMATION, AND SUPERCOOLED-ALLOY SOLIDIFICATION [518, 519, 523]

In [518] supercooling regularities are examined for the readily vitrifying alloy $Pd_{77.5}Cu_6Si_{16.5}$; we first give a brief exposition of the general concepts on thermally activated nucleation in a supercooled liquid. The activation energy ΔG^* required to produce a critical nucleus is dependent on the surface tension and on the free-enthalpy difference between the liquid and solid states. As $\Delta G^* > 0$, a certain supercooling $\Delta T = T_{mp} - T_n$ is required (T_{mp} and T_n are correspondingly the melting and nucleation points), before setting can begin. Homogeneous-nucleation theory predicts that the greatest supercooling has the limit $\Delta T/T_{mp} = 0.18$. ΔG^* in heterogeneous nucleation can be reduced by external-phase wetting in the bulk or at the container. The limiting values of $\Delta T/T_{mp}$ should then be less than 0.18. The temperature–time transition diagram can be calculated if one knows the nucleation rate i

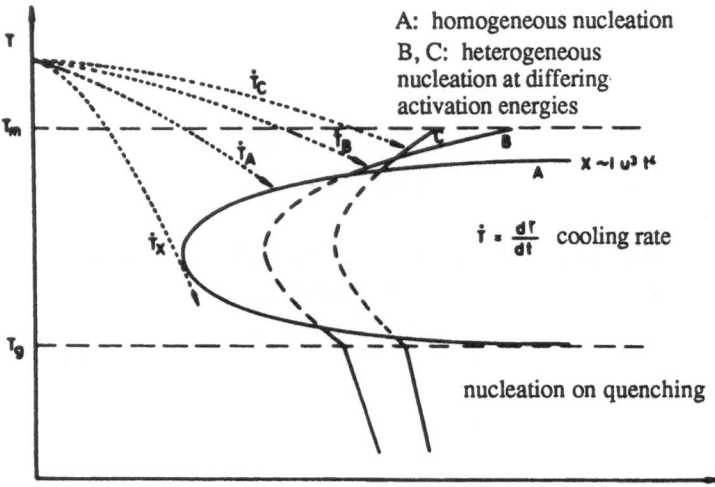

Fig. 60. Schematic temperature–time transition diagram on the assumption of homogeneous nucleation alone (A) and heterogeneous nucleation with differing activation energies (B, C) [518].

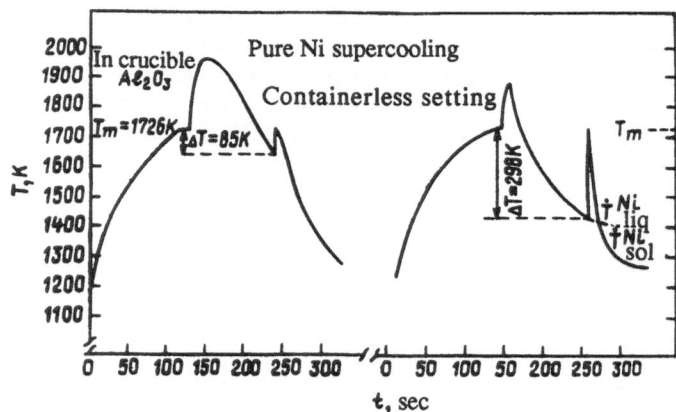

Fig. 61. Temperature–time profiles [519].

and the crystal growth rate U. Figure 60 shows such a diagram for homogeneous nucleation (curve A) and heterogeneous nucleation with different ΔG^* (curves B and C). The diagram enables one to determine the volume fraction X formed in a time t for a supercooling ΔT if I and U are known. It is evident from Fig. 60 that one can vary the cooling rate $\dot{T}$ to attain different points in the diagram. In the critical case $\dot{T} > \dot{T}k$, the cooling is so rapid that crystallization is prevented and the liquid attains a temperature below T_g (the glass temperature) and solidifies in a metastable amorphous (vitreous) state. Metal glasses are made in this way, which represents a new, promising class of materials with extraordinary properties, in particular great strength. This usually requires cooling rates on the order of 10^6°K/sec. However, in [518] papers are cited in which bulk metallic glasses have been obtained with rates of only about 1°K/sec by the elimination of heterogeneous nucleation. It is also stated that supercooling levels $\Delta T/T_{mp} > 0.50$ have sometimes been attained, i.e., much greater than 0.18 from homogeneous nucleation theory.

These conflicts on nucleation and transformation kinetics for supercooled metals give interest to fundamental studies, including those under microgravity with containerless treatment.

In [518] such studies were performed with $Pd_{77.5}Cu_6Si_{16.5}$, which includes the readily formed glass in the PdSi system, especially near the eutectic concentration $Pd_{82}Si_{18}$. A small amount of Cu reduces the crystallization rate by about an order of magnitude. The PdCuSi system can thus be used to examine supercooling. Measurements have been made on the ground and with reduced gravity, freely falling drops in tubes.

We cannot give all the details on the structure as related to temperature and setting conditions, and merely deal with the conclusions. It is shown that supercooled PdCuSi sets in a fashion determined by heterogeneous bulk nucleation, which gives rise to Pd_3Si dendrites. The bulk nucleation can be reduced by heating the liquid above the liquidus temperature, which increased the supercooling to $\Delta T = 310$ K. At that level, two types of metastable crystalline phase are formed, which have different microstructures. The transformation diagrams show that homogeneous-nucleation theory overestimates the nucleation rate in supercooled PdCuSi by more than two orders of magnitude. The diagram gives the critical cooling rate to prevent crystallization as 60°K/sec, i.e., the value providing metallic glasses. The transformation rates can be reduced further if one eliminates surface centers by chemical etching or ion bombardment.

In [519] Herlach et al. [518] examined effects from heterogeneous nucleation at the liquid–wall boundary by comparing temperature–time curves for melting and setting in pure Ni in He in an Al_2O_3 container and in a containerless experiment involving electromagnetic levitation. Figure 61 shows that the curves display heating retardation at $T = T_{mp} = 1726$ K [flat parts on the $T(t)$ curves], followed by a steep rise in $T(t)$ because of the resistance step at $T = T_{mp}$. The $T(t)$ curves on cooling differed considerably between the two cases. The containerless treatment led, first, to more rapid cooling, i.e., the heat was transferred by radiation and convection in the He. This eliminated the cooling resistance due to heat transfer through the substrate. Second, containerless Ni treatment increased the supercooling by about a factor of four (from 85 to 298°K), so that treatment is of great value in researching metal phase formation and vitrification. However, electro-magnetic levitation is applicable on the ground only for metals having $T_{mp} > 1000$°K, since levitation is possible only with a sufficient electromagnetic energy input.

Under microgravity, this limit can be reduced considerably, so there is good scope for examining phase kinetics in comparatively fusible alloys such as materials for making metallic glasses. Also, microgravity enables one to use larger specimens. Such experiments are planned for the future.

In [523] the conditions for metal-glass formation are examined from classical homogeneous-nucleation theory, where there is a survey of certain experimental studies. It is emphasized that the theory has been extended recently to multicomponent systems, which is particularly important, since all metallic glasses are alloys.

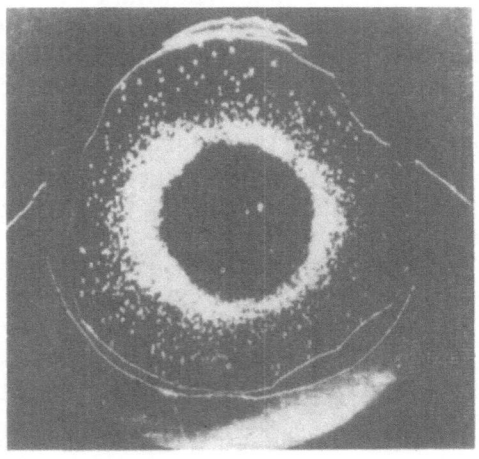

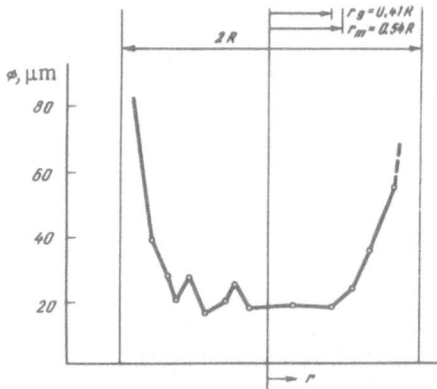

Fig. 62. Bubble distribution (top) and bubble sizes
(lower curve) in an orbital specimen [514].

The conclusions in [523] are analogous to the above, viz., the limiting factor in metal-glass formation is heterogeneous nucleation. Containerless processing can eliminate this, as can removing potential nucleation sources from the surface by etching or heating under vacuum, or else by the use of flux, when one can obtain fairly large volumes freed from nucleation centers. Microgravity may be useful in the flux technique, as large metal volumes can be kept surrounded by low-density low-fluidity flux, which is difficult under ground conditions. On the other hand, if heterogeneous centers occur in the bulk, one needs gravity to segregate them to the flux–metal interface. Microgravity treatment in the latter case may be unsuitable for producing large glass volumes.

To conclude, we may say that microgravity experiments with containerless treatment have already provided valuable information. Further research in that field appears promising.

4.2. GAS-BUBBLE FORMATION KINETICS IN GLASS SOLIDIFICATION [514, 522, 524]

In Section 1.6, we have considered theoretical papers on bubble formation and migration, and in Section 3.3 we have considered experiments on metals and model transparent systems. In [1] (p. 100) papers are cited on eliminating bubbles from glasses by outgassing or controlled displacement. In [1] it was also stated that gas diffusion coefficients in liquid glasses can be determined under microgravity from the time course of the bubble diameters.

Here we consider three papers on these topics appearing after those cited in [1].

In [514] microgravity effects were examined for inhomogeneities in molten glasses, where no differences were found between ground and space specimens apart from the bubble volume distributions. In the ground specimen, the bubbles were mainly evenly distributed over the cross section, while the space specimen (Fig. 62), first, had the bubbles clearly concentrating in an annular region around the axis and, second, there was a region almost free from bubbles at the center u, while, third, the mean diameter increased toward the edge (from 30 to 80 μm). In [514] an explanation is proposed for this unexpected phenomenon on the basis of the radiation pattern within a circular cylinder having a radiating and repeatedly reflecting side surface and containing a homogeneous absorbing material.

Under the conditions of [514] there is a radial temperature gradient having an annular concentric maximum, which under microgravity produces the observed bubble distribution by displacement.

The model of [514] indicates that such an annular distribution should be observed if: 1) the heat transfer within the specimen by radiation is comparable with that from conduction or exceeds it, 2) the thermal-radiation absorption coefficient and the reflection from the vessel walls are such that the heat rays are reflected at least once (better several times), and 3) the viscosity and the radial temperature difference allow the inhomogeneities to travel a distance comparable with the vessel size during the working time. On the whole, [514] shows that one needs to incorporate thermal-radiation effects from the vessel walls and possible focusing. Such phenomena can be used to clear liquids from bubbles and impurities.

In [522, 524] the time course of the bubble size was examined; in [522] observations were made on shrinking helium bubbles in glass under microgravity on STS-II (February 1984). Such observations had previously been made with brief weightlessness. In the STS-II experiment, the glass containing an artificial helium bubble at the center was melted in an IHF furnace and then kept at 1100°C for about 90 min. The bubble was photographed every minute, and the radius was recorded with an error of about 0.01 mm. Immediately after the glass melted, the bubble contracted sharply for several minutes because of pressure equalization, but then it expanded gradually for about 8 min on account of thermal expansion, after which it contracted almost linearly over about 80 min to a residual volume of about 10%. Ground experiments showed that the bubble began to grow again in the later stages, evidently from reverse diffusion into it by dissolved gases such as CO_2. This also explains why new bubbles form and grow. No bubble displacement from Marangoni convection was observed. The diffusion coefficient was found as $D = 3.9 \cdot 10^{-5}$ cm^2/sec at 1086°C or $D = 5.6 \cdot 10^{-5}$ cm^2/sec at 1126°C. However it is considered necessary to perform a further experiment at a higher temperature in order to detect convection effects. The MAUS apparatus used in [522] has proved suitable for such completely automatic experiments.

TABLE 4

Experiment	Temperature	Diffusion coefficient, cm^2·sec$^-$
Corrosion of SiO2 glass by fused $Na_2O \cdot 2SiO_2$	1160°C	$6 \cdot 10^{-7}$
Corrosion of SiO2 glass by fused $Na_2O \cdot 3SiO_2$	1160°C	$2.8 \cdot 10^{-7}$
^{22}Na diffusion in molten $Na_2O \cdot 3SiO_2$	1200°C	$2.5 \cdot 10^{-5}$
^{22}Na diffusion in a liquid of composition $0.5Na_2O \cdot 0.5Rb_2O \cdot 3SiO_2$	1200°C	$2.05 \cdot 10^{-5}$
He diffusion in molten $Na_2O–CaO \cdot SiO_2$	Temperature range 939–1036°C	$D_{He} = 8.95 \cdot 10^{-3} \exp (56.6$ kcal/mole$)/RT$

In [524] it is stated that differential equations have been drawn up and analyzed for dissolution or growth in one-component bubbles. However, the same problem for bubbles containing two or more components has received less attention, and many interesting aspects have not been examined. Consequently, [524] dealt with differential equations controlling bubble shrinkage or growth under microgravity in a model system containing multicomponent bubbles in molten glass. Not only are the complete equations derived there, but also approximate forms of them, which have been used to obtain a quantitative picture of the dissolution (dissipation) and growth. In particular, the radius may increase rapidly with time for certain values of the controlling parameters. In [524] there is a discussion of the conditions under which this can occur. Stationary bubbles are also considered, where the growth and shrinkage rates are equal. Paper [524] is thus an example of theoretical studies in this area.

4.3. REACTION KINETICS IN MOLTEN ALKALI GLASSES AND COMPONENT DIFFUSION

In [515] ES-307 results from SL-1 were given for alkali glasses.

It was planned that the microgravity experiments should cover diffusion in $Na_2O \cdot 3SiO_2$ and $Rb_2O \cdot 3SiO_2$ glasses, as well as the interactions of amorphous SiO_2 with alkali glasses. Faults in the IHF oven meant that only the first experiment could be performed partially and that there were considerable deviations from the planned course (set temperature–time conditions). Instead of the intended heating to 1200°C, stabilization for 30 min, and rapid cooling, the oven was heated only to 1180°C and was immediately switched off. This was followed by slow cooling. Nevertheless, it proved possible to estimate the diffusion for 1730 sec at 1180°C. The diffusion coefficient found for $0.5Na_2O \cdot 0.5Rb_2O \cdot 3SiO_2$ was $D = (1.1 \pm$

$0.3)\cdot10^{-5}$ cm$^2\cdot$sec^{-1}, which agreed with earlier values from brief weightlessness on rockets. Also, it was possible to relate the diffusion coefficient to the component concentrations, which agreed with the earlier Texus results (Table 4).

Although experiment ES-407 on SL-1 confirmed that these kinetic studies can be made on glasses, the accuracy is so far inadequate, and it will be necessary to repeat the entire experiment on D-1.

In [517] there is a brief survey of all the experiments on diffusion in glasses under microgravity with the Texus rockets and with the SL-1 and STS-11 flights. We give here the data from [517] on the microgravity diffusion coefficients.

In [517] it was concluded that microgravity experiments are useful not only in researching the diffusion and corrosion in glasses and in obtaining diffusion coefficients more accurate than ground ones, but also in other areas of high-temperature glass and ceramic technology.

4.4. CONCLUSIONS

Microgravity enables one to solve certain problems in glass science as enumerated in [1] (Section 4.1, pp. 93 and 94) and in this section, and it might seem that these should attract considerable attention to space experiments. Such experiments have been discussed in part above, but they are few and have so far been performed under prolonged weightlessness with universal systems. In order to produce perfect ultrapure glasses under microgravity, one needs specialized equipment similar to that previously proposed (see Section 4.2 in [1], pp. 94 and 95). It is necessary to develop such apparatus and to formulate new experiments for glass production in space.

Microgravity experiments similar to those described here (Section 4.1) should be directed to fundamental topics in nucleation for supercooled liquids, which are important in producing amorphous metals, which are interesting and promising materials.

The prospects for further experiments on glasses under microgravity are considered in a review [520], in which all previous experiments are discussed with emphasis on the prospects for research here. In particular, it is stated [520] that evidence on critical cooling rates under microgravity is important, for comparison with ground values, as well as the importance of information on the surface and diffusion phenomena in glasses at zero gravity. In [520] it is also stated that microgravity research on glasses requires new equipment providing higher temperatures, more rapid cooling, and stable levitation. It is also emphasized that previous microgravity glass research has constituted only a small fraction of research in materials science in space, although major and interesting programs are planned to solve problems in glass science.

Chapter 5

MICROGRAVITY GROWTH OF CRYSTALS
FROM AQUEOUS SOLUTION

In [526] information is given on the phenomenological theory of mass crystallization, which occurs when crystals grow from solution, and various experiments on it are discussed, including copper sulfate recrystallization [525].

In [1] (Chapter 5, pp. 101–108), papers [525, 526] have already been considered (although they were published after [1] appeared), and here we can supplement [1, 526] by considering three experiments on crystal growth from aqueous solution on SL-1.

5.1. GROWTH OF METALLIC-CONDUCTION
ORGANIC CRYSTALS [528]

In [528] experiment ES-322 is discussed, where it was planned that the SL-1 mission (see [1], p. 160) should examine the microgravity growth of TTF-TCNQ organic crystals showing metallic conductivity. It was proposed that counter-current diffusion should be used at 40 ± 0.1°C in a three-chamber reactor, with the chamber separated by valves opened at 0.3 mm a minute (without introducing perturbations) and then closed. Figure 63 shows the system. Above all three chambers, there was a rubber baffle, which eliminated the need to leave air bubbles (as envisaged in previous experiments) and which equalized the pressures. Figure 63 illustrates the concept. The two end chambers are filled initially with different solutions, and the valves are opened when microgravity is attained, so the solutions diffuse in the central growth chamber, where they react to produce a compound of lower solubility, which crystallizes. Two experiments on calcium tartrate crystal growth were particularly interesting, in one of which there were thin Teflon filters between the chambers, which eliminated convective perturbations on valve opening, while in the other there were no such filters. The two sets of results showed that the filters eliminate convection very effectively.

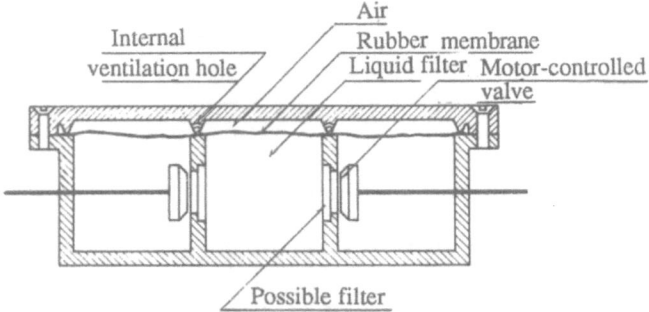

Fig. 63. Cross section of reactor showing elastic membrane eliminating the need for an air cavity and providing equal pressures even with the valves closed [528].

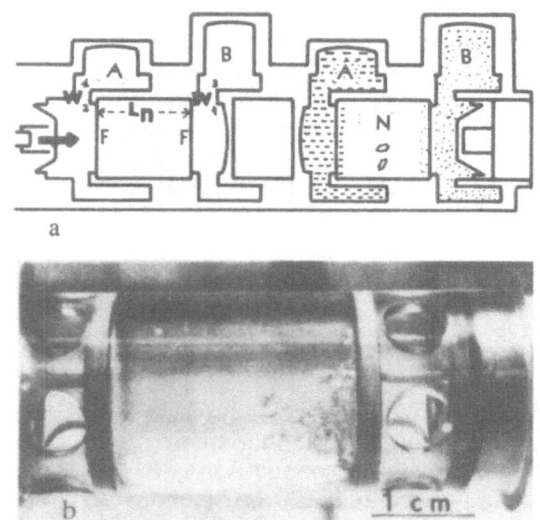

Fig. 64. Three-chamber space reactor: a) schematic drawing; b) central part containing brushite crystals [532].

In the discussion in [528] on further experiments on microgravity growth of crystals, it is stated that a new experiment is intended with TTF-TCNQ crystals of synthetic-metal type on the LDEF (long duration exposure facility), an unmanned module to be launched by NASA, as well as Eureka experiments. The LDEF experiment was undertaken before the results from the SL-1 were available, so no filters were placed between the chambers. On the other hand, the g perturbations in LDEF should [528] be less than on SL-1, while the run time is much longer and, therefore, it is possible that the LDEF experiment would be successful without filters.

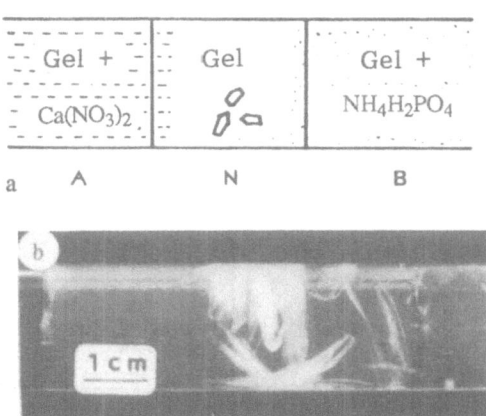

Fig. 65. Ground reactor: a) schematic drawing; b) brushite crystal grown in a gel in 2 months [532].

5.2. GROWTH OF SPARINGLY SOLUBLE CRYSTALS [532]

In [532] the ES-333 results are examined (see [1], p. 160) on microgravity production of sparingly soluble crystals on SL-1 (brushite $CaHPO_4 \cdot 2H_2O$ and lead monetite $PbHPO_4$) from aqueous solution, for comparison with ground results, but where there was growth from a gel medium ensuring no gravitational convection. The microgravity experiments, as in [528], employed a three-chamber reactor with elastic membranes and valves, so it was possible to react components A and B held in separate chambers and crystallize the product in the third (growth) chamber. When brushite is made, $A = Ca(NO_3)_2$, $B = NH_4H_2PO_4$, while with lead monetite $A = Pb(NO_3)_2$, $B = H_3PO_4$. Figure 64 shows the space reactor and a photograph of the middle containing a brushite crystal. Figure 65 shows the ground reactor in growing brushite from gels. The gel in the ground experiment was provided by polymerized tetramethoxysilane (TMS). A small amount of TMS is sufficient to prevent convection in the growth region. The TMS contents in the ground experiments varied from 2 to 10%.

The ground (gel) and space crystals were compared for morphology and defect structure by optical and x-ray topography methods; without going into the details, we give the main conclusions. The gel growth gave good crystals if high-purity material was not required, since the gel contaminated the crystals (was included in them mainly at the start). The convection was in fact inhibited by the gel. The space crystals were of good quality, but they had defects arising from the following factors. One of them is due to the retardation on the return of SL, where the crystallization continued in spite of the valves being closed. To avoid such defects, the crystals should be removed before the spacecraft returns. Other defects were steps on the platy crystals, which may be due to perturbations on valve closing. Finally, the third factor was defects associated with g perturbations during the flight. Optical holography should enable one to examine how these defects occur. In future experiments, attempts may be made to avoid all these three types of defects. In [532] it has thus been shown that countercurrent diffusion under

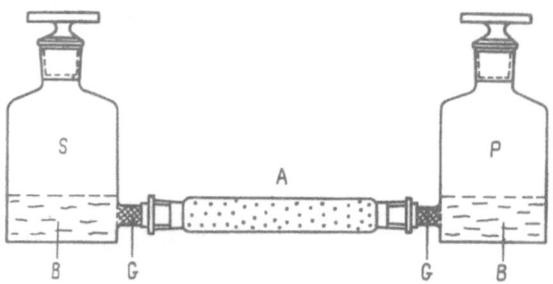

Fig. 66. Device for gel protein crystallization: P) protein solution in buffer B; S) salt solution in buffer; G) glass wool plug; A) polyglycine gel in buffer [530].

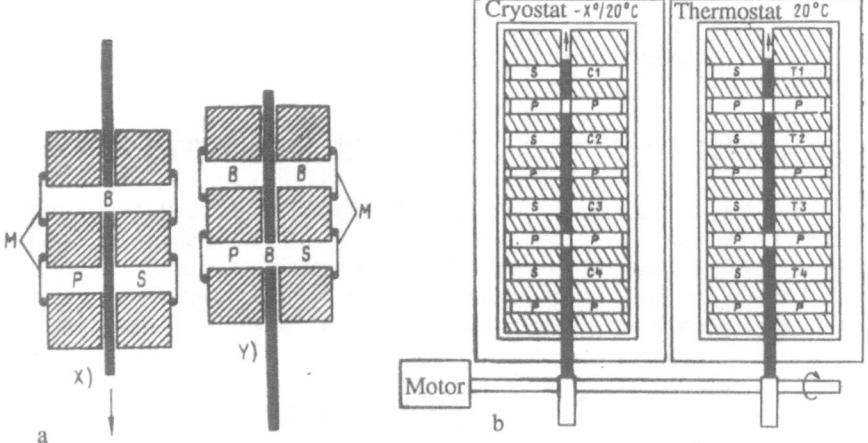

Fig. 67. a) Device for crystallizing protein in space free from convection: X) initial position; Y) final position for diffusion initiation; B) buffer solution; P) protein solution; S) salt solution; M) membrane, with the central black section representing the sliding valve. b) Space apparatus for examining the growth of lysozyme and β-galactosidase crystals: C1–C4, T_1, T_2) containers for the β-galactosidase; S) salt containers; T_3, T_4) lysozyme containers; P) buffer containers [530].

microgravity can give good crystals, which is particularly important when there is no gel system compatible with the material. Space then provides unique scope for making such single crystals.

5.3. PROTEIN CRYSTAL GROWTH [530, 531]

The last of the experiments on SL-1 concerns growing protein single crystals [530, 531] (see also [95]), experiment ES-334 (see [1], pp. 160–161), which was particularly successful and showed that there are good prospects for future use. Fairly large, perfect protein crystals (size up to 1 mm) are required for x-ray structure studies, which is a fundamental aspect of biophysics. On the ground, protein crystals can be grown by gel methods with an apparatus as shown in Fig. 66. However, the crystallization is extremely slow and gives only small crystals, which are not always free from contamination. Also, the brittle protein crystals, particularly those growing as filaments, are trapped in the tangled gel network and frequently break. Particular interest, therefore, attaches to growing protein crystals under microgravity without gels. This is the more so since preliminary experiments with brief weightlessness have shown an unexpected effect: a marked increase in crystallization rate under microgravity. It is suggested that this is due to molecular ordering regions at high protein concentrations (130 mg/cm^3). Such ordered clusters may aggregate to crystalline material if there is no turbulence. On the ground, gravitational convection may disrupt the clusters into smaller particles or even to individual molecules, which naturally require considerable time to crystallize. The increased protein crystal growth rates with brief weightlessness show the desirability of experiment ES-334 on SL-1.

Figure 67a, b shows the device for growing protein crystals free from convection on SL-1; as in [528, 532], the reactors have chambers separated by valves, which here are plates. When microgravity is attained, the sliding valves are displaced without vibration, which allows the protein solution P to interact with the salt solution S via the buffer B, as in Fig. 67a.

In [530, 531] two proteins were used with very different molecular masses: β-galactosidase, mass 465,000 daltons, and lysozyme, 14,307 daltons. The structure of β-galactosidase has so far not been examined because of the lack of crystals. Lysozyme was examined about 20 years ago.

Without going into details, we state that β-galactosidase crystals were made under microgravity 27 times larger in volume than those in the same apparatus on the ground, the factor for lysozyme being 1000. The space crystals were regular in shape, and the polarizing microscope showed that they had good quality.

Only further research can determine whether the method is suitable for other types of protein, although the results of [530, 531] are very promising, as is emphasized in [76], where there is a summary of all the experiments on materials science in space performed on SL-1.

5.4. CONCLUSIONS

It is necessary to formulate new experiments to resolve the problems associated with mass crystallization under microgravity from aqueous or other transparent solutions to judge from the evidence of [1] and the above. Experiments are desirable in which one can clearly and correctly trace the occurrence of nuclei and crystal growth *in situ*. One could use either existing equipment or new special designs for microgravity, where appropriate methods such as holography might be employed.

Simple, reliable apparatus and methods are required to meet the need for experiments on manned and unmanned vehicles, as well as to select the best systems providing answers to relevant problems, where the topic should be discussed in detail by specialists in each particular area, and the most suitable apparatus and methods and the most promising systems should be selected for use in space.

It is very promising to continue experiments on making protein crystals under microgravity larger in size and more perfect in structure than those grown on the ground. One needs not only select systems for x-ray diffraction research on protein crystals, but also to provide for testing the hypothesis of [530] on the reasons for the markedly accelerated protein crystal growth under microgravity. One should device methods of observing protein cluster formation (or possibly for model polymers) and the participation of such clusters in crystallization on seeds.

Chapter 6

APPARATUS AND METHODS FOR MICROGRAVITY
AND MATERIALS SCIENCE IN SPACE

In [1] the main specifications for space apparatus have been formulated, and a list was given of apparatus already used on Soviet space vehicles (Table 6 in [1], p. 109) as well as equipment prepared for Spacelab 1 (Table 7 in [1], pp. 111–113) and for the Eureka module (Table 8 in [1], pp. 114–115). Here we therefore consider only papers appearing after [1] giving additional information on the apparatus, including improved designs not mentioned in [1]. Also, this chapter covers in more detail than did [1] certain aspects of methods such as research on heater temperature patterns in Section 6.2, measuring accelerations on space vehicles in Section 6.5, holography under microgravity in Section 6.6, and levitation methods in Section 6.8. As [15] deals in more detail than did [1] with engineering equipment on Soviet vehicles, where many schemes and photographs are given, we use here only a little of the evidence presented in [15], such as the characteristics of the Pion equipment in highly condensed form. On the other hand, in [15] there were omitted some of the data supplementing [1], such as on the Soviet–Czechoslovak Kristallizator apparatus and on holographic equipment, which will be dealt with in somewhat more detail.

6.1. MICROGRAVITY HIGH-TEMPERATURE HEATERS

In [1] the main characteristics of most high-temperature space furnaces have already been given, while fuller characteristics have been given in [15] for the equipment used in the Salyut–Soyuz vehicles. So here we refer only to publications dealing with these devices operating in space, with measurements on the temperature patterns and descriptions of the control devices or other information not given in [1, 15].

In [549–552] there are descriptions of the Splav-01 apparatus together with methods of preparing for and conducting the experiments, which included ground development work, calculations, and measurements on the temperature pattern.

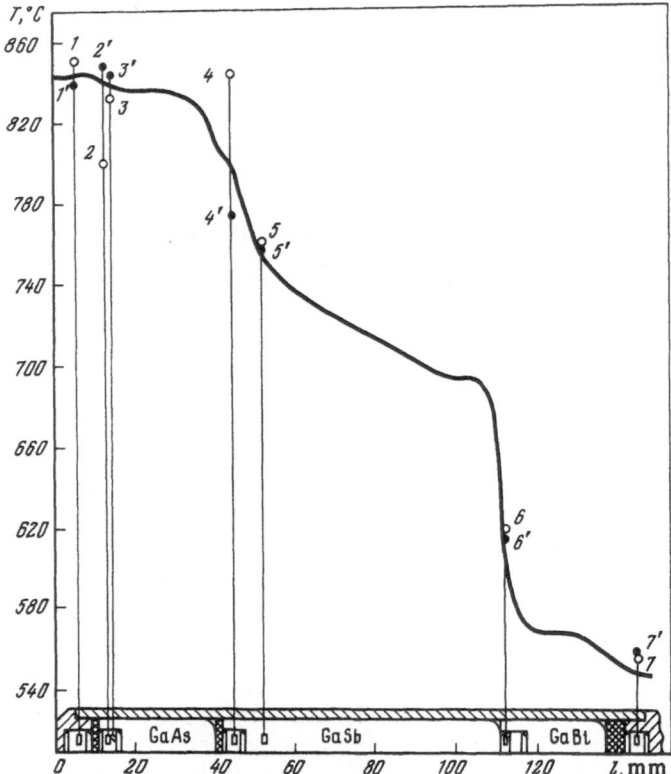

Fig. 68. Temperature distribution along the capsule axis, where the
solid line is from calculation, the circles being from ground tests and
the points from tests in flight [551].

Figure 68 compares the calculated distributions with those measured in space and
on the ground after the equipment had operated in space for more than 3 years. The
ground analog was substantially younger, but the temperature deviations in space
and on the ground did not exceed 11° as indicated by most sensors.

In [552] the temperature fluctuations in the working zone of the heating chamber were recorded with thermocouples at various points on the capsule simulator,
which provided a basis for modifying the equipment as regards the control-thermo-
couple location and working parameters.

In [539] the temperature pattern in the Kristall furnace was examined with
fusible wires (Imitator-1) and with thermocouples and a digital voltmeter (Imitator-
2). The temperature pattern in the probe corresponded to that in the tube containing
the molten semiconductor. The approach to the stationary pattern near the heater
was oscillatory with heavy damping and a long settling time (over 2 h). The
temperature initially exceeded the equilibrium value and then approached it slowly
(Fig. 69) in the region of container 1. This behavior must be borne in mind. Also,
there is a delay of up to 3 h if stationary conditions must be attained.

In [596] the temperature patterns in the Magma furnace were recorded with a
probe containing 10 thermocouples; the temperature patterns and heating and cool-

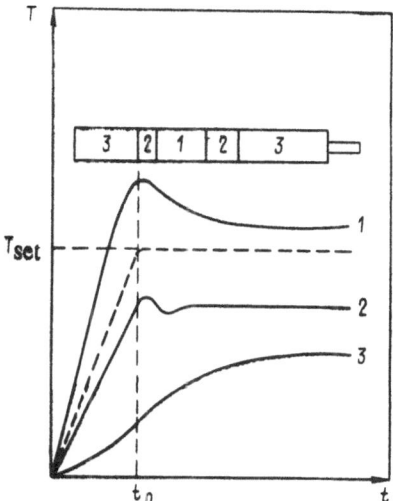

Fig. 69. Schematic representation of temperature profile settling in characteristic container parts (container in the initial position in the Kristall furnace). The temperature of the controlled thermoelement as a function of time is shown by the dashed line [539].

ing curves were similar to those given above for the Kristall apparatus. The superheating in the hottest part for a set temperature of 1000°C was 9°C, but it was over 20°C at 500°C, since the rate of approach to the steady state fell as the set temperature increased. When the furnace reached the steady state, the maximal temperature of 1014°C fell to 1009°C, while at the same time the temperature at the edges of the heater increased by 35–40°C. In the steady state, the fluctuations were ±3°C with a period of 1.5 min at a set 1000°C.

In [591–593, 630, 631] there are descriptions of an automatic recorder for parametric processes, which is used to measure thermocouple emfs (from 20 μV to 60 mV), with 16-point switching. The measurements are displayed and written to memory. The measurements may be compared with set values to provide remote monitoring and program control. Ground test results have been given, together with measurements on the temperature pattern in the Magma furnace by use of an Imitator-3 probe made from a ceramic whose thermal parameters simulated those of the working tubes, which was used under microgravity on Salyut-7. The thermal conditions (stationary and dynamic) differed as between the ground and space, which must be borne in mind in comparing synchronous measurements. It is therefore necessary to measure the specimen temperatures during the operations in order to distinguish effects due to temperature changes from those due to microgravity influencing the properties.

In [557] the temperature pattern was measured in the Magma-F apparatus under microgravity during the Soviet–French Kalibrovka-1 experiment, where it was concluded that there were differences in the temperature stabilization between the ground and space systems because of the lower voltage stability in the space supplies. The latter was subject to oscillations with a period of about 90 min corre-

Fig. 70. General view of the Kristallizator apparatus [605].

sponding to the vehicle being in the sunlit or shadow part of the orbit. The temperature stabilization deteriorated as the voltage increased. The temperature in the container in the stabilized state was 15–20° less than the set value, with the value dependent on the set temperature and settling time.

In [601] technical characteristics are given for the Kristall, Magma, and Korund furnaces, and various experiments performed with them are enumerated. Particular attention has been given in [601] and in [570, 571] to the automated space control for the Korund, where an analysis [570] indicated that it was desirable to build a multipurpose system with extensive control by microprocessors and microcomputers to ensure that the system could function properly, including on unmanned vehicles.

New equipment has been designed for Soviet space vehicles and automatic probes such as the Orion, Dyuna, Krater, and Menisk equipment, which implement programs for experimental manufacturing in space applied to semiconductor materials. In [602] this experiment has been characterized briefly, since no information on it has been given in [1, 15]. The specimen diameters were 25 mm with the Orion equipment, 40 with Dyuna, and 50 with Krater and Menisk. The Orion and Krater equipment have resistance heaters fitted with power supplies for conducting six processes automatically. The Menisk heater uses thermal tubes, while the Dyuna heater uses infrared radiation sources. It is proposed that this equipment should be operated on the space experimental manufacturing system on the scientific center (NTs) module.

The Kristallizator apparatus [605, 606] may be described in more detail, as it enables one to monitor and program processes for 19 different experiments without direct operator participation. Each experiment has its own temperature monitor and means of tube displacement and can be executed in any sequence. The following series of experiments (out of 19) may be repeated after the chamber containing the specimens and the cassettes with recordings have been changed. Figure 70 shows the general appearance.

The apparatus has been checked out on the Salyut-7 plus Soyuz orbiting station, which has led to preparations for a series of experiments in space. The Space Research Institute, Academy of Sciences of the USSR, has set up a system for ground development of experiments on materials science in space (KNOKOM) as in Fig. 71 to prepare for such experiments with the Kristallizator. KNOKOM is used in choosing the temperature and other conditions, to develop the details of technological processes, and thus to provide conclusions on the reliability in the space experiments. KNOKOM has been based on the Soviet–Czechoslovak Kristallizator apparatus, an Iskra-226 computer, and auxiliary equipment.

We now consider papers dealing with the heaters used on SL-1 and proposed for SL-D1 and under the Eureka program, where we note that the SL-1 flight, in spite of certain failures, largely confirmed the viability for all the apparatus made for it. In [661] there is a discussion of the reasons for control failure in the isothermal heating furnace (IHF) (see [1], p. 111) during the SL-1 flight, where measures were discussed for eliminating these deficiencies on D-1. In [611] there is a discussion of the behavior of the gradient heating furnace (GHF) on SL-1 (see [1], p. 111), where data are given on how the temperature profile varied with the microgravity level, for use with GHF. It has been concluded that GHF can be used again on the D-1 flight.

In [611, 608] there is a discussion of the mirror heating furnace (MHF) (see [1], p. 112), which is a two-ellipsoid furnace for growing crystals by zone melting or with a moving heater. During the SL-1 flight, the MHF operated successfully, although unfortunately the total available growth time could not be used to best advantage. There were difficulties with peripherals (the power supplies and cooling, together with unforeseen interference leading to accidental trip operation), so some experiments were unsuccessful. It was concluded that the MHF itself is viable under microgravity. In [608] it is stated that an improved design with a single ellipsoidal mirror (see Fig. 38 in [1]) will be used on D-1. In 1987 it is planned to use devices of this type on unmanned probes.

Mention is made in [15, 542] of the advantages of mirror furnaces, including developments in them for Soviet orbiting vehicles and automatic probes. Calculations have been made on the functions of such furnaces [583]. To this one can add that the Mir rockets have been used in the USSR to test solar furnaces with parabolic mirrors (see Section 7.4). The scope for using such furnaces on orbiting vehicles has been discussed in [541].

It is particularly important to calculate and measure the temperature distributions for all these furnaces for microgravity use; in [589] the patterns were simulated, and in [587] there was a discussion of methods for measuring such temperatures; in [552, 557, 591–593, 630, 631] there are descriptions of particular devices for such measurements and results obtained in space.

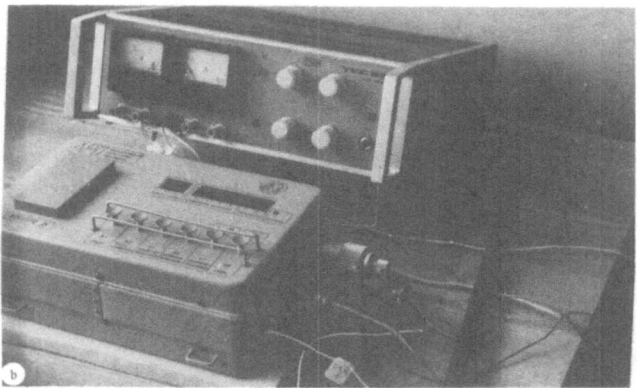

Fig. 71. Parts of the system for ground experiment development in materials science in space: a) Iskra-226 computer and control unit; b) automatic recording equipment with power supply; c) the PRARP apparatus with peripherals [593].

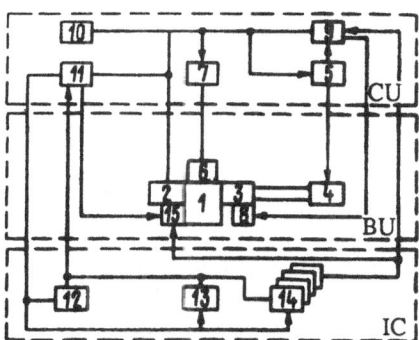

Fig. 72. Block diagram of the Pion apparatus: CU) control unit; BU) base unit; IC) interchangeable cells: 1) working chamber; 2, 3) illuminating and display parts of shadow apparatus; 4) movie camera; 5) movie camera control; 6) crystal-pulling system motor for Stepanov's method; 7) pulling system control; 8) time and temperature indicators; 9) temperature measurement and display; 10) power supply conversion and stabilization; 11) temperature regulator unit; 12) cell heaters; 13) interchangeable thermostat (up to 160°C); 14) interchangeable cells; 15) cold-junction thermostat [534].

One naturally tends to correct the temperature pattern as well as measure it. A recorder, including a microcomputer, has been built for thermostatic control under microgravity [581], which corrects for thermocouple nonlinearity and other errors, while in [559] an electrodynamic fluidized bed was used; in [582] a planar crystallization front was ensured in zone melting by adiabatic screens placed above and below the irradiated zone in a radiation furnace (mirror holders).

In [615, 616] a new multipurpose electromagnetic furnace was described for use on space vehicles such as the Shuttle. This [615] provides induction heating for materials with melting points up to 1600°C, or up to 1400°C according to [616] (evidently here with allowance for the container); the liquid is freely suspended and solidifies without container contact. A series of experiments beginning in 1985 was planned for this apparatus concerned with Ni–Sn alloy supercooling, particularly the effects of gravitation level on the crystallization.

On the whole, the data characterize advances in these furnaces for microgravity, temperature-pattern measurement, and acceleration-level recording. There has been a gradual tendency to automatic devices for unmanned craft.

6.2. APPARATUS FOR FLUID PHYSICS AND SOLUTION CRYSTALLIZATION [15, 31, 534–537, 548, 553, 556, 558, 560, 562, 564, 573, 576, 579, 594, 603, 604, 613, 619, 626, 639, 646]

In [1] the purposes and characteristics were described for the FPM fluid-physics module (p. 112, Fig. 41). In [619, 626] the use of FPM on the SL-1 flight was described, particularly some technical constraints found there, although in the main the apparatus worked normally.

In [626] an improved apparatus is described, viz., a second-generation FPM module, which was called the bubble, drop, and particle unit (BDPU). The BDPU is for research on hydrostatics, dynamics, thermodynamics, and electrodynamics, or in general two-phase mixture dynamics. The BDPU is intermediate between autonomous (dedicated) and multipurpose equipment. It consists of several modules, so standard units can be built up before the flight to give a system for any given experiment. The apparatus is assembled around a control panel, which provides electronic monitoring and recording and is fitted with power supplies, a startup service, and diagnostic facilities. The interchangeable diagnostic units employ lasers.

Fluid-physics research in the USSR has been based on the Pion equipment, which was not mentioned in [1] because the papers relating to the equipment have mainly appeared recently [15, 31, 534–537, 548, 573, 604]. Particularly detailed data have been given on the Pion and the uses in weightlessness-physics research in [15]. Therefore, only a few details are given here [534]. The Pion apparatus records processes in multiphase liquid media under microgravity as well as crystal growth from transparent liquids. The apparatus includes a basic unit, a control unit, a set of interchangeable cells containing substances, and measurement sensors (Fig. 72). The base unit is a shadow apparatus, which records optical inhomogeneities in transparent media and tracer displacements by cinematography or photography, where the working chamber accommodates the liquid cells and is fitted with illumination and display systems. The chamber is also fitted with a lateral illuminator and is set up on a baseplate, which provides for common optical-system adjustment (Fig. 73).

The control panel sets the operating modes and includes the following: supply-voltage conversion and stabilization, temperature regulators for the cells, the cold-junction thermostat, and the thermocouple sensors, as well as the movie camera control unit, etc. Basic information is provided by the movie recording, where the image also bears the time and the temperature at a selected point, which are recorded in digital form on the frame edge.

Figure 74 shows cells for Marangoni convection and bubble-drift experiments in a temperature gradient.

In [15] (see also [537]) there is a description of the upgraded Pion-M apparatus for model experiments on heat and mass transfer in liquids and in gas–liquid mixtures with various modes of heating, as well as for heat and mass transfer in single-crystal dissolution and growth in aqueous solution and research on Stepanov's crystallization method. It can also be used for making materials by radiative heating and in studying heat and mass transfer in gas-transport reactions.

In [548] the Pion apparatus was upgraded for use of model liquids used to study free-surface shaping and stability, as well as gravitational and other forms of convection. The apparatus allows one to fill the cell to a set volume, heat and cool the walls, rotate two disks independently, vibrate the cell, measure the liquid speed as well as the temperature and concentration patterns, and record the process on film. This apparatus thus provides for the experiments envisaged for the FLP and BDPU [619, 626].

In [31] it is stated that the Pion apparatus has been used in the Termogofr and Konvektsiya-U experiments on Salyut-7, which showed that Marangoni convection can be suppressed by adding surfactants. These recommendations on surfactant

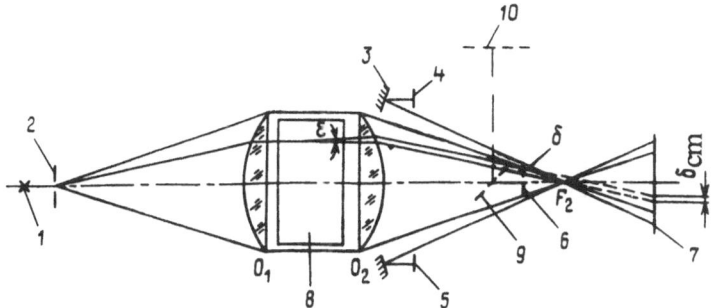

Fig. 73. Optical system: 1) light source; 2) stop; 3) system for displaying indicators and clock; 4) electronic clock; 5) temperature indicators; 6) display grid; 7) film plane; 8) object; 9) rotating mirror; 10) photographic plane; O_1, O_2) lenses [536].

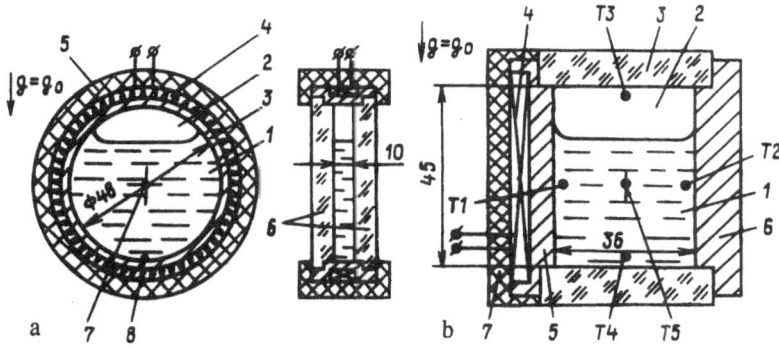

Fig. 74. Cells for convection experiment (a) and drift experiment (b): 1) working liquid; 2) bubble; 3) body; 4) heater. a. 5) Thermal insulator; 6) windows; 7, 8) thermocouples in liquid. b. 5) copper heat exchanger; 6) massive heat store; 7) heater insulator; T_1–T_5) thermocouples [534].

suppression of Marangoni convection can be used in certain processes for making semiconductor materials.

The Pion can also be used to examine crystal growth from transparent solutions, in particular aqueous ones. A precision thermostat for growing crystals from solution replaces the cell [556], and this stabilizes the temperature and enables one to follow the growth kinetics. The Textolite body is a 12 × 12 × 17 cm parallelepiped with a removable lid. The inner wall is made of thermally conducting material. The body contains the Plexiglas crystallizer. There are windows in the side walls for recording the crystal growth photographically. The heater power is 10 W. Under ground conditions, with an environmental temperature of 18–25°C, ammonium hydrogen phosphate can be crystallized on the following cycle: heating to 37 ± 0.5°C in 4–6 h followed by 280 h of crystallization at 30.011 ± 0.008°C.

The Biryuza apparatus [560] is intended for similar experiments and is analogous to the thermostat described in [562]. One of the instruments listed in [562] is also intended for examining aggregation in crystal growth from aqueous solution.

There are problems in adding and removing liquid (sampling) and eliminating air bubbles in research on fluid physics, electrophoresis, and growing crystals from solution. There are some special points here for microgravity. In [646] this topic is considered in detail, and recommendations are made on novel treatments for cell surfaces (e.g., coating with silicone oil by spraying), and also special cell designs. In [646] the design was proved effective on a Texus rocket flight. The design enables one to examine microgravity diffusion even when there are gas bubbles.

Apparatus has also been developed for simulating weightlessness in ground laboratories, where we may mention [553, 558, 564, 579] without dealing with the simulation methods used for microgravity research on liquids.

There are also recent papers [576, 594, 603, 613, 636, 637] on methods of studying liquid properties and motion. In [576] microgravity diffusion research is considered, where a multichannel automatic polarograph was used, which measured solute concentrations with high accuracy at set points. The apparatus was developed on the ground, and it was shown that it can measure concentration patterns in model liquids (solutions or melts) with an accuracy of 1%. A disadvantage is that the working zone contains indicator electrodes, which may interfere with the mass transfer.

In [594] the photochromic method is examined for studying convection in a liquid by means of color marks induced by UV irradiation. A collimated beam produces lines, points, and planes in a time of 10^{-8}–10^{-13} sec. The lifetime is determined by the concentration (10^{-4}–10^{-6}) and the radiation energy (0.05 J). This method has advantages over traditional particle or bubble tracers in that the measurements are made without contact and the reversible marks can be made repeatedly without lag.

In [603] there are descriptions of optical and probe monitoring methods for tracing flow structures arising from thermal or concentration convection, where exact characteristics can be recorded at set points. The pattern can be displayed and photographed. Pulse probing can be based on bridge circuits with radiofrequency generators, a frequency of 0.1–10 MHz, and a space-mark ratio up to 10^4; the method has been tested on the ground but is suitable for microgravity.

In [613] there is a method of measuring liquid thermal conductivity under various gravitational conditions by variable heating in a thin wire along the axis of a cylindrical container. The wire is used simultaneously as a heater and resistance thermometer. The heat transfer under gravity is different from that under microgravity; calculations have been made for various gravity levels and particular liquids.

Computer-controlled ground topographic equipment has been described [636, 637] for measuring infrared radiation from surfaces and internal layers in liquids in order to process and record data from experiments, which may be converted to temperature distributions. Temperature-distribution measurements are reported for a flat cell containing silicone oil. The apparatus could be used with minor modifications for space experiments.

In [15, 534] it is concluded that field experiments with model liquids need to be considered on the same basis as direct technological experiments and numerical calculations based on the mechanics of continuous media as a way of deriving recommendations on optimizing processes for making new materials under microgravity. This implies a need to improve equipment and methods in liquid physics, crystal growth from solution, making protein crystals, and the design of thermostats and cryostats, as mentioned in Section 5.5.

6.3. CUTTING, WELDING, AND COATING APPARATUS [15, 578, 588]

In [1] it is pointed out that great importance attaches to equipment developed at the Institute of Electrical Welding, Academy of Sciences of the Ukrainian SSR, for cutting and welding in space, which has been tested by Soviet spacemen; an appropriate bibliography is given. In [15, 588] additional information is given on the latest advances in this area (see also Section 3.9).

In [588] there is a description of a high-power miniature electron-beam apparatus with independent dc supply that had been developed, tested on the ground, and then tested in near-earth space. Such equipment has been used to eject electrons into space during the Soviet–French Arax experiment. Similar apparatus can be used for electron-beam welding and other operations in space. In [588] the design principle is described, and the characteristics of the main units are given that convert the dc from the primary power supply to ac, whose voltage is raised to the required level and then rectified to inject the electron beam. The units controlling the beam and converting the telemetry parameters to analog form are also described. Figure 75 shows the block diagram.

A novel method has been used to examine vacuum soldering under microgravity [506], which has already been mentioned in Section 3.9.

The Isparitel'-80 and Isparitel'-M are new equipment for depositing protective and hardening coatings in space and for major fundamental research on plasma coating and film crystallization from the vapor. These are not mentioned in [1], since the papers on the characteristics appeared only recently [15, 578]. Section 3.9 (see also [47]) deals with experiments performed recently with the Isparitel' equipment.

In [15] there are some test results obtained with this equipment in space, and here we need to add to what has been said in [578] only certain details not concerning the space test results.

The first form, the Isparitel'-80, provided the basis for the improved Isparitel'-M, since tests on the former for depositing films on Salyut-6 confirmed the design features. Aspects of the evaporation, condensation, and film formation under microgravity had been determined: reduced number of micropores, similarity in the condensate composition to the initial composition, etc.

The Isparitel'-M was designed for more detailed physicochemical research. The facilities were extended by including interchangeable units and scope for flexible programming and ongoing parameter monitoring, including the installation of several electron guns ranging from microfocused to completely defocused. The Isparitel'-M was installed on Salyut-7 and prepared for systematic research.

The conclusion from this section on cutting and welding, as well as on film deposition, is that such equipment is of specific practical significance and also that this is a good example of the ongoing improvement in equipment. A similar approach should be maintained in developing new generations of other equipment types.

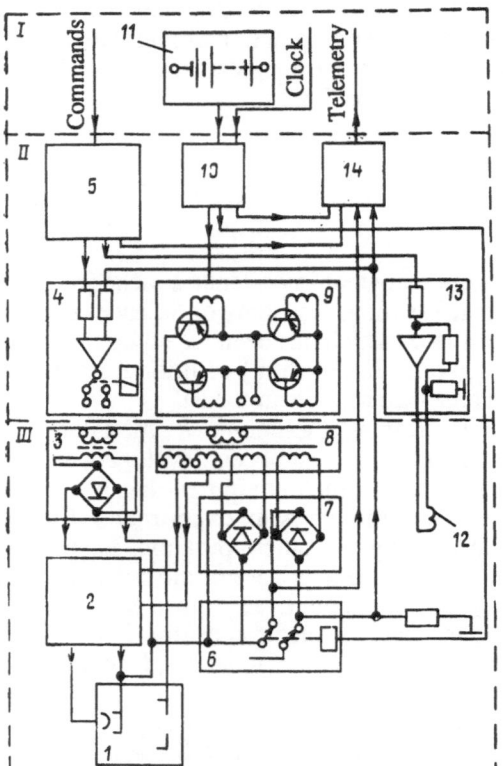

Fig. 75. Electron-beam apparatus block diagram:
1) electron gun; 2) cathode heater; 3) high-potential part of beam current regulator; 4) beam current regulator; 5) control unit; 6) high-voltage switch; 7) high-voltage rectifier; 8) high-voltage transformer; 9) inverter; 10) programmer; 11) storage battery; 12) deflecting coil; 13) deflecting-coil current regulator; 14) telemetry unit [474].

6.4. LEVITATION UNDER MICROGRAVITY
[15, 595, 609, 644, 645, 659]

Levitation is one of the methods requiring new developments for microgravity, as it provides for containerless treatment in space, which is necessary for future space manufacturing of new or improved materials.

In [1] papers were listed dealing with this, but the contents could be judged only from their titles. Only in the case of electrostatic levitation did [1] give an illustration (p. 125, Fig. 45), which was the limit of information on the detailed method. Here we deal briefly with several new papers on levitation without referring to earlier papers cited in [1], although some of them are important as regards priority.

In [15] there is a theoretical analysis of various levitation methods: electromagnetic, ion-plasma, and laser beam. The electromagnetic method has been characterized [15] as ineffective, while the electrostatic method involves electrical-strength problems. Nevertheless, there are many papers, such as [595, 609, 645, 659], on electrostatic levitation.

In [595] it has been shown that an insulating sphere immersed in a liquid under microgravity can be stabilized by an electrostatic field. Experiments and analytical studies on the stable equilibrium have been performed for a sphere coated with a spherical shell in axisymmetric electrostatic and uniform gravitational fields. This showed that a conducting sphere coated with an insulating shell can be levitated if $(R_i/R_0)^3 < (\varepsilon_m - \varepsilon_p)/(\varepsilon_m - 2\varepsilon_p)$, where R_i is the sphere radius, R_0 the shell radius, and ε_m and ε_p the dielectric constants of the medium and the shell. The experiments concerned the equilibrium condition for a Wood's-alloy sphere coated with polyethylene near a ring halfway between two grounded horizontal plates immersed in castor oil. A voltage of 2–15 kV was applied to the ring. Theory and experiment were compared.

In [609] specimens were weighed in an electrostatic field in a device having the following characteristics: a) the position was determined with a capacitance bridge, b) the suspended specimen was charged, and c) a configuration with feedback only on one axis was tested, with stability on the other two provided by connecting the corresponding electrodes to a constant potential. An analytic expression was obtained, where the electrostatic force was expressed as a function of the charge on the specimen, the voltages on the electrodes, the capacitance, and the dependence of it on the position. Theory and experiment agreed well.

In [645] a levitation method was proposed for solids and liquids in an electrostatic field, and it was emphasized that this differs from acoustic and electromagnetic methods in being universal: it can work in vacuum and the material does not have to be a conductor. Characteristics were described for conical, annular, and tetrahedron-shaped electrodes under various gravity conditions. A detailed description is given of the software for monitoring the field parameters. Experiments were performed on the ground with metallized spheres weighing 0.15 g. Under brief microgravity, $3 \cdot 10^{-2} \, g_0$, it was possible to support a 5-g specimen. It was assumed that a specimen of about 500 g could be kept in the suspended state during space flight.

The scope for electrostatic levitation under microgravity has also been pointed out in [659].

The views expressed in [595, 609, 645, 659] on electrostatic levitation are thus positive.

I now discuss other levitation methods mentioned in [15] (Section 5.6. pp. 218–223), but with condensed explanations to the figures given below. Figure 76 shows an electromagnetic levitator, which monitors the position of a conducting specimen on account of the additional currents arising from the small accelerations, which vary in magnitude and direction. The specimen is placed in a static magnetic field produced, for example, by solenoids carrying direct current. In [15] estimates were made of the forces in a particular case arising from interaction of the Foucault currents with the magnetic field. The method is suitable only when there are variable accelerations.

Figure 77 shows an ion-plasma levitator; the plasma flows are used to provide heating on a given program and for positioning in the heating zone. The magnetized-plasma flows in Fig. 77 are derived from two sources (usually out of three at

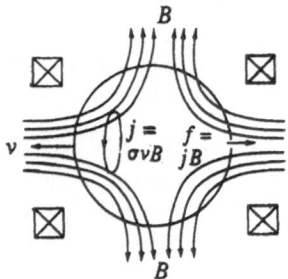

Fig. 76. Electromagnetic levitation provided by additional currents [15].

Fig. 77. Ion-plasma levitator: 1) plasma accelerators; 2) solenoids; 3) plasma flux [15].

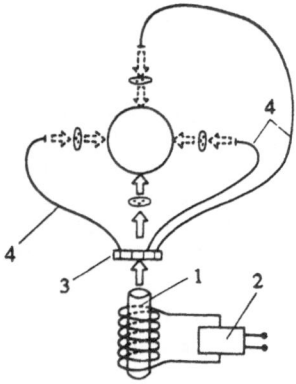

Fig. 78. Laser ablation levitator: 1) laser; 2) power supply; 3) distributor; 4) light guides [15].

angles of 120°) and propagate along appropriate field lines produced by solenoids. The flows provide heating, and the positioning is produced by controlled inequality between the flows. In [15] there is a discussion of levitation for a 170-g specimen, diameter 5 cm, with only 0.03% mismatch between the flows.

Figure 78 shows levitation produced by laser ablation; the material is heated by an independent source (not shown). The beams are directed by light guides. The evaporating material (the evaporation can also be produced by an electron beam) exerts a force in the reverse direction. The mass losses are only 0.1–0.5% of the specimen per hour, which is within the range caused by ordinary evaporation.

Other modes of levitation have also been discussed. In [659] there is a review of papers not only on electrostatic levitation, as indicated above, but also acoustic.

In [644] a simple method is described for levitating a set of drops or particles at once, which is based on gas flows. Ground experiments showed that the drops could be placed exactly by means of contactless manipulators. The method was described for manipulating drops under microgravity, together with the manipulator design.

This analysis shows that improvements in methods are important, as is the design of appropriate apparatus.

6.5. SPACE HOLOGRAPHIC EQUIPMENT (SHE)
   [540, 543–547, 555, 566, 567, 572, 597–599, 627, 641]

Holographic methods are important in space research generally and in developing materials science in space in particular, as is evident from the above citations, which deal with existing equipment and prospects for using it in space. Holography has advantages over ordinary photography. The image is more informative, since it is three-dimensional, and holography also provides for *a posteriori* analysis without loss of amplitude and phase information, which is particularly important for fast processes. An important advantage is that data can be transmitted from a transparent object via the phase structure, which provides information on heat and mass transfer, which is important in materials science in space.

Holography also provides higher resolution, because there are no distorting lenses, and also high sensitivity to change in state or position, as well as the scope for recording several objects or one object at different times on a single hologram.

These advantages and some others make holography convenient or sometimes irreplaceable under space conditions, but this has required completely new equipment: SHE, which has to meet specifications for space-station research equipment. The SHE should be strong, as light as possible, with good vibration stability, viable under microgravity, simple and reliable to use, consume as little energy as possible, and suitable for numerous purposes.

Equipment meeting these requirements has been built and tested on the Salyut-6 and Salyut-7 space stations [543–547, 566, 567, 572, 597–599].

The first form of small holographic equipment, the KGA-1, had dimensions 458 × 214 × 120 mm, mass about 5 kg, power drawn 60 W, supply voltage 27 V, exposure times from a fraction of a second to tens of seconds, recording on holographic films or plates, light source an LG-78 laser, and field of view 60 mm.

The KGA-1 employs the usual double-beam system (Fig. 79). The light from the source (1) is directed by mirrors (3) through lens (2) to the beam-splitting system, which is formed by semitransparent mirror (4) and mirror (5). The beam passing through the mirror and the object (6) and the reference beam are brought together in the film recording planes (7).

The second form, the KGA-2, was built to study electrophoresis (Fig. 80). The laser beam (1) is directed by mirror (2) to beam splitter (4). Between mirror (2) and splitter (4) there is the mobile shutter (3) to block off the beam, which is controlled by a cable. The lens (6) produces the object beam. The plane-parallel plate (7) is rotated to adjust the period of the finite-width fringes. The matt glass (9) between the electrophoretic column (8) (the object) and the hologram (11) serves to localize the interference pattern. The mirror (5) and lens (10) produce the reference beam. The floating aperture (12) enables one to record a series of double-

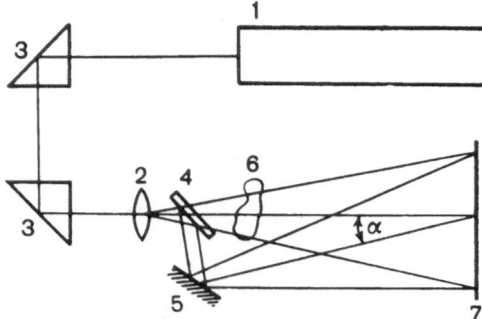

Fig. 79. Optical scheme for the KGA-1: 1) light source; 2) lens; 3) mirrors; 4, 5) beam-splitting system; 6) object; 7) film [572].

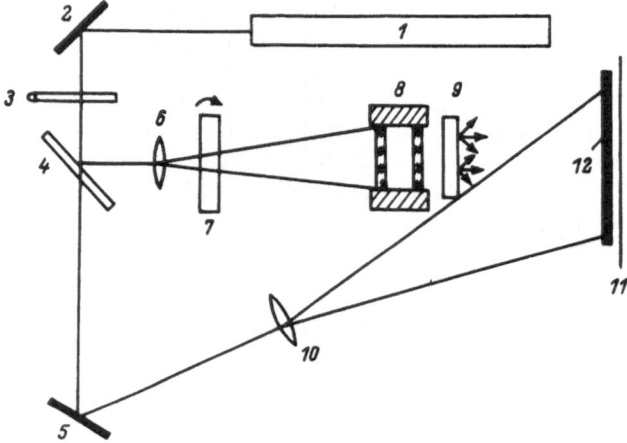

Fig. 80. Optical scheme for the KGA-2 [543].

exposure holograms on one frame. Each hologram on reconstruction gives an interference pattern, enabling one to measure the refractive-index differences. The series recorded at different times gives the displacement and fraction separations in a column (or other object). The fraction positions are determined from the fringe curvature. The KGA-2 is a holographic differential interferometer whose holographic volume is larger than that in the KGA-1 and is provided with a new device for recording double-exposure holograms.

Originally, the holograms were recorded (on the Salyut-6) photographically and were processed after the end of the flight, so it was impossible to monitor the course of the experiment, while in the Salyut-7 experiments TV transmission was used, i.e., to study physical processes accompanied by refractive-index changes. This extended the scope of this KGA equipment considerably by providing real-time operation, particularly in conjunction with TV display. The KGA-1 and KGA-2 have been used in various space experiments, which are described briefly below; these show that space technology has thus acquired a new, effective method.

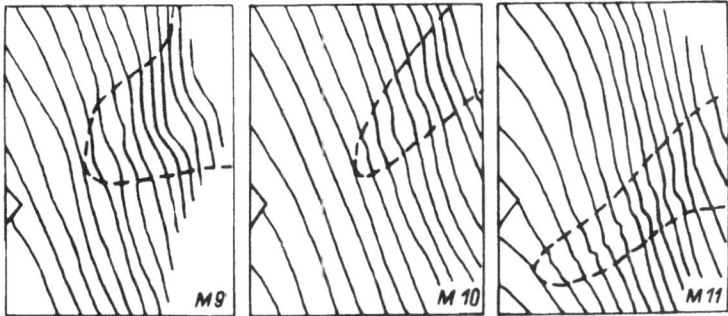

Fig. 81. Interferograms for 46 min, 52 min, and 1 h [543].

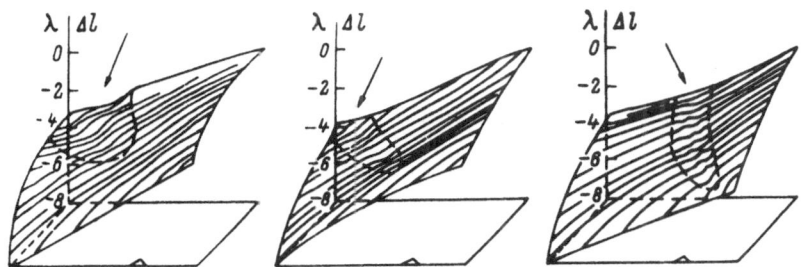

Fig. 82. Optical path-length variation front in isotachophoresis chamber at 46 min, 52 min, and 1 h [543].

The first holographic experiments were performed on the Salyut-6 on March 27, 1981, where the KGA-1 recorded the dissolution of an NaCl crystal during the flight of a Soviet–Mongolian team. The dissolution time under microgravity was 20 times that on the ground. Later, the Salyut-7 team performed tests on the KGA-2 in the Tavriya experiment on electrophoretic isolation of high-purity biologically active substances under microgravity [543], where estimates were made of the refractive index relative to the buffer solution. Also, it was possible to determine the refractive-index gradient from the change in fringe slope. A hologram for isotachophoresis under microgravity provided interference patterns for different instants (Fig. 81), where the dashed line shows the region of local changes associated with the biopreparation fractions. Figure 82 shows the changes in optical path length in this part of the chamber. The arrows indicate the regions of fraction localization. The fractions migrated at about 0.6 mm/min.

Holography can be applied to electrophoresis without introducing contrast materials that contaminate the preparation [543].

Equipment of this type can be used to examine heat transport [555]. Nonstationary convection was examined in a closed region arising from a line heat source in several liquids having Prandtl numbers from 4.6 to 34: water, ethanol, isopropanol, CCl_4, scipidar, and glycerol. This method with the KGA provided the temperature pattern at any time. A free-convection jet arose in the cavity. Figure 83 shows the temperature patterns at different times for a cavity filled with ethanol.

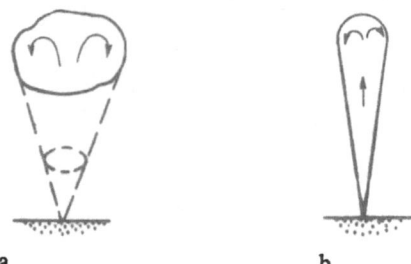

a   b

Fig. 83. Schematic representation of convective elements: a) thermal; b) floating jet [555].

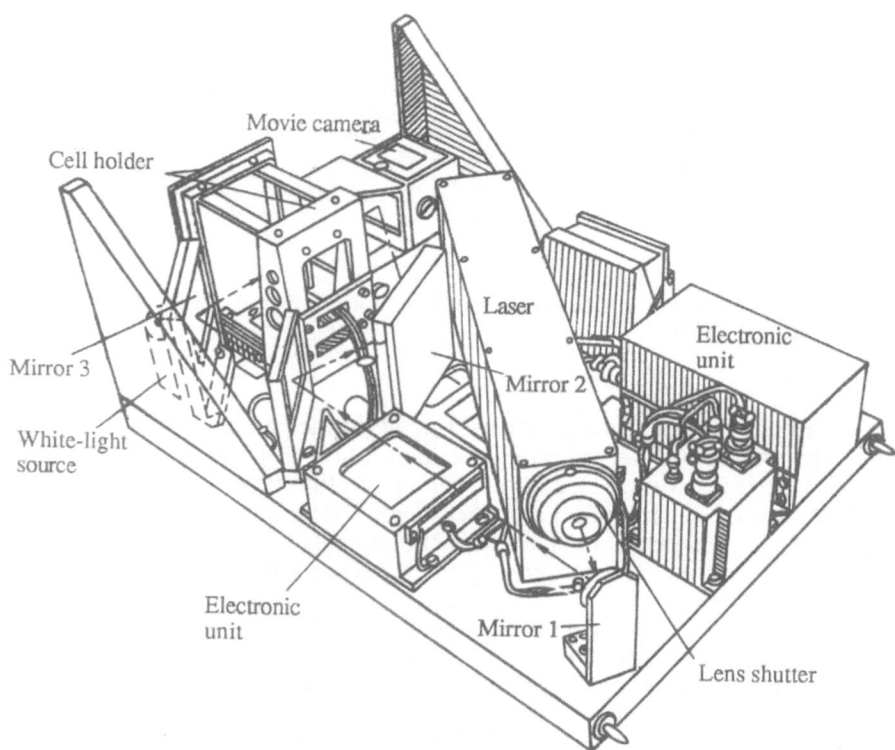

Fig. 84. The HOLOP holographic interferometer [627].

Such data give [555] the temperature profiles for various source powers and liquids, and also the temperature as a function of distance above the source along the symmetry axis. Without considering these data in more detail, we note that the results of [555] indicate that such equipment is effective with the methods developed for examining heat transfer in transparent media.

In the above, we have discussed experiments of the KGA type developed in the USSR and tested on Salyut-6 and Salyut-7. Holography in space experiments has also been examined abroad. In [627] it is stated that the HOLOP special holographic system has been built for a space laboratory on the D-1 mission. Figure 84 shows the HOLOP schematically. The HOLOP allows rapid adjustment to different experiments. The HOLOP is, in essence, a holographic interferometer for use with transparent media to record density changes. In [627] ground tests were reported for an experiment on Spacelab D-1, where the HOLOP is to be used to examine the density distribution in SF_6 (Frion) near the critical point. The specimen in a glass capillary, inside diameter 2.5 mm and length 70 mm, is cooled from 0.8°C above the critical point at $7 \cdot 10^{-3}$ °C/min. The ground experiment showed that the fringe separation was 10 times less than the calculated value (0.3 mm instead of 2.6 mm), which was due to gravitational convection, and this demonstrates that the HOLOP apparatus is of value for the same experiment under microgravity.

In [641] optical holography was applied to triglycine sulfate growth under microgravity on SL-3. The beam from a 107-Å He–Ne laser was used with two holographic systems, one for examining the specimen and the other for examining the light scattered by the crystal. The first object beam was split into two at the exit: one for recording on a film and the other for observation in a schlieren system in real time. The hologram was recorded on SO-253 film, width 70 mm. The resolution was about 20 μm. The holograms were recorded in two orthogonal spatial positions. Double exposure was used as mentioned above, which provides for observing the differences in the patterns in the crystal in accordance with temperature and concentration. The holograms were developed after the flight. The system had first been tested on a flight by the NASA KS-135 flying laboratory.

To complete this discussion, we may note the opinion expressed in [597–599] on the good prospects for holography in space research, particularly materials science. Holography can contribute to research on growth and dissolution dynamics under microgravity, as well as to studies on physicochemical and biophysical processes.

Holography can provide much in research on liquid behavior under microgravity, particularly as regards heat and mass transfer and gravitational or nongravitational convection.

In [31] it is stated that KGA-type apparatus enables one to examine microstructures with magnifications up to 100–300, so one can examine boundary layers in liquids at crystallization fronts, in particular in the proposal to perform synchronous experiments with the Pion and KGA-2 equipment.

Holography can also be applied to the topics mentioned in [1] and considered here, such as mass crystallization in transparent solutions and attempts to detect cluster formation in protein crystallization under microgravity.

6.6. MICROACCELEROMETERS AND RESEARCH ON GRAVITATION IN SPACE VEHICLES [15, 554, 569, 574, 577, 579, 584, 585, 610, 612, 632, 643]

The gravitational level and the time variations (g perturbations) must be known in analyzing microgravity experiments on aircraft, rockets, or spacecraft, so experiments on materials science in space must be accompanied by recordings of the microgravity levels and the time-varying microaccelerations.

In [1] results have been given from numerous microacceleration studies on rockets and spacecraft. However, research on this is continuing, and microacceleration meters are being improved [15, 569, 574], while measurements have been made on microgravity levels on aircraft, rockets, and spacecraft, in addition to theoretical calculations on accelerations in manned and unmanned spacecraft [290, 584, 585, 610, 643].

We first give some features of microaccelerometer design and sensitivity and then research results obtained with them on gravity levels in rockets and spacecraft. Section 1.10 has already dealt with theoretical calculations on gravity levels, so here we give only measurements.

In [569] (see also [15]) there is a brief description of accelerometers used in various experiments and then a more detailed description of a microaccelerometer used on the Salyut vehicles. This is a three-component precision pendulum accelerometer of electromechanical type with heavy feedback, which has an extended dynamic range. It includes three single-component sensors combined in a single unit. Figure 85 shows the sensor design. The inertial mass 1 is mounted on arm 2, which is rigidly coupled to mirror 3 and feedback coil 4 in the magnetic system 5. The moving part is connected to the body by torsion suspensions, which are also the current leads. The light source 6 works with a condenser 7 and stop 8, where the beam is reflected from mirror 9 and passes through lens 10 to mirror 3, which is on the moving element. The beam passes again through lens 10 and is reflected from the other face of mirror 9 and then passes through mask 11 to differential photocell 12. The unbalanced signal appearing when the mirror rotates is due to the acceleration. This signal is passed to the power amplifier 13, whose output is connected in opposition to the reverse electromagnetic converter 4. The feedback current produces a force that compensates for the acceleration. When the forces are in balance, the feedback current is proportional to the acceleration on the sensitivity axis, and the direction corresponds to the acceleration sign. Without going into details, we give the technical characteristics: measurement range $(10^{-6}-10^{-2})g_0$, threshold $10^{-8} g_0$, slope $2 \cdot 10^3 B/g_0$, time constant about 0.2 sec, power drawn 10 W, dimensions $150 \times 180 \times 140$ mm^3, and mass 5 kg.

This instrument has been used in a wide range of gravity studies on the Salyut vehicles. In [569] measurements were given on the accelerations a_x, a_y, a_z on three axes: 1) with the station unmanned (mothballed), $a_x \sim a_y \sim a_z \sim 10^{-5} g_0$; 2) with the usual team, $a_x = (2 \cdot 10^{-4}-1 \cdot 10^{-5})g_0$; $a_y = (10^{-3}-10^{-5})g_0$; $a_z = (10^{-3}-10^{-5})g_0$; and 3) during physical exercises under the training program, $a_x = 10^{-4} g_0$; $a_y = 8 \cdot 10^{-3} g_0$; $a_z = 8 \cdot 10^{-3} g_0$. In engineering experiments lasting more than 1 h, there were background accelerations of vibrational type having amplitudes of about $(0.8 \pm 1) \cdot 10^{-4}$ g_0 even under the quietest conditions.

This must be borne in mind in experiments on weightlessness and in prolonged material production. It is best, of course, to measure the acceleration level throughout the experiments, so much equipment for space use is fitted with instruments of this type.

An example is provided by [575], which deals with the accelerations on Salyut-7 during experiments with the Pion-M equipment. The sensors were rigidly attached to a special area in the Pion-M, and the readings were recorded during he Konvektsiya and Termogofr experiments. It proved possible to record bubble or

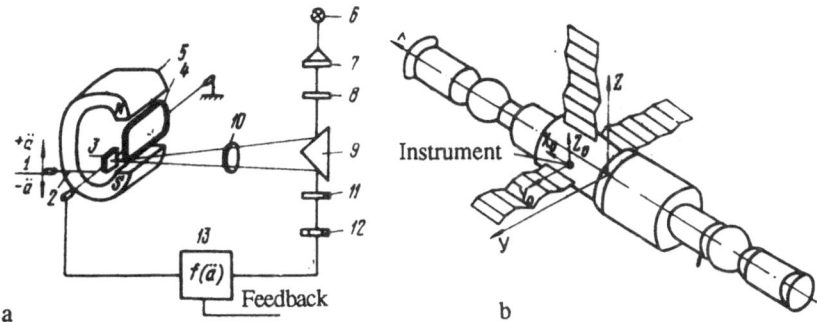

Fig. 85. The accelerometer (a) and instrument position (b) [569]; explanation in text.

label displacements in synchronism with the accelerations. There were damped vibrations when the movie camera operated together with certain low-frequency phenomena that affected the processes.

To supplement [1, 569] we give some brief details [632] on the acceleration levels (g/g_0) and microacceleration durations τ attained in various ways. Free-fall tubes have been used in the USA and France, diameter $d = 5$–30 cm, height $h \leq$ 100 m, which have given $g/g_0 = (10^{-6}$–$10^{-7})$ for several seconds, while towers ($d = 80$–150 cm, $h \leq 150$ m) have given $g/g_0 \sim 10^{-5}$, $\tau \sim 3$–6 sec. In aircraft in flight on ballistic paths (USA F 104 and KS 135, USSR IL 76: $g/g_0 \sim 10^{-2}$, $\tau \sim$ 15–30 sec) and on rockets $\sim g/g_0 \sim 10^{-4}$, $\tau \sim 5$–7 min (USA SPAR, Europe Texus, Japan TT-500, USSR Mir). Manned vehicles (USA Apollo, Skylab, and Spacelab, or USSR Salyut) have given $g/g_0 = 10^{-2}$,-10^{-4}, with τ from 8 days to 6 months. Unmanned satellites and special probes built in the USA and Europe for use at the end of the 1980s are proposed with levels of $g/g_0 = 10^{-5}$ for up to 6 months, with prospects for reuse.

In [554, 612] the discussions concern not only gravity levels on space stations but also certain other conditions that can influence the experiments, such as magnetic perturbations and atmospheric pollution in the vehicle.

6.7. CONCLUSIONS

The methods and equipment for weightlessness physics and materials science in space listed above are far from exhausting the topics covered in [534–661].

Here we merely enumerate some topics not mentioned in Sections 6.1–6.6.

In [538, 565, 607] there are discussions on automated microgravity experiments on unmanned probes and automatic equipment for space manufacturing. In [590] there is a discussion of choosing space-experiment apparatus.

In [568, 654] systems are described for producing vacuum under microgravity, and the general problem of vacuum on spacecraft is discussed, while in [561, 563, 635, 636, 640, 647] there are discussions on new methods for use under microgravity in studying physicochemical processes. Specific-heat discontinuities at the critical point, particularly in SF_6, may be examined [636] with the scanning differential calorimeter. In [604] droplet displacement in molten metals by means of x-ray topography was discussed. In [647] a new optical technique has been devised for simultaneous and independent determination of temperature and concentration gradients by measuring beam deflections at points of surface-tension gradient by means of the mirage effect (anomalous dispersion).

In [561, 563] it has been proposed to monitor melting and crystallization and to determine metal melting points and latent heats of fusion, specific heats, Peltier and Thomson coefficients, and other thermophysical parameters by recording the thermo-emf during melting and crystallization in a gradient furnace . The material is contained in the tube, with the ends fitted with platinum electrodes.

In [635] microgravity methods are proposed for examining phase transitions in liquid helium, which should provide high resolution near the λ transition, while in [617] there is a discussion of TV techniques in space experiments.

In [579, 580] there is a discussion of methods and apparatus for microgravity studies on physical processes in crystallization in transparent models (molten $NaNO_3$ and KNO_3), which simulate crystallizing metals. The method has been checked out on the ground and in an aircraft laboratory.

In [655, 658] methods of normal crystallization are presented, where the residual liquid is repelled and the parameters are measured: temperature, solidification front, and front speed, which involved periodic probe insertion.

In [652] experiments ES-303 and ES-325 are discussed, which were designed for SL-1 (see Table 12 in [1], pp. 153 and 156) for developing microgravity skin-technology methods. Texus-9 experiments showed that capillary forces can be used with a skin shell wetted by a melt and volume-change compensation to prevent the shell from being disrupted by the thermal expansion during heating in order to attain absolute shape stability. Unfortunately, ES-303, which would have confirmed the Texus results with the Ni/Ni_3Al–Mo system in an Al_2O_3 shell on SL-1, did not succeed because the IHF power supply system failed. Experiment ES-325 on melting iron in a skin shell was successful and confirmed that the skin technique is effective under microgravity.

In [621] a synchronous amplifier based on integrated circuits is described for detecting a Wheatstone-bridge unbalanced signal, where the bridge is part of a differential temperature regulator for an SL calorimeter. The purpose was to measure the specific heat at constant volume near the gas–liquid critical point under weightlessness. The sensors in the Wheatstone bridge were thermistors. The working frequency was 72.5 Hz. The device can detect temperature changes of ±5 mK. Tests showed a stability of 15 mK over 48 h or ±40 mK over several months.

This discussion of space equipment may be concluded by saying that research supports numerous experiments and preparations for industrial production of various materials in space.

In the conclusions in Sections 6.1–6.7, it is stated what the contributions are from the apparatus or methods to research on weightlessness physics, and what are the lines of advance. Here we would emphasize that these instruments and methods for space can also be very useful on the ground. In [597] this is claimed, for example, as regards space holographic equipment: "... previous instruments were suitable only for specially equipped laboratories. Space holographic equipment may find a place even in factories and schools and may give as striking results as do massive laboratory systems. When one compares the new holographic equipment with ordinary cameras as regards weight, we find that we are on the threshold of routine small-format holography. The effort devoted to designing instruments and methods for space will pay off not only in research on orbit but also in various ground tasks." This remark, particularly as applies to instruments of the KGA type, in fact indicates the assistance that is provided by developments in instruments and methods for space research as regards the specifications for the basic conditions in orbital stations and automatic probes, since these developments provide new methods and are, in general, useful for ground purposes.

Chapter 7

EXPERIMENTS ON SHORT-TERM WEIGHTLESSNESS: TOWERS, AIRCRAFT, AND ROCKETS

In [1] (Chapter 7, pp. 125–131) it was stated in discussing short-term weightlessness experiments that these serve not only to prepare for experiments with prolonged microgravity but also are sometimes of independent significance. It was emphasized [1] that numerous results have been obtained on weightlessness physics from short-term microgravity tests, and that this will remain so even when the opportunities for prolonged microgravity experiments are much extended. This is evident also from recent papers, which show that much evidence on materials science in space has been obtained recently from short-term weightlessness tests.

Part X in the bibliography is far from including all the papers on materials science in space involving short-term weightlessness, and as in other parts, many of the papers relate to rocket experiments. These cannot all be enumerated, and we merely characterize [662–688] briefly. First, in Section 7.1, we deal with papers where towers and aircraft have been used, and then with more numerous rocket studies, with separate emphasis (Section 7.2) on research in the USSR, since [1] did not reflect Soviet rocket experiments. Recently, [15] dealt with Soviet advances in materials science in space, where rocket studies were given appropriate attention, and so the survey is partly utilized here.

Section 7.3 deals selectively with certain experiments on foreign rockets, and Section 7.4 deals with apparatus used on rockets, since no particular emphasis has been given to this in Chapter 6.

7.1. EXPERIMENTS ON AIRCRAFT, TOWERS, AND BALLOONS [666, 669, 671, 675, 677, 684, 685]

In [675] it is stated that aircraft have been used in parabolic paths to simulate weightlessness briefly (for about 25 sec) in training spacemen and in various brief experiments, but the main significance of such flights is as a means of checking experiments intended for use on spacecraft. In [675] there is a discussion of experi-

TABLE 5

Material and specimen type	Material	No. of expts.	Main results
Semiconductors	Germanium, silicon	60	Specimen out of contact with tube walls, and production of few-dislocation crystals with uniform impurity distributions, growth rates anomalously high, up to 1 cm/min
Metals and alloys (magnetic and superducting)	Alloys based on iron, titanium, nickel, samarium-cobalt, niobium–silicon, niobium–gallium, and vanadium–gallium	25	Structures differing little from ground control specimens, but con-larger new-phase structures
Composites	Aluminum–graphite, aluminum–nickel–chromium carbide, copper–carbon, copper–chromium–carbon, iron–alundum	12	Homogeneity improved, immiscible-composition specimens markedly inhomogeneous, while wetting-improvement additives improve the homogeneity
Alloys with immiscibility ranges	Copper–chromium, aluminum–lead, aluminum–beryllium	8	Unmixing of immiscible components
Eutectic alloys	Aluminum–nickel, cadmium–zinc, copper–silver, lead–antimony	6	Specimens have a regular structure and finely divided structures between dendrites
Optical glass	Borophosphate glass	3	Specimens have zones of elevated transparency
Making shaped specimens by capillary shaping	Copper, silicon	4	Stepanov's method gives specimens of given shape, while a shape-control method has been tested involving varying the capillary pressure within the shaper
Contactless crystallization	Copper, silver, germanium	16	Spherical copper specimens have been made having regular dendritic structures; spherical silver specimens have large-block or single-crystal structures. The crystallization involved appreciable supercooling. Germanium crystallized from the center, with the heat lost through the liquid

ments for the liquid-physics module for the D-1 orbital laboratory under the ESA program, where results obtained on SL-1 are utilized. In December 1984 the ESA used two flights of an American KS-135 aircraft for the development, which was specially equipped for NASA programs. In [675] a list of the experiments is given, and the view is expressed that the testing purpose of the flights had been attained.

In the USSR, parabolic flights have been used [1] to test apparatus, particularly a flying laboratory (on an aircraft) for testing equipment for welding in space and for various other purposes.

In [666] results were presented on liquids moving under capillary forces during short-term weightlessness on an IL-76K laboratory aircraft. The acceleration level for 5–7 sec was $4 \cdot 10^{-4}$–$1 \cdot 10^{-3}$ m/sec^2; the perturbation frequency was 20–24 Hz. Cinematography was used to examine liquid filling in wedge devices. The flow mechanism confirmed the mathematical model. That study represented an extension of research on weightlessness hydromechanics.

The NASA program [684] involved experiments on iron solidifying in a furnace during short-term weightlessness on F-104 aircraft; the gravitational level was 0.03 g for 30 sec. A method has been proposed for preliminary development of weightlessness experiments on materials science.

Aircraft have been used in Sweden in preparatory experiments on materials science in space [677].

Towers have also been used in research on weightlessness physics. In [669] the scope for towers in developing space technology is discussed. Examples are considered of systems supporting the flight in space, and various facilities are compared. The paper gives experience accumulated over many years on a ground test system providing for simulating weightlessness and variable force fields for short periods. The recording and measuring equipment is described, with upgrading prospects and computerized processing. There is a survey of tower experiments in space technology and space-structure dynamics.

In [671] drop tubes have been considered as a supplement or alternative to space experiments. It is stated that such a tube 100 m high has been used at the Marshall Space Center in the USA and has shown that it is valuable for examining containerless solidification. The experiments are cheap, and one can process many specimens rapidly (usually 10 a day). Another advantage is that the supercooling can be tracked quite accurately, and one can readily record recalescence and multiple nucleation (see Sections 4–4.1). A disadvantage is that the specimen size is limited, as is the type of alloy that can be treated under microgravity. One can use specimens weighing from 50 to 500 mg [refractory alloys can be used in a vacuum of about 10^{-5} torr, while fusible alloys require an atmosphere (about 200 torr) to provide the necessary supercooling before collision with the bottom]. With long tubes, it has proved possible to develop superalloys, various steels, germanium–gold alloys, and niobium-based alloys.

Structure studies have given new information on the phase compositions, which indicates unique scope for such tube experiments on containerless solidification with considerable supercooling without quenching. Particularly successful experiments have been performed with refractory niobium-based alloys.

Here we may note [684] where rocket-type containers have been launched from a balloon at a height of 30 km. The time of free fall before the parachute opened was 10 sec. Test experiments were performed in 1980 and 1981. In September 1983, the first experiment was performed on the behavior of fish under microgravity by means of cinematography and telemetry.

7.2. MIR ROCKET EXPERIMENTS IN THE USSR [15, 663–668]

Between April 1975 and December 1982, the USSR launched Mir-2 rockets (VZAF-S high-altitude probes) seven times, where over 130 experiments were performed [15, 663–668]. An accelerometer and telemetry were used to measure the accelerations [668], which were found to be $(5-6)\cdot 10^{-3}$ g_0 during the standard operations (separating the probe from the launcher, opening the container, rapid stabilization-system operation), etc. The acceleration level on the rocket during the experiments was 1–2 orders of magnitude lower than on manned orbiting vehicles.

In [663–667] it is pointed out that microgravity lasting 10 min is not sufficient for some slow heat- and mass-transfer processes to attain steady states, which distinguished short-term weightlessness experiments from prolonged ones. The specimens on the Mir rockets were heated by exothermic reactions with burning temperatures up to 2400°C, which meant that materials with a melting point up to 1850°C could be melted in 10-mm molybdenum crucibles (foreign workers have used electrically heated furnaces, which have not provided similar results). Temperatures in the range 20–1800°C were recorded at 38 points with accuracy ±2%. The thermal-cycle time, including heating at 50–150 deg/min, did not exceed 560 sec. Table 5 lists the experiments performed on the Mir rockets. We cannot go into the details of the results from these over 130 experiments and merely indicate some general conclusions.

Rocket experiments confirm the prospects for space technology research on inorganic materials. Sometimes, space specimens have better characteristics. There are some unexpected effects: liquid detachment from tube walls, effectively containerless crystallization under nonequilibrium conditions, anomalously high crystallization rates for low-dislocation Ge and Si crystals (up to 10 mm/min), considerable supercooling before crystallization from the center, with heat transfer through the liquid, and impurity distributions along castings corresponding to diffusion-limited mechanisms (without convection).

Rockets have also tested design principles for microgravity instruments: exothermic-reaction furnaces, solar-heating furnaces with parabolic concentrators, which have been used to test containerless melting under space conditions, an apparatus containing shapers for growing crystals by Stepanov's method under microgravity, and other devices for containerless preparation of specimens such as spheres.

Another example is provided by research on the structures of spherical Ag and Cu specimens made by containerless crystallization under short-term weightlessness on the rocket [15, 667] (see [15], pp. 119–132, for a detailed description and photographs showing the structures). The surface was cellular in all cases, and there tended to be one or two crystallization centers, from which block single crystals frequently grew, the mechanism being dendritic. The cellular surface (Fig. 86) is due to the dendrite branches emerging, which in some cases have octahedron facets. The dendritic structure was evidently formed at high cooling rates, with considerable supercooling. However, it was difficult to judge the exact supercooling. The convection currents were probably slight, since they had no effect on the structure. In [667] it is stated that the research is being extended to Ag and Cu alloyed with Ni and Ge. These materials have differing distribution coefficients K_0: $K_{0_{Ni}} > 1$, $K_{0_{Ge}} < 1$, and they enable one to establish the front motion direction, i.e., to determine whether the crystallization is from the bulk or from the surface. This is only 1 out of 130 experiments under microgravity with the Mir rockets, and

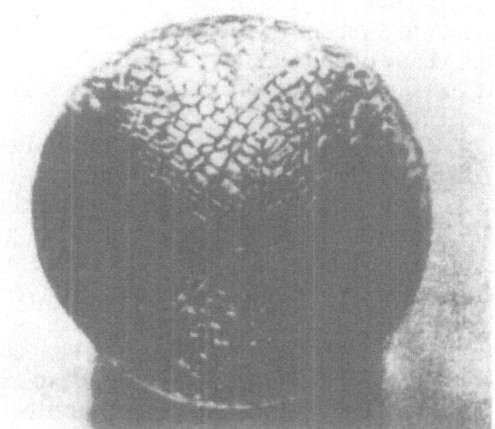

Fig. 86. Spherical copper specimen [15].

it should be indicated that [665] emphasizes that some of the results with these rockets have never before been obtained, such as contactless crystallization for Cu, Ag, and Ge, or making dislocation-free Ge and Si with anomalously high growth rates, testing capillary shaping under microgravity, studying liquid detachment from tube walls, supercooling, relaxation in heat and mass transfer, etc.

The general conclusions from all the studies in [15, 663–669] is that rocket experiments represent an important and independent class distinguished particularly by the fact that heat and mass transfer during melting and solidification may be nonstationary. The good performance and the hopeful results make it desirable to continue the experiments on accelerated crystallization on rockets.

7.3. SOME TEXUS-PROGRAM ROCKET EXPERIMENTS [672, 677, 679–683, 686–688]

The Mir rocket experiments in the USSR are exceeded in number by those performed in the USA with the SPAR rockets, in Europe with the Texus, and in Japan with the TT-500A.

However, my purpose here is not to list these experiments or to survey them even as briefly as in Section 7.2 for the Mir rocket experiments primarily because they are largely represented in [1] (see Chapter 7, pp. 125–131, and data on the SPAR experiments in Table 2 on pp. 50–54). We also give no data list on the experiments on materials science on rockets (rocket type, package load, weightlessness duration, launch method, number of launches, and so on) under the SPAR programs (up to 1980), Texus (up to 1983), and the Japanese ones (up to 1982) as these are given in [674]. we merely supplement the data of [674] by stating that most of the ESA program experiments on SL-1 represented here were prepared in experiments on rockets under the Texus program. In some cases, experiments have been performed on or are planned for the Texus rockets after the SL-1 flight. This shows that European progress in materials science in space continues to be based on these rocket experiments. In [676] it was proposed to survey some of the most interesting results obtained recently with the Texus rockets, but unfortunately the report of [676] is not yet available and there are no other publications on it.

Examples may be taken from papers on weightlessness physics experiments on the Texus rockets in recent years.

In the spring of 1984, the Texus-10 rocket carried two experiments of zone melting for 5-mm diameter Ge crystals; the liquid zones were displaced by 5 and 7 mm, respectively. The impurity segregation corresponded to laminary convection caused by surface-tension gradients, which was tested in ground experiments. In May 1985, the Texus-12 launch carried a repeat experiment, but with the specimen covered by a layer of glass to eliminate Marangoni convection. The results have been given in [672].

Papers [679–681, 686, 687] show that the rocket experiments continue for research in liquid physics; in [680–687] the TEM 06-4 experiment from the Texus-9 launch is reported, which was designed to determine floating-zone heat transfer. Estimates were made of the convection caused by Marangoni currents by measuring the total heat transferred to the lower disk by the floating zone. Flow conditions in these zones were compared between the ground and microgravity; during 6 min of microgravity, measurements were made under nonstationary conditions. There is a discussion of the calibration method, which was used to measure the heat transfer between the upper and lower disks. The flow speeds and Nusselt numbers are discussed, as obtained with computerized equipment in the laboratory. The frequencies and amplitudes in the flow-pattern oscillations have been measured, as was the temperature during the space experiment. The critical Peclet numbers were much larger in space than on the ground and in the stationary state.

In [687] experiments on liquid physics were reported from the Texus-5 and Texus-8 launches. Measurements were made on the transition from stationary to oscillatory thermocapillary convection in a liquid zone in sodium nitrate for cylindrical specimens 6–7 mm in diameter placed between heated graphite cylinders of the same diameter. The temperature difference provided a gradient in the surface tension. In the Texus-5 experiment, the transition to oscillatory convection occurred with a critical Marangoni number $Mg_c = 9600$ and an error of 20%. On Texus-8, $Mg_c = 8500$. A ground experiment gave $Mg_c = 10,200$. These differences are related to the stabilizing effects of heating from above on the ground.

In [679] the behavior of liquid bridges was reported (dimethyl silicone oil, surface tension 0.02 N/m, viscosity $5 \cdot 10^{-6}$ m^2/sec, density 920 kg/m^3) on Texus-12 (May 1985), with bridges between coaxial Al disks, which were of unusually large volume (diameter 30 mm and length 80 mm). The equipment was described, which may be used in Spacelab flights. Movie frames demonstrate the shape change up to bridge breakage. The results are briefly analyzed.

Superfluid helium [682, 683] has been used with short-term weightlessness (about 8 min) on rockets; the 8-liter cryostat contained 4 liters of He I and 4 liters of He II. The dynamic parameters and stability of He II were examined under varying microaccelerations, as well as the temperature homogeneity and stability in the He II vessel, the properties of thin superfluid films, the thermal destruction of these, and the dependence of film thickness on acceleration level. He II is fairly evenly distributed in the vessel under microgravity, and the initial motion rapidly dies away, and for Bond numbers $\leq 10^{-3}$, the situation is controlled by surface tension. An isothermal state in He II was attained very closely, with the temperature uniformity much better than on the ground, which was due to all the inner surfaces being coated with adsorbed He II. The layer thickness under microgravity was greater than that on the ground and also larger than that predicted by the van der Waals theory. Reliability tests were made on porous membranes and other cryogenic equipment; the cryostat was found suitable for microgravity experiments.

In [688] there is a survey of West German rocket experiments under microgravity as part of the Texus program since 1976. Seven launches are discussed 1) in December 1977, 2) in November 1978, 3) in April 1980, 4) in April 1981, 5) in May 1981, 6) in April 1982, and 7) in May 1982. All the launches apart from the third were successful, while in the third, rotation at 1 Hz produced an acceleration of $(0.1–0.2)g_0$. More than 90% of the experiments performed normally. The experiments are listed, and brief results are given: convection at interfaces, capillary phenomena, and metallurgical experiments (skin technology, peritectic reaction in MnBi, binary-system dispersion in the presence of immiscibility, composites, eutectic directional solidification, and directional solidification for doped semiconductors). The general conclusion was that rockets can carry out a wide range of experiments supplementary to the more expensive but more extensive Spacelab–Shuttle programs.

To judge from [674], no experiments have been performed on rockets in the USA since 1980. It is possible that since the start of the Shuttle launches, NASA has decided to stop rocket experiments, which are considered no longer desirable, but no direct statement of this has been found in the literature.

These examples demonstrate the interest in weightlessness physics experiments on rockets and the good scope for carrying them out.

7.4. APPARATUS FOR ENGINEERING EXPERIMENTS ON ROCKETS [16, 662, 670]

In [1] the Japanese TT-500A rocket and research equipment on it were described (pp. 127–131), which reflected the distinctive features by comparison with man–station equipment. There are, however, equipment specifications quite distinctive for rockets. First, the apparatus should operate automatically, and as the experiment is brief, furnaces have to heat up and cool very rapidly or else the specimen must be heated before the flight. Fluid-physics rocket modules also have to meet certain requirements, particularly for automatic control and recording. The rocket developments have assisted in another task, viz., fitting the GAS (get away special) modules, which are to be launched on Shuttle flights. In [677] this task has been considered in Sweden from experience accumulated with rocket equipment. Details are given of the TEM module used by the Swedish Space Corporation on five rocket launches under the Texus program, including 10 optical furnaces, 4 gradient furnaces, etc. Details are also given of the MURMEC rocket module, which included 4 isothermal furnaces with thermal tubes, 12 fast-heating systems for foam metals, 2 gradient furnaces, a three-axis accelerometer, and so on. In [677] there are also details of the GAS modules for space experiments in 1985–1986 on the effects of natural convection on metal solidification: four Sn–Pb specimens, with a total mass of about 8 kg, will be produced in three furnaces, with the heat transfer for directional solidification provided by helium.

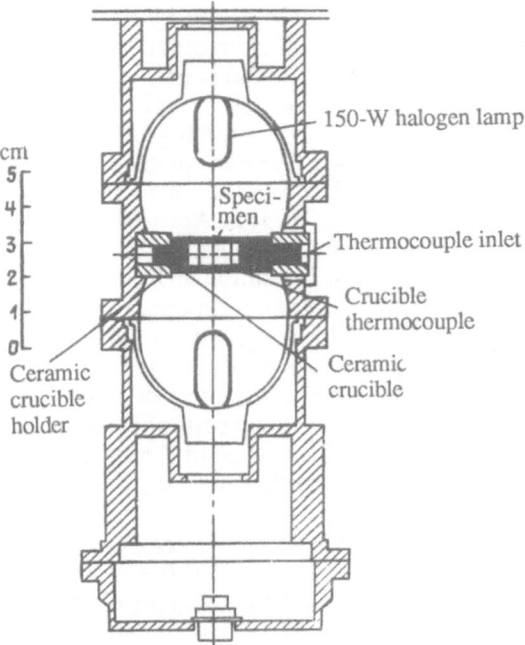

Fig. 87. Mirror furnace, cross section [677].

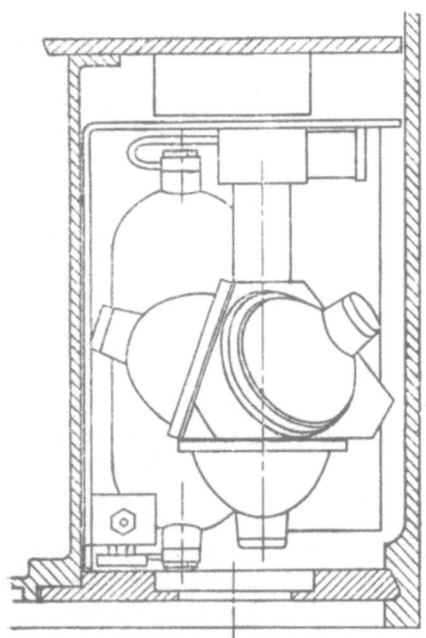

Fig. 88. Quadrupole mirror disposition in high-temperature furnace [677].

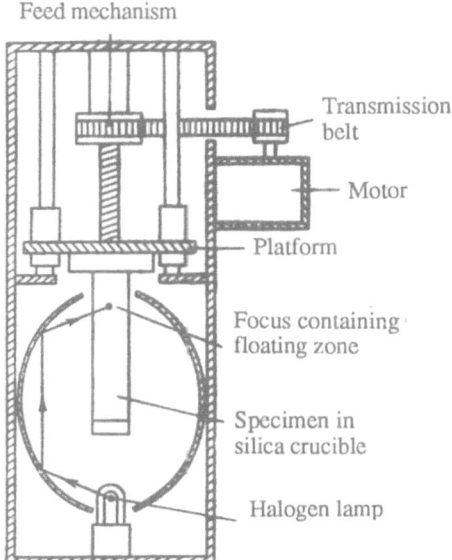

Fig. 89. Form of monoelliptic mirror furnace for crystal growing [677].

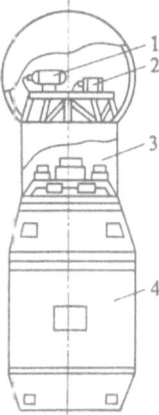

Fig. 90. High-altitude Mir rocket and research equipment location [15]: 1) Kamin radiation furnace; 2) high-sensitivity three-axis accelerometer; 3) Spring exothermic furnace; 4) service module.

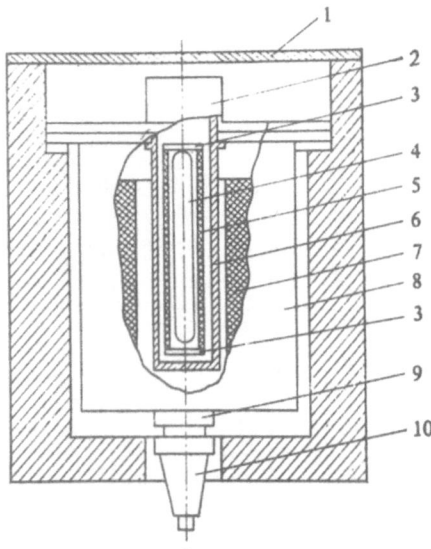

Fig. 91. Section of the VKT-15 exothermic furnace: 1) outside; 2) container shield; 3) insulating plug; 4) silica tube; 5) silica sheath; 6) molybdenum container; 7) exothermic mixture; 8) steel body; 9) ignitor for striking material; 10) lead zone sleeve.

We give here three figures from [677] representing a two-ellipsoid mirror furnace (Fig. 87), a form of quadrupole mirror disposition (Fig. 88), and a crystal-growing furnace (Fig. 89). In [677] several tables give details on the MURMEC and GAS apparatus parameters. It is planned to continue the program on aircraft and rockets (new furnaces will be developed for the TEM and MURMEC modules) not only under the Texus program but also under the new Swedish MASER (materials science experiment rocket) program, starting in 1986. Space experiments are planned in collaboration with ESA and NASA, and programs are being studied for setting up space platforms for making materials or for long-term experiments.

The apparatus used on the Mir rockets has been described in [15, 662]. In [15] there are detailed characteristics, so we merely give two figures from [15] and brief comments on them. Figure 90 shows the Mir rocket scheme with the equipment for processing materials. Inorganic materials may be melted and crystallized on these rockets by means of exothermic furnaces. Figure 91 shows the drawing of one of these. We have seen above, in Section 7.2, that such a furnace (burning temperature 2400°C) can melt materials of melting point up to 1850°C in a molybdenum container or up to 1300°C in silica tubes. Such a furnace is equivalent to a 100-kW electric furnace. The Sprint apparatus included 32 such furnaces (cells), a sensitive residual-acceleration meter, and a power supply, control, and monitoring system.

Section 7.2 deals with the experiments performed with this apparatus on the Mir rockets. In particular, such a furnace was used in 1978 in the first Kapillyar .experiment, which showed that Stepanov's method can be used to grow shaped crystals under microgravity. Later, the improved Forma experiment was performed on the Salyut–Soyuz combination [15].

In December 1982, a Mir rocket carried an experiment on melting material in space by concentrating the sun's rays; a special Kamin apparatus was used containing two parabolic mirrors, which concentrated the sun's radiation by a factor of about 300 in the visible region, while there was a system providing optimum orientation on the sun. The distance from the mirror vertex to the focus was 32 mm, the mean spot diameter 8 mm, and the length along the axis 10 mm. The spot contained the tube, which was made of glass having a high transmission for the solar spectrum. The tube contained 14 rods 1 mm in diameter and length 15 mm made of copper (melting point 1085°C), lead (327°C), zinc (420°C), and indium antimonide (525°C). Ground tests on the Kamin showed that the temperature in the spot attained 1000°C. The temperature attained in space was 750°C, so the sun's rays melted some of the specimens, and spheres 1 mm in diameter were found after the flight.

The Luch-1 solar furnace has also been developed in the USSR, but this has not yet been tested on a rocket and is meant for cutting and welding chrome–nickel and titanium alloy sheets (see [15], p. 152).

7.5. CONCLUSIONS

These data and those in Chapter 7 of [1] show that experiments with short-term weightlessness have played a considerable part in advancing weightlessness physics and have made a considerable contribution to materials science in space, including the development of automatic equipment.

Furthermore, one expects that future short-term microgravity experiments will be useful not only for preparing experiments for prolonged flights, but also for research on accelerated crystallization after heavy supercooling and for developing automatic-module equipment.

Chapter 8

CRYSTAL GROWTH AND ALLOY SOLIDIFICATION
UNDER ELEVATED GRAVITY

In [1] there is a survey of theoretical and experimental papers on alloy solidification and crystal growth under elevated gravity. The main purpose here is to examine the accentuated gravitational convection and the interaction between liquid and tube walls, which provides additional information on the relative roles of gravitational and capillary forces in crystal growth (by comparison with weightlessness). Also, elevated gravity should produce more layering for impurities or alloys differing considerably in density, which provides information on the reasons for banded impurity distributions on the ground and in space. Particular interest here attaches to the crystallization of alloys or highly doped materials, and in particular eutectics, under various conditions: microgravity, $g/g_0 \ll 1$, on the ground, $g/g_0 = 1$, and with elevated gravity, $g/g_0 > 1$, where g_0 is the acceleration due to gravity on the earth.

In [1] papers on this topic published up to 1983 were examined. Here this is supplemented by brief details of recent papers dealing with metal solidification under elevated gravity with methods somewhat different from those cited in [1]. The new methods are based on large centrifuges, particularly one fitted with the Medon mass growth apparatus, which can stand large forces.

Section 8.1 gives information on this centrifuge, which is installed at the Gagarin Spaceman Training Center, and on the Medon furnace, which was made by CNRS in France, while Sections 8.2, 8.3, and 8.4 give the first results obtained in this way. Section 8.5 discusses prospects for further research by this method.

8.1. CENTRIFUGE METHODS OF STUDYING SEMICONDUCTOR CRYSTAL GROWTH AND LIQUID SOLIDIFICATION [690–693]

8.1.1. Centrifuge Data

The centrifuge is a complicated, large-scale structure (total weight 305 tons), which is installed in a special large building (the centrifuge bay has an internal

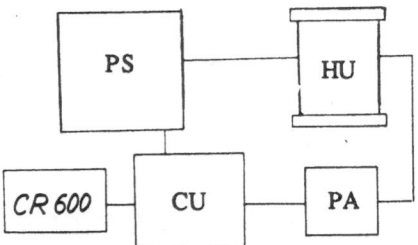

Fig. 92. Block diagram of the Medon apparatus.

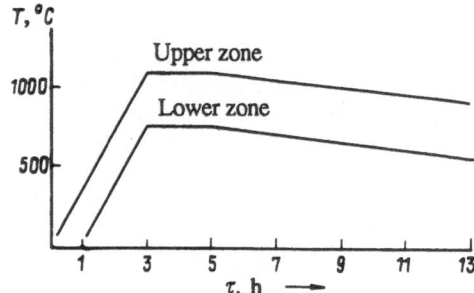

Fig. 93. Temperature profile in the heater unit in the Medon apparatus.

diameter of 45 m), where there is a special engineering staff capable of dealing with all the operating requirements. There is no need to give full details of the centrifuge here, and merely a few items on the unique systems are indicated.

The centrifuge has a radius of 18 m and provides a maximum acceleration of $50g_0$, with the maximum change in acceleration in 1 sec from $1g_0$ to $5g_0$.

Without giving the characteristics of the equipment in the centrifuge drive or in the cabin, rings, and so on, we merely note that the control system provides for setting and controlling g_0 factors, which can be varied, together with cabin orientation and oscillation, in a given fashion, where all the parameters can be controlled either manually from the control panel or by a programmer. The remote-control system and the telemetry serve to switch the equipment on and off in the cabin, including crystal-growing equipment, while signals are transmitted by cable from the cabin to the control panel. The sliding-contact system transmits electrical signals (for example, furnace thermocouple readings) to the panel. The TsF-18 can be used in growing crystals at various gravity levels, as the cabin can accommodate a special growth furnace designed for large forces. The Medon furnace was used in [690–693], which is described below. The design enables one to grow crystals up to 100 kg. The centrifuge facilities provide for orienting the axis in any direction, including along the total acceleration vector. The furnace is powered and controlled, and data are recorded on the temperature distribution by means of a transmission system working with the control panel via a rotating contact device.

8.1.2. The Medon Apparatus

The apparatus is intended to grow crystals with resultant accelerations up to $10g_0$; directional crystallization from the liquid is provided in a thermal-gradient zone by gradual temperature reduction. The furnace has two zones, which are heated by two platinum heaters, which give the maximum temperature in the hot zone up to 1600°C. The heater unit is enclosed in a steel cylinder filled with insulating material. Along the axis is a ceramic channel, on which the two heaters are mounted. The heater in the upper part provides the higher temperature, while the lower heater maintains the temperature gradient.

Figure 92 shows the block diagram, including the heater unit (HU), the control unit (CU), the preamplifier (PA), the power supply (PS), and the two-pen CR-600 pen recorder.

The apparatus provides programmed heating, stabilization, and controlled cooling.

Basic Characteristics

Maximum temperature	1600°C
Temperature gradient in crystallization zone	210°C/cm
Container length	30 cm
Tube length	9 cm
Crystallization zone length	4 cm
Furnace channel internal diameter	2 cm
Time to heat to set temperature	not more than 3 h
Controlled cooling time	99 h
Heater unit mass	45.6 kg
Supply voltage	220 V
Furnace power	300 W

Figure 93 shows the temperature profile in the heater unit.

The temperature is monitored, the processes are recorded, and the heating is controlled by four platinum–rhodium thermocouples, whose hot junctions are placed near the inner wall. The signal from the HU thermocouple passes to the PA and then to the CU. The last unit sets the maximum temperature in the upper heater, the temperature gradient, and the heating and cooling times. The two heaters are powered by the PS, which receives its controlled signals from the control unit. The CR-600 pen recorder provides for continuous furnace-temperature monitoring.

The heater unit and preamplifier are installed in a steel frame, which is mounted in a supporting frame installed in the cabin (Fig. 94); the other units are installed outside the cabin and are connected to the heater and amplifier via rotating contacts (sliding contacts).

We also give some information on the units shown in Fig. 92.

Heater Unit (HU). The crystallization furnace contains a ceramic channel, with an internal diameter of about 20 mm, in which one can install a tube 290 mm long. The ceramic tube is wound with upper and lower heaters. The tube is insulated and mounted in the jacket. The upper heater provides the higher temperature.

Fig. 94. Heater unit installed in the TsF-18 cabin.

Preamplifier (PA). The cabin contains the PA, which amplifies the signals from the thermocouples T_1 and T_2 in the HU, as shown on the scheme. The amplified signals pass through sliding contacts to the voltage amplifier in the CU, which is connected to the PS. The hot zone in the HU has two thermocouples T_1 and T_2. The amplified signal from T_1 passes to the temperature regulator in the HU, while the T_2 signal is used to drive the pen recorder.

The PA circuit is not given, and we merely note that it provides for amplifying the emf's from T_1 and T_2 by a factor of about 1000, so they are raised to the volt level at the output instead of millivolt, which is convenient in transmitting the readings via the sliding contacts.

Special balanced cables connect T_1 and T_2 to the amplifier, while the preamplifier itself is connected by ordinary leads to a voltage divider.

Power Supply (PS). The supply from the 220-V 50-Hz line is via the power supply as a separate unit placed in a case outside the cabin, which includes variac and transformers to supply the upper and lower heaters separately, as well as to adjust the supply voltage manually or via the control unit.

Control Unit (CU). The upper and lower heaters are controlled by this to give set temperature–time patterns.

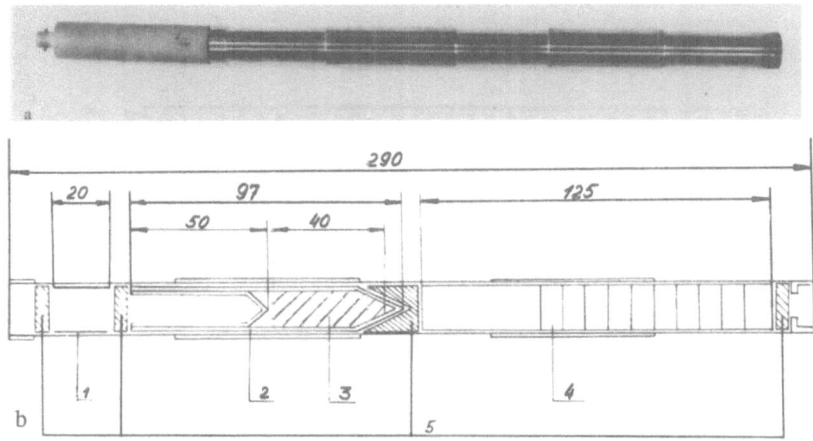

Fig. 95. The container: a) external view; b) internal design. 1) Silica sleeve; 2) tube containing material; 3) PbTe casting; 4) graphite insert; 5) silica-wool seal.

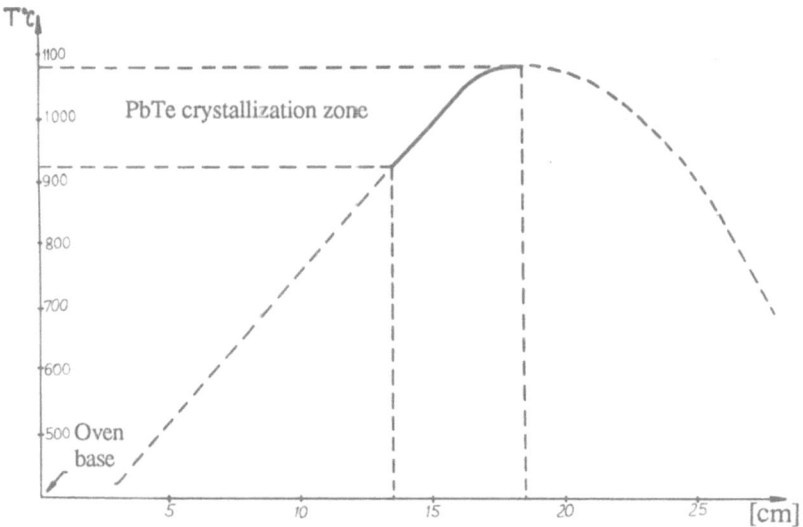

Fig. 96. Temperature profile in growing PbTe crystals.

8.2. PRODUCTION OF SILVER-DOPED LEAD TELLURIDE CRYSTALS UNDER ELEVATED GRAVITY [691, 693]

The Pb–Te specimens were 9 mm in diameter and 40 mm long and were sealed into silica tubes. A tube is placed in a stainless-steel container, with a wall thickness of 2 mm (Fig. 95), which is inserted in the heater channel, after which the centrifuge cabin is closed and the power supplies are switched on.

The Pb–Te crystals were grown with the temperature pattern shown in Fig. 93, with the temperature monitored by the CR-600; the material is heated to the melting point in 3 h, with the upper heater reaching 1095°C and the lower heater 755°C.

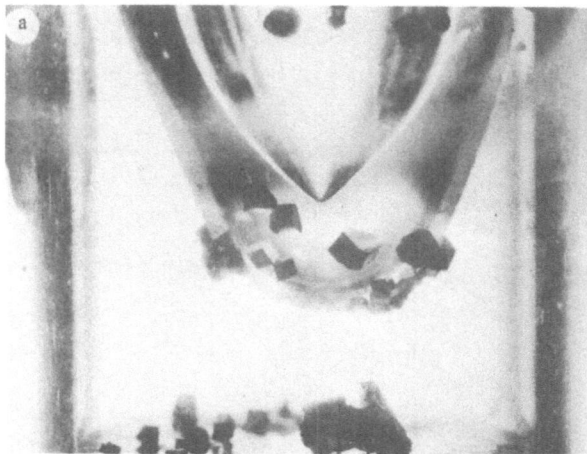

Fig. 97. Crystals at the end of the inner tube: a) cube; b) octahedra.

which correspond to a temperature gradient of 40°C/cm. Then the centrifuge is switched on and the material is kept at a given gravity in the molten state for 2 h. Then the programmed cooling is switched on, which provides a crystallization rate of about 4 mm/h. Figure 96 shows the temperature profile during PbTe growth.

Five experiments were performed with the same cycle but different $n = g/g_0$: 1, 2, 3, 4, and 5. In each experiment, the cabin was set with the resultant acceleration along the axis of the heater unit, i.e., along the specimen tube.

The centrifuge was stopped after the cycle was completed, the container was removed and cooled in air, and then the tube was extracted.

Examination before the tube was opened showed that the conical part of the crystal from which the growth began (the seed) had separated from the main part of the specimen, which evidently occurred after the end of the crystallization, during cooling. Visual examination also showed that bubbles were unevenly distributed along the specimen, with more in the upper part, while there are almost none at the

Fig. 98. SEM image from cube.

seed, while it was also evident that the distribution and sizes of these inclusions varied from specimen to specimen, i.e., were dependent on the gravity.

The most interesting effect seen before the tube was opened was that cubic PbTe crystals occurred at the end (Fig. 97).

The number and shape perfection of these increased with the gravity; these regular crystals are due to some feature of growth from the vapor at elevated gravity, since nothing such is observed on the ground or in space specimens. Usually, vapor deposition produces whisker PbTe, not cubic crystals.

The extracted crystals were examined in the scanning electron microscope. Figure 98 shows the surface of a cubic crystal. At this magnification, the surface is ideal. Figure 99 shows two cubic crystals intergrown as an octahedron. The surface of one cube has steps and growth spirals, in contrast to the previous.

In [691, 693] the structure and properties are given of the main castings, which were cut into parts and edges, where metallography revealed the features produced at the various gravity levels. The side surfaces contained many pores and shrinkage pits, whose number increased with the gravity. The shape and size also altered. This increase with gravity is due to increased interaction with the walls.

Metallography showed that the specimen made at $n = g/g_0 = 1$ consisted of several blocks, while there were three blocks for $n = 2$, and for $n = 3$–4, the specimens were in part single crystals, which they always were for $n = 5$. Some specimens, however, showed a mosaic structure; i.e., each large PbTe crystal consisted of a set of small slightly disoriented crystals (angles $\leq 1°$). The shapes and sizes of these grains varied. The mosaic structure was dependent on the slice number (see distance) and on the resultant acceleration. On the whole the perfection increased along the specimen and with the gravity, as the disorientation decreased, and for $n = 5$ one obtained a comparatively homogeneous single crystal.

The electrophysical measurements showed that Ag-doped PbTe has p-type conductivity. The mobility measured at 77–300°K was higher, and the majority-carrier concentration was lower, at high gravity, which is due to reduction in the structural defects.

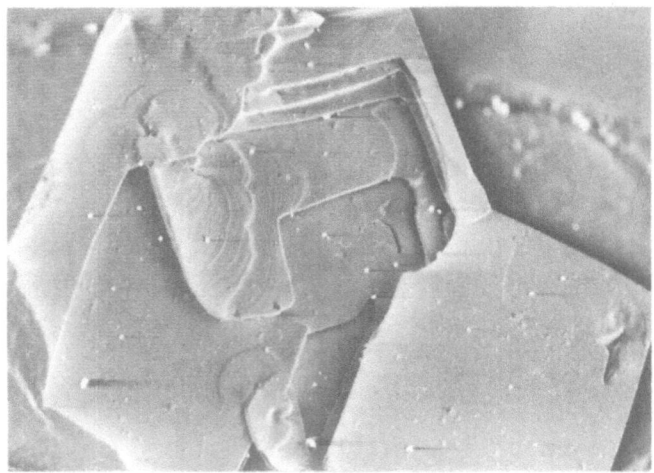

Fig. 99. SEM image from octahedron (two intergrown cubes).

These preliminary measurements on Ag-doped PbTe show that one can obtain better crystals by growing them in centrifugal forces, as proposed in papers by Müller and others cited in [1] (Chapter 8, pp. 132–149), which was derived from studies on Te-doped InSb.

The method of [690, 691, 693] should elucidate this important topic.

8.3. SOLIDIFICATION OF TELLURIUM–SILICON GLASS AND THE CRYSTALLIZATION OF TELLURIUM AND TELLURIUM–SELENIUM ALLOYS UNDER ELEVATED GRAVITY [390]

In [390] experiments are described on recrystallizing a tellurium single crystal, a tellurium polycrystalline billet, and a similar $Te_{0.92}Se_{0.08}$ one, as well as the solidification of vitreous $Si_{0.2}Te_{0.8}$ at $5g_0$ in a Ts-18 centrifuge. The electrical parameters were measured at 1.4–300°K. The elevated gravity affected the crystallization and the electrically active impurity distributions in tellurium and the tellurium–selenium alloy, as well as the solidification of vitreous tellurium–silicon. These studies were made in parallel with the ALKUTEST experiment on making these substances under microgravity [390].

Some details from [390] are given below.

8.3.1. Experimental Conditions

The specimen location in the furnace and the temperatures and times used varied from experiment to experiment.

In all the experiments, the cabin was oriented so that the centrifugal acceleration was along the furnace axis and thus that of the container. The solidification front moved against the centrifugal vector.

1. Complete Remelting for Tellurium and Tellurium–Selenium $Te_{0.92}Se_{0.08}$ Alloy. The tube was placed in the furnace zone heated above the melting point of tellurium (452°C); that zone was heated to 600°C over 2 h. After the temperature had been held for 30 min, the centrifuge was switched on and $5g_0$ was maintained for 30 min. Then the programmed cooling followed at about 130 deg/h (for tellurium) or 100 deg/h (for tellurium–selenium). The cooling was to 200°C, after which the centrifuge was switched off and the container was removed.

2. Incomplete Recrystallization of Tellurium Single Crystal. The tellurium crystal had been made by pulling from the melt, and the tube was placed in the container so that the melting front in the steady state passed through the middle. The upper part of the specimen (relative to the acceleration vector) melted, while the lower part remained crystalline and acted as a seed. During cooling, the crystallization began from the contact with the solid part, and conditions were favorable to making a single crystal. The upper part was heated to 500°C over 1.5 h, where it melted; after 30 min at a temperature to homogenize the liquid, the centrifuge was used to produce $5g_0$. The liquid was kept under these conditions for 30 min to equilibrate the temperature and impurity distributions. This was followed by cooling at about 70 deg/h. When the middle reached 200°C, the centrifuge was switched off and the container was removed.

3. Solidification and Vitrification in $Si_{0.2}Te_{0.8}$ Tellurium–Silicon Alloy. The process was analogous to case 1, but a higher cooling rate was used; i.e., a quenching effect was produced, with a high gravity maintained at 700°C, after which the furnace was switched off and the material cooled at about 300 deg/h.

These three recrystallized materials were carefully removed from the tubes and examined.

The electrical parameters were then checked (carrier mobility and concentration) in the intrinsic and impurity conduction ranges. The resistance and Hall effect were measured in a weak magnetic field at several points at helium temperatures (1.4–4.2°K), in liquid nitrogen (77.4°K), and near room temperature (290°K).

Decorative etching was used after the electrical measurements.

8.3.2. Results

1. Complete Tellurium Recrystallization. On one side, there were numerous bubbles (shrinkage pits), which were larger at the start; etching in 30% HNO_3 showed that there were numerous blocks of about the same size throughout the length.

The conductivity and Hall effect taken with the carrier concentration and mobility showed that the specimen had a comparatively low hole concentration in the impurity-conduction region ($p < 10^{15}$ cm^{-3} throughout the length). Near room temperature, the conductivity was inherent (n type, exponentially dependent on $1/T$). The electron concentration was then independent of the impurity level and was the same throughout the length. The Hall coefficient corresponded to that calculated for pure tellurium: $E_g = 0.325$ eV, ratio of electron and hole mobilities $\mu_n/\mu_p = 2$.

In the impurity conduction range (77°K and 4.2°K), the concentration decreased systematically toward the end, which is not characteristic of directional crystallization in tellurium and may be due to the impurity distribution arising in the liquid under elevated gravity. Estimates showed that $a = 5g_0$ was insufficient to produce any appreciable impurity-atom sedimentation by balancing between the diffusion fluxes and those produced by gravitational forces. The hole-concentration variation along the material was due to the difference between the convection fluxes under normal conditions and under elevated gravity when there are bubbles. This was evident from the uneven distribution of the bubbles by number and size, as they were larger and more numerous at the start. A control casting was made on an analogous pattern but with normal gravity, which showed the carrier concentration increasing toward the end: the impurities were rejected in the crystallization direction, as is widely used in making pure tellurium by zone melting.

The variation in hole concentration along the casting made at $5g_0$ did not alter at 4.2°K; the concentration merely fell slightly, although the temperature was reduced by almost a factor of 20, so the impurity-state activation energy is much the same along the casting and does not exceed 0.005 eV.

The hole mobility here was less than 1000 cm^2/V·sec, as would be expected, since the mobility in a polycrystalline specimen is determined by defects, which are individual in type. In the control specimen the hole mobility was somewhat higher, which was due primarily to a higher concentration of impurities that heal defects.

2. **Incomplete Recrystallization of Tellurium Single Crystal.** The remelted part had a uniform luster as characteristic of a single-crystal structure. Etching in 30% HNO$_3$ confirmed this. It also showed that a small part of the liquid had flowed into the seed region, where it had cooled rapidly and produced a poly-crystalline structure surrounding the single-crystal seed. The Hall coefficient and conductivity with the corresponding carrier concentration and mobility indicated low hole concentrations in the impurity range (4.2°K and 77°K) in the recrystallized part: $P_{(4.2-77)°K} = 1-2 \cdot 10^{14}$ cm^{-3}. At the end, the concentration increased to $P_{77} = 3 \cdot 10^{14}$ cm^{-3}. The fall in the measured concentration at the start of the recrystallized part was probably due to the factors encountered in complete recrystallization. The mobility in the recrystallized part was much higher than that in the polycrystalline materials and did not fall below 2500 cm^2/V·sec at 77°K. At 4.2°K, it increased to about 10^4 cm^2/V·sec, which shows that the material differed from polycrystalline in that the acoustic mechanism played a large part in hole scattering.

3. **Tellurium–Selenium Alloy Recrystallization.** The Te$_{0.92}$Se$_{0.08}$ casting after recrystallization is shown in Fig. 100; there were shrinkage pits (bubbles) and cracks. The surface, apart from the initial conical part 5 mm long, was dull, which is characteristic of tellurium and selenium alloys. The metallic lus-ter at the start indicates that here almost pure tellurium predominates, while the selenium level rises toward the end.

The Hall coefficient and conductivity were measured at various points at 291, 77.4, and 4.2°K, and the carrier concentration and mobility were calculated; the hole concentration was comparatively high: $P_{77°K} = (6-8) \cdot 10^{15}$ cm^{-3}, while the mobility was low: $\mu < 100$ cm^2/V·sec.

The Hall coefficients ($R > 0$) were of the same sign at 291, 77.4, and 4.2°K, which indicates hole conductivity throughout the range, with the hole concentration decreasing toward the end, as with complete tellurium recrystallization.

The tellurium–selenium system contains a continuous series of solid solutions, where the width of the forbidden band E_g varies smoothly, so the conductivity was examined between room and nitrogen temperatures (Fig. 101) to determine E_g and the impurity activation E_A. The high-temperature part showed that there were com-paratively deep impurity levels having $E_A \sim 70$ meV, which are absent from pure tellurium.

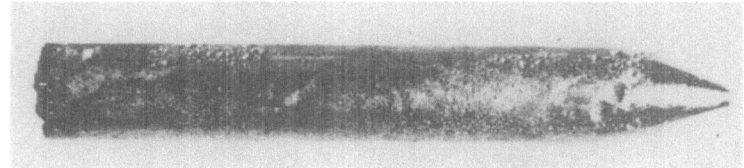

Fig. 100. A tellurium–selenium alloy specimen after recrystallization at $5g_0$.

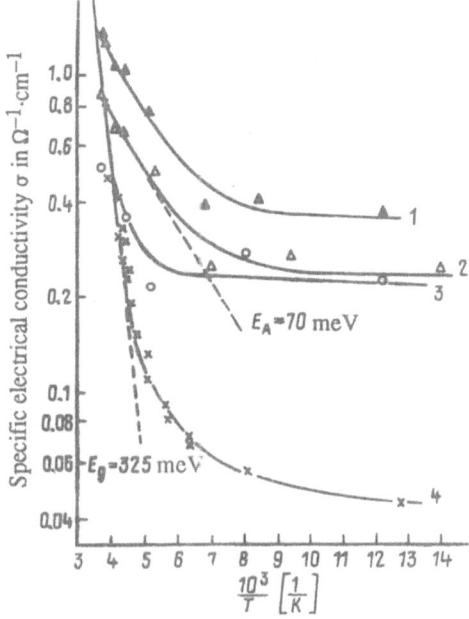

Fig. 101. Temperature dependence of the conductivity for $Te_{0.92}Se_{0.08}$ (curves 1–3) and for pure tellurium (curve 4).

For pure tellurium, the log $\sigma(1/T)$ dependence corresponds to theory for intrinsic conductivity with $E_g = 325$ meV.

Elevated gravity appears to cause the selenium to segregate as a second phase at the boundaries of the tellurium–selenium crystallites, which gives rise to a cellular structure; the reduction in the conductivity for the alloy throughout the temperature range then indicates thermal activation for the conductivity in the selenium-rich regions. This also explains the anomalously low hole mobility at low temperatures: 30 cm^2/V·sec at 4.2°K, which is due to the insulating character of the selenium-rich films at low temperatures.

Measurements made with temperature cycling were of poor reproducibility with this specimen, which was due to the microcracks and the thin layers between the crystallites.

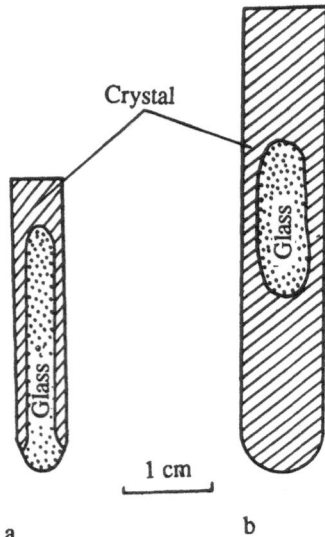

Fig. 102. Disposition of vitreous and crystalline regions in $Te_{0.8}Se_{0.2}$ solidified under elevated gravity (a) and in control specimen (b).

4. Vitrification in Remelting $Si_{0.2}Te_{0.8}$. The Si–Te phase diagram has a liquid of $Si_{0.2}Te_{0.8}$ composition (melting point 385°C) in the eutectic region on the tellurium side, where the following components coexist: $Si_2Te_3(SiTe_2 + Te)$; the densities differ: 4.39 g/cm^3 for Si_2Te_3 and 6.26 g/cm^3 for Te, so $5g_0$ should have an effect on the unmixing and vitrification. The $Si_{0.2}Te_{0.8}$ casting contained a considerable amount of glass, with the vitreous region inside (Fig. 102), where the figure also shows a controlled casting made on the same cycle but with normal gravity.

In both cases, the crystalline state occurred on the outside and the vitreous state within. This must be due to the cooling mode – the initial cooling rate was sufficient to produce a glass throughout the volume, but the subsequent rate below the melting point was insufficient to keep the material in the vitreous state: the $Si_{0.2}Te_{0.8}$ glass crystallized at $T > T_{cr} = 275$°C. The differences in position and content for the vitreous form in the two cases may be ascribed to the elevated gravity. Approximate resistivity measurements on the vitreous phase gave similar values in the two cases $(1-10)\cdot10^6$ $\Omega\cdot$cm at 293°K (the lower values occurred in zones adjoining the crystalline phase).

At the end of [390] the experiments are surveyed, and it is stated that the Te and Te–Se crystalline castings made at $5g_0$ show reduced concentrations of electrically active impurity centers in the crystallization-front direction, which differed from the distributions found in Te castings and single crystals grown under normal conditions. This distribution must be elucidated in additional experiments, which should include doping Te with various substances and using various conditions for keeping the liquid under elevated gravity. It is of particular interest to obtain purer Te and Te–Se by elevated-gravity recrystallization. The results of [390] have been presented in some detail here, apart from the numerous illustrations, in order to demonstrate the scope for the method described in Section 8.1 and to emphasize that the structure and properties of elevated-gravity materials raise questions requiring further centrifuge experiments.

8.4. SOLIDIFICATION OF Al–Cu EUTECTIC ALLOY UNDER ELEVATED GRAVITY [390, 692]

The purpose here is to elucidate how gravity affects directional crystallization in eutectic aluminum–copper alloys.

A special silica tube was devised, which enclosed a graphite container. The design prevented the molten material from entering the heater channel even if the silica tube is damaged.

The specimens were rods 70 mm long and 7 mm in diameter and contained a 67.7 wt. % Al and 33.3 wt. % Cu alloy, which corresponds to the Al–Cu eutectic. Each specimen was in a closed graphite container which, in turn, was placed in a silica tube evacuated to 10^{-3} mm Hg. The tube was placed in a stainless-steel container with a wall thickness of 2 mm.

The container was placed in the heater-unit channel, and then the centrifuge cabin was sealed and the power supply to the Medon system was switched on.

Figure 103 shows the tube position in the gradient field before crystallization started.

After a steady stage, the programmed cooling was initiated, which provided a crystallization rate of 2 cm/h or 65 deg/h.

The acceleration was along the furnace axis and thus along the container. The crystallization front moved against the acceleration vector.

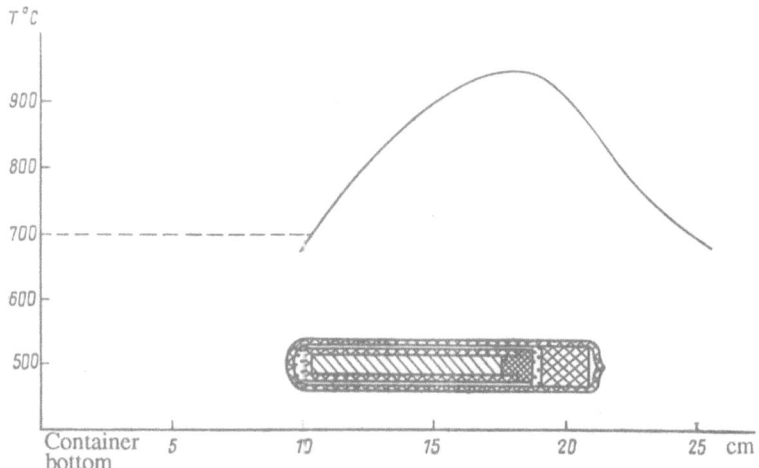

Fig. 103. Tube with specimen in temperature pattern in Medon equipment.

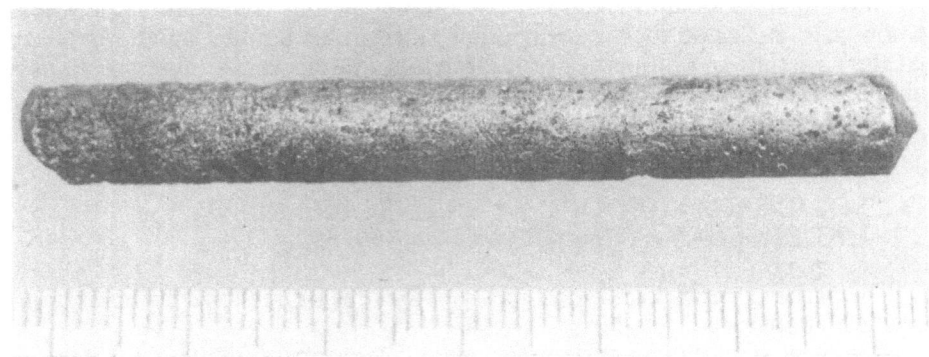

Fig. 104. Casting made with $1g_0 \rightarrow 2g_0 \rightarrow 5g_0$ steps, the surface showing necking on gravity switching (×3.5).

The centrifuge was halted and the container was removed and placed in a refractory vessel for cooling in air. Then the tube containing the Al–Cu specimen was removed and examined, which showed no external damage.

Five experiments were performed: at 1, 2, 5, and $7g_0$ and with the sequence $1 \rightarrow 2 \rightarrow 5g_0$ with one of the specimens. All the specimens were photographed. They had smooth surfaces, although flow streaks were evident at some points. Necking occurred in the specimen with stepped gravity, evidently because of the liquid detaching from the walls when the gravity level altered (Fig. 104).

The microstructure altered as the gravity increased; the expected regular platy structure would consist of Al and $CuAl_2$, but this was not obtained in all the specimens. In specimen A (gravity $1g_0$), there were a few highly disoriented eutectic grains at the middle. The structure was basically of two-phase type, but the second-phase segregations were regular in shape (Fig. 105a). Specimens B and C ($2g_0$ and $5g_0$, respectively) had the regular platy eutectic texture (Fig. 105b) throughout much of them, and in C there were virtually no deviations from the ideal texture. However, there was an interesting effect: the two-phase structure at the start was different from that at the end, as in the first case there was excess $CuAl_2$ segregation against the eutectic background (Fig. 105c), and in the second, Al against the eutectic background (Fig. 105d). This is probably because the elevated gravity enriched the head in copper and depleted it in the tail. One can similarly explain the absence of directional platy structure in D ($7g_0$), where there was the usual cellular microstructure, with the second-phase segregations along the aluminum grain boundaries.

8.5. PROSPECTS FOR STUDYING CRYSTAL GROWTH AND METAL SOLIDIFICATION UNDER ELEVATED GRAVITY

The data given in [1] and here (Sections 8.3 and 8.4) show that elevated gravity affects crystal structure and properties appreciably, and future centrifuge experiments should provide additional information on effects important in crystal grow-

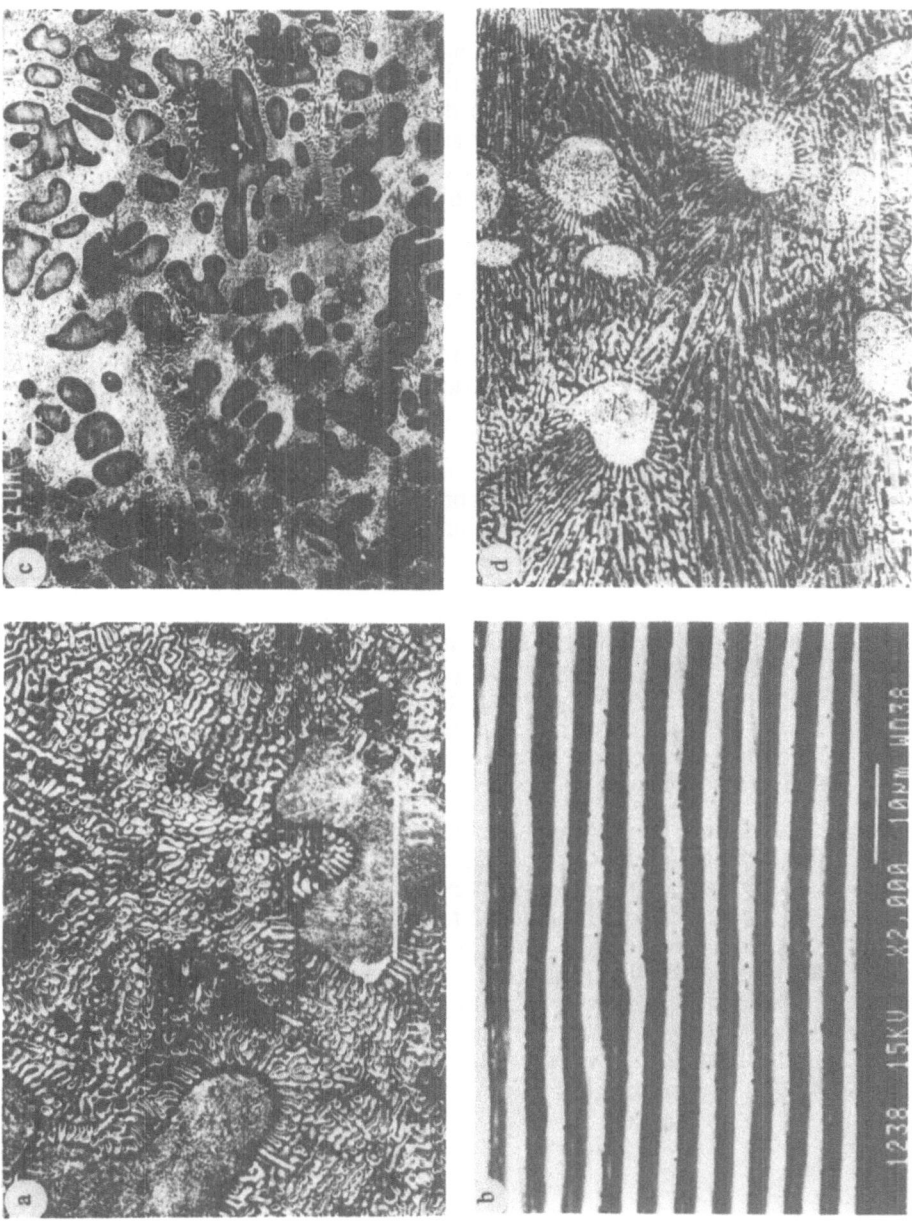

Fig. 105. Microstructures in Al–Cu specimens made at various gravity levels: a) $1g_0$; b) $2g_0$ and $5g_0$; c) excess $CuAl_2$ segregation against Al–Cu eutectic background; d) Al deposited against an Al–Cu eutectic background.

ing, such as the relation between temperature oscillations and impurity striation, gravitational convection, liquid interaction with tube walls, surface tension effects, bubble migration, and so on.

There is no need to use very large accelerations to examine such effects in Bridgman's method; it is sufficient to envisage $n = g/g_0 = 20$ or even less. The papers by Müller et al. cited in [1] indicate that it is sufficient for a Bridgman apparatus to be able to work at accelerations less than $n = 10$.

We conclude that it is desirable to develop crystal growth under elevated gravity, particularly by using a special furnace capable of withstanding considerable accelerations in a suitable centrifuge. The method described in Section 8.1 enables one to study crystal growth and metal solidification under elevated gravity and can provide new and important results.

The prospects involve certain theoretical and experimental tasks, where we may note particularly the following:

1. Theoretical studies on temperature oscillation and impurity stratification under elevated gravity should be used to select the most informative semi-conductor systems (with various dopants) for centrifuge experiments. It is desirable to predict the results, i.e., to provide a theoretical basis for these new experiments.
2. One should consider bubble distributions (number and size) in relation to gravity level and heating conditions. One can then make suggestions on special experiments for particular materials that include gas-producing substances.
3. There should be a theoretical discussion of the expected inclusion distributions, for example in dispersion-hardened composites (Al_2O_3 particles in Al, etc.) during melting and solidification in relation to the gravity level.
4. The possible effects of elevated gravity should be discussed for eutectic-phase segregations (spherical, filamentary, or platy) arising during solidification, where the most promising experiments should be selected.
5. New experiments should be done on the recrystallization of Te, Te–Se, and Te–Si to resolve some of the above problems.

These tasks should be accompanied by calculations on temperature patterns in crystals within heaters in relation to container wall thickness, as well as effects from thermal conductivity. Centrifuge experiments should be performed to check effects from changes in orientation of the acceleration vector relative to the temperature gradient, where the angles may be varied from 0 to 180°.

There are also some other particular tasks in elucidating the role of gravitation in crystal growth and metal solidification, as well as ones for glasses and other materials. When one discusses centrifuge studies, however, each experiment must have a proper basis, as is the case for reduced gravity.

In these cases, a preliminary theoretical analysis, as for microgravity, is not only useful but also necessary.

Chapter 9

MICROGRAVITY ELECTROPHORESIS

In [1] there is merely a list of papers dealing with electrophoresis under microgravity, with no separate section dealing with the topic. However, advances in electrophoretic research under microgravity have been very considerable, and such methods are being considered for the commercial production of drugs and biological materials in space. In [15] (Chapter 4, pp. 159–173), there is a survey of electrophoretic research under microgravity performed in the USSR. Here I also briefly present some of the evidence from [15] and from other recent publications [694–713]. As the topic was not analyzed at all in [1] (the present book is a continuation and supplement to [1]), I first give general information and a classification for electrophoretic processes in Section 9.1, then brief evidence on electrophoretic theory in Section 9.2, details of electrophoretic apparatus for microgravity in Section 9.3, and finally some electrophoretic results obtained recently in Section 9.4.

9.1. GENERAL ELECTROPHORESIS DATA [15, 699]

Classification of Electrophoresis Types [699]. In electrophoresis, electrically charged particles, cells, molecules, or ions migrate in an electric field between two electrodes (Fig. 106, the electrophoretic chamber 1). The electrodes 5 are placed in separate chambers 2 and 3 filled with buffer, while the electrophoretic chamber itself is filled with a buffer whose composition may vary over the volume and in time in accordance with the required separation. The purpose is to separate a mixture into fractions. The mixture (specimen or sample) usually occupies a zone in the chamber; if initially the sample is distributed throughout the volume, the zone arises during the electrophoresis. In [699] the following classification is based on zone behavior.

Zonal Electrophoresis. Here the components are divided into fractions, or zones, in the field in accordance with their electrophoretic mobilities; the zones ex-

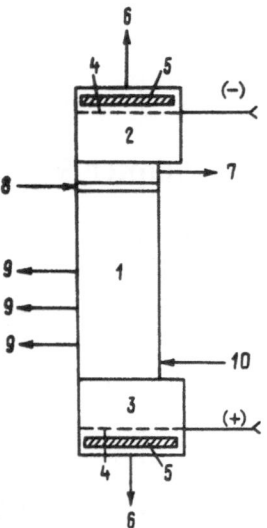

Fig. 106. Electrophoretic chamber [15]: 1) electrophoresis cell; 2, 3) electrode chambers; 4) electrodes; 5) gas stores; 6) gas tap-off systems; 7) solution drain system; 8) injection device; 9) fraction sampler; 10) buffer generation system.

pand by diffusion and also because of convection and the concentration dependence.

Isotachophoresis. This requires the buffer to contain two electrolytes (leader and terminator) having mobilities and dissociation constants such that all the components lie between them as regards transport rate. After a transient response, the mixture separates completely into sharply bounded zones not broadened by diffusion, whose widths are determined by the mobilities and concentrations. The entire system moves at a constant speed controlled by the leader.

Isoelectric Focusing. This is based on the mobilities for the amphoteric components being dependent on pH; each component has a pH at which the mobility becomes zero, which is called the isoelectric point pI. If the pH in the chamber varies over the range covering the pI for all the components, each will concentrate (be focused) around its isoelectric point. One gets a stationary state whose stability is determined by the stability of the pH distribution, while the zone widths are governed by the balance between diffusion and electromigration.

Free-Flow Electrophoresis. The above forms of electrophoresis are of stationary type, where there is no forced buffer displacement in the chamber. In free-flow electrophoresis, the buffer is passed through at a constant rate perpendicular to the migration direction. Usually, one employs the zone form, although there are forms of the method for isoelectric focusing and isotachophoresis. In all cases, convection interferes with the separation. Therefore, reducing convection by employing microgravity is useful in any method, but particularly in isoelectric focusing and isotachophoresis, since these give rise to the largest density inhomogeneities and temperature differences. However, zonal electrophoresis is widely used under microgravity because it is a good technology, particularly in the flow form.

9.2. THEORY OF MICROGRAVITY ELECTROPHORESIS [15, 698, 699, 701, 705, 706, 711]

See [698] for the mathematical theory of electrophoresis and [15] for a simplified analysis.

In [15] a solution is considered in which an electric field E is applied in the X direction and drives a particle of mass m and charge q by virtue of the force Eq with a velocity $v = dx/dt$. The liquid exerts a retarding force $F = f(dx/dt)$, where f is the resistance coefficient. Then the equation of motion is $m(d^2x/dt^2) = qE - f(dx/dt)$, and the solution is

$$v = \frac{qE}{f}\left[1 - \exp\left(-\frac{f}{m}t\right)\right].$$

(9.1)

As $f/m \gg 1$ for a particle of molecular size, (9.1) simplifies for short times t:

$$v = \frac{qE}{f}.$$

(9.2)

For a spherical particle with radius r, moving in a liquid having viscosity η, $f = 6\pi\eta r$, so the electrophoretic mobility is $u = (v/E) = q/6\pi\eta r$. We replace q/r by ε (ε is the dielectric constant, while ξ is the electrokinetic potential), which gives

$$u = \frac{v}{E} = \varepsilon\xi/6\pi\eta .$$

(9.3)

Then (9.3) shows that the mobility is independent of the size and is governed only by the viscosity and electrophysical parameters. The formula becomes more complicated than (9.3) when one incorporates the ion clouds around the colloidal particles in an electrolyte. The problem becomes even more complicated if one attempts to incorporate convection. In [15] one finds the basic equations for the hydrodynamics, electrodynamics, and thermodynamics of multicomponent liquids in electrophoretic separation under microgravity. In such a mixture, one has to consider the interaction between all the moving components and the interactions with the electromagnetic, gravitational, and other fields. The conservation equations for mass, energy, momentum, and entropy are written initially for the individual components. The equations for integral quantities applying to the mixture as a whole are obtained by summing the individual ones. Methods from nonequilibrium thermodynamics are used to solve the systems.

Here there is no need to consider these equations with all the details discussed in [698], or even in the condensed form given in [15]. We merely state that they can be solved, even when appropriate simplifications are made, only by numerical methods.

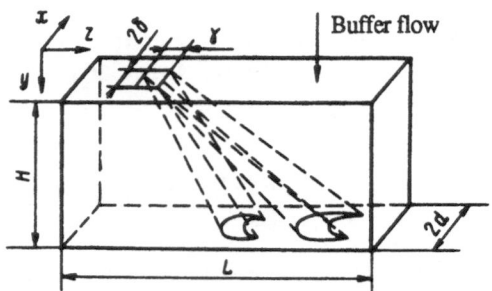

Fig. 107. Electrophoretic apparatus (flow electrophoresis model) [701].

Here we supplement what has been said about [15, 698] by indicating some short theoretical papers on electrophoresis. In [705, 706] states are evaluated where convection can arise during electrophoresis. In [705] convection was examined for electrolyte mixtures in zonal electrophoresis and isotachophoresis. Equations were derived for concentration-dependent and thermal forms of convection in an electric field for multicomponent chemically active electrolyte mixtures. These equations describe a specific effect, viz., a form of cross diffusion due to the joint effects of the electric field and instantaneous reactions. A model was given for convection during zonal electrophoresis in a buffer solution consisting of a weak acid and weak base with a supporting medium. Solutions have been found for various electrophoretic models corresponding to mechanical equilibrium.

In [706] it is emphasized that the choice of model for zonal electrophoresis for biological substances is dependent on the relative concentrations of the specimen and the buffer components. If the specimen concentration is low, the zone from the specimen moving in the electric field is symmetrical. As the concentration increases, it becomes unsymmetrical, and when the specimen and buffer concentrations become comparable, one side of the zone is a concentration step and the other is a spreading wave. In [706] there was a study of the resolving power in zonal electrophoresis. Numerical calculations have been confirmed by measurements on blood protein fractionation. Theoretical evaluations have been made on electrophoretic stability in the presence of convection. It has been found that the buffer system without specimen zone is unstable in relation to concentration inhomogeneities, and the zone makes this worse.

In [701] a model was proposed for flow electrophoresis under microgravity. Figure 107 shows the apparatus, with the chamber and the separation of a biological substance into two fractions differing in mobility. Various simplifications have been made in solving the equations, and the solution is split up into hydrodynamic and electrophoretic stages. This model forecast results from biotechnology experiments in space and provides for optimizing the electrophoretic equipment.

In [711] there is a discussion of the complications in free-flow electrophoresis, since the electric field generates an electroosmotic flow along the chamber walls. This interferes with the separation and disrupts the fraction collection. At the same time, it disrupts the temperature pattern. The field produces temperature gradients because of the Joule heating, which may affect the flow on account of thermal con-

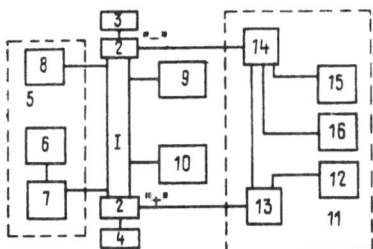

Fig. 108. The Tavriya apparatus [695]: 1) electrophoretic chamber (column); 2) electrode chamber; 3) hydrogen separator; 4) oxygen separator; 5) recirculation system and working solution supply; 6) displacement device; 7) buffer-mixture production device; 8) drainage device; 9) sample input; 10) fraction sampler; 11) power supply, control, and monitoring system; 12) sensor; 13) high-voltage unit; 14) current meter; 15) current indicator; 16) output unit.

vection. Therefore, convection is influenced by several parameters, and in [711] there are calculations on hydrodynamic phenomena in the chamber during free flow on that basis. Initially, the equations for the continuity of the motion and energy are given in general form, and then the general equations are adapted to free-flow electrophoresis and are solved by finite-difference methods subject to certain simplifying assumptions. The model enables one to identify the hydrodynamic perturbations arising during electrophoresis.

This ends the brief characterization of theoretical research on microgravity electrophoresis, where we conclude that the papers cited show that theoreticians have given serious attention to studying the topic, but the complexities require such calculations to be accompanied by microgravity electrophoresis experiments on model systems and on important materials.

9.3. MICROGRAVITY ELECTROPHORESIS APPARATUS [15, 694, 695, 699, 701, 702, 704]

Electrophoretic fractionation of biological material in space is complicated, where there are two main forms of scheme, in accordance with whether the vehicle is manned or not [694]. In the first, the apparatus and specimen are prepared on the ground, with fractionation during flight, and fraction return to the ground. The second is more complicated, where the apparatus is prepared on the ground and in space, with specimens and solutions brought up to the vehicle, the apparatus is serviced, the separated fractions are stored, and the material is returned. The apparatus, in general, should provide the following: 1) storing working solutions and initial material, 2) automatic solution and specimen supplied to chamber, 3) current-stabilized high-voltage power supply, 4) specimen fractionation, 5) thermostatic control, 6) separation and use of any gases produced, 7) automatic or semi-automatic control, and 8) data recording.

An example taken from [15] here is for the Tavriya apparatus, which meets the above requirements (Fig. 108). There is a rectangular chamber (Fig. 106) (thickness 3 mm, width 70 mm, and length 300 mm), which has two-sided cooling [702]. The specimen is supplied from a thermostatic container. The products are collected for transfer to the thermostatic collector on 60 channels; the transfer is by a peristaltic pump. The buffer solution in the collector section is at the same time drawn into the working chamber. The cooling system removes the heat dissipated, while the temperature-monitoring unit maintains the wall temperature at +4°C. The high-voltage source provides a field of 20–70 V/cm to suit the conditions and conductivity.

The monitoring unit outputs the data on the operation for telemetry.

The inner walls are coated with a fluoroepoxide material to provide high electrical strength and good heat transfer. A novel four-position automatic device [694] provides for flushing and servicing the working cavity and also for joining each of the 60 tubes in the fractionation pump to the corresponding fraction collectors in the thermostat, which contains the working solution and the spent-solution vessel. The storage unit enables one to use the same volume to store the solution and the separated fractions.

In [704] it is stated that multipurpose equipment has considerable advantages; for example, a free-flow apparatus can be used in a free liquid, or else in a free flow and for isoelectric concentration. This means that various fractionation methods can be used under microgravity and a single apparatus can be used for any experiments with various physicochemical processes.

In [699] there is a list of the technical problems and difficulties with electrophoresis. No matter what the type and the apparatus, there are difficulties both general and specific in electrophoresis under microgravity. For example, there is gas production at the electrodes. Even a small gas bubble entering the separation chamber can alter the process considerably, while the gas may also block the electrodes. There are many ways of overcoming these gas problems, e.g., with mechanical separators or by suitable electrode materials. In [699] papers are cited in which matrix electrodes made of porous platinized titanium were used, with a diaphragm in close contact. The diaphragm acts as a barrier layer, which does not allow the gases to pass through it and caused them to pass through the large pores in the electrodes. The electrolyte is drawn in through the small pores and supplied to the electrode.

A basic problem with stationary electrophoresis concerns sample input and fraction removal.[*] The resolving power in zone electrophoresis is substantially dependent on the initial zone width, in contrast to isoelectric focusing and isotachophoresis. Certain difficulties can arise if the chamber has a pH gradient, which has been discussed, for example, in [696] as well as in [695], where the Tavriya experiment methods are described. In that case, special equipment was used, where Fig. 108 shows that there are special units in addition to the chamber for storing the solutions and specimens, pumps for inputting the solutions and specimens, means of withdrawing the fractions, power supply, monitoring units, and stands for locating the equipment.

Under the Tavriya program, the following topics on methods have been handled:

---

[*]As stated in Section 6.2, the topic has also been considered in [646].

   a. Gas-separation principles have been tested, as well as means of introducing buffer solutions and substances to be separated, fraction sampling, and monitoring.

   b. A major line for future design improvement in the chamber should eliminate the factors that deform the fractions.

   c. The most reliable method of separating proteins, in spite of certain complications, is isoelectric focusing in a pH gradient, while the best ways of separating cells are by ordinary electrophoresis and by isoelectric focusing in a self-organizing pH gradient.

In column electrophoresis [696], there is a tendency to assemble equipment, in particular, involving special capillary collectors, electrobuffer circulation, and so on. There may be resolving-power loss on taking fractions from the chamber, so one needs an independent method of recording the process directly on the column or at its exit (without sampling). This can be based on visible or UV absorption as in [699], on thermometric or pH methods, or on holographic interferometry. In [701] it is stated that the KGA-2 holographic system (see Section 6.5) has been used in the Tavriya experiments to record isotachophoresis. Unstained biological preparations have been used with holography on the ground and under microgravity, which has shown that the KGA-2 can record displacement and fraction separation without sampling. The KGA-2 applications have been extended by means of a real-time holographic attachment, which outputs the data to a TV screen, as mentioned in Chapter 6.

9.4. RESULTS ON MICROGRAVITY ELECTROPHORESIS
[15, 695, 696, 699, 703, 707, 709, 712, 713]

In [1] no space electrophoresis experiments were examined, so here we use the table of [699] to characterize the main results obtained in the USA, the Federal Republic of Germany, and the USSR between 1971 and 1983 (Table 6).

In [699] emphasis is placed on the high proportion of unsuccessful experiments (in the initial cases) due to design faults and primitive recording systems. This experience has been utilized since 1982, as research has taken two lines: in the USSR, stationary electrophoresis in the forms of the Tavriya equipment and the Genom apparatus, and in the USA, with free-flow electrophoresis on an apparatus developed by McDonnell Douglas. Up to 1982, the electrophoresis parameters under microgravity were at best at the level of standard ground results. Starting in 1982, results were obtained that essentially could not be produced on the ground as regards resolving power and throughput. Electrophoresis had become a branch of space engineering.

The survey of [699] may be supplemented here with results from some recent papers on microgravity electrophoresis.

In [695, 696] Salyut-7 results under the Tavriya program were reported from several series of studies on separating biological test materials by the zone method. The table gives the materials and the type of electrophoresis already used in [695, 696], and here it may be added that the following essential advantages of microgravity electrophoresis are pointed out in [695, 696]: a) it is possible to fractionate cells, macromolecules, and fragments of them with a degree of homogeneity unattainable under ordinary (ground) conditions, and b) electrophoretic separation

TABLE 6. Microgravity Electrophoresis Experiments [699]

Space-craft	Country	Year	Electro-phoresis type	Materials	Recording	Results	Shortcomings
Apollo-14	USA	1971	Liquid zonal	Dyes, hemoglobin	Photography	Only dyes separated, performance as on ground	Unsatisfactory illumination
Apollo-16	USA	1972	Liquid zonal	Latex, blood cells	Photography	Separation as on ground, convection absent	Pronounced electroosmosis
Skylab	USA	1974	ITP	Ferritin, hemoglobin, blood cells	Photography	Only cells separated, process of zonal type	Leak from column, gas blocked electrodes
EPAS	USA	1975	Liquid zonal ITP	Erythrocytes, lympho-cytes, kidney cells	Photography, column freezing	Erythrocytes separated by isotachophoresis, process unsatisfactory	pH changed, gas blocked electrodes, cell viability low
EPAS	FGR	1975	Free flow	Rat bone-marrow and spleen cells, erythro-cytes	Optical density	Good separation	Excessive illumination, considerable loss of living material

Shuttle-3	USA	1982	Liquid zonal ITP	—	Photography, column freezing	Sharp separation indicated by photographs	Refrigerator failed on returning material to ground
Shuttle-4	USA	1982	Free flow	Three cell-culture specimens	Ground fraction analysis (197 fractions per specimen)	Extremely high performance, yield 400 times higher than on ground at 25% concentration	—
Salyut-7	USSR	1982	Isoelectric; ITP; liquid zonal	In ampholines and borate–polyol system; purified albumin, albumin, hemoglobin, dye, rat bone-marrow cells	Photography	High performance, as in gels on ground	—
Shuttle-6	USA	1983	Free flow	Albumin mixture, cell culture, hemoglobins, and polysaccharide	Fraction analysis	Product purity improved by a factor of 4, amount 700 times larger than on ground	—

Notes. A dash in the corresponding column means that this point is not dealt with in the papers. The table does not include experiments on Salyut-7 with the modified Tavriya equipment and the Pion in 1983 and 1984, nor Shuttle and Spacelab experiments in 1983 and 1984, as there are no descriptions of them in the papers.

TABLE 7. Potential Objects for Studying Electrophoresis under Microgravity [699]

Product	Applications [699]
Pure cell cultures producing insulin (1 year, 3.2 million patients)	Transplantation and pancreases for diabetic patients Overcoming transplant rejection by the use of pure cultures and selecting unrejected cells
Pure stem-cell cultures	Transplants in patients with bone-marrow neoplasia
Pure tumor-cell cultures	Fundamental research on cell membranes
Pure antihemophilia factors (1 year, 15,000 patients)	Treating hereditary hemophilia (blood clotting failure)
Pure B and T cell cultures	Use in anticancer therapy
Kidney cells producing urokinase	Antithrombotic agent
Growth hormones (1 year, 850,000 patients)	Stimulating bone growth in children and juveniles with physical deficiencies and treating yaws
Antitrypsin (1 year, 500,000 patients)	Retarding emphysema progress and accentuating the effects of anticancer preparations
Preparations for treating burns and wounds (1 year, 1.1 million patients)	Burn treatment
Peptide hormones made by solid-phase synthesis	Drug preparations, particularly requiring purification from analogs arising from incomplete reactions at various stages
Interferon (1 year, 20 million patients)	Antiviral and antitumor agent
Interferon and insulin made by genetic engineering;	Preparations suitable after purification from bacterial proteins
Monoclonal antibodies	Suitable after purification from host proteins
Pure immunoglobulins, four classes	Production of monospecific antibodies for diagnosing immune diseases (for example, agammaglobulinemia)
Isoenzymes in preparative amounts	Diagnosing various diseases
Pure virus antigens	Producing antiviral vaccines
DNA restricts	Genetic engineering

Note. The numerical data on the possible yields relate to equipment for free-flow electrophoresis used on the Shuttle and may be of advertising character.

can be applied to biological materials containing large particles such as cells, which under ordinary conditions cannot be fractionated because of mechanical hindrance provided by the supporting medium. A successful Tavriya experiment on electrophoresis (Table 6) is mentioned also in [709].

We may mention here also electrophoretic separation experiments for biological materials on Shuttle 8 (STS-8) [712, 713].

In [712] continuous-flow electrophoresis was used to separate specific secreting cells from suspensions of cultured human kidney cells and rat pituitary cells. The kidney cells were separated into more than 32 fractions. The mobility distribution in the space experiments was wider than in control ground experiments. All the cultured kidney-cell fractions produced urokinase (a kidney plasminogen activator used in medicine to lyse blood clots), with five or six fractions producing substantially more than others.

The pituitary cells were separated into 48 fractions, for each of which measurements were made on the growth hormone (GH) and prolactin (PRL). The fractions were grouped into eight mobility classes and were examined by immunocytochemical methods for various hormones. The hormone patterns showed that the specialized cells that produce GH and PRL are separable because of mobility differences.

In [713] individual fractions were used from primary kidney cells of human embryos obtained by electrophoretic separation on the STS-8 flight, which were successfully cultured after return from orbit. These were found to differ appreciably from one another and from the initial specimen in morphology, electrophoretic mobility, and urokinase type.

9.5. CONCLUSIONS: PROSPECTS FOR MICROGRAVITY ELECTROPHORESIS APPLIED TO PURIFYING AND SEPARATING BIOLOGICALLY ACTIVE COMPOUNDS [699, 704, 708, 710]

These papers deal with prospects for using electrophoresis in separating biological materials. In [708] foreign programs on making biomedical materials in space are examined; in [710] there is a survey of corresponding experiments in the FRG; and in [704] there is a discussion of using electrophoresis in microgravity. Here we summarize these discussions on promising materials for use in weightlessness, where in [699] it is stated that up to 1983 the studies were on models, but since 1983 there have been instances relevant to fundamental research and having value in biomedicine. Table 7 gives a list. In [699], however, it is emphasized that ground techniques in electrophoresis and other physicochemical methods are advancing rapidly, and some topics currently relevant for microgravity electrophoresis may soon be solved on the ground; i.e., space electrophoresis programs must be reconsidered in relation to ground progress.

In [699] it is stated that work is in hand on a large scale to choose electrophoretic materials for microgravity. In response to a notice published in 1978, NASA received about 360 suggestions for biological or biotechnical experiments on the Shuttles, which had a serious basis; e.g., the project from General Electric was oriented to immunobiology and involved the development of three different systems and the evaluation of economic aspects in all stages up to selling the finished diagnostic kits. McDonnell Douglas reported that materials made in space could be sold for $23 billion. It is clear that the manufacture of a commercial free-flow system for pharmacology means that the range of products will be subject to strict secrecy.

To conclude this discussion, it can be said with confidence that there is great scope for developing this section of materials science in space.

Chapter 10

RESEARCH IN MATERIALS SCIENCE
IN SPACE

Here we summarize research in recent years as surveyed in previous sections, and discuss the prospects for further progress in materials science in space.

In each chapter, and also in many sections, particular or final comments have been made and conclusions drawn that can be used in formulating general conclusions on the state of research in materials science in space and future prospects, particularly from the analysis of certain surveys included in Part II of the bibliography, i.e., [24–113]. These can be divided into:

1) papers discussing research on particular topics in microgravity physics and indicating further research lines;
2) review papers giving general surveys of research in materials science in space on various topics and various materials; and
3) papers dealing with general plans and programs for further fundamental research in various countries, as well as with prospects for commercial production in space.

These are discussed, respectively in Sections 10.1, 10.2, and 10.3. The conclusion Section 10.4 gives general conclusions on the entire book.

10.1. ASPECTS OF MICROGRAVITY PHYSICS AND MATERIALS SCIENCE IN SPACE [35, 39, 46, 47, 51, 52, 56, 59, 73, 84, 85, 95, 103, 104, 110]

The survey papers in Part II of the bibliography include some dealing with research on particular manufacturing topics. Most of these have already been cited and partly discussed in appropriate sections, so we restrict consideration to a brief enumeration and indication of the sections.

In Sections 1.1, 1.2, and 1.3 on the theory, there is a discussion of reviews [51, 52] on the convection and heat and mass transfer under weightlessness and ef-

173

fects particularly on crystal growth, while Section 1.4 deals, among other papers, with [30], which relates to controlling liquid surface stability by means of alternating fields, and in Section 1.9 there is a discussion of a review [46] dealing with methods of solving for capillary-liquid statics.

Chapter 2 deals with microgravity production technology for electronic materials, as considered in [56, 103], as well as with the crystallization of multicomponent semiconductors [59]. In [56] in addition to topics on semiconductor manufacturing, which is examined in Chapter 2, there is a discussion of apparatus design, which is considered in Chapter 6.

Making eutectic-structure alloys under microgravity is considered in reviews [59, 110], and is discussed in Chapter 3 (Section 3.4), while the review of [84] on the separation in immiscible alloys is dealt with in Section 3.1. In this same section there is a discussion of the particular topic of Ostwald coalescence in the separation of immiscible liquids, which is analyzed in review [74].

Microgravity growth of protein crystals is dealt with in review [95] and also in a reporter paper [77]; it is analyzed in Section 6.3. In [77] there is also mention of sparingly soluble crystal growth from solution, which is considered in Section 6.2.

In addition, review [104] deals with boundary-layer transport in crystallization from solution under microgravity, with reference to ice crystals growing from aqueous salt solutions. In [47] there is a discussion of making thin films under microgravity and their structure, microporosity, and properties.

Another review of this type is [85], which discusses one of the particular aspects of microgravity physics, viz., isochoric specific heats of liquid substances in the critical region (preparatory studies for Spacelab experiments).

10.2. RESEARCH RESULTS AND SURVEY PUBLICATIONS [15, 25, 26, 34, 38, 48, 54, 57, 58, 60, 61, 66, 67, 69, 75–78, 80, 86, 97, 102, 106, 108, 111, 112]

Survey publications deal with various microgravity experiments on materials science in recent years, where we may particularly note papers on experiments on Salyut-5, Salyut-6, and Salyut-7 not only under the national program but also under the Intercosmos program [15, 25, 26, 34, 37, 48, 54, 57], as well as reviews of experiments performed or prepared for execution on spacelab flights (SL-1, D-1, SL-3) [61, 69, 76–78, 80, 86, 102, 111]. These reviews include the Shuttle experiments and also experiments designed to test apparatus such as the MAUS apparatus on STS-7 and STS-11 [60, 66], or checking halogen-lamp furnaces on the GAS independent module [67]; there are also experiments with engineering apparatus designed in Japan on the Shuttle flights [112]. Furthermore, there are reviews of studies performed during short-term weightlessness with rockets in the USSR [15, 25] and elsewhere [108]. Attempts have been made to draw general conclusions from studies in the USSR and elsewhere in [58, 97] (quite apart from [1]). A survey paper [106] deals with the interest of metallurgists in microgravity research.

There is no need here to give more details on the experiments dealt with in these surveys, since the previous sections have dealt in fair detail with all these results. Also, the concluding sections to the chapters, or closing phrases in certain sections, reflect the main conclusions presented in these surveys.

For example, the conclusions from survey [58] are reflected in detail in Chapter 1 (Section 1.11). The conclusions from the reporter papers at a symposium [20] have also been incorporated above in the appropriate sections, e.g., the contents of [77] are considered in Chapter 5.

Recent space-manufacturing experiments are characterized at least partially in the collections [13, 20], as well as in [15]. The evidence in [13, 15, 20] is reflected in some detail in the present publication and largely characterizes what is new in space manufacturing and what has been attained in recent space experiments.

The bibliography and the contents of this book also show that the results of recent research here are far from limited to the evidence obtained with Salyut-6 [13] and Spacelab-1 [20], being now substantially supplemented by the experiments on Salyut-7 as well as on Spacelab-3 and D-1. Much new evidence has been obtained from rocket experiments, which is reflected in [15, 108] and has been incorporated into Chapter 7. Although the main conclusions from microgravity studies have already been given in the conclusions to Chapters 1–9, which correspond to the conclusions in the above surveys, we may complete this section by giving extracts from two such papers, one from studies in the USSR [15] and the other on experiments on Spacelab [76].

In [15] a survey of Soviet experiments in major fields of materials science in space leads to the conclusion that substantial advances have been made in making various materials and biological substances under microgravity, which demonstrate substantially improved properties. This applies particularly to materials for electronic engineering. Technologies have been developed that give some hope of economically producing certain specific materials in space.

Physics research under microgravity has indicated some new effects that can be used to improve future processes such as solidification in single-crystal materials at extremely high rates, supercooled-liquid solidification, means of monitoring thermocapillary convection, and electromagnetic methods of monitoring liquid-phase compositions. Working principles of engineering equipment have been analyzed (from reliable theoretical considerations) and optimization recommendations have been made.

Soviet space research has led [15] to the conclusion that the time has come for the next step in developing materials science in space, and the next stage has begun, viz., space production of major materials. The main purpose of this stage is to develop microgravity systems for manned and unmanned vehicles. Conclusions similar to those in [15] have been given in other surveys by Soviet researchers [25, 26, 38, 54, 57].

Here we may also give some extracts from an introductory paper [76] at the symposium [20] dealing with materials research in space on Spacelab-1. In [76] SL-1 experiments are considered as representing considerable scientific progress, in spite of the failures with certain experiments mentioned above.

The basic scientific results from SL-1 include [76] protein crystallization [530] (see Section 5.3). The experiments on growing protein crystals are very impressive, and the ESA has ordered a complicated apparatus for the Eureka flight with the agreement of outstanding biochemists, who will examine the crystallization of some new and important proteins. The marked increase in protein crystal growth rate in space is particularly intriguing from the viewpoint of the mechanism. It will be of interest in the future to examine molecular ordering in solution. It is possible that special ordering occurs under microgravity, which reduces the activation energy and increases the growth rate.

In [532] (see Section 5.2), the growth of highly perfect crystals from solution was reported. Around the crystals, there was a zone having a high dislocation density, which arose after return to gravity. In [76] it is emphasized that this result is good evidence for the harmful effects of gravitational convection on crystal perfection.

In [76] and in reporter paper [77] there is a discussion of a successful microgravity crystallization experiment for HgI_2 growing from the vapor and doped with Ar and styrene; this showed that gravity influences not only the growth but also the nucleation, which is greatly facilitated in space.

In [76] it is pointed out that microgravity can be used to examine second-order effects such as thermal diffusion in the absence of first-order effects caused by strong gravitational convection on the ground. This is evident from a successful experiment [485] on the thermal migration of Co in Sn (see Section 3.8). The conclusion is that an analogous experiment might be conducted on isotope separation in liquids. Such an experiment has been formulated for D-1 to elucidate the scope for isotope separation in space.

Successful experiments have also been performed on measuring the diffusion coefficient in molten tin [463] (see Section 3.8). Under microgravity, the diffusion coefficient is less than that on the ground by about a factor of 20 near 800°C; i.e., ground experiments are heavily affected by convection. Space is clearly the only place where one can make exact thermophysical measurements on melts.

The experiments of [376, 384] on microgravity crystallization of Si are also mentioned in [76] as early and interesting space experiments on a semiconductor more refractory than previous ones (see Section 2.1).

Finally, in [76] the experiments on fluid physics on SL-1 lead us to focus our attention [280] on free Marangoni convection, as well as [287] on liquid capillary properties (see Sections 1.3 and 1.9).

The general conclusion in [76] from a brief enumeration of certain successful SL-1 experiments is that although there are some difficulties and failures, on the whole the SL-1 flight provided interesting scientific results, some of which have applications.

In [76] some of these difficulties are enumerated, particularly financial problems, stringent limitations imposed by NASA on experiments with toxic substances, the need to modify the crystal-growing apparatus for use with solutions only a few months before the flight, and the prolonged run-up period before the flight because, for example, growing platinum compounds from solution was unsuccessful due to chemical decomposition. Other factors are that only incomplete data were available on the temperatures in certain experiments. It is important to overcome all these difficulties in subsequent flights. In particular, in order not to hold up developments in space manufacturing right at the start, it is necessary to device methods ensuring safety on operating in space with many materials so far not allowed by NASA.

In [76] it is also pointed out that European countries need to make efforts in order not to be left behind in space technology by the USA and Japan. The ESA members are only just beginning materials science in space with the SL-1 flight, which will be followed by others (SL-3, D-1, IML, Eureka, etc.), which will provide for prolonged experiments for manufacturing purposes. The space-station launch will provide the main scope for this. Section 10.3 deals with prospects and requirements here, including commercial returns.

10.3. FORECASTS, PROGRAMS, AND RESEARCH PROSPECTS IN MATERIALS SCIENCE IN SPACE [27–33, 36, 37, 40–45, 49, 50, 53, 55, 62–65, 68, 70–72, 76–79, 81–83, 87, 89–94, 96–101, 105, 107, 109, 111–113]

This section deals only with programs and development prospects for weightlessness physics and space manufacturing, and does not deal with previous research, although this is partly considered in the publications cited here.

In [27] the following specifications are drawn up for further microgravity research: equipment upgrading for pilot-plant production of certain materials (of the Korund type), and attention to research on heat and mass transfer, fluid mechanics, phase transitions, and so on, under weightlessness. This research should provide the scientific basis for making materials in space and for optimizing particular processes.

The subsequent stages in space manufacturing involve building flying power plants giving elevated power (tens of kilowatts) and independent automatic modules and production systems, periodically visited by spacemen to repair the equipment and bring back the materials.

In [31] the discussion of experiments on the Pion apparatus leads to refined specifications for equipment upgrading for use in weightlessness physics and for improvements in the methodology. One of the most important methodological aspects to be handled in the next stage concerns statistical confirmation. Various methods are available.

One of these is to use completely automatic equipment, which can be used in experiments on automatic spacecraft.

Another is to build multipurpose research apparatus maintained by skilled operators trained in space [75]. In that case, it is sufficient to formulate the purpose of the research, with the operator himself planning the detailed experiments and choosing the working conditions.

It is also possible to organize studies and manage experiments from the flight control center on the basis of data, including video data, supplied on-line from the spacecraft. One can then make extensive use of ground computing systems not only in experiments but also in managing engineering processes. Only properly planned mathematical studies and experiments on manned or automatic spacecraft can assist in careful ground development associated with making materials under microgravity in order to solve important problems in space technology.

In [50] it is suggested that soon there will be a transfer from long-lived stations of the Salyut type, which are staffed by rotating teams, to a multilink constantly manned orbital complex, which is a single system of large structures placed in orbits from 200 to 400,000 km and linked to the ground by transport vehicles. These will evidently contain specialized research laboratories, convenient living accommodations, high-power electricity sources, workshops, and possibly also constructional areas for making and installing typical components.

Large-scale economic tasks are in part formulated in [50], where the future orbital complex should have various facilities so far not available in space science. The sizes of some of these objects may be striking even for ground construction. For example, a parabolic antenna should have an effective aperture of 300–350 m. Also, in order to justify it economically, the design should be suitable for use for at least 15 years. Finally, there are exceptionally tight specifications for the manufacturing accuracy and for the stability in the geometrical and working characteristics throughout the life.

In [50] Soviet studies are enumerated that to some extent prepare for building such an orbital complex, in particular experiments on welding in space and depositing coatings, which have shown that structures can be repaired in open space and film coatings can be restored with a quality satisfying the requirements of the tightest ground standards. Ground tests on the future orbital complex may be accompanied by tests on small-scale ones based on general physical models representing orbital flight conditions. The behavior of these in open space may be examined over a period of 2–5 years, from which one can forecast reliably the states of the elements for 20 years or more. These models are represented by multilayer film composites, for which manufacturing and testing techniques have been under development on the ground in the USSR for some years. In [50] it is stated that related topics in the normal operation of orbital complexes cannot be handled without further fundamental studies, some of which are enumerated. These undoubtedly include studies in weightlessness physics and space manufacturing.

In light of [50], we may note [32], which discusses problems in making all such components. Here we restrict consideration to making liquid films on rigid frameworks under microgravity. Demonstration experiments on coating wire frameworks with liquid films were performed by American spacemen on Skylab; it is possible to make long-lived films under microgravity not only from soapy water but also from pure water, which is impossible on the ground. The microgravity films made from liquid metals and nonmetallic materials after stabilization and setting may constitute constructional elements in conjunction with the framework. A film construction on a cube framework shows that it can work properly in compression, torsion, and bending. In [32] two ways of making films on frameworks are indicated: 1) immersing the framework in a liquid (as on Skylab) and 2) stretching an initial body of liquid by expanding the framework.

We do not deal here with the conclusions on research on space manufacturing in surveys [28, 29, 30, 33, 37, 50] because, on the whole, they agree with the conclusions given in Section 10.2 from [15], as well as with the conclusions from [27, 31, 32]. Certain reviews [49, 53, 96, 97] also do not need to be considered because the evidence from them has been used not only in [1] (in particular pp. 150–152) but also because the topics are covered under the Intercosmos program and are dealt with in this book, since the author of [1] and of the present book is also an author or coauthor of [49, 53, 96, 97].

As regards [40, 41], it is sufficient to say that these deal with the principles of diffusion welding, where it is asserted that this is the space technology of the future.

We complete this survey of Soviet survey publications on this topic with a brief indication of long-range forecasts on industrial use of space and systems principles in space industrialization, which are discussed in [36, 42–45]. Those papers deal with the period from 1985 up to 3000, which is divided into 12 stages, for which the most reliable forecasts are for three stages whose realization requires about 25 years. We consider here the forecasts only on the first two stages, which have a bearing on space manufacturing. In stage I, 1985–1990, it is predicted that experimental production of improved materials will begin. In stage II, 1990–2000, it is suggested that space equipment and power plants giving 25–100 kW will represent a new generation of such systems, while there will be extensive use of space data systems and the industrial production of certain materials.

TABLE 8. Apparatus for the Recoverable EURECA Platform for Research in Materials Science in Space [101]

Multipurpose systems	Mass, kg	Area occupied on EURECA, mm	Height, mm	Mean power, W	Necessary data transmission rate, bit/sec
Automatic monoellipsoid mirror furnace facility (AMMFF)	200	1400×700	900	500 in treating specimens (including 150 for power supply and electronics)	300
Solution crystal growth facility (SGF)	90	700×700	350	From 10 to 60 in accordance with external temperature	100
Protein crystal growth facility (PCF)	100	700×700	1000	150	100
Multizone furnace suite (MFA)	150	700×700	1000	300	200
Automatic gradient furnace (AgHF)	160	1400×700	1000	400	200
Botanical facility (BF)	160	1400×700	90	150	100

We now consider plans and development forecasts for materials science in space made in Western Europe, Japan, and the USA on the basis of the foreign publications cited above [62–65, 68, 70–72, 76–79, 81, 82, 87, 89–94, 98–101, 105, 107, 109, 111–113].

We begin with the plans of the European Space Agency (ESA) on space-laboratory flights, which include manufacturing experiments [101]. The Spacelab-1 results have been discussed in detail here, and in accordance with ESA plans, the D-1 flight was performed in the fall of 1985. According to [72], out of 76 experiments planned for the D-1 (indicated partially in [1], Table 13, pp. 162–176), only one was unsuccessful. The D-1 results were discussed at a special symposium at the end of 1986. Several papers were also presented at the 36th Congress of the International Astronautical Federation [23]. The ESA plans for materials science in space have thus far been successful.

Following the D-1 flight in 1987, NASA collaborated with ESA in planning an international microgravity laboratory (IML) [89, 101] as well as the flight of a European retrievable carrier (EURECA), as mentioned in [1]. The experiments are being prepared in parallel with the development work on the platform, and according to [68], the test equipment for it should be available in the spring of 1987 from the firm of MBB/ERNO. In [101] the main characteristics are given for the EURECA and IML modules, with a list of the scientific equipment and the research purposes. In [62] there is a discussion of the payload prepared in the FGR. In [100, 101] there are also characteristics of the apparatus and experiments to be performed. We now give some information on these as regards materials science in space from [101].

We begin with the characteristics of the EURECA payload [81, 100, 101]. Although in [1] there is a partial description of this apparatus (see Table 8 on pp. 114 and 115, as well as Figs. 36–44 on pp. 119–124 in [1]), WE give some information from [101] (in [81] there is also some information on the flight load).

EURECA will be launched into a 500-km orbit for about 6 months and returned by Shuttle. In [101] the following advantages of the platform by comparison with the SL flights are enumerated:

1) 6 months' duration instead of about 1 week;
2) better microgravity conditions, viz., 10^{-5} g_0 instead of 10^{-3}–10^{-4} g_0 on SL;
3) purer atmosphere and better vacuum at 450–500 km;
4) simpler safety specifications, as the platform is unmanned;
5) vacuum seals are not required for the system;
6) toxic gaseous materials can more readily be used on recoverable platforms.

On the other hand, there are disadvantages:

1) automation complicates the systems, and
2) critical command intervention is impossible on a free-flying platform.

The DFVLR Agency in the FRG has organized a special ground laboratory for researchers preparing these experiments and has invited all the participants from the various West European countries to use these facilities. Here we may note that a ground laboratory for research in weightlessness physics and materials science in space was opened by NASA at the Lewis Research Center [87, 92].

West European materials scientists are also interested in the IML, for which the first launch is intended for mid-1987, with subsequent launches according to [101] occurring at intervals from 12 to 24 months. NASA has provided 50% of this recoverable vehicle for European equipment and proposes to combine the efforts of materials scientists in the USA and Europe in a world society on materials science in space.* NASA and ESA are discussing the IML statutes and program. The preliminary indication is as follows:

1) to support researchers throughout the world and to collaborate in international microgravity research;
2) to use Spacelab with the equipment in it made by participating countries and organizations;
3) NASA provides flight integration and costing;
4) NASA allows others to use 50% of the equipment it does not use;
5) the program includes at least three flights with intervals from 12 to 24 months. The first flight is in mid-1987;
6) if NASA is not interested in an apparatus, the participants will refund to NASA the costs arising from the Shuttle payment principles;
7) ESA will supply a double rack concerned with fluid physics and a biological rack (BIORAK–ANTHRORACK); and

*The European Low Gravity Research Association (ELGRA) is a union of European scientists interested in using microgravity for materials science in space (see [91] for the structure, statutes, and plans of ELGRA, which is supported by ESA), which is intended to link together all the scientific bodies in various countries concerned with microgravity physics into an international federation or union on microaccelerations (IFMA or IUMA).

8) a specialist on the payload may be among the crew if the equivalent equipment in the double rack is supported.

ESA has made the following proposals: 1) to consider EURECA in the same framework, 2) to reserve part of IML for industrial users, and 3) to use IML for student experiments.

These plans indicate that launching reusable platforms will be important not only for research purposes but also commercial purposes. In [79] there is a discussion of using such platforms launched and recovered by Shuttle for processing materials (plans up to 2010 involving the creation of a space factory). The payload in the orbital platforms can be increased to 3.5 tons, power supply to 15 kW, and flight duration up to 200 days. It should be possible to reduce the cost of making materials on such platforms considerably if they are supported by future orbital stations.

NASA has already launched an unmanned module for prolonged experiments, the long-duration experimental facility (LDEF), on which various experiments have been performed, in particular on synthetic metals [528], as mentioned in Section 5.1.

In [101] the description of the EURECA and IML programs is followed by a discussion of ESA microgravity plans designed to extend research in materials science in space in Europe with the maximum possible use of existing apparatus, which should support fresh Spacelab and rocket flights, particularly for new ESA members.

Future plans include building a space station in the 1990s for materials research, where prolonged experiments could be performed, e.g., on crystal growth from vapor, and also ultimately for industrial production in space.

Here we return again to [76] (see the end of Section 10.2), in which SL-1 results led to a discussion of preparations for future research on space stations.

In [76] it is emphasized that materials scientists are not yet ready to deal with topics that need to be resolved before the station launch. The space-station project according to [76] will advance science and technology immensely. A real problem envisaged by materials scientists is commercialization. At present, there are few projects that can be used for production in space. This stage has not yet been reached even with the very interesting SL-1 results, since it is stated that we will not be ready to transfer to a space station in less than 10 years. Many space-materials scientists in the USA and Europe fear that the agency during this period will spend the money mainly on equipment and will give little support to ground research for new experiments or testing fundamental aspects involved in transfer to space production. This danger cannot be overestimated and should be strongly emphasized. The extent of the ground research should be sufficient for proper use on the station. Otherwise, new experiments will not be properly prepared when the station is ready, nor will there be serious space manufacturing projects. There is a considerable danger that the agency will consider as serious materials-science proposals only those that have already been proposed from time to time, whereas in fact materials science is a vast field of activity with enormous potential. The reasons why this potential has not yet been realized according to [76] are as follows:

1. The lack of necessary financial support to ground research.
2. The lack of good power supplies in space. So far, research has been based only on fusible model materials and small amounts. It is proposed that the space station should use 60 kW for materials science.

3. Major parameters such as temperature and pressure are not stable in space. The best experiments on crystal growth and nucleation require a temperature stability of about 0.01°C, which should be borne in mind in designing the space-station modules. In [76], the following conclusions are drawn:

   1) First, if the governments of the various countries wish to produce materials in space, they must increase the budgets for applied and ground research decisively.

   2) Second, when one designs materials science modules for the space station, one should use the high-grade systems employed on SL-1. However, it must not be forgotten that these were based on the requirements of the 1970s, while the space station should meet the needs of science in the 1990s and later years. Consequently, the space-station module design should be different.

The ESA materials science work group has formulated the following recommendations on the design of this module for a future space station. The final words [76] are: materials science in space is in its infancy. As in any new discipline, it is encountering skepticism from established sciences, particularly because space equipment costs so much. The early SL-1 results are very hopeful and provide an impetus to policy for developing this science. The success will be dependent on how far materials science in space is supported in the near future.

In addition to [76], the data from the symposium [20] on the SL-1 results included [71] (see also [72]), which contained some interesting statements from participants in various areas of materials science and condensed-state physics on the current and future research in microgravity. There is no point in reproducing these statements in detail here, but they indicate some definite progress in research on microgravity physics and the need for further detailed studies in this area as already reflected in this book but not mentioned here (for example, the proposal to use microgravity for studying plastic strain near the melting point). There are also general arguments and, in particular, the proposal to organize a microgravity society, the need to organize links with scientific societies generally, and even with one's colleagues, i.e., communication problems, as communication is obligatory in a large area such as microgravity science, where one needs to link together not only scientists but also funding sources and politicians, who need to be convinced of the value of research conducted in space.

This survey of papers on plans and development prospects may be completed by enumerating certain publications without characterizing the contents in detail because they largely reproduce what has been said above, particularly to emphasize the extensive interest in this area in various countries.

We have noted above the activities of ESA members in the area of materials science in space, and here we may also note publications on the activities of DFVLR in the FRG [88], especially the expansion of the MAUS* program [107], and a paper [109] on results and research plans with rockets, studies by French researchers under the MEPHISTO program [70, 94], and surveys by Italian researchers [63, 90, 91, 98]. Swedish plans have been mentioned above (see [677] and Chapter 7). Interest in space research and development prospects there in Japan can be judged from [82, 83, 93, 105].

---

*See the abbreviations list in [1], p. 8.

All the West European and Japanese plans for expanding research in materials science in space (apart from experiments with short-term weightlessness) are closely related to the plans of NASA in the USA, as is evident from the above description of the EURECA and IML projects, as well as preparations for the future space station. Examples are provided not only by the links of ESA and NASDA with NASA but also by contracts signed between European firms and NASA [65]. In [113] an example is given of an agreement of a US firm with NASA. Other examples of this type have been given in Chapter 9.

In all the surveys without exception included in Part II in the bibliography [24–113], as in all the papers cited here, the general view is expressed that further fundamental research is required on the physics of weightlessness with manned and unmanned vehicles (orbital stations and platforms), as well as with short-term weightlessness on rockets, aircraft, and towers. Almost all statements on the prospects for space manufacturing are optimistic, i.e., on space industrialization.

10.4. CONCLUSIONS

The general conclusion is that the 3–4 years covered by the papers cited here have seen many new studies in weightlessness physics and materials science in space.

The conclusions section in each chapter deals with the results of these studies. The following are more briefly the conclusions chapter by chapter.

1. Theoretical developments in materials science are dealt with in Chapter 1 (see the section list for that chapter) and demonstrate considerable advantages. On the other hand, theoreticians should be involved more in discussing future experiments under microgravity, and they should take an active part in suggesting experiments that would advance the theoretical principles of materials science in space and the solution of fundamental problems in condensed-state and weightlessness physics.

2. Experiments on growing semiconductor crystals under microgravity (Chapter 2) have confirmed some previous conclusions (have improved the statistical basis) and have provided some new and interesting results:

1) They have confirmed that $A^{III}B^V$, $A^{II}B^{VI}$, and $A^{IV}B^{VI}$ crystals having improved impurity distributions and less defectiveness can be grown under microgravity, in particular GaAs.
2) They have shown that convection can be suppressed to reduce nonstoichiometry as well as dopant inhomogeneity.
3) They have shown that bubble formation can be prevented for certain materials growing under microgravity (GaP).
4) They have provided new results on ternary-crystal growth under microgravity, which confirm the promise from further research designed to produce crystals of this type of considerable practical importance in forms more perfect than those attainable on the ground.

Other important results include the formulation of certain unresolved or debatable aspects for which further work is required. These include elucidating the following:

1) the effects of nongravitational convection on dopant distribution in doped-crystal growth, as well as on composition homogeneity and stoichiometry deviations in binary and ternary semiconductor compounds;

2) features of nonequilibrium phenomena (in particular, supercooling) under microgravity, which may adversely affect the homogeneity;

3) detailed effects from partial contact with tube walls when the liquid does not wet them; and

4) various unresolved and debatable aspects concerning details of crystal growth from the vapor under microgravity, particularly when it is necessary to consider physical and chemical gas transport.

These fundamental problems and others, if solved, will provide a basis for the commercial production of some semiconductors in space, as well as for improving techniques for making them on the ground.

3. Some new and interesting results have also been obtained on metal, alloy, and composite solidification under microgravity (Chapter 3). These new microgravity experiments have been used in:

1) detailed analysis of immiscible-alloy separation mechanisms and the scope for reducing or even eliminating macroscopic separation under microgravity;

2) in research on composite melting and crystallization under microgravity (in new experiments), where a difference from previous experiments is that direct *in situ* observation has been used in space with droplets and bubbles, which has provided valuable information on composite solidification under space conditions;

3) in producing foam metals in space, where it has been shown that microgravity favors bubble nucleation, which reduces the bubble size, reduces the formation energy, and increases the rate. It is concluded that future research should deal with detailed processes in nucleation, growth, and agglomeration, as well as bubble elimination;

4) in research on eutectic crystallization, various and sometimes conflicting conclusions have been reached, which may be explained as due to differences in impurity composition in the materials used by the different workers. Unambiguous and statistically sound conclusions required new experiments on single materials having definite impurity compositions;

5) an earlier conclusion has been confirmed that microgravity has an appreciable effect on the structures and physical properties of superconducting and magnetic materials;

6) it has been concluded that advances in fundamental concepts on dendritic and cellular crystallization for alloys require experiments under microgravity;

7) it has been found that capillary forces play a particular part under microgravity, especially in shaping methods involving crystallization with free surfaces such as Stepanov's method;

8) it has been found that microgravity experiments give valuable information on the physical constants in mass transfer such as diffusion, thermal diffusion, and electrical transport. Such information is difficult or impossible to obtain under ground conditions, so future use of microgravity is required to make measurements in the absence of gravitational convection on ther-

mophysical and physical parameters, particularly to test theories of heat and mass transfer; and

9) considerable advances have been made in welding and soldering in space from studies on the fundamental trends under microgravity. Thermal evaporation and condensation methods have been used to produce numerous unique coatings under microgravity, whose properties indicate that repairs can be carried out in orbit. Methods have been devised for producing special thin-walled welded structures, and the reasons for burns during welding have been established.

4. In new research on glass solidification under microgravity (Chapter 4), an important part is played by solidification in supercooled alloys, nucleation, and metallic-glass formation. Microgravity has provided new and valuable information on these processes as well as on diffusion in silicate glasses. Further space experiments should provide evidence on differences in critical cooling rates under microgravity by comparison with ground conditions, as well as information on surface and diffusion phenomena in glasses without interference from gravity. Advances in glass research under microgravity require new equipment providing higher temperatures, more rapid cooling, and stable levitation. Major interesting programs are being executed or planned to resolve substantial problems concerning glasses.

5. The new results on crystal growth from aqueous solution under microgravity (Chapter 5) include some interesting results on organic crystals having metallic conductivity, where a novel three-chamber reactor was used with separation by valves with special filters, which eliminate convection when the valves open. These synthetic-metal crystals have been grown on an unmanned NASA module for prolonged experiments (the LDEF). Successful experiments have also been performed on making sparingly soluble crystals in a three-chamber reactor.

Growing protein crystals under microgravity has been particularly successful. In space, it is possible to make much larger and more perfect protein crystals, which are also purer, and much more rapidly at that. Further growth experiments will therefore undoubtedly be undertaken.

Fundamental research will also be undertaken soon on the reasons for the rapid protein crystallization under microgravity (the cluster hypothesis). Also, mass-crystallization mechanisms and kinetics may be examined *in situ* by direct methods such as holography.

Crystal growth under microgravity, whether from solutions or from molten semiconductors and metals, has thus provided not only new interesting and hopeful results, but has also raised some problems that need attention in future theoretical and experimental studies.

6. Apparatus for research on the physics of weightlessness and materials science in space (Chapter 6) requires ongoing improvement. It is already necessary to envisage automatic equipment and subsequent generations of methods for unmanned platforms and new orbital stations, as well as apparatus requiring higher power inputs and for organizing the production of certain materials in space. Considerable advances have been made recently here, and results have been obtained on the following lines:

1) from detailed studies on the characteristics of existing high-temperature furnaces (temperature patterns) under microgravity, where new designs had been developed. There is a gradual transfer to automatic heating devices capable of operating on unmanned modules, as well as higher-power

systems providing higher temperatures and operating with larger specimens, which approach industrial sizes;

2) further improvements to apparatus and methods for research on fluid physics under microgravity on the basis that experiments with this apparatus not only serve to resolve fundamental problems, but also provide recommendations on optimizing processes for making new materials under microgravity. Improvements are also being made to apparatus for growing crystals from solution, including special thermostats and cryostats for making protein crystals; new types of three-chamber reactors and valves for them are being devised, and new methods of observing crystal nucleation and growth in solution;

3) there are ongoing improvements in equipment for cutting and welding, and also for depositing films in space, where new generations of equipment are being designed, and new types of devices are being proposed;

4) there is ongoing development in theoretical principles and equipment for levitation under microgravity, such as electrostatic, electromagnetic, acoustic, ion-plasma, and ablation types, or by means of gas flows. There has recently been considerable progress here;

5) new small and light holographic systems for space experiments, which have already been used several times under microgravity and which have been recommended for further use in fluid physics, crystal growth, and other space experiments;

6) improved microacceleration meters have been designed which have been used in numerous studies on gravitational conditions on space vehicles; and

7) there have recently been discussions and developments in other methods of examining structure and property changes in condensed media under microgravity: spectroscopic, x-ray topographic, and optical, including the use of TV techniques, as well as other methods relevant to materials science in space, such as the use of vacuum on vehicles, skin technology, and so on.

Advances in apparatus and methods for materials science in space make it possible to carry out many experiments and can support the production of materials in space. Some methods devised for space can be useful also on the ground in new technology.

7. The short-term weightlessness provided by towers, aircraft, and (especially) rockets (Chapter 7) has continued to be used in numerous experiments not only in preparing for experiments under prolonged microgravity on space vehicles, orbital stations, and unmanned platforms, but also for purposes of independent significance in materials science in space. Results published recently on short-term microgravity conditions confirm the interest in experiments on weightlessness physics under such conditions. In particular, the Mir and Texus rockets have been equipped with new apparatus providing good scope for space materials research.

Experiments on short-term weightlessness play a considerable part in advancing our concepts of weightlessness physics and have made a substantial contribution to general advances in materials science in space, including the development of automatic equipment.

One expects that in the future experiments under short-term microgravity will be useful in spite of the increasing scope for prolonged experiments in space on manned or unmanned vehicles.

8. Here, as in [1], data have been given on new research not only on microgravity but also elevated gravity as regards crystal growth and alloy solidification (Chapter 8). The significance of this research is that gravitational convection and wall-interaction effects should be substantially increased at elevated gravity, which provides additional information on how gravitational and capillary forces are involved in crystal formation.

During the last 2–3 years, new methods have been devised for studying crystal growth and metal setting under elevated gravity, in which simple but massive growth apparatus has been installed on a large centrifuge giving accelerations up to $30\ g_0$. The method has been tested, and the first experiments have been performed on crystallizing Ag-doped PbTe as well as Te, Te–Se, and Te–Si materials, in addition to the solidification of Al–Cu eutectic alloys at various gravity levels. Unexpected results have been obtained on crystallization and dopant distribution changes occurring as gravity increases, which should be the subject of future, more detailed research.

There are several experimental and theoretical problems requiring elucidation in which this method of altering the gravity level can be used to examine effects on crystal growth and the solidification of metals, composites, and glasses.

9. New and interesting results have recently been obtained, with prospects for industrial use, in electrophoresis during weightlessness (Chapter 9). New equipment efficient for space use has been devised. Electrophoresis theory for microgravity has also been developed.

Up to 1982, the results on electrophoretic separation in space were not better than ground results, but, since 1982, results have been obtained essentially unavailable on the ground as regards resolving power and throughput. Electrophoresis has become a branch of space engineering. Up to 1983, the studies were conducted on models, but since 1983, they have been done on materials of interest to fundamental research as well as materials of value in biomedicine. However, ground electrophoresis techniques are advancing rapidly, and various aspects currently important for microgravity electrophoresis may soon be resolved on the ground; i.e., space electrophoresis programs should be reconsidered in light of ground advances. Nevertheless, there are forecasts that preparations made in space will find a market and will be economically producible.

On the whole, it seems clear that there are good prospects for this division of materials science in space.

10. Points 1–9 above deal briefly with new research in materials science in space and with arguments advanced by materials scientists on the prospects for advances in weightlessness physics and materials science (Chapter 10, particularly Section 10.3), which lead on the whole to optimistic forecasts on the future of this science.

There are the following trends in materials science in space at present:

1) vigorous developments in fundamental research on weightlessness physics accompanied by comprehensive preparations for the commercial production of certain materials;

2) major trends to performing long-term experiments on unmanned platforms, together with shorter studies on manned vehicles, as well as on rockets, aircraft, and towers;

3) preparations for work with new orbiting stations, in which materials research will be supported by higher available power levels in experiments and in future industrial systems;

4) in some countries, special ground research laboratories have been set up for materials science in space not only to prepare experiments for space but also to undertake synchronous ground experiments. The main specifications for such ground research have been formulated, but more attention should be paid to them, including financial support; and

5) there is a tendency to wide-ranging discussion of the desirability of new space experiments based on involving researchers from many branches of science, both theoretical and experimental.

We have thus discussed the new features in materials science in space as reflected in recent papers and have pointed out many major results obtained in the last 3–4 years, where we may emphasize that there are many new problems waiting for research in weightlessness physics and materials science in space, which need to be solved in the next few years.

This book shows that there is a need for further strengthening of research on weightlessness physics and materials science in space, as these two provide good examples of the peaceful and fruitful use of recent advances in the use of space. If one combines the efforts of researchers in various countries in this area and sets up appropriate international societies, the result may be considerable acceleration of progress in this branch of science.

It is clear that there will be ongoing advances in materials science in space and in ground laboratories on the basis of future research on weightlessness physics and materials science in space.

REFERENCES

I. BOOKS, COLLECTIONS, AND CONFERENCE PROCEEDINGS

1. L. L. Regel', *Materials Science in Space: Advances in Science and Technology, Space Research* [in Russian], Vol. 21, Izd. VINITI, Moscow (1984); L. Regel, *Sciences des materiaux dans l'espace. Theorie–experiences–technologie: technique et documentation* [French translation], Lavoisier, Paris (1985); *Materials Science in Space: Theory, Experience, Technology*, Halsted Press (1987).
2. Collection: *Space Research in the Ukraine* [in Russian], Naukova Dumka, Kiev, No. 16 (1982); No. 17 (1983); No. 18 (1984); No. 19 (1985).
3. Collection: *K. É. Tsiolkovskii and Space Manufacturing Problems: Proceedings of the 16th Meeting Dealing with the Scientific Consequences and the Development of Tsiolkovskii's Ideas*, Kaluga, September 14–17, 1981, Section: K. É. Tsiolkovskii and Space Manufacturing Problems [in Russian], IIET AN SSSR, Moscow (1982).
4. Collection: *Space Prospects and Problems: Proceedings of the 17th Meeting Dealing with the Scientific Consequences and the Development of Tsiolkovskii's Ideas*, Kaluga, September 15–17, 1982, Section: K. É. Tsiolkovskii and Space Manufacturing Problems [in Russian], IIET AN SSSR, Moscow (1983).
5. Collections: *Tsiolkovskii's Ideas and Space Manufacturing Problems: Proceedings of the 18th Meeting Dealing with the Scientific Consequences and the Development of Tsiolkovskii's Ideas*, Kaluga, September 13–17, 1983, Section: K. É. Tsiolkovskii and Space Manufacturing Problems [in Russian], IIET AN SSSR, Moscow (1984).
6. Collections: *Tsiolkovskii's Ideas and Space Manufacturing Problems: Proceedings of the 19th Meeting Dealing with the Scientific Consequences and the Development of Tsiolkovskii's Ideas*, Kaluga, September 1984, Section:

K. É. Tsiolkovskii and Space Manufacturing Problems [in Russian], IIET AN SSSR, Moscow (1985).

7. Collection: *Fluid Mechanics and Transfer Processes under Conditions of Weightlessness* [in Russian], UNTs AN SSSR, Sverdlovsk (1983).

8. Collection: *Engineering Experiments under Conditions of Weightlessness* [in Russian], UNTs AN SSSR, Sverdlovsk (1983).

9. Collection: *Fluid Mechanics and Space Research* [in Russian], Nauka, Moscow (1985).

10. *Abstracts of the 3rd All-Union Seminar on Fluid Mechanics and Heat Transfer under Conditions of Weightlessness* [in Russian], Chernogolovka (1984).

11. Collection: *The Gagarin Lectures on Space Flight and Aviation, 1983 and 1984* [in Russian], Nauka, Moscow (1985).

12. Collection: *The Gagarin Lectures on Space Flight and Aviation, 1985* [in Russian], Nauka, Moscow (1986).

13. Collection: *Salyut 6–Soyuz: Materials Science and Technology* [in Russian], Nauka, Moscow (1985).

14. V. S. Avduyevsky, S. D. Grischin, L. V. Leskov, V. I. Polezhayev, and V. V. Savitchev, *Scientific Foundations of Space Manufacturing: Advances in Science and Technology in the USSR, Technology Series* [in Russian], Mir, Moscow (1984).

15. V. S. Avduyevsky, S. D. Grishin, L. V. Leskov, V. V. Savitchev, and V. T. Khryapov, *Manufacturing in Space: Processing Problems and Advances. Advances in Science and Technology in the USSR, Technology Series* [in Russian], Mir, Moscow (1984).

16. *Abstracts of Papers of the 25th Plenary Meeting of COSPAR*, Graz, Austria, June 25–July 11, 1984.

17. *Abstracts of Papers of the 26th Plenary Meeting of COSPAR*, Toulouse, France, June 30–July 11, 1986.

18. *The Effect of Gravity on the Solidification of Immiscible Alloys: Proceedings of an RIT/ESA/SSC Workshop*, Jarva Krog, Sweden, January 18–20, 1984.

19. *5th European Symposium: Material Sciences under Microgravity. Results of Spacelab-1, Abstracts of Papers*, Schloss Elmau, West Germany, November 5–7, 1984.

20. *5th European Symposium: Material Sciences under Microgravity. Results of Spacelab-1*, Schloss Elmau, West Germany, November 5–7, 1984, Paris (1984).

21. *Abstracts of Papers of the 35th Congress of the International Astronautical Federation*, Lausanne, October 8–13, 1984, IAF-84.

22. *Abstracts of Papers of the 36th Congress of the International Astronautical Federation*, Stockholm, Sweden, October 7–12, 1985.

23. *Abstracts of Papers of the 37th Congress of the International Astronautical Federation*, Innsbruck, Austria, October, 1986.

II. REVIEW PAPERS, EXPERIMENT PROGRAMS, AND GENERAL ASPECTS OF MATERIALS SCIENCE IN SPACE

24. V. S. Avduevskii, "Space should be peaceful," *Zemlya Vselennaya*, No. 5, 6–11 (1984).

25. V. S. Avduevskii, S. D. Grishin, and L. V. Leskov, "Ten years of space engineering," in *The Scientific Creativity of K. É. Tsiolkovskii and the Development of His Ideas* [in Russian], Moscow (1984), pp. 120–125.
26. V. S. Avduevskii, V. V. Gorbatko, S. D. Grishin, V. M. Zholobov, A. N. Lobachev, and V. V. Savichev, "Engineering experiments on the Salyut-5 station," in *Tsiolkovskii's Ideas and Modern Scientific Problems* [in Russian], Nauka, Moscow (1984), pp. 44–49.
27. V. S. Avduevskii and L. V. Leskov, "Problems and prospects in space technology," in *The Gagarin Lectures on Space Flight and Aviation, 1983 and 1984* [in Russian], Nauka, Moscow (1985), pp. 28–33.
28. V. S. Avduevskii, "Space engineering and technology," in *Hydroaeromechanics and Space Research* [in Russian], Nauka, Moscow (1985), pp. 24–31.
29. V. S. Avduevskii, "Basic purposes of research on fluid mechanics and on heat and mass transfer under conditions of weightlessness," *Izv. Akad. Nauk SSSR, Ser. Fiz.* **49**, No. 4, 627–634 (1985).
30. V. S. Avduevskii and L. V. Leskov, "The technology in setting up new engineering processes: the technology for making materials in space," in *The Scientific Principles of Advanced Engineering and Technology* [in Russian], Mashinostroenie, Moscow (1985), Chapter 3, pp. 134–155.
31. V. S. Avduevskii, M. S. Agafonov, S. D. Grishin, V. P. Levtov, and L. V. Leskov, "Problems in experimental research on the fluid mechanics of weightlessness," in *The Gagarin Lectures on Space Flight and Aviation, 1985* [in Russian], Nauka, Moscow (1986), pp. 39–45.
32. I. V. Barmin, A. V. Egorov, V. G. Maslennikov, and A. V. Korovin, "Making materials under microgravity conditions," in *Tsiolkovskii's Ideas and Space Manufacturing Problems: Proceedings of the 18th Meeting Dealing with the Scientific Consequences and the Development of Tsiolkovskii's Ideas*, Kaluga, September 13–17, 1983, Section: K. É. Tsiolkovskii and Space Manufacturing Problems [in Russian], IIET AN SSSR, Moscow (1984), pp. 136–143.
33. I. V. Barmin, N. A. Bezdenezhnykh, V. A. Briskman, G. Z. Gershuni, E. M. Zhukhovitskii, V. G. Kozlov, A. P. Lebedev, V. I. Polezhaev, G. F. Putin, A. F. Pshenichnikov, and A. S. Senchenkov, "The program for experiments on an apparatus for examining hydrodynamic phenomena under conditions of weightlessness," in *Abstracts of the 3rd All-Union Seminar on Fluid Mechanics and Heat Transfer under Conditions of Weightlessness* [in Russian], Chernogolovka (1984), pp. 121–124; *Izv. Akad. Nauk SSSR, Ser. Fiz.* **49**, No. 4, 698–707 (1985).
34. J. Bartel and R. Kuhl, "The contribution of the GDR to the Interkosmos program of materials research in space with the Salyut-6 orbiting station," *Kosm. Issled.* **23**, No. 5, 783–791 (1985).
35. N. A. Bezdenezhnykh, V. A. Briskman, A. A. Cherepanov, and M. G. Sharov, "Controlling the surface stability of a liquid by means of alternating fields," in [7], pp. 37–56.
36. S. D. Grishin and L. V. Leskov, "The basic stages in the development of space manufacturing," in [4], pp. 3–7.
37. S. D. Grishin, L. V. Leskov, and V. V. Savichev, "Problems in the physics of weightlessness and prospects for space manufacturing," in [3], pp. 18–22.

38. V. S. Zemskov, L. I. Ivanov, E. M. Savitskii, M. R. Raukhman, B. P. Mikhailov, V. N. Pimenov, I. N. Belokurova, R. S. Torchinova, M. I. Bychkova, and V. N. Meshcheryakov, "Basic results from experiments under conditions of weightlessness and some problems in materials science in space," in [10], pp. 115–117; *Izv. Akad. Nauk SSSR, Ser. Fiz.* **49**, No. 4, 673–680 (1985).

39. V. S. Zemskov, M. R. Raukhman, and E. A. Kozitsyna, "Features in the crystallization of multicomponent alloys under conditions of weightlessness," in [6].

40. A. F. Kazakov and A. G. Braun, "Diffusion welding: a future space technology," in [6].

41. N. F. Kazakov, S. P. Rusin, and V. A. Kazakov, "The prospects for using diffusion welding in space," in [6].

42. L. V. Leskov, "Forecasting industrial uses of space," in [5], pp. 3–13.

43. L. V. Leskov, "A systems principle for analyzing prospects for industrializing space," in [6].

44. L. V. Leskov, "Physics prospects and the industrial utilization of space," in [6].

45. L. V. Leskov, "Space industrialization: the next millenium," *Nauka Zhizn'*, No. 6, 24–28 (1985).

46. A. D. Myshkis and L. A. Slobozhanin, "Methods of solving problems in the statics of a capillary liquid," in [7], pp. 6–27.

47. V. P. Nikitskii and G. V. Zhukov, "Thin films in space engineering and technology," in [5], pp. 78–81.

48. Yu. A. Osip'yan and L. L. Regel', "Some results from research in materials science in space," in [13], pp. 5–11.

49. Yu. Osip'yan and L. Regel', "Establishing the physics of weightlessness," *Pravda*, November 12, 1985.

50. B. E. Paton, "On future orbits," in [11], pp. 33–36.

51. V. I. Polezhaev, "Hydrodynamics and heat and mass transfer in crystal growth," in *Mechanics of Liquids and Gases, Advances in Science and Technology* [in Russian], VINITI (1983), pp. 18, 198–268.

52. V. I. Polezhaev, "Research on convection and on heat and mass transfer under conditions of weightless," in [10], pp. 4–6; *Izv. Akad. Nauk SSSR, Ser. Fiz.* **49**, No. 4, 635–642 (1985).

53. L. L. Regel', *State of the Art and Development Prospects for Materials Science in Space* [in Russian], Institute of Space Research, Academy of Sciences of the USSR, Preprint No. 960 (1984).

54. V. T. Khryapov, E. V. Markov, E. P. Prokop'ev, V. M. Biryukov, E. T. Solomin, N. A. Kul'chitskii, S. I. Rozhkov, V. V. Smirnov, and V. A. Tatarinov, "Research on making semiconductor materials in space performed with the Kristall apparatus," in [4], pp. 24–31.

55. V. T. Khryapov, E. V. Markov, N. A. Kul'chitskii, V. M. Biryukov, E. T. Soloman, V. I. Seliverstov, and Yu. M. Zhelannyi, "State of the art, development prospects, and problems in extending semiconductor materials science in space," in [5], pp. 64–71.

56. V. T. Khryapov, N. A. Kul'chitskii, and E. V. Markov, "Research on technological processes in making semiconductor materials in space and setting up an apparatus system," in [10], pp. 113–115.

57. V. T. Khryapov, N. A. Kul'chitskii, V. M. Biryukov, and E. V. Markov, "Some results obtained on semiconductor materials science with the Kristall, Magma, and Korund apparatus on the Salyut-6 and Salyut-7 space stations," in [11], p. 240.

58. A. A. Chernov, "Crystallization under conditions of weightlessness," in [10], pp. 6–8.

59. Yu. V. Cheshlya, "Prospects and problems in manufacturing alloys with controlled quasieutectic structure under conditions of weightlessness," in [5], pp. 72–77.

60. D. Baum, H. Stolze, and P. Vits, "First flight data from MAUS-payloads on STS-7 and STS-11," in [18], p. 154.

61. K.-W. Benz, "Einkristallzuchtungen in der Schwerlosigkeit," *Lab. Prax.* (1984–5); Labor 2000, pp. 16–19.

62. J. Bock, B. Haase, J. Schawer, A. Tegtmeier, and G. Wieczorek, *Deutsche Nutzlastbeistellungen zu EURECA-1*, Fachinformationszentrum, Eggenstein–Leopoldshafen (1985).

63. F. Borlasta, F. Giani, and V. Guarnieri, "Current and prospective Italian activities in microgravity research and application from a space industry viewpoint," in [23].

64. Ch. Bulloch, "Materials processing in space: plenty of prophets, but what about profits?," *Interavia* **39**, No. 7, 679–683 (1984).

65. Deere and Co., "Designing metallurgical tests aboard Shuttle," *COSPAR Inf. Bull.*, No. 100, 55–56 (1984).

66. "Erfolgreicher Einsatz von MAUS-Nutzlasten in Weltraum," *DFVLR-Nachr.*, No. 40, 12–14 (1983).

67. "ESA first 'Get Away Special payload'," *COSPAR Inf. Bull.*, No. 102, 61 (1985).

68. "EURECA: contract signed," *Spaceflight* **27**, No. 11, 391–392 (1985).

69. A. Eyer, H. Leiste, M. Schuhmacher, and H. Walcher, *Entwicklung von Kristallisationsexperimenten für Spacelab: Silizium (FSLP), CdTe'DI), ZnS(DI). Bau und Erprobung einer Monoellipsoidspiegelheizanlage. Phase 2*, Fachinformationszentrum, Eggenstein–Leopoldshafen (1984).

70. I. I. Favier and A. Rouzaud, "Experimental study of morphological and convective instabilities: The Mephisto space program," in [17], p. 388.

71. B. Feurbacher, "Present and future of microgravity research," in [20], pp. 451–454.

72. B. Feurbacher, "Ergebnisse wissenschaftlicher Experimente im Weltraumlabor Spacelab," *Stahl Eisen* **105**, No. 25–26, 21–29 (1985).

73. J. F. Garibotti, W. E. Davis, and N. R. Adsit, "Materials and structures for space applications," *AIAA/NASA Space Syst. Technol. Conf.*, Costa Mesa, Calif., June 5–7, 1984. Collect. Techn. Pap., New York (1984), pp. 50–58.

74. M. Harr and D. Langbein, *Untersuchung der Ostwald–Reifung von Ausscheidungen in Flüssig–Flüssig-Systemen unter Schwerlösigkeit*, Fachinformationszentrum, Eggenstein–Leopoldshafen (1984).

75. "JPL scientist in space," *Space Flight* **27**, No. 9–10, 355–356 (1985).

76. E. Kaldis, "Material sciences research on board Spacelab-1: an overview," in [20], pp. XVII–XIX.

77. E. Kaldis, "Crystal growth from solutions and nucleation from the vapor phase," in [20], pp. 439–441.

78. E. Kaldis, "The view of the microgravity scientific community," *Earth–Orient. Appl. Space Technol.* **5**, No. 1–2, 27 (1985).
79. D. E. Koelle, "Reusable commercial space processing platforms," *Space Manuf. 1983: Proc. Conf.*, Princeton, N. J., May 9–12, 1983, San Diego, Calif. (1983), pp. 119–133.
80. W. Köhler, "Im Spacelab forschen bei Mikroschwerkraft," *VDI-Nchr.*, **37**, No. 49, 10–11 (1983).
81. W. A. Kral, "EUREKA – Freifliegende, wiedervervendbare Platform," *Astronautic* **22**, 35–36 (1985).
82. Akira Kubusono, "Present space laboratory toward future space factor," *Kagaku Keuku. Chem. Educ.* **32**, No. 6, 496–498 (1984).
83. I. Kudo and H. Fujisada, "Space semiconductor processing factory," *Space Solar Power Rev.* **5**, No. 2, 189–195 (1985).
84. D. Langbein, "On the separation of alloys exhibiting a miscibility gap," in [18], pp. 3–12.
85. R. Lange and J. Straub, *Die isochore Wärmekapazität fluider Staffe im kritischen Gebiet-Voruntersuchungen zu einem Spacelab–Experiment*, Fachinformationszentrum, Eggenstein–Leopoldshafen (1984), p. 136.
86. P. Langereux, "Le vol Spacelab-1 a eu une excellente vateur scientific," *Air Cosmos* **22**, No. 1007, 36–37 (1984).
87. "Lewis Research Center opens microgravity materials science lab," *NASA Activ.* **16**, No. 10, 6–7 (1985).
88. D. Manski, "Analyse und Optimierung kleiner Space-Shuttle Antriebsplatformen," *DFVLR, Forschungsbericht 84–28*, Wissenschaftliches Berichtwesen der DFVLR, Cologne (1984).
89. "Microgravity," *ESA Bull.* No. 43, 49–51 (1985).
90. R. Monti, "Space processing," *Earth-Orient. Appl. Space Technol.* **5**, No. 1–2, 129–138 (1985).
91. L. G. Napolitano, "Present and future prospects of microgravity," *Atti 25 Conv. Int. Sullo Grazio*, Rome, March 26–28, 1985, pp. 265–286.
92. Th. A. Nobbe, "Bringing deep space experiments down to earth," *Mach. Des.* **57**, No. 23, 32–34 (1985).
93. Tatsozou Obayashi, "Construction of space base," *Nikon kikai gakkaim, J. Jpn. Soc. Mech. Eng.* **88**, No. 83, 1133–1134 (1985).
94. C. Potard, "Solidification processes in microgravity," *J. Brit. Interplanet. Soc.* **39**, No. 2, 71–74 (1986).
95. *Protein Single-Crystal Growth under Low Gravity*: Proceedings of ESA–DFVLR Workshop, Freiburg, March 19–20, 1984, ESA SP-1067, Paris (1984), pp. VIII and 70.
96. L. L. Regel, "Current state and perspectives of space material science," in [18], p. 151 (see translation, [53]).
97. L. L. Regel, "The analysis of new results in the field of material science in space," *IAF-86-281*, in [23].
98. "Ricerca previsioni technologiche la fabrica va nello spazio," *Ing. Mecc.* **34**, No. 11–12 (1985).
99. H. P. Schmidt and K. Wittmann, "Microgravity user Support Centre for EUREKA-1," *Z. Fluguiss. Weltraumforsch.* **10**, No. 1, 6–12 (1986).
100. G. Seibert and A. Hahne, "First payload for the European retrievable carrier EURECA," in [14], p. 336.

101. D. J. Shapland, "ESA and microgravity research," *Space: Dev. Roll. Eur. 18th Eur. Space Symp.*, London, June 6–9, 1983, San Diego, California (1984), pp. 19–34.
102. "Spacelab preliminary scientific results 'excellent'," *Space Age Times* 12, No. 3–4, 28–29 (1985).
103. H. Suemune and K. Kinoshita, "Technology of preparing materials for electronic engineering parts in space," *Dzaupe Kagaku, J. Mater. Sci. Soc. Jpn.* 20, No. 16, 308–315 (1984).
104. C. Tiby, H. Erdmann, D. Langbein, and H. Behret, *Transportvorgange in der Grenzschicht bei Lösungskristallisation unter Mikrogravitation: Wachstum von Eis in wäszrigen Salzlösungen*, Fachinformationszentrum, Eggenstein–Leopoldshafen (1984).
105. T. Umeda, "Experiments on new materials in space," *Sokeidzai* 26, No. 10, 23–25 (1985).
106. "Une nouvel le métallurgie en microgravité spatial," *Air Cosmos* 22, No. 1001 (1984).
107. P. Vits, H. Stolze, G. Fechner, V. Groth, H. Anderle, B. Haase, and G. Börchers, *Untersuchung zur Erweiterung der MAUS-Ressourcen*, Fachinformationszentrum, Eggenstein–Leopoldshafen (1984).
108. H. U. Walter, "Results of materials-science experiments with sounding rockets," *ESA J.* 7, No. 3, 235–246 (1983).
109. H. U. Walter, "Kristallzüchtung im Weltraum," *Z. Flugwiss. Weltraumforsch.* 7, No. 6, 372–384 (1983).
110. W. R. Wilcox, G. F. Eisa, V. Baskaran, and D. C. Richardson, *Study of Eutectic Formation*: Final Report on Contract NASA 8-34887, August 1984 (Prepared for G. C. Marshall Space Flight Center NASA).
111. K. Wittman, "Wiederflug des Werkstofflabors in der deutschen Spacelab-Mission D1," *DFVLR-Nachr.*, No. 42, 1–3 (1984).
112. Tatsuo Yamanaka, "Manned scientific role in space," *Nihon Kikai Gakkaishi, J. Jpn. Soc. Mech. Eng.* 88, No. 803, 1135–1140 (1985).
113. "Zero-gravity laboratory," *Space World*, V-2-255, 34 (1985).

III. THEORY OF HEAT AND MASS TRANSFER IN SPACE; LIQUID PHYSICS, THEORY, AND EXPERIMENT

114. V. S. Avduevskii, L. V. Leskov, and V. V. Savichev, "Features of heat and mass transfer in full-scale experiments with model substances," in [11], pp. 121–126.
115. M. S. Agafonov and L. V. Leskov, "Contact phenomena at liquid–solid boundaries under conditions of weightlessness," in [10], pp. 168–170.
116. M. S. Agafonov, V. N. Golubev, L. D. Kizim, V. V. Romanov, and V. A. Solov'ev, "Measurements on the behavior of a liquid–gas system in a cylindrical layer in the presence of a radial temperature difference," in [12], pp. 208–209.
117. M. S. Agafonov, V. N. Golubev, L. D. Kizim, V. V. Romanov, and V. A. Solov'ev, "Simulating convection currents at low Prandtl numbers," in [12], p. 210.

118. V. K. Andreev, A. A. Rodionov, and V. A. Sapozhnikov, "Stability in the planar and axisymmetric thermocapillary motion of a liquid," in [10], pp. 49–50.

119. N. Yu. Anisimov and L. V. Leskov, "Physical features of directional crystallization under conditions of weightlessness," in [8], pp. 124–139.

120. L. K. Antanovskii, "TPCP methods in research on thermocapillary convection and dynamic contact-angle problems," in [10], pp. 50–51.

121. N. A. Anfimov, S. V. Ermakov, and A. I. Feonychev, "A numerical study of the flow and heat transfer in a liquid on the Pion apparatus under conditions of weightlessness," in [10], pp. 170–172.

122. Yu. V. Apanovich and V. I. Klykov, "Simulating the symmetry of capillary forces in graphoepitaxy from solutions," in [10], p. 170.

123. L. G. Badratinova, "The stabilizing effects of gas pressure on the stability of the equilibrium of a liquid in a closed vessel for small Bond numbers," in [10], pp. 87–88.

124. L. G. Badratinova and O. M. Lavrent'eva, "The equilibrium form of a molten semiconductor material in a tube and its stability," in [10], pp. 88–90.

125. I. V. Barmin and A. S. Senchenkov, "Solving the Laplace capillary equation for a two-coupled axisymmetric surface by linearization and application to crystallization under microgravity conditions," in [8], pp. 115–123.

126. I. V. Barmin, B. E. Vershinin, I. G. Levitina, and A. S. Senchenkov, "The shape and stability of a rotating liquid zone," in [10], pp. 90–91; *Izv. Akad. Nauk SSSR, Ser. Fiz.* **49**, No. 4, 661–666 (1985).

127. I. V. Barmin, V. A. Briskman, A. L. Zuev, A. F. Pshenichnikov, and A. S. Senchenkov, "Thermocapillary phenomena in liquid films and droplets," in [11], p. 241.

128. I. V. Barmin and A. S. Senchenkov, "The effects of growth conditions on crystal shape in crucible-free zone melting," in [12], pp. 206–207.

129. N. A. Besdenezhnykh, V. A. Briskman, D. V. Lyubimov, A. A. Cherepanov, and M. T. Sharov, "Controlling the stability of a liquid interface by means of vibrations and of electric and magnetic fields," in [10], pp. 18–20.

130. I. V. Belova and A. L. Ovsyannikova, "An analysis of the effects from thermoelastic stresses on crystallization of a sphere under conditions of weightlessness," in [10], pp. 172–173.

131. V. S. Berdnikov, A. G. Zabrodin, and V. A. Markov, "Thermal gravitational-capillary convection in a rectangular cavity," in [7], pp. 136–151.

132. V. S. Berdnikov, A. G. Zabrodin, and V. A. Markov, "Thermal gravitational capillary convection in a horizontal layer of liquid heated from below," in [10], pp. 27–28.

133. V. S. Berdnikov and V. I. Panchenko, "The structures of thermal gravitational capillary and mixed forms of convection in growing single crystals by pulling from the melt," in [10], pp. 203–204.

134. A. B. Barkov, M. K. Bologa, and V. L. Solomyanchuk, "Use of an electrodynamically fluidized bed for controlling heat transfer under conditions of weightlessness," in [10], pp. 62–64.

135. R. V. Birikh, V. I. Chernatynskii, and A. N. Sharifulin, "Vibrational convection in a cylindrical layer of infinite or finite length with a constant force component," in [10], pp. 28–29.

136. M. K. Bologa, I. A. Kozhukhar', O. I. Mardarskii, and I. V. Kozhevnikova, "Electroconvective heat transfer," in [7], pp. 106–115.

137. M. K. Bologa, I. A. Kozhukhar', I. V. Kozhevnikov, A. V. Malakhov. and S. A. Usov, "Electroconvective phenomena and prospects for using them under conditions of weightlessness: review," in [10], pp. 21–23.

138. N. K. Bologa, A. B. Didkovskii, S. M. Klimov, A. N. Maiboroda, L. M. Moldavskii, I. K. Savin, and F. M. Sazhin, "Fluid mechanics and heat transfer in two-phase media in the presence of electric fields under conditions of simulated microgravitation," in [10], pp. 65–66.

139. N. D. Borisov and V. I. Smirnov, "The stability of the equilibrium in a conducting liquid in an electrostatic field," in [10], pp. 91–93.

140. L. M. Braverman, G. Z. Gershuni, E. M. Zhukhovitskii, A. K. Kolesnikov, and V. M. Shikhov, "New results from research on vibrational and convectional instability," in [10], pp. 11–13.

141. Yu. K. Bratukhin and A. L. Zuev, "Thermal capillary drift of an air bubble in a horizontal Hele–Shaw cell," in [10], pp. 51–52.

142. V. A. Briskman and A. L. Zuev, "Ground simulation of thermocapillary bubble drift under conditions of weightlessness," in [8], pp. 95–100.

143. Yu. A. Buevich, "Thermocapillary convection under conditions of weightlessness," in [10], pp. 16–18.

144. Yu. A. Buevich and V. V. Mansurov, "Dynamic instability and self-excited oscillations in a crystallization front," in [10], pp. 173–175.

145. A. V. Bune, V. L. Gryaznov, K. G. Dubovik, S. A. Nikitin, A. I. Prostomolotov, and V. I. Polezhaev, "A software package for examining technological-process parameters under conditions of weightlessness," in [11], pp. 247–248.

146. Yu. V. Val'tsiferov, "A numerical study of nonstationary thermal convection in a thin-walled cylindrical vessel having hemispherical ends under conditions of weightlessness," in [10], pp. 29–31.

147. N. A. Berezub and V. I. Polezhaev, "An analysis of the factors influencing liquid epitaxy under conditions of weightlessness and on the ground," in [10], pp. 167–168.

148. A. F. Voevodin and O. N. Goncharova, "Calculations on free convection in a variable gravitational field," in [10], pp. 31–32.

149. A. F. Voevodin and N. A. Leont'ev, "The effects of a dopant on the melting and crystallization of a sphere under conditions of weightlessness," in [10]. pp. 176–177.

150. A. F. Voevodin, N. A. Leont'ev, A. G. Petrova, and V. V. Pukhnachev. "Thermal diffusion in the melting and crystallization of a sphere under conditions of weightlessness," in [11], p. 234.

151. Yu. M. Gel'fgat and M. Z. Sorkin, "Magnetohydrodynamic effects influencing the convective-transport parameters in experiments on space engineering,' in [11], pp. 289–290.

152. G. Z. Gershuni and E. M. Zhukhovitskii, "Vibrational thermal convection under conditions of weightlessness," in [7], pp. 86–105.

153. G. Z. Gershuni, E. M. Zhukhovitskii, and V. M. Shikhov, "Stability of the vibrational convective flow of a liquid in a planar layer," *Izv. Akad. Nauk SSSR, Ser. Fiz.* **49**, No. 4, 643–648 (1985).

154. A. P. Grigin, V. A. Petrov, and N. V. Pet'kin, "A study of the effects from a gravitational field on the diffusion in a cell having cylindrical electrodes," in [10], pp. 69–70.

155. V. L. Gryaznov, "A numerical study of thermal concentration-convection and the formation of layered structures in melts under conditions of normal and reduced gravity," in [10], pp. 32–34.

156. A. V. Gudzovskii, V. A. Gushchin, T. V. Kondranin, and A. A. Serebrov, "Simulating the three-dimensional flow of a viscous incompressible liquid under conditions of weightlessness," in *Aerophysical and Space Research* [in Russian], Moscow (1984), pp. 21–26.

157. A. A. Gudzovskii, V. A. Gushchin, T. V. Kondranin, and A. A. Serebrov, "Simulating the displacement of a liquid from a vessel by another liquid under conditions of weightlessness," in [11], pp. 246–247.

158. G. A. Dolgikh and A. N. Feonychev, "A numerical analysis of the MA-060 experiment performed under the Soyuz–Apollo program," in [10], pp. 177–179.

159. K. G. Dubovik, "Simulating the interaction between thermocapillary and natural convection and between thermocapillary and capillary concentration-dependent convection," in [10], pp. 52–53.

160. E. A. Eremin and A. K. Kolesnikov, "Stability of stationary states of heat and mass transfer in a layer of mixture under conditions of weightlessness with an exothermic reaction at the boundary," *Izv. Akad. Nauk SSSR, Ser. Fiz.* **49**, No. 4, 649–654 (1985).

161. V. I. Ermakov, "Stability of a planar boundary between liquids in an electric field," in [10], pp. 93–95.

162. S. V. Ermakov and A. I. Feonychev, "A numerical study of convection in an annular region having one or two free boundaries under conditions of weightlessness," in [12], pp. 215–217.

163. M. P. Zavarykin, S. V. Zorin, and G. F. Putin, "Measurements on vibrational-thermal convection," in [10], pp. 34–36.

164. A. A. Zaitsev and D. A. Kazenin, "Effects of wetting angle on thermal capillary convection and bubble detachment in boiling under conditions of weightlessness," in [10], pp. 53–55.

165. V. S. Zemskov, I. N. Belokurova, and D. M. Khavisu, "The impurity distribution in the cross section of a crystal produced by directional crystallization under conditions of weightlessness," in [12], pp. 204–205.

166. S. M. Zen'kovskaya, "Convection under conditions of weightlessness with an oscillating gravitational field," in [10], pp. 36–37.

167. D. A. Ivanova and V. G. Kozlov, "Measurements on heat transfer under conditions of vibrational convection," in [10], pp. 38–39.

168. I. I. Ievlev and A. B. Isers, "Stability in the equilibrium of a spherical drop suspended in a liquid insulator by means of an electric field," in [10], pp. 95–97.

169. É. L. Kalyazin, "Dynamic characteristics of gridded liquid-collection devices under conditions of weightlessness," in [10], p. 111.

170. A. G. Kirdyashkin, "Thermocapillary and thermal-gravitational convection in a horizontal layer of liquid," in [7], pp. 126–135.

171. A. G. Kirdyashkin, "Periodic thermocapillary flows," in [10], p. 13.

172. A. G. Kirdyashkin, V. I. Polezhaev, and A. I. Fedyushkin, "Thermal convection in a horizontal layer with lateral heat input," in [9], pp. 170–187.

173. Yu. A. Kirichenko and G. M. Gladchenko, "Some microscopic characteristics of bubble boiling in weak gravitation," in [7], pp. 72–85.

174. Yu. A. Kirichenko, G. M. Gladchenko, and K. V. Rusanov, "The relationship between the microscopic characteristics of bubble boiling in oxygen and the conditions for the occurrence of a heat-transfer crisis in a weak gravitational field," in [10], pp. 79–81.

175. Yu. A. Kirichenko, G. M. Gladchenko, and K. V. Rusanov, "States of deteriorating heat transfer in the bubble boiling of oxygen under conditions simulating those of weak fields," in [10], pp. 81–83.

176. B. K. Kopbosynov and V. V. Pukhnachev, "Thermocapillary motion in a thin layer of liquid," in [7], pp. 116–125.

177. B. K. Kopbosynov, "Calculating one-dimensional thermocapillary motion in a gas–liquid mixture," in [10], pp. 55–57.

178. A. V. Korol'kov, V. S. Kuptsova, and V. V. Savichev, "Thermal gravitational convection within a cylindrical vessel in the presence of a variable gravitational-acceleration vector," in [12], pp. 218–219.

179. N. D. Kosov, Yu. I. Zhavrin, and S. M. Belov, "Convective instability during diffusion in some ternary mixtures in closed volumes at elevated pressures," in [10], pp. 39–41.

180. V. V. Kuznetsov, "Application of boundary-layer theory to the velocity and concentration patterns in growing single crystals by directional crystallization and crucible-free zone melting," in [10], pp. 179–180.

181. O. M. Lavrent'eva, "The optimal damping of the free surface of a liquid under conditions of weightlessness," in [10], pp. 97–98.

182. A. P. Lebedev, "Some nonlinear effects in the oscillations of a drop bounded by a substrate," in [10], pp. 99–100.

183. D. V. Lyubimov, N. I. Lobov, and A. A. Cherepanov, "The equilibrium boundary between liquids in a high-frequency vibrating field," in [10], p. 100.

184. T. P. Lyubimova, B. I. Myznikova, and E. L. Tarunin, "Simulating free convection in a liquid in a gap between coaxial cylinders," in [10], pp. 41–43.

185. E. D. Lyumkis, B. Ya. Martuzan, and É. N. Martuzane, "Interaction between flows caused by thermocapillary convection and rotation in zone melting: effects on impurity distribution," in [8], pp. 163–178.

186. E. D. Lyumkis, B. Ya. Martuzan, and É. N. Martuzane, "Simulation of thermal convection caused by rotation in zone melting under conditions of reduced gravity," in [10], p. 208.

187. S. A. Maslyaev, G. S. Birova, I. P. Sasinovskaya, and V. N. Pimenov, "Solid–liquid interaction for metals with normal gravitation and under conditions of weightlessness," in [12], p. 207.

188. V. M. Myznikov, "The effects of temperature inhomogeneity on convective processes under conditions of weak gravitation," in [10], pp. 43–44.

189. B. I. Myznikova and E. L. Tarunin, "Numerical simulation of free convection," in [7], pp. 152–160.

190. Nguyen Thanh Nghi, L. L. Regel', and T. A. Cherepanova, "A study of heat and mass transfer under conditions using growing GaP crystals from solution," in [13], pp. 122–124.

191. A. A. Nepomnyashchii and I. B. Simanovskii, "A numerical study of thermocapillary convection in a two-layer system," in [7], pp. 161–166.

192. A. A. Nepomnyashchii and I. B. Simanovskii, "A study of thermocapillary convection in a two-layer formulation for weightlessness conditions," in [10], pp. 57–58.

193. V. G. Nefedov, O. S. Ksenzhek, V. G. Serebritskii, A. V. Palamarguko, V. P. Nikitskii, and G. V. Zhukov, "Features of the gas release mechanism in the electrolysis of water under conditions of weightlessness," in [11], p. 243.

194. S. A. Nikitin, V. I. Polezhaev, and A. I. Fedyushkin, "Convection and impurity distribution in crystals formed by directional crystallization under conditions of weightlessness," in [8], pp. 140–150.

195. S. A. Nikitin, "A numerical study of the hydrodynamics and impurity distribution in directional crystallization under conditions of weightlessness," in [10], pp. 181–182.

196. A. G. Petrova, "Minor oscillations in the pulling rate as a source of nonuniformity in the impurity distribution during crystallization in weak force fields," in [10], pp. 182–183.

197. V. K. Polevikov, "An iterative difference method of calculating the equilibrium forms for the free surface of a liquid," in [10], pp. 100–103.

198. V. K. Polevikov and I. V. Nikiforov, "A numerical study of convection in a liquid having a curved free boundary," in [10], pp. 44–46.

199. V. I. Polezhaev and N. A. Verezub, "A numerical study of liquid epitaxy under conditions of weightlessness," in [11], pp. 235–239.

200. V. I. Polezhaev and N. A. Verezub, "A parametric study of liquid epitaxy," in [12], p. 208.

201. C. I. Polezhaev and K. G. Dubovik, "Interaction between thermocapillary and gravitational forms of convection for ground processing by zone melting," in [12], pp. 219–220.

202. V. I. Polezhaev and A. I. Fedyushkin, "A study of thermal concentration-dependent convection in a square region," in [12], p. 218.

203. V. I. Popov, "A study of MHD phenomena at a liquid–gas boundary occurring when a two-phase system goes over to weightlessness," in [12], p. 221.

204. L. N. Popova, "Nonaxisymmetric forms for the free surface of a liquid in a vessel," in [10], pp. 103–105.

205. A. I. Prostomolotov, "A numerical study of impurity-distribution mechanisms under conditions of normal and reduced gravity in Czochralski's method," in [10], pp. 204–205.

206. V. V. Pukhnachev, "Models for thermocapillary motion," in [10], pp. 14–16.

207. A. F. Pshenichnikov, "The stationary forms of a thin nonisothermal layer of liquid having a free surface," in [10], pp. 58–60.

208. V. Ya. Rivkind, "Hydrodynamics of a rotating nonisothermal drop under conditions of weightlessness," in [10], p. 13.

209. V. A. Saranin, "The hydrodynamics of thermoelectric effects in melts under conditions of weightlessness," in [10], pp. 71–73.

210. A. S. Senchenkov, "Some problems in the deformation and stability of a two-sided liquid film under conditions of weightlessness," in [10], p. 105.

211. Yu. B. Sklevskii, "A parametric study of thermocapillary convection in a rectangular channel," in [10], pp. 60–62.

212. L. A. Slobodkin, "The stability of molten material in zone melting," in [4], pp. 18–23.

213. L. A. Slobozhanin, "Equilibrium shapes for a rotating drop under conditions of weightlessness," in [7], pp. 28–36.

214. L. A. Slobozhanin, "A study of hydrostatic problems simulating zone melting," in [10], p. 25; *Izv. Akad. Nauk SSSR, Ser. Fiz.* **49**, No. 4, 652–660 (1985).

215. L. A. Slobozhanin, "Formulations on the equilibrium and stability of a system having a line of contact between three liquids," in [10], pp. 107–110.

216. L. A. Slobozhanin, "Stability of the equilibrium of a liquid having an uncoupled free surface in a closed system," in [10], pp. 110–111.

217. E. L. Tarunin, "Electroconvection in a closed region under conditions of weightlessness," in [10], pp. 73–75.

218. V. A. Tatarchenko, "Features of crystallization in capillary shaping under conditions of weightlessness," in [8], pp. 101–114.

219. V. A. Tatarchenko, "Capillary shaping in crystallization under conditions of weightlessness," in [10], pp. 9–10.

220. A. I. Fedonenko and A. I. Zhakin, "Use of electroconvective flows for accelerating heat transfer under conditions of weightlessness," in [10], p. 71.

221. A. I. Fedyushkin, "A numerical study of thermal concentration-dependent convection in a closed region," in [10], pp. 46–47.

222. A. I. Feonychev and G. A. Dolgikh, "The effects of crystallization rate and conditions at the phase boundary on heat transfer and impurity distribution in a cylindrical tube with directional crystallization," in [11], p. 229.

223. V. T. Khryapov, E. P. Prokop'ev, K. V. Markov, V. M. Biryukov, and N. A. Kul'chitskii, "Application of Landau's boundary-layer theory to chemical hydrodynamics in a single-crystal growth under conditions of weightlessness," in [10], p. 147.

224. Yu. D. Chashechkin, V. I. Neklyudov, and V. S. Belyaev, "Laboratory simulation of layered structures involving thermal concentration-dependent convection with normal and reduced gravity," in [10], pp. 47–49.

225. T. A. Cherepanova and V. V. Lyukhin, "Crystallization mechanisms and kinetics with reduced gravitation," in [8], pp. 152–162.

226. T. A. Cherepanova and A. L. Ushkans, "Theoretical aspects of making foam materials under microgravity conditions," in [10], pp. 185–186.

227. P. S. Chernyakov, "Dynamics of vapor bubbles and heat transfer in the bulk of boiling underheated or superheated liquid under conditions of weak mass forces," in [10], pp. 85–87.

228. V. P. Shalimov, "Gas inclusions in a liquid under conditions of weightlessness," in [10], pp. 159–160.

229. V. P. Shalimov, "Flow of a liquid through a capillary under conditions of weightlessness indicated by the Romanian–USSR Kapillyar experiment," in [13], pp. 143–150.

230. I. M. Yavorskaya, Yu. N. Belyaev, and N. I. Fomina, "Simulating centrally symmetric convection by means of an alternating electric field," in [9], pp. 188–200.

231. D. Avnir, A. Bewersdorff, and M. Kagan, "Chemical instabilities at liquid interfaces," G. 1.3.4, in [17], p. 336.

232. J. K. Baird, "Application of the theory of Ostwald ripening to microgravity experiments," in [19], p. 70; in [20], pp. 319–321.

233. C. Barta, A. Triska, Z. Pokorna, and J. Trnka, "Crystallization of KNO_3 under conditions of diffusive and convective regime," G. 1.1.5, in [17], p. 333.

234. H. F. Bauer, "Transient thermal Marangoni convection in a liquid bridge," Z. Flugwiss. Weltraumforsch. 7, No. 2, 120–133 (1983).

235. H. F. Bauer, "Liquid surface oscillations induced by temperature fluctuation," Z. Flugwiss. Weltraumforsch. 7, No. 4, 274–278 (1983).

236. H. F. Bauer, "Theoretical study about the Marangoni convection in a liquid column in zero gravity," *Acta Astronaut.* **11**, No. 6, 301–311 (1984).

237. H. F. Bauer, "Oscillation of a viscoelastic liquid drop in zero gravity," in [19], p. 65; in [20], pp. 265–270.

238. H. F. Bauer, "Free surface and interface oscillation of an infinitely long viscoelastic liquid column," *Acta Astronaut.* **13**, No. 1, 9–22 (1986).

239. D. Beysens, "Stability of critical fluid mixtures: experimental simulation of microgravity conditions," in [21], p. 161.

240. C. Bisch, "Existence et position de cercles nodaux sur une calotte sphérique liquide oscillant en microgravité a sa fréquence principale de résonance," *Acta Astronaut.* **11**, No. 3–4, 173–177 (1984).

241. C. Bisch, "Axial and tangential vibration modes for semi-free liquid spheres in microgravity: nodal circles and nodal points position," in [21], p. 155.

242. R. Boudreault, "Numerical simulation of convections in the microgravity environment," in [19], p. 62; in [20], pp. 259–264.

243. D. Camel, P. Tison, and J. J. Favier, "Marangoni flow regimes in liquid metals," in [22], p. 214.

244. F. M. Carlson and A. H. Eraslan, "Convection phenomena in Bridgman–Stockbarger crystal growth," in [22], pp. 204–205.

245. Ch.-H. Chun, "Numerical study of thermal Marangoni convection and comparison with experimental results from the TEXUS-rocket program," *Acta Astronaut.* **11**, No. 3–4, 227–232 (1984).

246. Ch.-H. Chun, "Verification of turbulence developing from oscillatory Marangoni convection in a liquid column," in [20], pp. 271–280.

247. Ch.-H. Chun, "Interferometric study of the thermal Marangoni convection around a drop and a bubble near a heated horizontal wall," G. 1.4.1, in [17], p. 336.

248. Liang Lai Chun and Ti Chai An, "Surface temperature distribution along a thin liquid layer due to thermocapillary convection," in [22], IAF-85-223.

249. S. R. Coriell, "Double diffusive convection during directional solidification," G. 1.3.3, in [17], p. 335.

250. V. Delitzsch, H. Eckelmann, and W. Wuest, "The influence of thermocapillarity on the migration of droplets in a liquid possessing a uniform temperature gradient," in [19], p. 89, in [20], pp. 245–249.

251. H. M. Dürr, "Numerical studies of fluid oscillation problems by boundary integral techniques," in [19], Programm, p. 68.

252. A. H. Eraslan, F. M. Carlson, and Zen Sheu Ji, "Modeling of convection phenomena in Bridgman–Stockbarger crystal growth," IAF-86-286, in [23].

253. P. V. Farrel and B. D. Peters, "Droplet vaporization in supercritical pressure environments," in [22], Preprint IAF-85-193.

254. W. N. Gill, C. C. Hsu, N. D. Kazarinoff, M. A. Noack, and J. I. Verhoeven, "Thermocapillary and buoyancy driven convection in supported and floating zone crystallization," in [16], pp. 328–329; *Adv. Space Res.* **4**, No. 5, 15–22 (1984).

255. J. M. Haynes, "Preliminary observations on fluid physics experiments I-ES 327 and 339," in [19], p. 19, in [20], pp. 43–46.

256. J. M. Haynes, "Assessment of SL-1 results: IES-327: spreading kinetics and IES-339: capillary hysteresis," *Earth-Orient. Appl. Space Technol.* **5**, No. 1–2, 67–68 (1985).

257. V. V. Iliukhin, V. P. Shalimov, S. J. Budurov, P. D. Kovachev, and S. A. Toncheva, "On the conditions of gas inclusions formation in melts under zero gravity state (after Pirin experiment)," *Acta Astronaut.* **11**, No. 9, 585–592 (1984).

258. M. Jurisch and J. Barthel, "Oscillatory thermocapillary convection in floating zones of refractory metals under normal gravity," G. 1.2.6, in [17], pp. 334–335.

259. V. M. Korovin, "Hydrodynamic aspects of electromagnetic vibrational treatment of molten metals under microgravity conditions," *Acta Astronaut.* **11**, No. 3–4, 207–212 (1984).

260. V. M. Korovin, "Eddy current induced migration of droplets in electrically conducting fluid in microgravity environments," in [22], p. 212.

261. C.-L. Lai and A.-T. Chai, "Surface temperature distribution along a thin liquid layer due to thermocapillary convection," in [22], Programm, p. 68.

262. D. Langbein and W. Heide, "The separation of liquids due to Marangoni convection," in [16], p. 329; *Adv. Space Res.* **4**, No. 5, 27–36 (1984); *Z. Flugwiss. Weltraumforsch.* **8**, No. 3, 192–199 (1984).

263. D. Langbein, "Liquids and interfaces," in [20], pp. 443–447.

264. D. Langbein and W. Heide, "Entmischung von Flussigkeiten auf-grund von Grenzflächenkonvektionen," *Z. Flugwiss. Weltraumforsch.* **8**, No. 3, 192–199 (1984).

265. D. Langbein, "Study of convective mechanisms under microgravity conditions," G. 1.2.1, in [17], pp. 333–334.

266. G. Lebon, "Recent developments in surface-tension driven instabilities," *Acta Astronaut.* **11**, No. 7–8, 353–359 (1984).

267. J. C. Legros, M. C. Limbourg-Fontaine, and G. Petre, "Influence of a surface tension minimum as a function of temperature on the Marangoni convection," *Acta Astronaut.* **11**, No. 2, 147 (1984).

268. J. C. Legros, G. Petre, and M. C. Limbourg-Fontaine, "Study of the Marangoni convection around a surface tension minimum under microgravity conditions," in [16], pp. 329–330; *Adv. Space Res.* **4**, No. 5, 37–41 (1984).

269. J. H. Lichtenbelt, "Marangoni convection and mass transfer from the liquid to the gas phase," G. 1.3.2, in [17], p. 335.

270. M. C. Limbourg, J. C. Legros, and G. Petre, "The influence of a surface tension minimum on the convective motion of a fluid in microgravity (D1 Mission Results)," G. 1.2.4, in [17], p. 334.

271. T. Maekawa and I. Tanasawa, "Onset of Marangoni convection in an infinite layer of an electrically conducting liquid under magnetic field," G.1.2.5, in [17], p. 334.

272. T. Maekawa, I. Tanasawa, J. Ochiai, K. Kuwahara, M. Morioka, and S. Enya, "Two-dimensional Marangoni and buoyancy convection related to crystal growth techniques in space," in [14], p. 333; *Adv. Space Res.* **4**, No. 5, 63–66 (1984).

273. I. Martinez, "Results of an experiment on liquid column stability aboard Spacelab-1," in [19], p. 17; in [20], pp. 31–36.

274. J. Meseguer, "Stability of long liquid columns," in [19], p. 67; in [20], pp. 297–300.

275. R. Monti, L. G. Napolitano, and G. Russo, "Experimental study of thermal Marangoni flows in silicon oil floating zones," *Acta Astronaut.* **11**, No. 7–8, 369–378 (1984).

276. R. Nahle, "Bubble dynamics in a fluid medium with temperature gradient-D-1 results," in [23], IAF-86-285.
277. L. G. Napolitano, "Marangoni convection in space microgravity environments," *Science* **225**, No. 4658, 197–198 (1984).
278. L. G. Napolitano and G. Russo, "Similar axially symmetric Marangoni boundary layers," *Acta Astronaut.* **11**, No. 3–4, 189–198 (1984).
279. L. G. Napolitano and C. Golia, "Effects of gravity levels on Marangoni–Stokes flows in Plateau configurations," *Acta Astronaut.* **11**, No. 3–4, 213–225 (1984).
280. L. G. Napolitano, R. Monti, and G. Russo, "Results of the Marangoni free convection experiment," in [19], p. 15; in [20], pp. 15–22.
281. L. G. Napolitano, R. Monti, and G. Russo, "Marangoni convection in low gravity: experiment IES-328," *Earth-Orient. Appl. Space Technol.* **5**, No. 1–2, 69–82 (1985).
282. L. G. Napolitano, C. Golia, and A. Viviani, "Numerical simulation of unsteady Marangoni flows," in [19], p. 90; in [20], pp. 251–258.
283. L. G. Napolitano, R. Monti, and G. Russo, "Marangoni experiment on Spacelab-1," in [16], p. 329.
284. L. G. Napolitano, C. Golia, and A. Viviani, "Effect of variable transport properties on thermal Marangoni flows," in [22], Programm, p. 68.
285. L. G. Napolitano, "Recent development of Marangoni flows: theory and experimental results," G. 1.2.3, in [17], p. 334.
286. J. Ochiai, K. Kuwachara, M. Morioka, S. Enya, T. Maekawa, and I. Tanasawa, "Experimental study of Marangoni convection," in [19], p. 66; in [20], pp. 291–295.
287. J. P. Paday, U. Merbold, and B. Lichtenberg, "Capillary forces in a low gravity environment," in [19], p. 14; in [20], pp. 9–14.
288. G. Petre and D. Wozniak, "Measurement of the variation of interfacial tension with temperature between immiscible liquids of equal density," in [22], Programm, p. 68; Preprint IAF-85-198.
289. G. Petre, M. Cl. Limbourg-Fontaine, and J. Cl. Legros, "Preliminary results of TEXUS-8 experiments on effects of surface tension minimum," *Acta Astronaut.* **12**, No. 3, 203–206 (1985).
290. V. I. Polezhaev, A. P. Lebedev, and S. A. Nikitin, "Mathematical simulation of disturbing forces and material sciences processes under low gravity," in [19], p. 84; in [20], pp. 237–243.
291. V. I. Polezhaev and K. G. Dubovik, "Instability secondary structures and unsteady regime of Marangoni convection," G. 1.2.2, in [17], p. 334.
292. S. J. Robertson and L. W. Spradley, "Effect of enclosure shape on natural convection velocities in microgravity," in [22], p. 215.
293. H. Rodot and C. Bisch, "Oscillations of semifree liquid volume in space: experiment 326 in Spacelab-1," in [19], p. 16; in [20], pp. 23–29.
294. G. Russo, "Order of magnitude analysis of unsteady Marangoni convection," in [21], Preprint IAF-84-208.
295. G. Russo, "Influence of imposed pressure on surface tension gradient driven convection," in [22], p. 213.
296. D. Schwabe and A. Scharmann, "Measurement of the critical Marangoni number of the laminar oscillatory transition of thermocapillary convection in floating zones," in [20], pp. 281–289.

297. D. Schwabe and A. Scharmann, "Microgravity experiments on the transition from laminar to oscillatory thermocapillary convection in floating zones," in [16], p. 330; *Adv. Space Res.* **4**, No. 5, 43–47 (1984).

298. D. Schwabe and A. Scharmann, "Messung der kritichen Marangonizahl für den Übergang von Stationarer zu oszillatorischer thermokapillarer Konvection unter Microgravitation: Ergebnisse der Experimente in den ballistischen Raketen TEXUS 5 und TEXUS 8," Z. *Flugwiss. Weltraumforsch.* **9**, No. 1, 21–28 (1985).

299. O. Schwabe, "Crystal growth and Marangoni convection in space," G. 1.1.1, in [17], p. 332.

300. P. J. Sell, E. Maisch, and J. Siekmann, "Hydrodynamic forces resulting from liquid motion in capillary tubes," in [16], p. 330; *Adv. Space Res.* **4**, No. 5, 49–52 (1984).

301. P. J. Sell, E. Maisch, and J. Siekmann, "Fluid transport in capillary systems under microgravity," *Acta Astronaut.* **11**, No. 9, 577–583 (1984).

302. P. J. Sell, E. Maisch, and J. Siekmann, "Experimental study of fluid transport in capillary systems," in [21], p. 160.

303. J. Straub and K. Nitsche, "The anomaly of the specific heat at constant volume (Cv) of sulfur hexafluoride SF_6 at critical state," in [19], p. 73.

304. R. S. Subramanian, "Capillary driven motion of bubbles and droplets – some new results," in [16], p. 334.

305. A. E. P. Veldman and M. E. S. Vogels, "Axisymmetric liquid floating under low-gravity conditions," *Acta Astronaut.* **11**, No. 10–11, 641–649 (1984).

306. D. Villers and J. K. Platten, "Rayleigh–Benard instability in systems presenting a minimum in surface tension," in [19], p. 68; in [20], pp. 301–305.

307. J. P. Vreeburg, "Axisymmetric free surface behavior of moving liquid in a cylinder in microgravity," in [19], p. 18; in [20], pp. 37–42.

308. J. P. B. Vreeburg, "Overview of Spacelab experiment IES-330," *Earth-Orient. Appl. Space Technol.* **5**, No. 1–2, 83–89 (1985).

309. J. P. B. Vreeburg, "Liquid motions in partially filled containers: preliminary results of the D-1 Mission," G. 1.4.5, in [17], p. 337.

310. M. Wadih and B. Roux, "Natural convection for supercritical conditions in oscillatory microgravity environment (0-jitter)," G. 1.2.7, in [17], p. 335.

311. R. West, "Effect of gravity on the solidification of binary alloys – a numerical solution," in [22], pp. 216–217.

312. B. Zappoli and F. Elie, "On boundary conditions for hydrodynamic equations in reduced gravity," in [21], p. 162.

313. B. Zappoli, "Oscillatory growth regime: D-1 numerical nonlinear stability analysis," IAF-86-287, in [23].

314. M. Zell, J. Straub, and A. Weinzierl, "Nucleate pool boiling in subcooled liquid under microgravity: results of TEXUS experimental investigations," in [19], p. 72; in [20], pp. 327–333.

315. E. Zimmerman and J. Siekmann, "Experimental investigation of the dynamic contact angle," G. 1.4.6, in [17], p. 337.

316. I. M. Yavorskaya, N. I. Fomina, and Yu. N. Belyaev, "A simulation of central-symmetry convection in microgravity conditions," *Acta Astronaut.* **11**, No. 3–4, 179–189 (1984).

IV. GROWTH OF SEMICONDUCTOR CRYSTALS
FROM THE MELT UNDER MICROGRAVITY

317. O. V. Abramov, A. S. Okhotin, Zh. Yu. Chashechkina, I. P. Kazakov, O. I. Rakhmatov, and S. A. Zver'kov, "Results on $Pb_{1-x}In_xTe$ crystals made by Bridgman's method with the Kristall apparatus under conditions of the orbiting Salyut-6 station," in [8], pp. 47–58.

318. O. V. Abramov, G. E. Ignat'ev, S. A. Zver'kov, and Zh. Yu. Chashechkina, "Distributions of Te–Se solid solution components in the Kristall apparatus," in [10], pp. 161–162.

319. I. V. Barmin, Yu. L. Volkov, Yu. G. Gromakov, A. V. Egorov, V. A. Kazymov, É. S. Kopeliovich, N. S. Lagozina, V. V. Rakov, and I. G. Filatov, "Crucible-free zone melting for semiconductors under microgravity conditions," in [4], pp. 8–17.

320. I. V. Barmin, V. S. Zemskov, E. A. Abramova, V. N. Kondrat'ev, M. R. Raukhman, and A. S. Senchenkov, "Impurity distributions in an InSb crystal grown at zero gravity with a gap between the ampul and the growing crystal," in [10], pp. 165–166.

321. I. V. Barmin, V. S. Zemskov, I. T. Krutovertsev, N. P. Konovalova, A. V. Laptev, and M. R. Raukhman, "Impurity segregation in InSb crystals grown by directional crystallization as affected by growth conditions," in *Doped Semiconductor Materials* [in Russian], Nauka, Moscow (1985), pp. 127–131.

322. N. A. Bul'enkov, V. V. Savichev, M. S. Agafonov, A. F. Belyanin, and V. P. Martovitskii, "Crystallization of a doped germanium liquid under conditions of brief weightlessness," in [11], pp. 224–226.

323. N. A. Verezub, É. S. Kopeliovich, V. I. Polezhaev, and V. V. Rakov, "Heat and mass transfer in melts of certain $A^{III}B^V$ semiconductor compounds under conditions of zero gravity," in [8], pp. 79–84.

324. N. A. Verezub, I. N. Zubritskaya, A. V. Egorov, É. S. Kopeliovich, L. V. Nazarova, V. V. Rakov, B. A. Shekhtman, E. S. Yurova, and I. M. Yur'eva, "Studies on making some semiconductor systems with the Splav apparatus," in [10], pp. 120–121; *Izv. Akad. Nauk SSSR, Ser. Fiz.* **49**, No. 4, 687–690 (1985).

325. Yu. L. Volkov, É. S. Kopeliovich, N. S. Lagozina, V. V. Rakov, E. S. Yurova, I. M. Yur'eva, and V. V. Fedorov, "Inhomogeneities in semiconductor materials made under natural conditions," in [10], p. 167.

326. Yu. L. Volkov, I. N. Zubritskaya, É. S. Kopeliovich, L. V. Nazarova, A. S. Senchenkov, I. G. Filatov, and B. A. Shekhtman, "Laboratory studies on zone melting for semiconductor materials in a crucible-free zone melting apparatus," in [11], p. 206.

327. R. R. Galonzka, Yu. Aulyaitner, T. Varminski, I. Bonk, A. Endzheichak, Z. Furmanik, É. Mizera, K. Godvud, A. Shcherbakov, A. S. Okhotin, I. A. Zubritskii, I. V. Barmin, A. A. Il'in, R. P. Borovikova, V. T. Khryapov, and E. V. Markov, "The crystallization of $Hg_{1-x}Cd_xTe$, $Hg_{1-x}Cd_xSe$, and $PbSe_xTe_{1-x}$ solid solutions under conditions of zero gravity: the Sirena-1, -2, and -3 experiments," in [13], pp. 164–175.

328. P. Dias, E. Vichil, D. Romero, E. Puron, S. De Rookes, F. Sanches, V. M. Andreev, S. G. Konnikov, and T. B. Popova, "The growth of epitaxial GaAlAs–GaAs films under zero-gravity conditions," in [13], pp. 129–132.

329. V. S. Zemskov, M. R. Raukhman, I. V. Barmin, A. S. Senchenkov, I. G. Shul'pina, and L. M. Sorokin, "Results on doped indium antimonide crystals grown on the Salyut-6 and Soyuz space vehicle combination," in [3], pp. 38–56.

330. V. S. Zemskov and M. R. Raukhman, "Component liquation in an indium–antimony–bismuth melt caused by gravitation," in [3], pp. 57–64.

331. V. S. Zemskov, M. R. Raukhman, A. V. Laptev, I. V. Barmin, A. S. Senchenkov, I. G. Smirnova, and S. M. Pchelintsev, "The tellurium distributions in indium antimonide crystals grown under zero-gravity conditions," in [3], pp. 87–104.

332. V. S. Zemskov, I. V. Barmin, M. R. Raukhman, and A. S. Senchenkov, "Impurity distributions in semiconductor crystals grown from melts under conditions of zero gravity," *The Gagarin Lectures on Cosmonautics and Aviation*, April 5–9, 1982 [in Russian], Moscow (1984), pp. 127–133.

333. V. S. Zemskov, I. V. Barmin, M. R. Raukhman, and A. S. Senchenkov, "An experiment on growing doped indium antimonide under conditions in the orbit of the Salyut-6 and Soyuz space vehicle combination," in [8], pp. 30–46.

334. V. S. Zemskov, I. N. Belokurova, I. L. Shul'pina, and A. N. Titkov, "The structures of germanium–silicon solid-solution crystals made under zero-gravity conditions," in [13], pp. 241–243.

335. V. S. Zemskov, M. R. Raukhman, A. V. Laptev, I. V. Barmin, A. S. Senchenkov, I. G. Smirnova, and S. M. Pchelintsev, "The tellurium distributions in indium antimonide crystals made under zero-gravity conditions," in *Doped Semiconductor Materials* [in Russian], Nauka, Moscow (1985), pp. 132–138.

336. V. S. Zemskov, M. R. Raukhman, and E. A. Kozitsyna, "The crystallization of multicomponent alloys under zero-gravity conditions," *Fiz. Khim. Obrab. Mater.*, No. 5, 44–49 (1985).

337. V. S. Zemskov, M. R. Raukhman, B. P. Mikhailov, V. N. Pimenov, and I. N. Belokurova, "Technological developments in processing as a major condition for space experiments," in [12], pp. 203–204.

338. V. S. Zemskov, M. R. Raukhman, E. A. Kozitsyna, I. V. Barmin, A. I. Antipov, A. S. Senchenkov, I. T. Krutovertsev, B. A. Shekhtman, and N. K. Karaseva, "Features of ground experiments on directional crystallization for indium antimonide," in [12], pp. 205–206.

339. H. Zusman, K. Stecker, W. Eichler, S. Langhammer, N. H. Kuen, N. V. Sung, I. V. Vyung, C. Z. Hoai, V. M. Truzhenikov, and M. B. Shcherbina-Samoilova, "Results from the Halong-2,3 experiments on the controlled crystallization of $Bi_{1-x}Sb_xTe_3$ solid solutions in the Kristall apparatus on board the Salyut-6," in [13], pp. 37–51.

340. É. S. Kopeliovich, M. G. Mil'vidskii, and V. V. Rakov, "Features of making semiconductor materials under special conditions," in [3], pp. 32–37.

341. E. Lendvay, M. Harsy, G. Gereg, I. Gyuro, F. Koltai, J. Gyulai, T. Lohner, F. Pasti, G. Mezery, E. Kotiai, M. Ranky, L. L. Regel', V. T. Khryapov, and N. A. Kul'chitskii, "The Eötvös experiment: growing GaSb crystals under microgravity conditions," in [13], pp. 79–89.

342. E. Lendvay, I. Gyuro, M. Harsy, F. Koltai, L. L. Regel', and N. A. Kul'chitskii, "History of the structure in GaSb grown under zero-gravity conditions," in [13], pp. 98–100.

343. T. Lohner, L. Varga, G. Mezey, F. Pasti, E. Kotiai, J. Gyulai, L. L. Regel', N. A. Kul'chitskii, and V. T. Khryapov, "A study on GaSb and GaAs crystals grown under microgravity conditions by OP and PIXE methods," in [13], pp. 90–94.

344. W. Lupei, D. Toma, D. Prunariu, I Brydus, A. Popa, P. I. Antonov, D. D. Dryuchenko, V. P. Shalimov, I. L. Shul'pina, L. I. Popov, G. V. Zhukov, N. A. Kul'chitskii, and A. V. Korovin, "Capillary shaping for germanium crystals under the Romania–USSR Kapillyar-1 program," in [13], pp. 133–142.

345. Nguyen Thanh Nghi, L. L. Regel', T. A. Cherepanova, and A. L. Ushkans, "The formation of pores in the crystallization of the Bi_2Te_3–Bi_2Se_3 system under microgravity conditions," in [13], pp. 125–128.

346. L. L. Regel', R. V. Parfen'ev, V. V. Popov, and I. Gyuro, "Low-temperature studies on transport phenomena in gallium antimonide made under microgravity conditions," in [10], pp. 150–152.

347. L. L. Regel', O. I. Rakhmatov, R. V. Parfen'ev, and N. A. Red'ko, "The electrophysical and thermal properties of PbTe made under microgravity conditions," in [10], pp. 156–159.

348. L. L. Regel', O. N. Rakhmatov, N. A. Red'ko, and R. V. Parfen'ev, "A study of transport in single-crystal p–PbTe grown under zero-gravity conditions," Fiz. Tverd. Tela, **26**, No. 4, 1242–1245 (1984).

349. L. L. Regel' and Nguyen Thanh Nghi, "The growth of Bi_2(Te, Se)$_3$ and GaP under microgravity conditions," in [10], pp. 160–161.

350. L. L. Regel' and Nguyen Thanh Nghi, "Flux crystallization of GaP under microgravity conditions," in [13], pp. 116–121.

351. L. L. Regel' and Nguyen Thanh Nghi, "The segregation of Bi_2Te_3 in growing Bi_2(Te, Se)$_3$ solid solutions under zero-gravity conditions," in *Doped Semiconductor Materials* [in Russian], Nauka, Moscow (1985), pp. 142–145.

352. L. L. Regel', Nguyen Thanh Nghi, R. V. Parfen'ev, N. A. Red'ko, and V. V. Sologub, "Low-temperature transport phenomena in Bi_2(Te, Se)$_3$ crystals made under microgravity conditions on the Salyut-6 station," Kosm. Issled. **23**, No. 4, 651–654 (1985).

353. V. A. Tatarinov, A. S. Pashinkin, V. T. Khryapov, and N. A. Kul'chitskii, "A study of the effects from thermal capillary convection on the crystallization of germanium under microgravity conditions," in [10], pp. 136–137.

354. V. A. Tatarinov, V. T. Khryapov, T. V. Kul'chitskaya, E. V. Markov, and N. A. Kul'chitskii, "The impurity distributions in germanium single crystals grown under microgravity conditions," in [10], pp. 137–139.

355. V. A. Tatarinov, V. T. Khryapov, N. A. Kul'chitskii, E. V. Markov, L. A. Fokina, and V. P. Nikitskii, "Container-free crystallization of a semiconductor material from a melt under microgravity conditions," in [10], pp. 139–141.

356. V. A. Tatarinov, V. T. Khryapov, A. S. Pashinkin, and N. A. Kul'chitskii, "The effects of gravity on liquid mixing and impurity segregation for germanium," in [10], pp. 141–143.

357. V. A. Tatarinov, N. A. Kul'chitskii, A. S. Pashinkin, and V. T. Khryapov, "Technology for growing bulk and ribbon single crystals from a liquid under zero-gravity conditions," in [11], pp. 286–287.

358. V. M. Truzhenikov, M. B. Shcherbina-Samoilova, H. Zusman, K. Stecker, S. Langhammer, N. H. Kuen, H. V. Sung, N. V. Byung, and C. Z. Hoai, "Crystallization of $(Bi_{1-x}Sb_x)_2Te_3$ solid solutions under microgravity conditions as indicated by the Halong-2,3 experiments," in [10], pp. 155–156.

359. V. M. Truzhenikov, E. V. Markov, V. M. Biryukov, and V. T. Khryapov, "Container-free growth of gallium arsenide single crystals under zero-gravity conditions," in [11], p. 287.

360. V. T. Khryapov, V. A. Tatarinov, T. A. Kul'chitskaya, N. A. Kul'chitskii, E. V. Markov, and R. S. Krupyshev, "Growing bulk germanium crystals by directional crystallization under zero-gravity conditions," in [8], pp. 59–71.

361. V. T. Khryapov, E. V. Markov, V. V. Smirnov, V. M. Biryukov, T. I. Markova, and V. M. Truzhenikov, "Preliminary results of crystallizing GaAs under microgravity conditions," in [10], pp. 145–146.

362. V. P. Shalimov, D. D. Dryuchenko, V. Lupey, and D. Toma, "Features of capillary shaping for germanium under zero-gravity conditions: results from the Soviet–Romanian Kapillyar experiment," in [10], p. 154.

363. K. Somogyi and T. Görog, 'Some electrophysical parameters of GaSb grown under zero-gravity conditions," in [13], pp. 95–97.

364. I. L. Shul'pina, L. M. Sorokin, A. S. Tregubova, G. N. Mosina, V. S. Zemskov, M. R. Raukhman, and E. A. Kozitsyna, "Structures in doped indium antimonide crystals containing p–n junctions grown under zero-gravity conditions," in Doped Semiconductor Materials [in Russian], Nauka, Moscow (1985), pp. 138–142.

365. P. G. Barber and R. K. Crouch, "Electrochemical etching of the semiconductor lead–tin telluride," J. Electrochem. Soc. 131, No. 12, 2803–2805 (1984).

366. I. V. Barmin and A. S. Senchenkov, "Stability diagram of a liquid isorotating zone of an arbitrary volume," in [21], pp. 164–165.

367. K. W. Benz and G. Nagel, "GaSb semiconductor crystals under microgravity: experiment ES-323 in Spacelab-1," in [19], p. 46; in [20], pp. 157–161.

368. K. W. Benz, "Die Herstellung von Halbleitereinkristallen während der 1 Spacelab-Mission," Metall(W. Berlin) 39, No. 12, 1138–1140 (1985).

369. M. Bloom, "Space-grown GaAs crystals promise performance," Microwaves RF, No. 5, 45–49 (1985).

370. E. D. Bourret, J. J. Derby, R. A. Brown, and A. F. Witt, "Segregation effect during growth of pseudobinary systems with large liquid–solidus separation,' Acta Astronaut. 11, No. 3–4, 163–171 (1984).

371. T. Carlberg, "A preliminary report on floating zone experiments with germanium crystals in a sounding rocket," in [19], p. 85; in [20], pp. 367–373.

372. R. K. Crouch, A. L. Fripp, W. J. Debnam, I. O. Clark, and F. M. Carlson, "Optimization studies for the growth of $Pb_{1-x}Sn_xTe$ in space," Adv. Ceram. 5, No. 10, 186–194 (1983).

373. R. K. Crouch, A. L. Fripp, W. J. Debnam, I. O. Clark, P. G. Barber, and F. M. Carlson, "Experimental investigation of the effects of gravity on thermosolutal convection and compositional homogeneity in Bridgman grown compound semiconductors," in [21], pp. 152–153.

374. R. K. Crouch, A. L. Fripp, W. J. Debnam, I. O. Clark, P. G. Barber, and F. M. Carlson, "Experimental investigation of the effects of gravity on thermosolutional convection and compositional homogeneity in Bridgman grown

compound semiconductors," *Acta Astronaut.* **12**, No. 11, 923–929 (1985).

375. P. Diaz, E. Vigil, R. Romera, E. Puron, S. Roux, F. Sancher, V. M. Andreev, S. G. Konnikov, and T. B. Popova, "Homo y heteroepitaxia de GaAl As–GaAs en condiciones de microgravidez," *Cienc. Tec. Fis. Mat.*, No. 4, 23–28 (1984).

376. E. Eyer, H. Leiste, and R. Nitsche, "Floating zone growth of silicon under microgravity in Spacelab-1," in [19], p. 49; in [20], pp. 173–182.

377. A. L. Fripp, R. K. Crouch, W. J. Debnam, I. O. Clark, F. M. Carlson, and P. G. Barber, "Preparation for microgravity science investigation of compound semiconductor crystals," in [22], pp. 188–189.

378. A. L. Fripp, R. K. Crouch, W. J. Debnam, Y. Clark, and I. O. Clark, "Effects of supercooling in the initial solidification of PbTe–SnTe solid solutions," *J. Crystal Growth* **73**, No. 3, 304–310 (1985).

379. "Gallium arsenide development in space studied," *Aerospace Daily*, **136**, No. 28, 223 (1985).

380. H. C. Gatos, J. Lagowski, L. M. Pawlowicz, F. Dabkowski, and C.-J. Li, "Crystal growth of GaAs in space," in [19], p. 82; in [20], pp. 221–225.

381. H. C. Gatos and J. Lagowski, "Growth of GaAs – stoichiometry and defect structure," G. 1.1.3, in [17], p. 332.

382. I. Gyuro, E. Lendvay, T. Görög, M. Harsy, I. Pozsgai, K. Somogyi, F. Koltai, T. Lohner, J. Gyilai, M. Ranky, L. Varga, J. Giber, L. Bori, L. L. Regel, N. A. Kulchitsky, and V. T. Khryapov, "Crystal growth of GaSb under microgravity conditions," *Acta Astronaut.* **11**, No. 7–8, 361–368 (1984).

383. D. T. J. Hurle, "Melt and solution growth experiments involving semiconductors (ES 319, ES 321, ES 322, ES 323, ES 324)," in [20], p. 437.

384. H. Kolker, "Crystallization of a silicon sphere," in [19], p. 48; in [20], pp. 169–172.

385. E. Lendvay, M. Harsy, T. Görög, I. Gyuro, I. Pozsgai, F. Koltai, J. Gyulai, T. Lohner, G. Mezey, E. Kotai, F. Paszti, V. T. Khryapov, N. A. Kulchitsky, and L. L. Regel, "The growth of GaSb under microgravity conditions," *J. Crystal Growth* **71**, 538–550 (1985).

386. G. Nagel and K. W. Benz, "Traveling-heater growth of GaSb under reduced gravity during FSLP," in [16], p. 329; *Adv. Space Res.* **4**, No. 5, 23–26 (1984).

387. Nguyen Thanh Nghi and Nguyen Hoc, "Heterogeneous bubble nucleation during crystal growth from solution under microgravity conditions," in [16], p. 335.

388. Nguyen Thanh Nghi, "Influence of gravity on the nucleation and solidification of binary semiconductor alloys," G. 1.5.4, in [17], p. 338.

389. L. L. Regel and Nguyen Thanh Nghi, "On basic results of the joint Soviet–Vietnamese program 'Halong' on space semiconductor technology," *Acta Astronaut.* **11**, No. 3–3, 155–162 (1984).

390. L. L. Regel, R. V. Parfenjev, I. V. Vidensky, I. I. Farbstein, A. V. Mihailov, A. M. Turchaninov, N. K. Shulga, and B. T. Melch, "Semiconductor Te crystals. Te–Se, Te–Si alloys and directed crystallization of eutectic Al–Cu alloys under zero gravity conditions," IAF-86-283, in [23].

391. H. Rodot and O. Tottereau, "Lead telluride crystal growth under microgravity conditions," in [19], p. 40; in [20], pp. 135–139.

392. A. Rouzaud, D. Camel, and J. J. Favier, "Thermal and thermosolutional convective effect of orientation relative to gravity," in [16], p. 328; *Adv. Space Res.* **4**, No. 5, 3–8 (1984).

393. D. Schonholz, R. Dian, and R. Nitsche, "Solution growth in cadmium telluride under microgravity in Spacelab-1," in [19], p. 47; in [20], pp. 163–167.

394. V. S. Zemskov, I. N. Belocurova, I. L. Shulpina, and A. N. Titkov, "The structural features of germanium–silicon solution crystals obtained under microgravity," *Adv. Space Res.* **4**, No. 5, 11–14 (1984).

395. V. S. Zemskov, I. N. Belokurova, D. M. Khaizhu, A. N. Titkov, and I. L. Schulpina, "Microgravity effects on the segregation of components in growing germanium–silicon solid solution single crystals," G. 1.5.3, in [17], p. 338.

V. CRYSTAL GROWTH FROM THE VAPOR–GAS STATE UNDER MICROGRAVITY CONDITIONS

396. A. S. Drobyshev and A. G. Karpushin, "The effects of gravitational direction on the dynamics of gas–solid transition," in [10], pp. 77–79.

397. V. S. Zemskov, I. N. Belokurova, B. S. Vasilina, O. I. Mavrin, A. N. Titkov, and I. L. Shul'pina, "Crystallization of germanium–silicon solid solutions from the vapor state under zero gravity," in [11], p. 228.

398. V. S. Zemskov, I. N. Belokurova, B. S. Vasilina, O. I. Mavrin, A. N. Titkov, and I. L. Shul'pina, "The distribution of dopant silicon in germanium crystals grown from the vapor under zero gravity," in *Doped Semiconductor Materials* [in Russian], Nauka, Moscow (1985), pp. 145–149.

399. S. A. Kornev, N. A. Kul'chitskii, E. V. Markov, and V. T. Khryapov, "CdSe mass transfer kinetics in growth from the vapor phase under ground conditions and with microgravity," in [10], pp. 126–128.

400. S. A. Kornev, N. A. Kul'chitskii, E. V. Markov, V. T. Khryapov, and I. V. Lubashevskaya, "Techniques in growing large cadmium selenide crystals from the vapor by physical deposition under zero-gravity conditions," in [11], p. 286.

401. E. V. Markov, E. P. Prikop'ev, E. T. Solomin, V. D. Tarasov, L. A. Fokina, and V. T. Khryapov, "Simulating the growth and doping of single-crystal semiconductor epitaxial films from the vapor state under conditions of zero gravity," in [3], pp. 105–109.

402. H. Opperman and I. V. Barmin, "The effects of convection on the chemical transport of Ge by iodine," in [13], pp. 29–36.

403. S. I. Rozhkov, N. A. Kul'chitskii, V. M. Biryukov, V. T. Khryapov, and E. V. Markov, "The chemical transport of germanium by iodine under microgravity and under ground conditions for various directions of transport with respect to the gravitational field," in [10], pp. 129–130.

404. V. V. Smirnov, E. V. Markov, V. T. Khryapov, and Yu. N. Kuznetsov, "A study of mass transfer in the $ZnO-H_2-(Ar)$ gas-transport system," in [10], pp. 131–133.

405. V. V. Smirnov, E. V. Markov, V. T. Khryapov, and Yu. N. Kuznetsov, "Mass transfer in the $ZnO-H_2$ gas-transport system in experiments with the Kristall apparatus," in [10], pp. 133–135.

406. V. T. Khryapov, S. I. Rozhkov, N. A. Kul'chitskii, V. M. Biryukov, and E. V. Markov, "A study of germanium transfer in the Ge–GeI$_4$ system under microgravity," in [8], pp. 72–78.

407. V. P. Chegnov, V. M. Yazov, E. V. Markov, and V. T. Khryapov, "The growth of lead and tin telluride crystals from the vapor," in [10], p. 148.

408. M. Piechotka and E. Kaldis, "Effect of foreign gases on the vapor growth of α-HgI$_2$ crystals," G. 1.1.7, in [17], p. 333.

409. F. Rosenberger, "The absence of convective stability in incongruent crystal growth on earth," in [16], p. 333.

410. H. Wiedemeier and R. C. Whiteside, "Transport phenomena and crystal growth of the GeSe–Xenon system for different gravitation conditions," in [16], p. 328; *Adv. Space Res.* **4**, No. 5, 1 (1984).

411. H. Wiedemeier and S. B. Trivedi, "Chemical and physical vapor transport of Ge-chalcogenides under terrestrial and microgravity conditions," G. 1.1.6, in [17], p. 38.

VI. SOLIDIFICATION OF METALS, ALLOYS, COMPOSITES, AND EUTECTICS UNDER MICROGRAVITY

412. C. Barta and A. Triska, "Major problems in the Morava 1 and 2 experiments," in [13], pp. 12–20.

413. S. I. Budurov, P. D. Kovachev, S. A. Toncheva, N. G. Nenchev, F. R. Khashimov, V. T. Khryapov, V. A. Tatarinov, and A. S. Okhotin, "The formation and crystallization of an iron–zinc alloy under microgravity," in [13], pp. 60–63.

414. S. I. Budurov, P. A. Petrov, P. D. Kovachev, V. T. Khryapov, F. R. Khashimov, and V. A. Tatarinov, "Determining wetting angles under microgravity," in [13], pp. 64–66.

415. S. I. Budurov, S. A. Toncheva, P. D. Kovachev, A. S. Okhotin, I. G. Filatov, V. V. Ilyukhin, V. P. Shalimov, A. L. Ushkans, and T. A. Cherepanova, "Making foam aluminum under microgravity conditions," in [13], pp. 73–78.

416. G. M. Vlasov, "Making catalysts in zero gravity," in [10], pp. 76–77.

417. Yu. M. Gel'fgat, "Decomposition and component sedimentation in liquid-metal systems with immiscible regions in zero gravity and microgravity," in [7], pp. 57–71.

418. Yu. M. Gel'fgat, L. A. Gorbunov, and M. Z. Sorkin, "Forced homogenization in a multicomponent alloy in the absence of natural convection," in [10], pp. 23–25.

419. Yu. M. Gel'fgat and L. A. Gorbunov, "Liquid homogenization in multicomponent systems in zero gravity," *Izv. Akad. Nauk SSSR, Ser. Fiz.* **49**, No. 4, 667–672 (1985).

420. V. G. Zabotin, A. I. Osipov, A. I. Kosenko, and A. I. Pervyshin, "Features of gas cutting under vacuum," in [3], pp. 76–81.

421. V. V. Ilyukhin, V. P. Shalimov, S. I. Budurov, P. D. Kovachev, and S. A. Toncheva, "The occurrence of gas inclusions in melts under zero gravity," in [13], pp. 67–72.

422. V. A. Kazakov and S. P. Rusin, "Features of diffusion in a two-component alloy on heating under vacuum," in [12], p. 211.

423. V. F. Lapchinskii, "Some problems in welding thin sheet metal in space," in [2], Issue 18, pp. 9–14.

424. B. P. Mikhailov and N. A. Palii, "Results from ground trials on recrystallization under conditions of brief weightlessness for niobium–gallium alloys," in [11], p. 289.

425. B. E. Paton, D. A. Dudko, and V. F. Lapchinskii, "Welding processes in space," in *Welding and Special Electrometallurgy* [in Russian], Kiev (1984), pp. 121–129.

426. V. N. Pimenov, S. A. Maslyaev, I. P. Sasinovskaya, G. S. Birova, S Ya. Betsofen, and E. B. Rubina, "Features of crystallization in space for aluminum-base alloys," in [11], p. 246.

427. L. L. Regel', A. M. Durachenko, I. V. Videnskii, and L. Oyuunbiling, "The effects of microgravity on the effective diffusion coefficients in metallic liquids," in [10], pp. 152–153.

428. L. L. Regel', I. V. Videnskii, C. Potard, P. Morgand, and G. Nomen, "Surface morphology of immiscible Al–In alloys made under microgravity," in [10], pp. 154–155.

429. L. L. Regel', I. V. Videnskii, V. E. Morozov, T. V. Zhukov, and Shriramamurti Ramachandra Rao, "The structure and properties of silver–germanium alloys made under microgravity in the Soviet–Indian supercooling experiment with the Isparitel' M apparatus," in [12], pp. 210–211.

430. E. M. Savitskii, B. P. Mikhailov, M. I. Bychkova, R. S. Torchinova, and I. D. Giller, "The effects of crystallization under zero gravity on the structure and properties of superconducting and magnetic materials," in [3], pp. 23–31.

431. E. M. Savitskii, B. P. Mikhailov, N. A. Palii, B. A. Shekhtman, and V. S. Martyushov, "Making crystals from a supersaturated solution in the niobium–tin system," in [11], p. 228.

432. I. P. Sasinovskaya, S. A. Maslyaev, V. N. Pimenov, and S. Ya. Betsofen, "Crystallization in the copper–aluminum system under zero gravity," in [11], pp. 226–227.

433. V. A. Tatarchenko, S. K. Brantov, L. V. Leskov, V. L. Levtov, and M. S Agafonov, "Crystallization of indium by Stepanov's method under microgravity," *Izv. Akad. Nauk SSSR, Ser. Fiz.* **49**, No. 4, 708–710 (1985).

434. R. S. Torchinova, A. S. Ilyushin, A. A. Nikolaev, and I. A. Nikanorova, "The crystal structure of an intermetallide crystallized under zero gravity," in [11], pp. 243–244.

435. E. G. Fuks, A. Roos, and G. Buza, "Metallurgical research in space under the Bealuts program," in [13], pp. 101–115.

436. T. A. Chernyshova, V. N. Meshcheryakova, L. I. Kobeleva, and M. P. Arsent'eva, "Wetting in the carbon–silumin system under earth conditions and under zero gravity," in [10], pp. 149–150.

437. T. A. Chernyshova, V. N. Meshcheryakova, L. I. Kobeleva, and M. P. Arsent'eva, "The wetting of carbon materials by aluminum alloys under zero gravity," in [11], p. 250.

438. Yu. V. Cheshlya, A. A. Babareko, and V. V. Romanov, "Structure of a eutectic silver–copper alloy melted under zero gravity," in [11], p. 223.

439. Yu. V. Cheshlya, "Microstructures of eutectic alloys crystallized under zero gravity and on the ground in capillaries," in [11], p. 245.

440. "Electron-beam technology in open space," *Avtom. Svarka*, No. 12, 1–2 (1984).

441. H. Ahlborn and J. Hogel, "Influences affecting separation in monotectic alloys under microgravity," in [18], pp. 37–40.

442. H. Ahlborn and K. Lohberg, "Influences affecting separation in monotectic alloys under microgravity," in [19], p. 25; in [20], pp. 55–61.

443. W. H. M. Alsem, T. Luyendijk, and H. Nieswag, "ES-325 unidirectional solidification of cast iron," in [19], p. 28; in [20], pp. 79–85.

444. F. Barbier, C. Patuelli, P. Gonndi, and R. Montanari, "Melting and solidification in 0 *g* of sintered alloys. Experiment ES 311 A and B: bubble reinforced materials," in [19], p. 31; in [20], pp. 101–107.

445. F. Barberi, P. Gondi, R. Montanari, and C. Patuelli, "Experiment ES-311: bubble reinforced materials," *Earth-Orient. Appl. Space Technol.* **5**, No. 1–2, 57–62 (1985).

446. A. Borgman, H. Fredriksson, and H. Shahani, "On the mechanism of the coalescence process in immiscible alloys," in [22], p. 45.

447. J. Bert, D. Henry, P. Layani, G. Chuzeville, J. Dupuy, and B. Roux, "Space experiment on thermal diffusion: preparation and theoretical analysis," in [19], p. 75; in [20], pp. 347–352.

448. B. Billia, H. Jamgotchian, J. J. Favier, and D. Camel, "Cellular morphologies in lead–thallium alloys," in [20], pp. 409–412.

449. B. Billia, H. Jamgotchian, J. J. Favier, and D. Camel, "Morphological instabilities during unidirectional solidification of Pb–Ti alloys: influence of the gravity field orientation," in [16], p. 331.

450. D. Camel, I. I. Favier, and M. D. Dupouy, "Preliminary results of the D-I-WL-GHF-04 experiment on dendritic solidification of Al–Cu alloys," G. 1.5.6, in [17], pp. 338–339.

451. P. D. Caton and W. G. Hopkins, "The influence of microgravity on the solidification of Al–Pb alloys: results of SL-1 experiment IES-309," in [19], p. 24.

452. P. F. Clancy and W. Heide, "Acoustic mixing of an immiscible alloy (Pb–Zn) in microgravity," in [18], pp. 73–77.

453. T. W. Clyne and P. J. Goodhew, "Formation and retention of the aluminide dispersions in aluminum," in [18], pp. 71–72.

454. A. Deruytterre and L. Froyen, "Melting and solidification of metallic composites," in [18], pp. 65–67.

455. A. Deruyterre and L. Froyen, "Melting and solidification of metallic composition in space," G. 1.5.1, in [17], pp. 337–338.

456. A. M. Durachenko and L. L. Regel, "Crystallization of model alloys in microgravity and uniform compression conditions," in [22], pp. 202–203.

457. J. J. Favier and J. De Goer, "Directional solidification of eutectic alloys," in [19], p. 39; in [20], pp. 127–133.

458. J. J. Favier, "FSLP experiments in metallurgy," in [20], pp. 435–436.

459. H. F. Fischmeister, A. Kneissl, R. Pfefferkorn, and W. Trimmel, "Solidification and Ostwald ripening of near monotectic Zn–Pb alloys: Spacelab experiment IES 313," in [18], pp. 41–42.

460. H. F. Fischmeister and H. E. Exner, "Ostwald ripening, interfacial energies, and diffusion coefficients in monotectic systems," in [18], pp. 69–70.

461. H. Fredriksson, "Space results on the solidification of immiscible alloys," in [18], pp. 25–34.

462. H. Fredriksson, "The effect of the temperature gradient on the solidification of immiscible alloys on earth and in space," in [16], p. 330.

463. G. Frohberg, K.-H. Kraatz, and H. Weber, "Self-diffusion of Sn^{112} and Sn^{114} in liquid tin," in [19], p. 56; in [20], pp. 201–205.

464. L. Froyen and A. Deruytterre, "Melting and solidification of metal matrix composites under microgravity," in [19], p. 27; in [20], pp. 69–78.

465. L. Froyen and A. Deruytterre, "Melting and solidification of metallic composites in Spacelab," *Phys. Mag.* **6**, No. 2, 133–141 (1984).

466. E. G. Fuchs, A. Roosz, and G. Buza, "Das werkstofftechnologische Weltraumexperiment BEALUCA," *Z. Metallk.* **75**, No. 3, 185–195 (1984).

467. S. H. Gelles and A. J. Markworth, "Space Shuttle experiments on Al–In liquid phase miscibility gap (LPMG) alloys," in [19], p. 86; in [20], pp. 417–422.

468. S. H. Gelles and A. J. Markworth, "Space Shuttle experiments on Al–In liquid phase miscibility gap (LPMG) alloys," in [16], p. 330.

469. M. E. Glicksman, "Implementation of space flight experiments concerning dendritic growth," in [16], p. 331.

470. H. K. Henson and K. E. Drexler, "Method for processing and fabricating metals in space," US Patent No. 480,677, November 6, 1984.

471. W. G. Hopkins, "Solidification of Al–Pb alloys under microgravity in TEXUS-7: preliminary report," in [18], pp. 83–86.

472. Huang Tao, Lu Deyang, and Zhou Yaohe, "Diffusion convection effects on constrained dendritic growth in dilute alloys," IAF-86-280, in [23].

473. M. Kelley and E. Ethridge, "Containerless solidification of NiAl," *Microgravity Science and Application Program Tasks*, NASA T.M. 87568, May 1985, p. 66.

474. H. Klein, H. U. Walter, A. Bewersdorf, and J. Pottschke, "Observation of processes in a solidifying dispersion," in [16]; *Adv. Space Res.* **4**, No. 5, 57–62 (1984).

475. H. Klein, R. Nähle, and K. Wanders, "Transport processes in directional solidification during space shuttle flight," in [19], p. 87; in [20], pp. 375–378.

476. H. Klein, R. Nähle, and K. Wanders, "Metal-like solidification of a multiphase dispersion in low gravity during a space shuttle flight," *Z. Flugwiss. Weltraumforsch.* **9**, No. 1, 14–20 (1985).

477. A. Kneissl and H. F. Fischmeister, "Schmelzen und Estarren von übermonotektischen Zink-Blei-Legierungen unter Schwerlosigkeit," *Metall (W. Berlin)* **38**, No. 9, 831–837 (1984).

478. A. Kneissl and H. Fischmeister, "The behavior of hypermonotectic zinc–lead alloys during melting and resolidification under microgravity," in [19], p. 26; in [20], pp. 63–68.

479. V. Laxmanan, "Dendritic solidification in a binary alloy melt. Steady state versus morphological stability theories," in [16], p. 80; in [20], pp. 403–407.

480. "Lotexperiment der BAM beim Spacelab-Flug durchgeführt," *TÜ* **25**, No. 3, 100 (1984).

481. R. Lück and B. Predel, "Thermodynamics of the decomposition of liquid alloys," in [18], pp. 13–23.

482. T. Luyendijk, H. Nieswaag, and W. H. M. Alsem, "Unidirectional solidification of cast iron," *Science* **225**, No. 4658, 200–202 (1984).

483. T. Luyendijk and H. Nieswaag, "Serste resultaten van net gietijzer experiment in Spacelab," *Lastechniek* **51**, No. 2, 22–26 (1985).

484. Y. Malmejacc and J. P. Praizey, "Nucleation of eutectics (IES 312)," in [19], p. 33.

485. Y. Malmejac and J. P. Praizey, "Thermomigration of cobalt in liquid tin (IES 320)," in [19], p. 42; in [20], pp. 147–152.

486. B. P. Michailov, R. S. Torchinova, and M. I. Bychkova, "The results of space technological experiments performed with the superconducting and magnetic alloys," in [22], p. 201.

487. P. Morgand and C. Potard, "Immiscible alloys experiment aboard Salyut 7: results and interpretation of aluminum–indium emulsion solidified in microgravity," in [16], p. 331; *Adv. Space Res.* **4**, No. 5, 53–56 (1985).

488. G. Müller and P. Kyr, "Directional solidification of InSb–NiSb eutectic," in [19], p. 41; in [20], pp. 141–146.

489. G. Otto, "Stability of metallic dispersions," in [19], p. 69; in [20], pp. 379–388.

490. G. H. Otto, "First results of a MAUS experiment to investigate the stability of a metallic dispersion," in [18], pp. 43–46.

491. T. Persson, "An *in situ* technique for the determination of electrotransport in liquid metallic alloys," in [16], p. 335; *Adv. Space Res.* **4**, No. 5, 81–84 (1984).

492. V. N. Pimenov, V. N. Kubasov, and L. I. Ivanov, "Influence of zero gravity state on the crystallization of metallic materials," *Acta Astronaut.* **11**, No. 10–11, 687–690 (1984).

493. G. Poletti and D. Cambiaghi, "Surface forces in contacting solids," in [19], p. 74; in [20], pp. 341–346.

494. C. Potard, "Filtration-theory approach to immiscible-alloys solidification," in [18], pp. 79–82.

495. C. Potard and P. Morgand, "Directional solidification of a vapor emulsion aluminum–zinc in microgravity: first analysis results," in [19], p. 38; in [20], pp. 121–125.

496. J. P. Praizey, "Thermomigration in liquid metallic alloys," G. 1.3.1, in [17], p. 335.

497. L. L. Regel, I. V. Vidensky, M. Potard, and P. Morgand, "Surface structure of Al–In alloy obtained under microgravitation," in [21], p. 156.

498. H. Riepl, "Surface examination of nickel catalysts processed under microgravity," in [19], p. 78; in [20], pp. 359–363.

499. F. Rossito and G. Chersini, "Experiment IES-340, 'Adhesion of metals': analysis of results," in [16], p. 57; in [17], pp. 207–211.

500. A. K. Sample and A. Hellawell, "The mechanisms of formation and elimination of channels segregation during alloy solidification," in [16], p. 338.

501. K. Sasabe and E. Siegfried, "Flow process in capillary gaps during brazing," *Schmäss. Schneid*, No. 11, E192–E195 (1985).

502. E. M. Savitsky, R. S. Torchinova, A. S. Ilushin, A. A. Nikolaev, and I. A. Nikanorova, "Some peculiarities of the crystal structure and magnetic properties of the intermetallic Gd_3Co compound crystallized in microgravity conditions," in [21], pp. 157–158.

503. E. M. Savitsky, R. S. Torchinova, and S. A. Turanov, "Influence of crystallization in microgravity conditions on the structure and magnetic properties of GdCo and $(Gd_{0.2}Tb_{0.8})_3Co$ compounds," *Acta Astronaut.* **11**, No. 3–4, 185–188 (1984).

504. E. M. Savitsky, R. S. Torchinova, A. S. Ilushin, A. A. Nikolaev, and I. A. Nikanorova, "X-ray analysis of the crystal structure and thermal expansion of the intermetallic compound $(Gd_{0.2}Tb_{0.8})_3Co$ crystallized in space," *Acta Astronaut.* **11**, No. 10–11, 691–694 (1984).
505. R. J. Schaeffer and S. R. Coriell, "Solid–liquid interface distortions due to convection," in [16], p. 311.
506. E. Siegfried, K. Frieler, and R. Stickler, "Vacuum brazing under microgravity. Experiment IES 304/305," in [19], p. 30; in [20], pp. 95–99.
507. D. M. Stefanescu and J. C. Hendrix, "Low-gravity effects on the solidification of flake and spheroidal graphite cast irons," in [16], p. 332.
508. S. Takahaschi, "Melting and hardening of metal under weightlessness," *Zairyo Kagaku, J. Mater. Sci. Soc. Jpn.* **20**, No. 6, 305–308 (1984).
509. R. S. Torchinova, "The summary of results on crystallization in microgravity conditions of the R_3Co intermetallic compounds," IAF-86-282, in [23].
510. I. Vera, "The casting and mechanism of formation of semipermeable membranes in a microgravity environment," G. 1.4.2, in [17], p. 336.
511. M. D. Vignon, D. Camel, and J. J. Favier, "Order of magnitude analysis of convective effects in dendritic solidification," *Acta Astronaut.* **12**, No. 4, 257–263 (1985).
512. H. U. Walter, "Preparation of dispersion alloys – component separation during cooling and solidification of dispersions of immiscible alloys," in [18], pp. 47–64.
513. H. U. Walter, "FSLP experiments on composite materials," in [20], pp. 425–433.

VII. GLASS SOLIDIFICATION UNDER MICROGRAVITY

514. K. Gert, K. Hilbert, D. Unangst, A. S. Okhotin, and I. V. Barmin, "The effects of microgravity on inhomogeneity distributions in molten glass," in [13], pp. 21–28.
515. M. Braedt, V. Braetsch, and G. H. Frishat, "Interdiffusion in the glass melt system $(Na_2O + Rb_2O)\cdot 3SiO_2$," in [19], p. 32; in [20], pp. 109–112.
516. R. H. Doremus, "Crystallization and melting of fluoride glasses with and without a container," in [16], p. 335.
517. G. H. Frishat, "Microgravity research in glasses and ceramics," *J. Brit Interplanet. Soc.* **39**, No. 2, 90–91 (1986).
518. D. M. Herlach, F. Gillessen, and R. Willnecker, "Undercooling investigations on the easy glassformer Pd–Cu–Si," in [19], p. 77; in [20], pp. 389–397.
519. D. M. Herlach, R. Willnecker, and F. Gillessen, "Containerless undercooling of Ni," in [20], pp. 399–402.
520. N. Y. Kreidel, "Glass experiments in space," *J. Non-Crystalline Solids* **80**, No. 1–3, 587–593 (1986).
521. Moriya Ioshiro, "Processing glass and ceramic materials in space," *Zairyo Kagaku, J. Mater. Sci. Soc. Jpn.* **20**, No. 6, 300–305 (1984).
522. V. Rosenkrantz and G. H. Frischat, "Shrinking of a gas bubble in a glass melt under microgravity conditions," in [19], p. 76; in [20], pp. 353–357.
523. F. Spaepen, "Crystal nucleation and glass formation in metallic alloy melts," in [20], pp. 215–219.

524. M. C. Weinberg, "A two-component gas bubble in a glass melt in low gravity," in [16], p. 334.

VIII. CRYSTAL GROWTH FROM AQUEOUS SOLUTION UNDER MICROGRAVITY

525. A. A. Vedernikov, I. V. Melikhov, D. G. Berdonosova, M. A. Prokof'ev, V. V. Ilyukhin, and V. I. Lisoivan, "Surface morphology and internal defects in $CuSO_4 \cdot 5H_2O$ crystals grown under microgravity," in [10], p. 160.
526. A. A. Vedernikov and I. V. Melikhov, "Mass crystallization under microgravity," in [13], pp. 141–157.
527. "Crystal clear," *Space World*, No. V-9-261 (1985).
528. G. Galster and K. F. Nielsen, "Crystal growth from solution," in [19], p. 54; in [20], pp. 189–191.
529. V. V. Ilykhin, L. D. Ishkhakova, V. F. Komarov, I. V. Melikhov, M. A. Prokofiev, V. K. Trunov, and A. A. Vedernikov, "Peculiarities of crystallization process in the solution under microgravity (Erdenet experiment)," *Acta Astronaut.* **11**, No. 10–11, 651–658 (1985).
530. W. Littke and Ch. John, "Protein single-crystal growth under microgravity," in [19], p. 53; in [20], pp. 185–188.
531. W. Littke and Ch. John, "Protein single-crystal growth under microgravity," *Earth-Orient. Appl. Space Technol.* **5**, No. 1–2, 63–66 (1985).
532. M. C. Robert, F. Lefaucheux, and A. Authier, "Simulation and results of a Spacelab growth experiment. Growth and characterization of brushite and lead monetite," in [19], p. 55; in [20], pp. 193–199.
533. M. C. Robert and F. Lefaucheux, "Solution growth under microgravity," G. 1.1.2, in [17], p. 332.

IX. APPARATUS FOR SPACE EXPERIMENTS, METHODS, LEVITATION DEVICES, AND METHODS OF SIMULATING ZERO-GRAVITY PROCESSES ON THE EARTH

534. V. S. Avduevskii, M. S. Agafonov, S. D. Grishin, V. L. Levtov, L. V. Leskov, V. V. Romanov, and V. V. Savichev, "Measurements in fluid mechanics and heat and mass transfer under zero gravity made by means of the Pion apparatus," in [8], pp. 15–29.
535. V. S. Avduevskii, M. S. Agafonov, N. A. Anfimov, S. V. Ermakov, V. L. Levtov, V. V. Romanov, V. V. Savichev, and A. I. Feonychev, "Results on fluid mechanics and on heat and mass transfer obtained with the Pion system on the Salyut-6 station," in [11], pp. 248–249.
536. V. S. Avduevskii, M. S. Agafonov, S. D. Grishin, V. L. Levtov, L. V. Leskov, V. V. Romanov, and V. V. Savichev, "Experimental studies on the fluid mechanics of zero gravity," in [10], pp. 117–119; *Izv. Akad. Nauk SSSR, Ser. Fiz.* **49**, No. 4, 681–686 (1985).
537. M. S. Agafonov, S. D. Grishin, V. N. Golubev, L. A. Dedyukov, V. A. Korolev, V. L. Levtov, V. A. Novikov, V. V. Romanov, and A. F. Frolov, "An apparatus for examining features of zero gravity," in [11], pp. 279–280.

538. M. S. Agafonov, I. R. Kuznetsov, and V. V. Savichev, "A promising automatic apparatus for space manufacturing," in [6].

539. W. Eichler, K. Stecker, H. Zussmann, R. Kuhl, H. Spetke, V. M. Truzhenikov, M. B. Shcherbina-Samoilova, A. I. Kotenkov, G. V. Zhukov, A. A. Il'in, N. V. B'yung, N. Kh. Kuen, N. V. Sung, and Ch. S. Hoai, "The Imitator 1 and 2 experiments on temperature profiles in the Kristall oven," in [13], pp. 52–59.

540. P. F. Ar'kov, A. L. Barannikov, N. M. Ganzherli, S. B. Gurevich, V. N. Golubev, V. B. Konstantinov, V. A. Korolev, I. A. Maurer, B. F. Ryadinskii, V. N. Sobolev, N. S. Cheberyak, and D. F. Chernykh, "Operation of the KGA-2 apparatus under laboratory conditions and in the Salyut-7 station," in [12], pp. 213–214.

541. P. B. Bairamov, N. R. Korneev, and O. A. Luchev, "Solar furnaces in space engineering," in [11], p. 283.

542. P. B. Bairamov, N. R. Korneev, and V. V. Shokin, "A mirror furnace on a space vehicle," in [11], pp. 283–284.

543. A. L. Barannikov, N. M. Ganzherli, S. B. Gurevich, V. B. Konstantinov, I. A. Maurer, S. A. Pisarevskaya, B. F. Ryadinskii, A. A. Serebrov, V. N. Sobolev, V. M. Stolovitskii, M. S. Cheberyak, and D. F. Chernykh, "Holography in electrophoresis research on Salyut 7," Pis'ma Zh. Éksp. Teor. Fiz., No. 11, 659–662 (1985).

544. A. L. Barannikov, N. M. Ganzherli, S. B. Gurevich, V. B. Konstantinov, I. A. Maurer, S. A. Pisarevskaya, B. F. Ryadinskii, A. A. Serebrov, V. P. Savinnykh, V. N. Sobolev, M. S. Cheberyak, and D. F. Chernykh, "Use of holography in space," in [5], pp. 122–126.

545. A. L. Barannikov, N. M. Ganzherli, S. B. Gurevich, V. B. Konstantinov, B. F. Ryadinskii, V. N. Sobolev, O. D. Ustimenko, M. N. Tsaplin, M. S. Cheberyak, and D. F. Chernykh, "Holographic TV methods in research on heat and mass transfer," in [10], pp. 186–188.

546. A. L. Barannikov, N. M. Ganzherli, S. B. Gurevich, V. B. Konstantinov, B. F. Ryadinskii, V. K. Samsonov, V. N. Sobolev, M. S. Cheberyak, and D. F. Chernykh, "Holographic TV methods in research on physical processes under zero gravity," in [11], pp. 292–293.

547. A. L. Barannikov, N. M. Ganzherdi, S. B. Gurevich, V. B. Konstantinov, I. A. Maurer, B. F. Ryadinskii, Yu. P. Semenov, V. N. Sobolev, O. D. Ustimenko, N. M. Tsaplin, M. S. Cheberyak, and D. F. Chernykh, "Holographic TV methods in research on heat and mass transfer," Izv. Akad. Nauk SSSR, Ser. Fiz. 49, No. 4, 711–714 (1985).

548. I. V. Barmin, V. A. Briskman, B. I. Burshtein, V. K. Butoshchin, V. A. Bushuev, V. P. Zharov, A. S. Senchenkov, and A. A. Shevchenko, "An apparatus for examining hydrodynamic phenomena under zero gravity," in [10], pp. 188–189.

549. I. V. Barmin, A. V. Egorov, A. V. Korovin, V. S. Martyushov, M. N. Smirnova, and B. A. Shekhtman, "Ground developments of experiments to be performed with the Splav-01 apparatus," in [11], p. 288.

550. I. V. Barmin, L. V. Leskov, A. V. Egorov, M. G. Bulychev, A. M. Dolbin, G. M. Luk'yanova, and A. N. Kurilov, "Methodological aspects of technological experiments," in [6].

551. I. V. Barmin, A. V. Egorov, and A. S. Senchenkov, "Some results obtained with the Splav-01 apparatus," in [13], pp. 158–163.

552. I. V. Barmin and S. N. Gavritskii, "Temperature fluctuations in an electrically heated chamber with automatic control," in [21], p. 220.

553. N. A. Bezdenezhnykh, B. I. Burshtein, G. V. Puzanov, and T. M. Sharov, "Laboratory simulation of the equilibrium values and fluctuations in the free surface of a liquid under zero gravity subject to vibrational loading," in [10], pp. 189–191.

554. M. Yu. Belyaev, S. G. Venevtsev, V. M. Stazhkov, and V. P. Teslenko, "The physical conditions on research space vehicle," in [4], pp. 18–26.

555. V. S. Berdnikov, N. M. Ganzherli, S. B. Gurevich, and I. A. Maurer, "Free convection in a closed cavity examined by real-time holographic interferometry," Preprint FTI im. A. F. Ioffe AN SSSR, No. 996, 18 (1986).

556. N. L. Bondarenko, A. A. Dubova, V. A. Novikov, A. F. Frolov, V. F. Parvov, and E. V. Kartashov, "A precision thermostat for growing crystals from solution," in [11], p. 281.

557. A. V. Bochkarev, L. L. Regel', V. A. Tatarinov, and V. P. Shalimov, "Results from the Soviet–French Kalibrovka experiment (calibration)," in [12], p. 212.

558. V. A. Briskman, B. I. Burshtein, M. P. Zavarykin, S. B. Zorin, G. F. Putin, A. F. Pshenichnikov, and G. V. Yastrebov, "Instruments and methods for earth simulation of nongravitational heat and mass transfer mechanisms," in [11], p. 296.

559. Sh. A. Vakhidov, V. V. Vishnevskii, Yu. G. Reutov, Yu. E. Tetelya, V. P. Usenko, and A. N. Kulashov, "Use of an electrodynamic fluidized bed for thermostatic control under zero gravity," in [10], pp. 67–68.

560. Sh. A. Vakhidov, V. Kh. Gataullin, and I. Kh. Tukhvatullin, "The Biryuza and Analiz instruments for examining physical processes under zero gravity," in [10], p. 191.

561. Sh. A. Vakhidov and V. Kh. Gataullin, "A gravitationally sensitive method of examining melting and crystallization," in [10], p. 191.

562. Sh. A. Vakhidov, V. Kh. Gataullin, V. P. Nikitskii, I. Kh. Tukhvatullin, V. I. Smyshlyaev, G. V. Zhukov, M. N. Mukhsinov, and V. S. Dmitriev, "An apparatus suite for examining physicochemical processes under zero gravity," in [11], p. 284.

563. Sh. A. Vakhidov, V. Kh. Gataullin, V. P. Nikitskii, I. Kh. Tukhvatullin, and G. V. Zhukov, "A method of monitoring and measuring thermophysical properties of materials during melting and crystallization under zero gravity," in [11], p. 292.

564. L. M. Visloguzov, M. P. Elagin, A. L. Lebedev, and V. I. Polezhaev, "Methods and apparatus for ground simulation of hydrodynamic processes under zero gravity," in [11], pp. 294–295.

565. P. N. Voz'milov, A. V. Solovkin, E. E. Solomonova, and V. T. Khryapov, "A standardized hardware and software suite for automating experiments in space semiconductor materials science," in *Proceedings of the 3rd All-Union Symposium on Modular Data-Acquisition Systems*, February 16–18, 1982, Topic: Microcomputers and Microprocessors in Automated Systems [in Russian], Moscow (1983), pp. 181–184.

566. N. M. Ganzherli, S. B. Gurevich, I. A. Maurer, and D. F. Chernykh, "Devising faster development methods for real-time holographic interferometry," in *State of the Art and Development Prospects in Optical Methods of*

Data Transmission, Storage, and Processing [in Russian], LFTI AN SSSR (1984), pp. 198–204.

567. N. M. Ganzherli, S. B. Gurevich, V. B. Konstantinov, I. A. Maurer, S. A. Pisarevskaya, D. F. Chernykh, and M. S. Chebryak, "Facilities of holographic apparatus of KGA type," in [11], p. 285.

568. G. I. Glikman, V. E. Pashkov, and V. P. Sborets, "Evacuation systems," in [11], pp. 281–282.

569. S. D. Grishin, V. B. Dubovskoi, S. S. Obydennikov, and V. V. Savichev, "A study of microaccelerations on board the Salyut 6 orbiting space station," in [8], pp. 6–14.

570. N. I. Gul'ko, E. T. Solomin, and A. B. Shagin, "A multipurpose control system for technological processes operated under special conditions," in [11], p. 277.

571. N. I. Gul'ko, E. T. Solomin, V. D. Tarasov, and A. V. Shagin, "The controlled system in the Korund-1 MNTs multifunctional technological plant," in [11], p. 278.

572. S. B. Gurevich, V. B. Konstantinov, and D. F. Chernykh, "Holography in space research," *Vestn. Akad. Nauk SSSR*, No. 3, 44–53 (1983).

573. L. A. Dedyukov, L. A. Gritsaenko, V. A. Korolev, V. A. Novikov, A. F. Frolov, V. N. Golubev, A. G. Barannikov, and N. L. Bondarenko, "An apparatus for model experiments on heat and mass transfer under zero gravity," in [10], pp. 191–193.

574. V. B. Dubovskoi, V. I. Leont'ev, S. S. Obydennikov, V. A. Savichev, and A. B. Sevast'yanov, "The physical conditions on board a space vehicle," in [3], pp. 65–75.

575. V. B. Dubovskoi, V. I. Leont'ev, S. S. Obydennikov, and V. V. Savichev, "Measuring accelerations during engineering experiments with the PION-M apparatus on board the Salyut 7," in [12], pp. 212–213.

576. M. P. Elagin, A. P. Lebedev, V. G. Petukhov, V. I. Polezhaev, A. V. Shmelev, and É. T. Chekunov, "Measuring salt concentrations in model liquids by an oscillographic method: suggestions for the Diffuziya experiment on earth satellites," in [11], pp. 293–304.

577. V. I. Zhimailov, V. P. Nikitsii, S. B. Ryabukha, M. S. Khlystunov, V. V. Lebedev, and A. A. Serebrov, "Vibrational acceleration measurements on Salyut 6 and Salyut 7," in [11], p. 299.

578. G. V. Zhukov, A. A. Zagrebal'nyi, V. F. Lapchinskii, L. O. Neznamova, V. P. Nikitskii, and V. V. Stesin, "Apparatus and technology for depositing thin films for various purposes in space," in [11], p. 300.

579. M. P. Zavarykin, K. G. Kostarev, G. F. Putin, and A. F. Pshenichnikov, "Laboratory simulation of convection in weak gravitational fields," in [10], pp. 193–195.

580. A. A. Zagrebal'nyi, V. F. Lapchinskii, V. V. Stesin, and Yu. M. Chernitskii, "Research on melt crystallization under zero gravity by means of transparent models," in [2], No. 18, pp. 14–17.

581. D. A. Kolpakov, E. T. Solomin, and A. V. Shagin, "A precision multichannel temperature meter," in [11], pp. 278–279.

582. N. R. Korpeev and O. A. Luchev, "Formation of a planar crystallization front in zone melting in a radiative furnace," in [11], p. 247.

583. N. R. Korpeev, O. A. Luchev, and V. V. Shokin, "Functions of mirror furnaces in technology," in [11], pp. 282–283.

584. A. P. Lebedev and V. I. Polezhaev, "Calculations on microaccelerations on satellites and recommendations on reducing them," in [11], pp. 245–246.
585. A. P. Lebedev and V. I. Polezhaev, "Perturbing accelerations on an orbital station and their effects on impurity distribution in liquids," in [10], pp. 180–181.
586. L. V. Leskov and V. V. Romanov, "Hydrooptic systems under zero gravity," in [10], pp. 195–197.
587. E. T. Makrushin and É. F. Reut, State of the art and development prospects for temperature measurements in space plants," in [11], p. 291.
588. B. E. Paton, V. D. Shelyagin, O. K. Nazarenko, Yu. N. Lankin,, V. K. Mokhnach, Yu. V. Neporozhnii, E. N. Baishtruk, V. I. Kirienko, V. I. Pekker, Yu. I. Drabovich, and G. F. Pazeev, "A high-power miniature electron-beam apparatus for technological operations and physics experiments under space conditions," in [2], No. 18, pp. 3–9.
589. Yu. P. Perfil'ev, A. S. Senchenkov, I. G. Smirnova, and S. M. Pchelintsev, "Simulating the temperature patterns in systems," in [11], pp. 249–250.
590. Yu. S. Pronin, "A method of choosing equipment structure for space-technology experiments," in [4], pp. 32–38.
591. L. L. Regel', M. B. Shcherbina-Samoilova, A. V. Bochkarev, V. P. Shalimov, G. V. Zhukov, A. I. Kotenkov, V. D. Leonov, V. N. Truzhennikov, V. A. Tatarinov, H. Kwass, F. Winkler, G. Kell, M. Günter, R. Kuhl, H. Spetke, and H. Zussmann, "The ARP apparatus: methods and experiments, with the first results obtained on the ground and on board the Salyut 7 plus Soyuz orbital vehicle," in [12], p. 209.
592. L. L. Regel', M. B. Shcherbina-Samoilova, V. P. Shalimov, G. V. Zhukov, A. I. Kotenkov, V. D. Leonov, V. N. Truzhenkikov, V. A. Tatarinov, R. Kuhl, H. Zussmann, H. Kwass, G. Kell, F. Winkler, H. Spetke, and V. Khailiger, "Tests on an automatic recorder for parameters under ground and space conditions," Preprint No. 1058, IKI AN SSSR, Moscow (1986).
593. L. L. Regel', V. P. Shalimov, D. D. Dryuchenko, I. V. Videnskii, A. I. Golubkov, A. V. Mikhailov, A. M. Turchaninov, and M. B. Shcherbina-Samoilova, "A ground development system for materials science in space based on the Kristallizator apparatus," Preprint No. 1101, IKI AN SSSR, Moscow (1986).
594. Yu. S. Ryazantsev, V. N. Yurechko, V. F. Mandzhikov, V. A. Barachevskii, I. A. Vysotskii, and A. V. Semenov, "The motion of a liquid in a closed volume examined by photochromic visualization," in [10], pp. 197–198.
595. V. A. Semenov, "Levitation of a sphere in an electrostatic field," in [10], p. 184.
596. V. M. Truzhenikov, V. A. Tatarinov, Yu. M. Zhelannyi, V. S. Kazachok, V. I. Seliverstov, and V. P. Nikitskii, "Analyzing temperature patterns in the Magma electric heating apparatus," in [10], pp. 143–145.
597. V. Tuchkevich and S. Gurevich, "Holography enters space," *Pravda*, December 1, 1983.
598. V. M. Tuchkevich, Yu. P. Semenov, and S. B. Gurevich, "Holography colonizes space," *Zemlya Vselennaya*, No. 3, 17–24 (1984).
599. V. M. Tuchkevich and S. B. Gurevich, "Holography in earth orbit," *Nauka SSSR*, No. 3, 35–39 (1985).
600. F. L. Fal'kon and P. V. Peres, "Zone melting with a temperature gradient in the water–sucrose system under zero-gravity conditions," *Kosm. Issled.* **23**, No. 3, 388–492 (1985).

601. V. T. Khryapov, E. V. Markov, N. A. Kul'chitskii, V. T. Solomin, P. N. Voz'milov, V. I. Seliverstov, and Yu. M. Zhelannyi, "Technology in semiconductor materials science in space and the creation of specialized engineering equipment," *Elektron. Promst.*, No. 1, 77–80 (1983).

602. V. T. Khryapov, V. I. Seliverstov, and Yu. M. Zhelannyi, "Multipurpose and specialized apparatus for an NTs module," in [11], p. 277.

603. Yu. D. Chashechkin, "Optical and probe methods of monitoring the parameters of process models under ground and orbital conditions," in [11], pp. 290–291.

604. V. S. Avduevsky, S. D. Grishin, and L. V. Leskov, "Research of heat capillarity convection features in nongravitation," in [21], p. 159.

605. C. Barta, A. Triska, J. Trnka, and L. L. Regel, "Experimental facility for materials research in space (CSL-1)," in [19]; in [20], pp. 413–415.

606. C. Barta, A. Triska, J. Trnka, and L. L. Regel, "Experimental facility for materials research in space, CSR-1," in [16], p. 337; *Adv. Space Res.* **4**, No. 5, 95–98 (1984).

607. D. Baum, G. Otto, and P. Vits, "MAUS – a flight opportunity for automated experiments under microgravity conditions," *Acta Astronaut.* **11**, No. 3–4, 239–245 (1984).

608. P. Behrmann and M. Eyb, "Spiegelofen für das Labor im Weltraum," *Lab. Prax, Sonderpubl.; Labor 2000*, 8–11 (1984–5).

609. A. Bernard, J. P. Canny, R. Juillerat, and P. Toubol, "Electrostatic suspension of samples in microgravity," in [21], p. 163.

610. R. Boudreault, "Atmospheric induced aerodynamic accelerations onto a free flying microgravity platform," in [19], p. 61.

611. G. Cambon, B. Castets, O. Boutemy, J. P. Gicquel, R. Aubron, M. Boddaert, and J. Benoit, "Gradient heating facility behavior," in [19], p. 37; in [20], pp. 115–119.

612. H. K. F. Ehlers, S. Jacobs, and L. J. Leger, "Space shuttle contamination measurements for flights STS-1 through STS-4," *J. Spacecraft Rockets* **21**, No. 3, 301–308 (1984).

613. M. Eklof and T. Persson, "A gravity-sensitive *in situ* technique for studies of the thermal conductivity of fluids," in [16], p. 337.

614. F. L. Falcon, P. V. Perez, and G. F. Znukov, "Fusion zonal congradiente de temperatura en el sistems H_2O–sacorosa en condiciones de microgravider,' *Cienc. Tec. Fis. Mat.*, No. 4, 11–16 (1984).

615. M. C. Flemings, R. I. Frost, and J. J. Szekley, "Electromagnetic containerless undercooling facility for STS," in [16], p. 337.

616. R. T. Frost, M. C. Flemings, J. Szekely, N. El-Kaddah, and Y. Shiohara, "Electromagnetic containerless undercooling facility and experiments for the Shuttle," *Adv. Space Res.* **4**, No. 5, 99–103 (1984).

617. T. Fukuda, S. Tanaka, S. Fujiwara, and K. Onozuka, "Dedicated data recording video system for Spacelab experiments," *Acta Astronaut.* **11**, No. 3–4, 199–205 (1984).

618. O. K. Garrioff and D. B. Debra, "A simple microgravity table for an orbital space station," *Earth-Orient. Appl. Space Technol.* **5**, No. 3, 161–163 (1985).

619. A. Gonfalone, "The fluid physics module. Technical description," in [19], p. 13; in [20], pp. 3–7.

620. "Gradientenofen für Weltraumexperimente," *Elektrowarme Int.* **844**, No. 1, 51 (1986).

621. M. Grubic and K. Kemmerle, "A precision lock-in amplifier for temperature control in a Spacelab colorimetric experiment," *J. Phys. E: Sci. Instrum.* **18**, No. 7, 572–574 (1985).

622. H. Grunditz, "Experimental equipment for metallurgy and fluid science studies under microgravity," IAF-86-266, in [23].

623. H. Hamacher, R. Jilg, and U. Merfold, "The microgravity environment of the D1-mission," IAF-86-268, in [23].

624. D. I. Jones and R. G. Owens, "A microgravity isolation mount," IAF-86-270, in [23].

625. K. Kawasaki, T. Shimada, T. Yamanaka, and H. Azuna, "Floating furnace for supercooling processing," IAF-86-267, in [23].

626. J. Kingdon and A. Gonfalone, "The bubble drop and particle unit (BDPU): a second generation fluid physics facility," in [14], p. 336; *Adv. Space Res.* **4**, No. 5, 91–94 (1984).

627. H. Klein and K. Wanders, "Holographic interferometry near gas/liquid critical point. Ground-based study of a D-1 experiment," in [19], p. 71; in [20], pp. 323–325.

628. H. Klein and K. Wanders, "Holographic interferometry near gas/liquid crystal points: Results of Spacelab D-1," IAF-86-284, in [23].

629. W. E. Knabe and F. L. Bremen, "Microgravity characteristics and their measurement in orbital systems," IAF-86-269, in [23].

630. R. Kuhl, H. Suessman, and L. L. Regel, "Results of thermal conditions during crystal growth processes in space and on earth," in [22], Program, p. 45.

631. R. Kuhl, H. Quaas, and H. Suessman, "ARP – a multipurpose instrumentation for experiments in material sciences in space," IAF-86-260, in [23].

632. "La microgravite: au sol, en avion et en satelite," *Air Cosmos* **22**, No. 1001 (1984).

633. R. Lange and F. Straub, "Die isochore Warmekapazitat fluider Stoffe im kritt-Gebiet. Voruntersuchungen zu einem Spacelab–Experiment," *Forschungsber. Bundesmin. Forsch. Technol.* **136**, No. 34 (1984).

634. S. Lioy, V. Defilipp, and R. Monti, "BDPU – Bubble drop and particle unit design characteristics," IAF-86-265, in [23].

635. J. A. Lipa, "Prospects for high resolution measurement near the lambda-transition in space," 75th Jubilee Conf. Helium-4, St. Andreus, Aug. 1–6, 1983, Singapore (1983), pp. 208–209.

636. R. Monti and G. P. Russo, "Nonintrusive methods for temperature measurements in liquid zones in microgravity environments," *Acta Astronaut.* **11**, No. 9, 543–551 (1984).

637. R. Monti and R. Fertezza, "Nonintrusive techniques for thermal measurements in microgravity material science experiments," G. 1.4.3, in [17], p. 336.

638. R. Morera, J. Lodos, and E. Casanova, "Crecimiento de monocristales de sacarosa en el espacio cosmico," *Cienc. Tec. Fis. Mat.*, No. 4, 3–9 (1984).

639. K. Nitsche, J. Straub, and R. Lange, "Isochoric specific heat of sulfur hexafluoride at the critical point. A Spacelab experiment for the German D-1 mission in 1985," in [19], p. 45; in [20], pp. 335–340.

640. G. H. Otto, "First results of a MAUS experiment to investigate the stability of a metallic dispersion," in [18], pp. 43–46.

641. R. B. Owen and R. L. Kroes, "Holography on the Spacelab 3 mission," *Opt. News* **11**, No. 7, 12–16 (1985).
642. J. C. Perron, P. Chretien, C. Gachier, and N. Lecando, "G 300, the first French GAS microgravity measurements of fluid thermal conductivity," IAF-86-264, in [23].
643. V. I. Polezaev, A. P. Lebedev, and S. A. Nikitin, "Mathematical simulation of disturbing forces and material science processes under low gravity," in [16], p. 333.
644. C. Potard and P. Dussere, "Contactless positioning manipulation and shaping of liquids by gas bearing for microgravity applications," in [16], p. 337, *Adv. Space Res.* **4**, No. 5, 105–108 (1984).
645. W. K. Rhim, M. Collender, M. T. Hyson, W. T. Simms, and D. D. Elleman, "Development of an electrostatic positioner for space material processing," *Rev. Sci. Instrum.* **56**, No. 2, 307–315 (1985).
646. J. Richter, J. Hermanns, and W. Merkens, "Entrüften und Befüllen einer Diffusions Meszzeibe unter verminderter Schwerigkeit," *Z. Flugwiss. Weltraumforsch.* **8**, No. 6, 415–419 (1984).
647. F. Rosenberger and M. R. Banish, "Anomalous dispersion mirage effort: simultaneous concentration and temperature measurements in fluid mixtures," in [16], pp. 337–338.
648. C. Roulle, D. Valentian, and W. A. Biemann, "GHF – a modular facility for the GAS," IAF-86-263, in [23].
649. H. Sainct, F. Jamin-Changeart, and J. P. Praizey, "Metallurgy laboratory for COLUMBUS," IAF-86-261, in [23].
650. "Schwerkraftlose Experimente der deutschen Spacelab-Mission," *Materialprüfung* **27**, No. 12, 369–370 (1985).
651. Joshioki Sorimachi, Ryushi Kuwano, Masahiro Kudo, Hiromi Ono, Yotsuo Kamimiyota, Yoshito Narimatsu, Jun Tanu, and Hitomi Okazaki, "Get-away special space experimental system on board the space shuttle," *NEK Ruxo* **38**, No. 13, 73–77 (1985).
652. H. Sprenger, "Skin casting of alloys and composites: Results of SL-1 and TEXUS experiments," in [19], p. 29; in [20], pp. 87–94.
653. J. Szekely and N. El-Kaddah, "The mathematical description of the steady state and transient temperature and velocity fields in levitated metal droplets both under normal and microgravity," in [16], p. 333.
654. S. Takahashi, "Ultrahigh vacuum and weightlessness in space," *C. Pamukkycy, Ceram. Jpn.* **19**, No. 8, 644–650 (1984).
655. H. M. Tensi, H. Fuchs, P. F. Harmathy, and J. Schmidt, "Normalkristallisation mit Abschrecken der Restschmelze unter Weltraumbedingungen. Teil 1: Ausgeführte Kristallisationanlagen," *Aluminium (BKD)* **60**, No. 7, 499–502 (1984).
656. H. M. Tensi, H. Fuchs, P. F. Harmathy, and J. Schmidt, "Normalkristallisation mit Abschrecken der Restschmelze unter Weltraumbedingungen. Teil II: Experimentelle Möglichkeiten der Versuchseinrichtungen," *Aluminium (BKD)* **60**, No. 8, 614–617 (1984).
657. I. Vera, J. N. Cassanto, P. Todd, Z. R. Korszun, and L. S. Rock, "Automated canister for microgravity biotechnology experiments," G. 1.1.4, in [17], pp. 332–333.
658. P. Vits and S. Walter, "Facilities for microgravity research," IAF-86-262, in [23].

659. T. G. Wang, E. Trinh, W.-K. Rhim, D. Kerrisk, M. Barimatz, and D. D. Elleman, "Jet propulsion (JP. L.). Containerless processing technologies at the Jet Propulsion Laboratory," *Acta Astronaut.* **11**, No. 3–4, 233–237 (1984).

660. "'Wassertropfen-Zange' der Unit Konstanz mit Spacelab," *Tech. Heute* **39**, No. 1, 13–14 (1986).

661. K. Wittmann, "The isothermal heating facility: technical description and performance," in [19], p. 23; in [20], pp. 49–54.

X. EXPERIMENTS WITH BRIEF WEIGHTLESSNESS ON TOWERS, AIRCRAFT, AND ROCKETS

662. M. S. Agafonov, S. D. Grishin, L. A. Dedyukov, V. L. Levtov, and V. A. Novikov, "An apparatus for performing technological experiments under conditions of brief weightlessness," in [11], pp. 280–281.

663. M. S. Agafonov, V. L. Levtov, D. V. Leskov, and V. V. Savichev, "Technological experiments under conditions of brief weightlessness," in [5], pp. 106–115.

664. M. S. Agafonov, V. L. Levtov, L. V. Leskov, and V. V. Savichev, "Technological experiments on high-altitude rockets," *Zemlya Vselennaya*, No. 5, 27–34 (1984).

665. M. S. Agafonov, V. L. Levtov, L. V. Leskov, and V. V. Savichev, "Nonstationary technological processes under conditions of brief weightlessness," in [10], pp. 162–164; *Izv. Akad. Nauk SSSR, Ser. Fiz.* **49**, No. 4, 691–697 (1985).

666. L. G. Aleksandrov, S. A. Kiselev, I. R. Kuznetsov, S. S. Obydennikov, T. Yu. Rodionova, V. V. Savichev, and D. M. Shinakov, "The dynamics of a liquid in a capillary subject to small accelerations on board a laboratory aircraft," in [12], p. 214.

667. A. A. Babareko, I. N. Belokurova, N. F. Bogdanova, S. N. Gorin, V. S. Zemskov, M. Z. Mukhoyan, and V. V. Savichev, "Features of metal crystallization on high-altitude probe rockets," in [11], p. 226.

668. V. B. Dubovskii, S. S. Obydennikov, V. V. Savichev, and A. B. Sevast'yanov, "Perturbing accelerations on a high-altitude probe," in [11], p. 227.

669. É. L. Kalyazin, A. G. Mednov, and V. V. Naryshkin, "Simulating hydrodynamic processes on a zero-gravity tower," in [11], p. 295.

670. V. L. Levtov, L. V. Leskov, and V. V. Savichev, "Methodological features of crystallization processes under conditions of brief weightlessness," in [11], pp. 287–288.

671. R. J. Bayuzick, N. D. Evans, W. P. Hofmeister, K. R. Johnson, and M. B. Robinson, "A review of drop tubes as a supplement/alternative to space experiments," in [16], p. 336; *Adv. Space Res.* **4**, No. 5, 85–90 (1984).

672. T. Carlberg, "Floating zone experiments with germanium in sounding rockets," in [22], p. 200.

673. N. P. Evans, R. I. Bayuzick, and E. A. Kenik, "Metastable structures in drop tube processed niobium-based alloys," G. 1.5.5, in [17], p. 338.

674. B. Franks, "Technical aspects of material science experiments on sounding rockets," *6th ESA Symp. Rocket and Balloon, Programs and Related Res.,* Interlaken, April 11–15, 1983, Paris (1983).

675. D. Frimout and A. Gonfalone, "Parabolic aircraft flights – an effective tool in preparing microgravity experiments," *ESA Bull.,* No. 42, 58–63 (1985).

676. W. Herfs, "Outstanding recent TEXUS sounding rocket experiments," in [22], Program, p. 45.

677. R. Johnson, S. Wallin, P. Holm, and E. Soderdah, "Materials science experiment in Sweden. Sounding rocket – GAS, Space Shuttle," *6th ESA Symp. Eur. Rocket and Balloon, Programs and Related Res.,* Interlaken, April 11–15, 1983, Paris (1983).

678. J. C. Legros, H. C. Limbourg, J. H. Lichtenbelt, and D. Frimout, "Some results from parabolic flights," G. 1.4.7, in [17], p. 337.

679. I. Martinez and A. Sanz, "Long liquid bridges aboard sounding rockets," *ESA Journal* **9**, No. 3, 323–328 (1985).

680. R. Monti, L. G. Napolitano, and G. Mannara, "TEXUS flight results on convective flows and heat transfer in simulated floating zones," in [19], p. 83; in [20], pp. 229–236.

681. R. Monti and G. Mannara, "TEXUS experiment on the convective heat transfer induced by Marangoni flows," in [21], Preprint.

682. M. Murakami, N. Nakaniwa, S. Hayakawa, T. Matsumoto, H. Murakami, K. Noguchi, K. Ugama, and H. Nagano, "Zero-gravity experiment on superfluid helium," *Inst. Space Astronaut. Sci. Rep.,* No. 604, 23 (1983).

683. M. Murakami, N. Nakaniwa, S. Hayakawa, T. Matsumoto, H. Murakami, K. Noguchi, K. Ugama, and H. Nagano, "Zero-gravity experiment on superfluid helium aboard a rocket," *Cryogenics* **25**, No. 3, 154–161 (1985).

684. M. Namiki, S. Ohta, T. Yamagami, Y. Koma, H. Akiyama, H. Hirosawa, and J. Nishimura, "Microgravity experiment system utilizing a balloon," *Adv. Space Res.* **5**, No. 1, 83–86 (1986).

685. "NASA flies zero-g. F-104 flight series for John Deere iron study," *NASA Activ.* **15**, No. 9, 6 (1984).

686. G. Petre, M. C. Limbourg-Fontaine, and J. C. Legros, "Preliminary results of TEXUS-8 experiments on effects of surface tension minimum," in [23], Preprint, IAF-86.

687. D. Schwabe and A. Scharmann, "Messung der kritischen Marangonizahl für den Übergang von stationerer zu oszillatorischer thermokapillarer Konvektion unter Mikrogravitation: Ergebnisse der Experimente in den ballistischen Raketen TEXUS 5 und TEXUS 8," *Z. Flugwiss. Weltraumforsch.* **9**, No. 1, 21–28 (1985).

688. H. U. Walter, "Scientific results and accomplishments of the TEXUS program," *Acta Astronaut.* **11**, No. 10–11, 659–672 (1984).

XI. CRYSTAL GROWTH UNDER ELEVATED GRAVITY

689. S. A. Kornev, T. V. Kul'chitskaya, V. M. Biryukov, N. A. Kul'chitskii, V. T. Khryapov, and E. V. Markov, "Growth of epitaxial germanium films under conditions of microgravity and with elevated gravity in a centrifuge," in [10], pp. 124–126.

690. L. L. Regel', G. V. Sarafanov, and A. M. Turchaninkov, "State of the art and development prospects for research on crystal growth and metal solidification under conditions of elevated gravity," *Inst. Kosm. Issled. Akad. Nauk SSSR*, Preprint No. 907 (1984).
691. L. L. Regel', G. V. Sarafanov, I. V. Videnskii, A. M. Turchaninov, and H. Rodot, "Making lead telluride crystals doped with silver under conditions of elevated gravity," *Inst. Kosm. Issled. Akad. Nauk SSSR*, Preprint No. 908 (1984).
692. L. L. Regel', I. V. Videnskii, I. M. Safonova, A. M. Turchaninov, A. V. Mikhailov, and H. Rodot, "Solidification of a eutectic alloy under conditions of elevated gravity," *Inst. Kosm. Issled. Akad. Nauk SSSR*, Preprint No. 1049 (1985).
693. H. Rodot, L. L. Regel, G. V. Sarafanov, H. Hamidi, I. V. Videnski, and A. M. Turchanonov, "Cristaux de telluride de plomb elaborés en centrifugeuse," in *Program and Abstr. ICCG-8*, July 13–18, York, England, PSI/243; *J. Cryst. Growth* **79**, 77–83 (1986).

XII. ELECTROPHORESIS UNDER MICROGRAVITY

694. B. A. Adamovich, A. A. Elatorunskii, V. S. Poleshchuk, and D. Ch. Peregud, "Apparatus design and fractionation technology for biological preparations based on electrophoresis in a free flow in a space vehicle," in [11], p. 297.
695. G. Yu. Azhitskii, L. A. Vavirovskii, I. A. Gus'kov, A. I. Kiselev, A. N. Lebyazh'ev, A. A. Lepskii, O. V. Mitichkin, G. V. Troitskii, and V. E. Fokin, "Biotechnology experiments on Salyut 7," in [5], pp. 91–99.
696. G. Yu. Azhitskii, G. V. Troitskii, O. V. Mitichkin, and G. K. Sharaeva, "Liquid electrophoresis, isoelectric focusing, and isotachophoresis under microgravity conditions," *Dokl. Akad. Nauk Ukr. SSR Ser. B*, No. 4, 56–60 (1984).
697. A. A. Aksenov, A. V. Gudzovskii, T. V. Kondranin, and A. A. Serebrov, "Hydrodynamic aspects of producing an artificial pH gradient in an electrophoretic column under zero-gravity conditions," in [12], p. 215.
698. V. G. Babskii, M. Yu. Zhukov, and V. I. Yudovich, *Mathematical Theory of Electrophoresis*, Consultants Bureau (1989).
699. V. G. Babskii, "The state of the art in electrophoresis under zero gravity," in [10], p. 228; *Izv. Akad. Nauk SSSR, Ser. Fiz.* **49**, No. 4, 6, 724–730 (1985).
700. A. L. Barannikov, N. M. Ganzherli, S. B. Gurevich, V. B. Konstantinov, I. A. Maurer, S. A. Pisarevskaya, B. F. Ryadinskii, A. A. Serebrov, V. N. Sobolev, M. S. Cheberyak, and D. F. Chernykh, "Holographic studies on electrophoresis," in [11], p. 240.
701. M. S. Bello and V. I. Polezhaev, "A mathematical model for flow electrophoresis under zero-gravity conditions," in [10], pp. 232–234.
702. V. M. Galkin, N. B. Panernaya, D. P. Peregud, and V. S. Poleshchuk, "An apparatus for fractionating biological materials by electrophoresis in a free flow of liquid under zero gravity," in [10], pp. 235–237.

703. A. V. Gudzovskii, T. V. Kondranin, and A. A. Serebrov, "Physical features of liquid electrophoresis under microgravity, data from the Tavriya experiment," in [5], pp. 100–105.

704. B. B. Egorov and V. A. Peredkov, "Development prospects for purifying biologically active compounds under zero gravity," in [11], p. 298.

705. M. Yu. Zhukov and V. I. Yudovich, "Convective stability in electrophoresis," in [10], pp. 229–231.

706. L. E. Korol' and M. Yu. Zhukov, "A theoretical and experimental study of convection in free zone-column electrophoresis," in [10], pp. 237–238.

707. L. I. Kurlanov and O. F. Gorbachev, "Electrophoretic separation of mixtures containing nitrogen tetroxide," in [10], pp. 238–240.

708. A. A. Lepskii, G. M. Luk'yanova, O. V. Mitichkin, and A. A. Milenkov, "A study of foreign programs and proposals for making biomedical preparations on space vehicles," in [11], pp. 233–234.

709. O. V. Mitichkin, L. A. Vavirovskii, G. Yu. Azhitskii, S. E. Savitskaya, A. A. Serebrov, and V. E. Fokin, "Results from the Tavriya biotechnology experiment on board Salyut 7," in [11], pp. 230–233.

710. H. Binnenbruck, M. Roth, and W. Steinborn, "Biochemistry/biotechnological processing – a new discipline of the German microgravity program first results and future aspects,' in [22], Program, p. 45.

711. B. Riscons and J. Bertrand, "Some aspects of continuous flow electrophoresis in microgravity," in [22], pp. 210–211.

712. D. R. Morrison, G. H. Barlow, C. Cleveland, M. A. Farrington, R. Grindeland, J. M. Hatfield, W. C. Hymer, J. W. Lanham, M. L. Lewis, D. S. Nachtwey, P. Todd, and W. Wilfinger, "Electrophoretic separation of kidney and pituitary cells on STS-8," in [16], pp. 334–335; *Adv. Space Res.* **4**, No. 5, 67–76 (1984).

713. D. R. Morrison, M. L. Lewis, G. H. Barlow, P. Todd, M. E. Kunze, B. E. Sarnoff, and Z. Li, "Properties of electrophoretic fractures of human embryonic kidney cells separated on Space Shuttle flight STS-8," in [16], p. 335; *Adv. Space Res.* **4**, No. 5, 77–79 (1984).

XIII. MISCELLANY: COMBUSTION, BIOLOGICAL PROCESSES, AND POLYMER SEPARATION AND SYNTHESIS IN SPACE

714. V. G. Babskii, "The role of mass transfer in microorganism growth at zero gravity," in [10], pp. 241–243.

715. E. N. Baulina and I. D. Anikeeva, "The mutation frequency in *Chlorella* as influenced by the orientation of the cell with respect to the gravitational vector," in [10], pp. 243–245.

716. A. K. Kolesnikov, "Hysteresis in chemically active mixtures under zero gravity," in [10], pp. 83–85.

717. N. I. Kon'shin, V. G. Man'ko, and V. A. Kordyum, "Instruments and methods for cultivating microorganisms under zero gravity," in [10], p. 243.

718. V. B. Librovich and G. M. Makhviladze, "A theoretical study of combustion processes under zero gravity," in [10], p. 13.

719. A. I. Merkis, R. S. Laurinavichus, D. V. Shvyagzhdene, and Yu. V. Darginavichene, "Gravitational sensitivity and plant growth under zero gravity," in [10], pp. 226–228.

720. L. R. Pal'mbakh and N. K. Slashcheva, "Features of early embryonic development of fish in gravitational fields," in [10], pp. 245–248.

721. G. P. Parfenov, "Realizing biological processes under zero gravity," in [10], p. 228.

722. K. A. Trukhanov, A. N. Zaikin, S. É. Shnol', and N. M. Kirgizova, "Prospects for research on hydrodynamic and heat- and mass-transfer problems in the biophysics of weightlessness," in [10], pp. 231–232.

723. A. L. Berlad and N. Josi, "Gravitational effects on extinction conditions for premixed flames," in [21], pp. 166–167.

724. D. E. Brooks, S. Bamberger, J. M. Harris, and J. M. Van Alstine, "Rationale for two-phase polymer system microgravity separation experiments," in [19], p. 79; in [20], pp. 315–318.

725. L. Dintenfass, P. Osman, and H. Maguire Jedrzejczyk, "Experiment on aggregation of red cells under microgravity on STS 51-C," G. 1.4.4, in [17], p. 337.

726. S. S. Dosanjh, J. Peterson, A. C. Fernandez-Pello, and P. J. Pagni, "The effect of buoyancy on smoldering combustion," in [22], pp. 208–209.

727. P. V. Farell and B. Peters, "Droplet vaporization in supercritical pressure environments," in [22], pp. 206–207.

728. M. Gieras, R. Klemens, and S. Wojcicki, "Ignition and combustion of coal particles at zero gravity," in [21], p. 174.

729. "How a gas burns in space is target of zero-g flights by Lewis Lear Jet," *NASA Activ.* **15**, No. 12, 15–16 (1984).

730. K. Isao, "Synthesis of high-molecular-weight composites in space," *Zairyo Kagaku, J. Mater. Sci. Soc. Jpn.* **20**, No. 6, 295–299 .

731. S. Okajima, H. Kahno, and S. Kumagai, "Combustion of emulsified full droplets under microgravity," in [21], pp. 170–171.

732. O. D. Ronney, "Effect of gravity on halogenated hydrocarbon flame retardant effectiveness," in [21], pp. 175–176.

733. J. M. Van Alstine, J. M. Harris, R. S. Snyder, P. A. Curreri, S. Bamberger, and D. E. Brooks, "Results of initial two-phase polymer system microgravity separation experiments," in [21], p. 88; in [20], pp. 309–313.

734. J. W. Vanderhoff, M. S. El-Aasser, F. J. Micale, E. D. Sudol, C. M. Tseng, A. Silwanowisz, D. M. Kornfeld, and F. A. Vicente, "Preparation of large-kinetics and process development," *J. Dispers. Sci. Technol.* **5**, No. 3–4, 231–236 (1984).

735. M. Vedha-Nayagam and R. A. Altenkirch, "Gravitational effects on flames spreading over thick solid surface," in [21], pp. 172–173.

736. F. A. Williams, "Ignition and burning of single liquid droplet," in [21], pp. 168–169.

737. R. S. Young, "Gravity and the living organism," in [16], p. 334.

INDEX

**DMV Seminar
Band 26**

Geometry of Higher Dimensional Algebraic Varieties

Yoichi Miyaoka
Thomas Peternell

Springer Basel AG

Authors:

Yoichi Miyaoka
RIMS
Kyoto University
606-01 Kyoto
Japan

Thomas Peternell
Mathematisches Institut der
Universität Bayreuth
95440 Bayreuth
Germany

1991 Mathematical Subject Classification 14Jxx

A CIP catalogue record for this book is available from the Library of Congress, Washington D.C., USA

Deutsche Bibliothek Cataloging-in-Publication Data
Geometry of higher dimensional algebraic varieties / Yoichi
Miyaoka ; Thomas Peternell. - Basel ; Boston ; Berlin :
Birkhäuser, 1997
  (DMV-Seminar ; Bd. 26)
  ISBN 978-3-7643-5490-9   ISBN 978-3-0348-8893-6 (eBook)
  DOI 10.1007/978-3-0348-8893-6
NE: Miyaoka, Yoichi; Peternell, Thomas; Deutsche Mathematiker-
  Vereinigung: DMV-Seminar

© 1997 Springer Basel AG
Originally published by Birkhäuser Verlag in 1997
Camera-ready copy prepared by the author
Printed on acid-free paper produced from chlorine-free pulp. TCF ∞
Cover design: Heinz Hiltbrunner, Basel

ISBN 978-3-7643-5490-9

Contents

Preface

This book is based on lecture notes of a seminar of the Deutsche Mathematiker Vereinigung held by the authors at Oberwolfach from April 2 to 8, 1995. It gives an introduction to the classification theory and geometry of higher dimensional complex-algebraic varieties, focusing on the tremendeous developments of the subject in the last 20 years. The work is in two parts, with each one preceeded by an introduction describing its contents in detail. Here, it will suffice to simply explain how the subject matter has been divided. Cum grano salis one might say that Part 1 (Miyaoka) is more concerned with the algebraic methods and Part 2 (Peternell) with the more analytic aspects though they have unavoidable overlaps because there is no clearcut distinction between the two methods. Specifically, Part 1 treats the deformation theory, existence and geometry of rational curves via characteristic p, while Part 2 is principally concerned with vanishing theorems and their geometric applications.

Part I
Geometry of Rational Curves on Varieties

Yoichi Miyaoka

RIMS
Kyoto University
606-01 Kyoto
Japan

Introduction: Why Rational Curves?

This note is based on a series of lectures given at the Mathematisches Forschungsinstitut at Oberwolfach, Germany, as a part of the DMV seminar "Mori Theory". The construction of minimal models was discussed by T. Peternell, and my task was to give an overview of various aspects of the study of rational curves on algebraic varieties, including the following topics:

(a) Techniques which enable us to find rational curves on certain classes of varieties;

(b) Characterization of uniruled varieties (varieties that carry sufficiently many rational curves) in terms of canonical divisors;

(c) Generic semipositivity of the cotangent bundle of non-uniruled varieties and its application to the "abundance conjecture" in dimension three;

(d) Decomposition of a given variety into the "non-uniruled part" and the "rationally connected part",

(e) Application of the techniques above to the theory of Fano varieties.

Rational curves are, by definition, images of the projective line $\mathbb{P}^1$ by non-constant morphisms, and thus very special objects in algebraic geometry. In fact, a "very general algebraic variety" will not carry any rational curve on it. Still the study of embedded rational curves is of great importance in algebraic geometry by at least two reasons.

First of all, it has a close connection to the minimal model program (MMP). This relationship is discussed in detail in Peternell's lectures.

Secondly, the varieties that contain sufficiently many rational curves constitute a distinguished closed class in the set of algebraic varieties, as will be illustrated in Lectures II and V. What we mean by "varieties with sufficiently many rational curves" are *uniruled varieties* or *rationally connected varieties*, the latter being special cases of the former. Roughly speaking, a variety X is said to be uniruled if X is set-theoretically a union of rational curves on it. If, more strongly, two arbitrary points on X can be joined by a rational curve on X, X is called *rationally connected*. Uniruled (or rationally connected) curves are nothing but rational curves. In dimension two, a uniruled [resp. rationally connected] surface is a ruled [resp. rational] surface.

The theory of surfaces would give a good illustration of the peculiarity of uniruled varieties. If you look at the Enriques classification of algebraic surfaces, you will find that rational and ruled surfaces form a very special class, which require treatments completely different from those for the other classes.

To begin with, ruled surfaces have no absolute minimal model in the sense of Zariski, while the existence of absolute minimal models plays a crucial role in detailed structure theory for surfaces in other classes. Instead, each relatively minimal ruled surface has a very simple biholomorphic structure (the projective plane $\mathbb{P}^2$ or a $\mathbb{P}^1$-bundle over a curve). Furthermore, you can tell whether a given surface is ruled or not by checking a single invariant P_{12}; in other words, it is quite easy to separate ruled surfaces from other surfaces.

Ruled surfaces have another singular property. Namely, there is no good theory of moduli for ruled surfaces. To see this, consider the Hirzebruch surface $F_{2d} = \mathbb{P}(\mathcal{O} \oplus \mathcal{O}(2d))$, $d = 0, 1, \dots$ As a topological space, this surface is just $S^2 \times S^2$, but F_{2d} and F_{2e} are not mutually isomorphic if $d \neq e$. Any deformation of F_{2d} is of the form F_{2e}; indeed, if $d \leq e$, there is a one-parameter family of surfaces X_t, $t \in \mathbb{C}$, such that $X_0 \simeq F_{2e}$ and $X_t \simeq F_{2d}$, $t \neq 0$. Thus the moduli space of ruled surfaces homeomorphic to $S^2 \times S^2$, if any, would be a weird space. Set-theoretically, it must be identical with $\{0, 1, 2, 3, \dots\}$; however, as a topological space, a positive integer e is contained in the closure of any smaller non-negative integer $d < e$. Such a space cannot be a good moduli space.

These phenomena also occur in higher dimension, thus leading us to the following general principles (or, perhaps, working hypotheses):

— The class of uniruled varieties is a sort of minority group in the universe of algebraic varieties, clearly distinguished from the other classes in terms of simple invariants.
— General theory (moduli etc.) does not well fit to these varieties.
— Instead, a uniruled variety has comparatively simple (birational or biregular) structure.
— For non-uniruled varieties, construction of good birational models (a minimal model) would provide powerful machinery for finer general theory.

With these principles in mind, S. Mori and other people proposed an overall strategy in the classification theory. The strategy, called the *minimal model program*, could be summarized into the following four steps:

Step 1. Given an algebraic variety X, find bad curves (usually rational) which have "bad effect" on the analysis of X.

Step 2. Contract bad curves found in Step 1 and reiterate this process as long as possible; we often need, however, to insert a new procedure called "flip" to continue this step.

Step 3. Study the structure of the resulting variety Y, on which there is no "bad" rational curve any more; if Y is birational to X (i.e. Y is a minimal model of X), then everything would be fine from the view point of birational geometry.

Step 4. When Y is not birational to X, then study the fibres of the projection $X \to Y$, which are necessarily uniruled and expected to have simple structure.

The subjects of my lectures are related to the steps 1 and 4, as well as some part of Step 3. Step 2, which is the minimal model program in a stricter sense, is discussed in Peternell's lectures.

Deformation theory is an omnipresent powerful machinery at work in these lectures. Varieties we deal with are, in principle, defined over the complex numbers. However, our argument heavily depends on geometry in characteristic $p > 0$

through mod p reduction technique. We often go to positive characteristics in order to get more room for maneuvers, and then come back to characteristic zero.

In what follows, I tried to make the exposition as accessible as possible. Readability has priority over the coherence of logic. Required background knowledge is, hopefully, minimal. What I assume are:

(a) Classical theory of curves and algebraic surfaces; for the second material, [Beauville 1] will be a nice survey work.

(b) Very basic knowledge of schemes, which would be provided by [Hartshorne 2].

(c) A little portion of terminology from category theory like contravariant functors.

Fundamental materials needed in the proofs (Hilbert schemes, Chow schemes, deformation theory, foliations, etc.) will be explained, if not fully, in the lectures. In order to help the reader get better understanding of the topic, I included scores of examples and exercises, which the reader is advised to check and solve.

Lecture I
Deformations and Rational Curves

Overview

In this lecture, we give a brief account on the fundamental notions such as Hilbert schemes, Chow schemes, deformation of morphisms, and intersection pairing between curves and divisors, along with their application to the construction of rational curves on a projective variety.

1   Deformation with base points produces rational curves

1.1   Let X, Y be complete varieties over an algebraically closed field k, and T a connected k-scheme with a marking $o \in T$ by a closed point. Consider a k-morphism $f : Y \to X$. A deformation of f parametrized by T is, by definition, a k-morphism $\tilde{f} : T \times Y \to X$ such that $\tilde{f}|_{\{o\} \times Y}$ is identical with f. When T is reduced, we often employ a parametric representation, $\tilde{f} = \{f_t : Y \to X\}_{t \in T}$.

Given a closed subscheme $B \subset X$, we have the notion of deformation of f with *base point set* B; namely, a deformation $\tilde{f} : T \times Y \to X$ such that $\tilde{f}|_{T \times B}$ is a trivial deformation. More precisely, $(\tilde{f}|_B)^* : \mathcal{O}_X \to \mathcal{O}_T \otimes \mathcal{O}_B$ satisfies $(\tilde{f}|_B)^*(\alpha) = 1 \otimes (f|_B)^*(\alpha)$ for every $\alpha \in \mathcal{O}_X$. (Intuitively it amounts to the equality $\tilde{f}|_{t \times B} = f|_B$ for every $t \in T$.)

The main topic in this lecture is the following

1.2 Theorem   *Let X be a projective variety over an algebraically closed field, H an ample divisor on it, Y a smooth projective curve, and $B \subset Y$ a zero-dimensional reduced subscheme of degree d. Assume that a non-constant morphism $f : Y \to X$ has a non-trivial deformation $\tilde{f} : T \times Y \to X$ with base point set B, parametrized by a one-dimensional variety T. If $f(Y)$ is not a rational curve, then X contains a rational curve C with the following two properties:*

$$(C, H) < \frac{2 \deg f^* H}{b}. \tag{1.2.1}$$

$$C \text{ meets } f(B) \text{ at least at one point.} \tag{1.2.2}$$

Varieties in positive characteristics or non-Kähler manifolds often provide counterexamples to statements which are true for projective complex manifolds. When we work in the category of k-schemes, Theorem 1.2 holds in any characteristic without reservation. It also stays true in analytic category if we replace X, Y and $\tilde{f} : T \times Y \to X$ by a compact Kähler manifold, a compact Riemann surface and a non-trivial deformation with base points of an analytic mapping. However, once the Kähler condition is dropped in this case, (1.2) is no more true. In fact, we have the following example.

1.3 Example   Let Y be a compact Riemann surface of genus ≥ 1. Fix a closed point $y_0 \in Y$, and consider the vector bundle $\mathcal{E} = \mathcal{O}_Y \oplus \mathcal{O}_Y(y_0)$, where we view y_0 as an effective divisor on Y. The group $\mathbb{C}^\times$ naturally acts on $\mathcal{E}^\circ = \mathcal{E} \setminus (0\text{-section})$ as the fibrewise multiplication. Fix a constant $\alpha \in \mathbb{C}^\times$ such that $|\alpha| < 1$. Then the infinite cyclic subgroup $\langle \alpha \rangle \subset \mathbb{C}^\times$ acts properly discontinuously on $\mathcal{E}^\circ$. The quotient space $X = \mathcal{E}^\circ / \langle \alpha \rangle$ is a compact complex manifold equipped with a natural projection $\phi : X \to Y$. Every fibre of ϕ is isomorphic to the Hopf surface $S_\alpha = (\mathbb{C}^2 - \{(0,0)\})/\langle \alpha \rangle$, with $\pi_1(S_\alpha) \simeq H_1(S_\alpha, \mathbb{Z}) \simeq \mathbb{Z}$. Since the first Betti number of a compact Kähler manifold is always even, S_α is not Kähler, so that the ambient space X cannot be Kähler, either.

Let us construct a non-trivial deformation $\{f_s : Y \to X\}$, $s \in \mathbb{C}$, with non-empty base point set $\{y_0\}$. Let 1 denote the canonical section of $\mathcal{O}_Y$ as well as its image in $\mathcal{O}_Y(y_0)$ via the natural inclusion map $\mathcal{O}_Y \hookrightarrow \mathcal{O}_Y(y_0)$. Then $(1, s1)$, $s \in \mathbb{C}$, defines sections $Y \to \mathcal{E}^\circ$, thereby inducing a family of sections $f_s : Y \to X$ parametrized by $S = \mathbb{C}$. The morphism

$$\tilde{f} : \mathbb{C} \times Y \to X, \quad \tilde{f}(s, y) = f_s(y)$$

is thus a deformation of $f = f_0$ with parameter space $\mathbb{C}$. Noting $1 \in \mathcal{O}(y_0)$ has a unique zero at y_0, we see that $f_s(y_0) = f_0(y_0)$ for every $s \in \mathbb{C}$. Namely the closed point $\{y_0\}$ is a (unique) base point of the deformation $\tilde{f}$.

Despite a quite similar setting as in Theorem 1.2, this non-Kähler manifold X does not contain any rational curve. Indeed, if $C \subset X$ were a rational curve, $\phi(C) \subset Y$ must be a point because $g(Y) \geq 1$. Hence C would be contained in a fibre $\simeq S_\alpha$, whose universal covering is $\mathbb{C}^2 - \{(0,0)\} \subset \mathbb{C}^2$. Since $\mathbb{P}^1$ is simply connected, the normalization $\mathbb{P}^1 \to C$ would then induce a non-trivial analytic map $\mathbb{P}^1 \to \mathbb{C}^2 \setminus (0,0) \subset \mathbb{C}^2$, or equivalently, two holomorphic functions on $\mathbb{P}^1$, at least one of which being non-constant. This is clearly impossible.

1.4 What is going on in this example?

As will be shown below, the rational curves in Theorem 1.2 are constructed as irreducible components of *limit cycles* of $\{f_s(Y)\}$. In Example (1.3), however, the one-parameter family of 1-cycles $\{f_s(Y)\}$ does not have any limit cycle. Indeed, if the family had a limit $f_\infty(Y)$, which must be a compact 1-cycle, then the volume of $f_s(Y)$ (with respect to a hermitian metric on X, arbitrary but independent of s) would be uniformly bounded, because $\mathbb{C} \cup \infty = \mathbb{P}^1$ is compact. But you can easily check that the volume of $\{f_s(Y)\}$ tends to infinity as $|s|$ grows, so you cannot find a limit cycle of $\{f_s(Y)\}$ on X.

In case X carries a Kähler metric ω, the situation is completely different. Cycles with respect to ω have constant volume if they form a continuous family and their fundamental classes define the same homology class of X (cf. Exercise 1.5 below). In this case, we can expect that a family of cycles could be compactified. Indeed that is always the case on (compact) Kähler complex varieties or on projective varieties over an arbitrary algebraically closed field. Namely the parameter space of closed subschemes or compact analytic subspaces is a disjoint union of projective schemes or compact Kähler spaces according as X is projective or Kähler. Such a parameter space is called the *Hilbert scheme* (projective case) or the *Douady space* (Kähler case).

1.5 Exercise   (1) Let M be a compact differentiable manifold and ω a smooth 2-form on M. Consider the functional $\Gamma \mapsto \int_\Gamma \omega$ on the space of 2-cycles on M. Prove that this functional factors through $H_2(M, \mathbb{R})$ if and only if ω is d-closed.

(2) Let the notation be as in (1.3). Show that $\mathbb{P}(\mathcal{E}^*) = \mathcal{E}^\circ / \mathbb{C}^\times$ is a projective variety with natural projections $\pi : \mathbb{P}(\mathcal{E}^*) \to Y$, $\rho : X \to \mathbb{P}(\mathcal{E}^*)$. Define a one-parameter family of sections $\overline{f}_s : Y \to \mathbb{P}(\mathcal{E}^*)$ by $\overline{f}_s = \rho \circ f_s$. Check that the limit cycle $Z_\infty = \lim_{|s| \to \infty} \overline{f}_s(Y)$ exists on $\mathbb{P}(\mathcal{E}^*)$. ($Z_\infty$ is a union of the section of π corresponding to the section $(0,1) : Y \to \mathcal{E}$ plus the fibre $\simeq \mathbb{P}^1$ of π over y_0.)

(3) Given a locally free sheaf $\mathcal{E}$ of rank r on a k-scheme S, define the *projective bundle* $\pi : \mathbb{P}(\mathcal{E}) \to S$ as follows. Over a point $s \in S$, the fibre of π is the set of the one-dimensional quotients (not subspaces) of the r-dimensional $k(s)$-vector space $k(s) \otimes_{\mathcal{O}_S} \mathcal{E}$. Show that, if you fix a local basis $e_1, \ldots, e_r$ of $\mathcal{E}$ on $U \subset S$, this construction gives a local identification $\mathbb{P}(\mathcal{E})|_{\pi^{-1}(U)} \simeq U \times \mathbb{P}^{r-1}$.

A given global rational section $e \in H^0(U, \mathcal{E})$ determines a family of hyperplanes in the fibres $\simeq \mathbb{P}^{r-1}$ of π, and thus a divisor H_e on $\mathbb{P}(\mathcal{E})$. Prove that the linear equivalence class of H_e does not depend on the choice of e. (H_e, or its linear equivalence class, is called the *tautological divisor* and denoted by $\mathbf{1}_\mathcal{E}$.)

Check that $\mathbb{P}(\mathcal{E})$ is nothing but $\mathrm{Proj}(\mathrm{Sym}\mathcal{E})$ as an S-scheme, where $\mathrm{Sym}\mathcal{E}$ is the symmetric tensor algebra $\oplus_{i \geq 0} S^i \mathcal{E}$ over $\mathcal{O}_S$.

2   Hilbert schemes and Hom schemes

2.1 By a *polarized k-scheme* we mean a pair (X, H) of a projective k-scheme and an ample Cartier divisor H on it. Easy induction on dimension tells us the Euler characteristic $\chi(X, \mathcal{O}_X(tH))$ of the line bundle $\mathcal{O}_X(tH)$, viewed as a function in $t \in \mathbb{Z}$, is a polynomial $h(X, t) = h((X, H), t) \in \mathbb{Q}[t]$. $h(X, t)$ is called the *Hilbert polynomial* of (X, H). The dimension n of X over k is equal to the degree of the polynomial $h(X, t)$, while the degree H^n of X is given by the leading coefficient of $h(X, t)$ multiplied by $n!$.

Let X be a projective S-scheme with relatively ample divisor H. (Namely, there is a closed embedding of X into a projective bundle $\mathbb{P}(\mathcal{E})$ over S, with the pullback of the tautological line bundle being linearly equivalent to H.) Given a point $s \in S$, let X_s denote the associated fibre $k(s) \otimes_{\mathcal{O}_S} X$. Then the correspondence $s \mapsto h(X_s, t)$ defines a function on S. It is well-known that this function is locally constant whenever X is flat over S (for a proof, see e.g. [Mumford 3, page 50]). Put in another way, the Hilbert function stays constant under projective, flat deformation.

Fix a polynomial $h = h(t) \in \mathbb{Q}[t]$. Let us construct the Hilbert scheme $\mathrm{Hilb}^h(X)$ as a projective S-scheme. First, we introduce the Hilbert functor $\mathcal{H}ilb_X^h$.

Let S be a scheme and X a projective S-scheme, with a relatively ample divisor H. For a given S-scheme V, define $\mathcal{H}ilb^h_X(V)$ as the set

$$\{ Z \subset V \times_S X \mid \mathrm{pr}_V : Z \to V \text{ is proper, flat, } \chi(Z_v, \mathcal{O}(tH)) = h(t), \forall v \in V \}.$$

Given an S-morphism $V \to W$, the correspondence $Z \mapsto V \times_W Z$ defines a map $\mathcal{H}ilb^h_X(W) \to \mathcal{H}ilb^h_X(V)$. In other words, $\mathcal{H}ilb^h_X$ is a contravariant functor from the category of S-schemes to the category of sets. A theorem of A. Grothendieck tells us that $\mathcal{H}ilb^h_X$ is representable by a projective scheme $\mathrm{Hilb}^h(X)$:

2.2 Theorem [FGA] *Let X be a projective S-scheme. Then there exists a projective S-scheme $\mathrm{Hilb}^h(X)$ together with a subscheme $\mathcal{Z}^h \subset \mathrm{Hilb}^h(X) \times_S X$ characterized by the following universal property:*

— If $Z \subset V \times_S X$ sits in $\mathcal{H}ilb^h_X(V)$ (i.e. if Z is proper flat over V, with fibres having Hilbert polynomial h), then there is a unique S-morphism $V \to \mathrm{Hilb}^h(X)$ such that $Z = V \times_{\mathrm{Hilb}^h(X)} \mathcal{Z}^h$.

By the universal property, $\mathrm{Hilb}^h(X)$ is unique up to isomorphisms, and is called the *Hilbert scheme* of (X, H) attached to the Hilbert polynomial h. The subscheme $Z^h \subset \mathrm{Hilb}^h(X) \times_S X$ is said to be the *universal subscheme* because any flat family of closed subschemes with Hilbert polynomial h on X is obtained from Z^h.

2.3 Rough idea for the construction of Hilbert schemes

You may skip this paragraph if you can believe in the existence of Hilbert schemes.

Step 1 (Boundedness). For simplicity, assume that $S = \operatorname{Spec} k$, the spectrum of a field, and that H is a hyperplane section of an embedding $X \hookrightarrow \mathbb{P}^N$. Let Z be a closed subscheme of X. Let $h_Z = h_Z(t)$ be the Hilbert polynomial $\chi(Z, \mathcal{O}_Z(tH))$. The degree of h is equal to the maximum dimension of the irreducible components. For example, if $Z \subset X$ is a zero-dimensional subscheme, then h_Z is the constant function $\deg Z$.

Fix a polynomial $h \in \mathbb{Q}[t]$. We show by induction on $\deg h$ that there is a constant $m(h)$, depending only on h, such that
(a) the restriction map $H^0(\mathbb{P}^N, \mathcal{O}(tH)) \to H^0(Z, \mathcal{O}(tH))$ is surjective and that
(b) the higher cohomology $H^i(Z, \mathcal{O}(tH))$, $i \geq 1$ vanishes
for every $Z \subset X$ with $h_Z = h$ and $t \geq m(h)$.

If the degree d of zero-dimensional subschemes Z is fixed, then we can find a constant $m = m(d)$, independent of Z (in this case, $m = d$) such that the restriction map $H^0(\mathbb{P}^N, \mathcal{O}(tH)) \to H^0(Z, \mathcal{O}(tH))$ is surjective for $t \geq m$. Thus, if $h = h(t)$ is the constant function d, then $m(h) = d$.

Given Z of dimension n, let Z_H be a general hyperplane section of dimension $n - 1$. Then you calculate $h_{Z_H}(t) = h^*(t) = h_Z(t) - h_Z(t-1)$. Consider the cohomology exact sequence

$$
\begin{aligned}
0 \;\longrightarrow\; & H^0(Z, \mathcal{O}((t-1)H)) \;\longrightarrow\; H^0(Z, \mathcal{O}(tH)) \;\longrightarrow\; H^0(Z_H, \mathcal{O}(tH)) \\
\longrightarrow\; & H^1(Z, \mathcal{O}((t-1)H)) \;\longrightarrow\; H^1(Z, \mathcal{O}(tH)) \;\longrightarrow\; H^1(Z_H, \mathcal{O}(tH)) \\
\longrightarrow\; & \qquad\qquad \cdots
\end{aligned}
$$

Since the Hilbert polynomial $h_{Z_H} = h^*(t)$ is fixed, the induction hypothesis tells you that there is a constant $m(h^*)$ such that

(1) $H^0(\mathbb{P}^N, \mathcal{O}(tH)) \to H^0(Z_H, \mathcal{O}(tH))$, and hence
     $H^0(Z, \mathcal{O}(tH)) \to H^0(Z_H, \mathcal{O}(tH))$ as well, is surjective and that

(2) $H^i(Z_H, \mathcal{O}(tH)) = 0$ ($i > 0$) whenever $t \geq m(h^*)$.

Hence if $i \geq 1$ and $t \geq m(h^*))$, then $H^i(Z, \mathcal{O}(tH)) \simeq H^i(Z, \mathcal{O}((t+1)H)) \simeq \ldots = 0$, so that $\dim H^0(Z, \mathcal{O}(tH)) = h(t)$. Let $I_Z \subset \mathcal{O}_{\mathbb{P}^N}$ [resp. $I_{Z_H} \subset \mathcal{O}_H$] denote the ideal sheaf of $Z \subset \mathbb{P}^N$ [resp. $Z_H \subset H \simeq \mathbb{P}^{N-1}$] and observe the three exact sequences

$$
\begin{array}{ccccccccc}
0 & \longrightarrow & I_Z(tH) & \longrightarrow & \mathcal{O}_{\mathbb{P}^N}(tH) & \longrightarrow & \mathcal{O}_Z(tH) & \longrightarrow & 0, \\
0 & \longrightarrow & I_{Z_H}(tH) & \longrightarrow & \mathcal{O}_H(tH) & \longrightarrow & \mathcal{O}_{Z_H}(tH) & \longrightarrow & 0, \\
0 & \longrightarrow & I_Z((t-1)H) & \longrightarrow & I_Z(tH) & \longrightarrow & I_{Z_H}(tH) & \longrightarrow & 0.
\end{array}
$$

Assume that $t \geq m(h^*)$. By our induction hypothesis, you easily deduce:

(1) $H^i(I_{Z_H}(tH)) = 0$ for $i \geq 1$;
(2) if $H^0(\mathbb{P}^N, I_Z(tH)) \to H^0(H, I_{Z_H}(tH))$ is surjective, so is $H^0(\mathbb{P}^N, I_Z((t+1)H)) \to H^0(H, I_{Z_H}((t+1)H))$;
(3) $H^1(\mathbb{P}^N, I_Z(tH)) \to H^1(\mathbb{P}^N, I_Z((t+1)H))$ is surjective;
(4) $H^1(\mathbb{P}^N, I_Z(tH)) \to H^1(\mathbb{P}^N, I_Z((t+1)H))$ is not injective if and only if $H^0(\mathbb{P}^N, I_Z(tH)) \to H^0(H, I_{Z_H}(tH))$ is not surjective.

From (4), it follows that $\dim H^1(I_Z(tH))$ is strictly decreasing for the t such that $H^0(\mathbb{P}^N, I_Z(tH)) \to H^0(H, I_{Z_H}(tH))$ is not surjective. On the other hand, $\dim H^1(I_Z(m(h^*))$ is bounded from above by $\dim H^0(Z, \mathcal{O}(m(h^*)H)) = h(m(h^*))$, a constant depending only on h. Hence we see that $\dim H^1(\mathbb{P}^N, I_Z(tH))$ will be constant (and hence 0) if $t \geq m(h) = m(h^*) + h(m(h^*))$. This shows that $H^0(\mathbb{P}^N, \mathcal{O}(tH)) \to H^0(Z, \mathcal{O}(tH))$ is surjective and that $H^i(Z, \mathcal{O}_Z(tH) = 0$, $i \geq 1$ if $t \geq m(h)$. □

Step 2 (Construction). Fix $m \geq m(h)$ as in Step 1. Then we have a natural surjection $H^0(\mathbb{P}^N, \mathcal{O}(tmH)) \to H^0(Z, \mathcal{O}(tmH))$, $t = 1, 2, \ldots$ In particular, the quotient vector space $H^0(Z, \mathcal{O}(mH))$ defines a point $[Z]$ of the Grassmann variety $\mathrm{Grass}(H^0(\mathbb{P}^N, \mathcal{O}(mH)), h(m))$. (Here we mean by $\mathrm{Grass}(V, e)$ the set of the quotient vector spaces of dimension e of a fixed vector space V. Given an e-dimensional quotient space Q of V, we get the one-dimensional quotient $\wedge^e Q$ of $\wedge^e V$. This correspondence gives an embedding, i.e. the *Plücker embedding*, of $\mathrm{Grass}(V, e)$ into $\mathbb{P}(\wedge^e V)$.) It is easy to see that the correspondence $Z \mapsto [Z]$ is injective. Conversely, given a $h(m)$-dimensional quotient space Q of $H^0(\mathbb{P}^N, \mathcal{O}(mH))$, let $Z(Q)$ denote the subscheme $\subset \mathbb{P}^N$ defined by the homogeneous polynomials sitting in $\mathrm{Ker}(H^0(\mathbb{P}^N, \mathcal{O}(mH)) \to Q)$. You can check that the $Q \in \mathrm{Grass}(H^0(\mathbb{P}^N, \mathcal{O}(mH)), h(m))$ such that $Z(Q)$ has Hilbert polynomial $h(t)$ form a closed set $\mathrm{Hilb}^h(\mathbb{P}^N)$. It is again a closed condition on $Q \in \mathrm{Hilb}^h(\mathbb{P}^N)$ that $Z(Q) \subset X$, and hence such Q's form a closed subset $\mathrm{Hilb}^h(X) \subset \mathrm{Hilb}^h(\mathbb{P}^N) \subset \mathrm{Grass}(H^0(\mathbb{P}^N, \mathcal{O}(mH)), h(m))$. □

We call $\mathrm{Hilb}(X) = \coprod_h \mathrm{Hilb}^h(X)$ the *Hilbert scheme* of X (or, more precisely, of (X, H)); it is a disjoint union of countably many projective schemes $\mathrm{Hilb}^h(X)$.

2.4 Exercise (1) Show that $\mathrm{Hilb}^1(\mathbb{P}^1)$ attached to the constant function 1 is isomorphic to $\mathbb{P}^1$ by considering $H^0(\mathbb{P}^1, \mathcal{O}(1))$. What do you get, if you consider $\mathrm{Grass}(H^0(\mathbb{P}^1, \mathcal{O}(2)), 1) \simeq \mathbb{P}^3$ instead?

(2) Show that $\mathrm{Hilb}^n(\mathbb{P}^1) \simeq \mathbb{P}^n$ by considering $\mathrm{Grass}(H^0(\mathbb{P}^1, \mathcal{O}(n)), n)$.

(3) Show that $\mathrm{Hilb}^1(X)$ is canonically isomorphic to X, a projective k-scheme.

(4) Show that a closed subscheme $Z \subset \mathbb{P}^N$ is equal to a line if and only if the Hilbert function is $t+1$. Hence $\mathrm{Hilb}^{t+1}(\mathbb{P}^N)$ is the Grassmann variety $\mathrm{Gass}(N+1, N-1)$, identified with the space of 2-dimensional vector subspaces of a fixed $(N+1)$-dimensional vector space.

(5) Put $h = 2t + 2$. Show that a subscheme in $\mathbb{P}^3$ with Hilbert polynomial h is either a disjoint union of two lines or a plane quadric *plus* a single embedded point. Show that plane quadrics form a projective variety of dimension 8. Prove that the disjoint unions of two lines form a 8-dimensional open subset U of $\mathrm{Hilb}^{2t+2}(\mathbb{P}^3)$. Thus $\mathrm{Hilb}^{2t+2}(\mathbb{P}^3)$ consists of two irreducible components of dimension 11 and 8, respectively. What is the closure of U in $\mathrm{Hilb}^{2t+2}(\mathbb{P}^3)$?

2.5 Let us observe some implications of the existence Hilbert schemes. In what follows, only the existence counts; you could forget the construction in order to get the results.

(a) Consider a complex projective variety X. Any k-vector space is flat over a field k, so a closed complex subscheme $Z \subset X$ is flat over $\mathrm{Spec}\,\mathbb{C}$. Hence there is a unique morphism $\mathrm{Spec}\,\mathbb{C} \to \mathrm{Hilb}(X)$ such that Z is the pull-back of $\mathcal{Z}$. Of course the morphisms $\mathrm{Spec}\,\mathbb{C} \to \mathrm{Hilb}(X)$ correspond to the points on $\mathrm{Hilb}(X)$ viewed as a complex analytic space. In this sense $\mathrm{Hilb}(X)(\mathbb{C})$ is the set of all the closed $\mathbb{C}$-subschemes on X.

(b) More generally, consider a ring homomorphism $\mathcal{O}_S \to k$ to a field k, i.e. a k-valued point on S. Any k-module is flat over k, so that there is a one-to-one correspondence between the k-subschemes of $\mathrm{Spec}\,k \times_S X$ and the k-rational points on $\mathrm{Spec}\,k \times_S \mathrm{Hilb}(X)$.

(c) As a useful application of Hilbert schemes, we show a sort of Hasse principle on subschemes, which are repeatedly used in the subsequent lectures.

Any projective scheme $\overline{X}$ over a field k can be viewed as a scalar extension of a scheme X defined over a ring R, finitely generated over $\mathbb{Z}$. For instance given a complex projective variety $\subset \mathbb{P}^N$ defined by homogeneous polynomials $f_1, \ldots, f_r$, let R be the ring $\mathbb{Z}[\alpha_1, \ldots, \alpha_m]$, where α_i's are the coefficients appearing in the defining equations. Thus X is defined over R, and we can think of X as a projective R-scheme.

Suppose that R is of characteristic zero. Take p, an arbitrary prime number, and $\mathfrak{p}$ a minimal prime ideal $\supset pR$ (there are finitely many choices of such $\mathfrak{p}$ for given p). Then $R/\mathfrak{p}$ is an algebra over a finite field $\mathbf{F}_q$ of characteristic p, and its field of fractions k is of characteristic p. The fibre product $k \otimes_R X$ is a projective k-scheme, called a *reduction of X modulo p*.

2.6 Proposition   *Fix finitely many polynomials $h_i \in \mathbb{Q}[t]$, $i = 1, \ldots, r$. Let R be a ring of characteristic zero, finitely generated over $\mathbb{Z}$. For a projective R-scheme X, suppose that modulo p reductions $\bigcup_i \mathrm{Hilb}^{h_i}(k \otimes_R X)$ are non-empty for infinitely many primes p. Then $\bigcup_{i=1}^r \mathrm{Hilb}^{h_i}(K \otimes_R X) \neq \emptyset$, where K is the field of fractions of R, or, equivalently, the generic point of $\mathrm{Spec}\, R$. Conversely, if $\bigcup_i \mathrm{Hilb}^{h_i}(K \otimes_R X) \neq \emptyset$, then $\bigcup_i \mathrm{Hilb}^{h_i}(k \otimes_R X) \neq \emptyset$ for every modulo p reduction.*

This easily follows from the facts that (a) a projective (or, more generally, a noetherian) S-scheme is flat over an open dense subset of S and that (b) a projective morphism is proper (and hence universally closed). It is essential that h_i runs through only finitely many polynomials.

2.7 Exercise   (1) Since K, the fraction field of R, is not algebraically closed, the non-emptiness of $\mathrm{Hilb}^h(K \otimes X)$ does not imply the non-emptiness of the K-rational points on it. Find an example of a $\mathbb{Z}$-scheme X such that $\mathrm{Hilb}^1(X) \simeq X$ carries $\mathbb{F}_p$-rational points for each prime number p, but has no $\mathbb{Q}$-rational points. (*Hint:* Choose X as a certain quadric $\subset \mathbb{P}^2$.)

(2) Let $C \subset \mathbb{P}^2$ be a plane curve of genus $g \geq 2$ defined over $\mathbb{Z}$. Assume that C contains the $\mathbb{Z}$-valued point $P = (0, 0, 1)$. Let H be the ample divisor $\mathrm{pr}_1^*(P) + \mathrm{pr}_2^*(P)$ on $C \times_{\mathbb{Z}} C$. Let $\Gamma_p \subset \overline{\mathbb{F}}_p \otimes C$ be the graph of the Frobenius $\overline{\mathbb{F}}_p$-morphism $C \to C$ (see Lecture II), defined over $\mathbb{F}_p$. (Γ_p is well-defined at least for the p such that $C \bmod p$ is smooth.) Compute the Hilbert polynomial h_p of Γ_p. Prove that $\bigcup_p \mathrm{Hilb}^{h_p}(\mathbb{Q} \otimes_{\mathbb{Z}} X) = \emptyset$.

2.8   One of the most important applications of Hilbert schemes will be the construction of moduli spaces. Of course it is impossible to treat the theory of moduli in this series of lectures, but I would like to give you a rough idea what a moduli space is.

Consider some meaningful condition on algebraic varieties X (e.g. X has a given topological type, etc.), and the set $\mathcal{M}$ of the isomorphism classes of the varieties subject to this condition. It often happens that $\mathcal{M}$ has structure of an algebraic variety, scheme, or algebraic space (as we saw in Introduction, $\mathcal{M}$ may have no such structure if the condition imposed is not good enough). When this is the case, $\mathcal{M}$ is called the *coarse moduli space* of the varieties with the specified condition. In some special cases, the situation is even better in the sense that there is a universal family $\mathcal{X} \to \mathcal{M}$. That means: if $X \to S$ is an S-scheme with each closed fibre in the set of isomorphism classes $\mathcal{M}$, then there is a unique morphism $S \to \mathcal{M}$ such that X is (S-isomorphic to) the fibre product $S \times_{\mathcal{M}} \mathcal{X}$. $\mathcal{M}$ is called the *fine moduli space* when a universal family exists.

The Hilbert scheme $\mathrm{Hilb}(X)$ can be viewed as the fine moduli space of closed subschemes, while the Chow scheme $\mathrm{Chow}(X)$ (see Section 3 below) is the coarse moduli space of effective cycles on X.

2.9 Examples   (1) The isomorphism classes of the compact Riemann surfaces of genus one can be identified with $\mathbb{C}$ via the j-invariant. Namely, two Riemann surfaces of genus one are mutually isomorphic (over $\mathbb{C}$) if and only if their j-invariants are the same. $\mathbb{C}$ is thus the coarse moduli space of Riemann surface of genus one.

(2) The set of the complex elliptic curves with level two structure is identified with $\mathcal{M}_{1,2} = \mathbb{C} \setminus \{0, 1\}$. (Level n structure on an abelian variety A of dimension g is, by definition, an identification of the group of the n-division points $_n A$ with $(\mathbb{Z}/n\mathbb{Z})^{2g}$.) The Legendre canonical form

$$y^2 = x(x-1)(x-\sigma), \quad \sigma \neq 0, 1, \infty$$

gives the universal family $\mathcal{C}_{1,2} \subset \mathcal{M}_{1,2} \times \mathbb{P}^2$ of such curves parametrized by $c \in \mathcal{M}_{1,2}$. More precisely, if $\mathcal{X} \to S$ is a proper smooth morphism, with fibres being curves of genus one, and there are four disjoint sections $g_i : S \to \mathcal{X}$ such that $2(g_i(s)) \sim 2(g_j(s))$ on each fibres X_s, then there is a unique morphism $S \to \mathcal{M}_{1,2} = \mathbb{P}^1 \setminus \{0, 1, \infty\}$ which makes $\mathcal{X}$ the fibre product $S \times_{\mathcal{M}_{1,2}} \mathcal{C}_{1,2}$ on which the four sections are the pull-backs of the sections $\sigma \mapsto (\sigma, \infty), (\sigma, 0, 0), (\sigma, 1, 0), (\sigma, \sigma, 0)$. Thus $\mathcal{M}_{1,2} = \mathbb{A}^1 - \{0, 1\}$ is the fine moduli space of elliptic curves with level two structure.

(3) Let A be an abelian variety of dimension g and assume that an ample divisor H on A satisfies $H^g = g!$. Then the pair (A, H) is called a *principally polarized abelian variety*. The coarse moduli space, or equivalently the set of isomorphism classes, $\mathcal{A}_g$ of the principally polarized abelian varieties exists [Mumford-Fogarty]. Over the complex numbers, $\mathcal{A}_g$ is identified with the quotient $\mathbf{Sp}(g, \mathbb{Z}) \backslash \mathcal{H}_g$, where $\mathcal{H}_g$ is the Siegel upper halfspace $\{A \in M_g(\mathbb{C}) \mid {}^t A = A, \operatorname{Im} A \gg 0\}$. If we add level n structure ($n \geq 3$), there exists the fine moduli space $\mathcal{A}_{g,n}$, provided the characteristic of the ground field is coprime with n.

(4) D. Mumford constructed the coarse moduli space of the smooth curves $\mathcal{M}_g$ of genus ≥ 2 [Mumford-Fogarty]. By the Torelli theorem, $\mathcal{M}_g$ is a subset of $\mathcal{A}_g$ through the correspondence $C \mapsto J(C)$, where $J(C)$ is the Jacobian with a principal polarization coming from the theta divisor. Putting some additional structure (level structure, say), we can construct the fine moduli for curves with the structure, and $\mathcal{M}_g$ is realized as a finite quotient of this fine moduli space.

2.10 A pair (X, H) of a projective variety X and an ample divisor H is called a *polarized variety*. An isomorphism between two polarized varieties (X_1, H_1) and (X_2, H_2) are defined to be an isomorphism $f : X_1 \tilde{\to} X_2$ such that $f^* H_2 \sim H_1$. The principle of the construction of the moduli space of polarized varieties is this:

Step 0. Select the condition you pose on the polarized varieties (X, H).

Step 1. Check that there are only finitely many choices of the possible Hilbert polynomials $\chi(X, \mathcal{O}_X(tH))$ under your condition chosen in Step 0. Fix one of the possible Hilbert polynomials.

Step 2. Take a sufficiently large m such that $|mH|$ defines an embedding of X into $\mathbb{P}^N$, with N independent of X. Then the set of the triples (X, H, ι) is identified with a subset $F \subset \text{Hilb}(\mathbb{P}^N)$, where ι is an embedding such that mH is linearly equivalent to a hyperplane section.

Step 3. Finally in order to kill the choice of the embeddings, divide out F by the action of $\mathbf{PGL}(N+1)$.

There are some difficulties to clear Step 1. Once it is cleared, Step 2 causes no problem thanks to the theory of Hilbert schemes. The hardest part is usually Step 3, which is so delicate you cannot expect there is a good quotient space in general. However, if you have proceeded up to Step 1, then, in principle, you can list up all the possible polarized variety (X, H) subject to the condition in Step 0. When this is the case, we say the polarized varieties (X, H) are *bounded*. Since $\text{Hilb}^h(\mathbb{P}^N)$ is projective and hence of finite type, it follows that there are only finitely many flat deformation invariants taken by the $[X] \in \mathcal{M}$.

2.11 Examples   (1) Let X be a smooth projective variety of dimension n such that the canonical divisor K_X is ample. The pair (X, K_X) is called a (smooth) canonical n-fold. Any isomorphism $f : X_1 \xrightarrow{\sim} X_2$ induces $K_{X_1} \sim f^* K_{X_2}$ so the isomorphisms between the canonical n-folds are identical with the isomorphisms in the ordinary sense. Fix the dimension n and the degree $d = K_X^n$, Then we want to show that the possible Hilbert polynomials for the (X, K_X) are finitely many. In view of the Riemann-Roch theorem (see [Hirzebruch], [Borel-Serre])

$$\chi(X, \mathcal{O}_X(tH)) = \frac{t^n}{n!} H^n - \frac{t^{n-1}}{(n-1)!} H^{n-1} K_X + \ldots,$$

this follows from:

Theorem [Kollár-Matsusaka] (characteristic zero)   *Given a triple (n, d, a) of positive integers, there are constants $b_1, b_2, \ldots, b_n$ with the following property:*

— If the Hilbert polynomial $\chi(X, \mathcal{O}_X(tH)) = \frac{1}{n!} a_0 t^n + a_1 t^{n-1} + a_2 t^{n-2} + \ldots + a_n$ of an n-dimensional polarized variety satisfies $a_0 = d$, $a_1 \leq a$, then $|a_i| \leq b_i$, $i = 1, \ldots, n$.

Put in another way: The Hilbert polynomial of a polarized n-fold is determined up to finitely many possibilities if you bound the first two coefficients from above.

Caution: The statement is valid only for polarized varieties (i.e. irreducible, reduced schemes). For instance, by adding components of dimension $\leq n - 2$, you can arbitrarily increase the lower term without affecting the top two coefficients.

Thus the canonical smooth n-folds of bounded degree are bounded. In particular, there are only finitely many possibilities for an arbitrary discrete topological invariant (which is clearly invariant under smooth deformation). The coarse moduli spaces of canonical n-folds were constructed by H. Popp [Popp] as algebraic spaces. For $n = 1, 2$, the moduli spaces are quasi-projective [Mumford-Fogarty][Gieseker 2].

(2) A Fano manifold X is, by definition, a smooth projective variety with $-K_X$ ample, which is naturally viewed as a polarized variety. The theorem of Kolár-Matsusaka cited above then tells us:

The Fano manifolds of dimension n are bounded if and only if there exists a constant $d = d(n)$ such that $(-K_X)^n \le d$ for an arbitrary n-dimensional Fano manifold X.   □

On a compact Kähler variety, there is a counterpart of the Hilbert scheme, called the *Douady space*.

2.12 Theorem [Douady]   *Let X be a compact Kähler variety. Then there exist a disjoint union* $\mathrm{Douad}(X)$ *of countably many compact Kähler varieties* $\mathrm{Douad}^\alpha(X)$ *and a closed analytic subspace* $\mathcal{Z} \subset \mathrm{Douad}(X) \times X$ *such that for every complex space V and for every closed analytic subspace $Z \subset V \times X$ flat over V, there is a unique analytic mapping $V \to \mathrm{Douad}(X)$ with $Z = V \times_{\mathrm{Douad}(X)} \mathcal{Z}$.*

2.13   Let S be a scheme and let X and Y be S-schemes. Given an S-scheme V, let $\mathcal{H}om_S(Y, X)(V)$ denote the set of S-morphisms $\{f : V \times_S Y \to X\}$. $\mathcal{H}om_S(Y, X)(V)$ is naturally a contravariant functor (S-schemes) $\to$ (Sets), by attaching the morphism $f \circ (g, \mathrm{id}_Y) : W \times_S Y \to X$ to given $f : V \times_S Y \to X$ and $g : W \to V$. Associating to f the graph $\Gamma_f = \{(y, f(y)) \in Y \times_S X\}$, which is a closed subscheme of the projective variety $Y \times_S X$, we obtain a natural inclusion $\mathcal{H}om_S(Y, X)(V) \hookrightarrow \mathcal{H}ilb(Y \times_S X)(V)$. Thus the contravariant subfunctor $\mathcal{H}om_S(Y, X)(\cdot)$ is expected to be representable by a subset of $\mathrm{Hilb}_S(Y \times X)$, which is indeed the case as we see below.

Let H_X and H_Y be S-ample divisors on X and Y, respectively. Then $H = \mathrm{pr}_X^* H_X + \mathrm{pr}_Y^* H_Y$ is an ample divisor on $X \times_S Y$. Let us calculate the Hilbert polynomial of Γ_f with respect the polarization H. If you recall the Riemann-Roch theorem, then you trivially check that the Hilbert polynomial $h_{\Gamma_f}(t)$ is uniquely determined by the data $(f^* H_X)^i H_Y^j c(Y)^K$, where $K = (k_1, k_2, \ldots)$ is a multi-index and $c(Y)^K$ is the Chern monomial $c_1(Y)^{k_1} c_2(Y)^{k_2} \cdots$ and the indices run over the triples $(i, j, K) \in \mathbb{Z}_{\ge 0}^{2+\dim Y}$ such that $i + j + k_1 + 2k_2 + \ldots = \dim Y$. Thus the Hilbert polynomial is naturally determined by $f^* H_X$ and H_Y.

On the other hand, the set $\{[Z] \in \mathrm{Hilb}_S(Y \times_S X) \mid \mathrm{pr}_Y : Z \to Y \text{ is an isomorphism}\}$ is an open subset in $\mathrm{Hilb}_S(Y \times_S X)$.

2.14 Proposition   *Let X and Y be projective S-schemes. Then $\mathrm{Hom}_S(Y, X)$ is naturally an open subset of $\mathrm{Hilb}(Y \times_S X)$ via the correspondence $\gamma : [f] \mapsto [\Gamma_f]$. Let $\mathcal{Z} \subset \mathrm{Hilb}(Y \times_S X)$ be the universal subscheme. The restriction $\mathcal{Z}|_{\mathrm{Hom}_S(Y,X)}$ of the universal subscheme $\mathcal{Z}$ to the open subset $\mathrm{Hom}_S(Y, X)$ is a subscheme $\subset \mathrm{Hom}_S(Y, X) \times_S Y \times_S X$ which is canonically identified with $\mathrm{Hom}_S(Y, X) \times_S Y$ via $\mathrm{pr}_{\mathrm{Hom}(Y,X) \times_S Y}$. The projection $\mathrm{pr}_X|_{\mathcal{Z}}$ thus induces a morphism $\Phi : \mathrm{Hom}_S(Y, X)$*

$\times_S Y \to X$, given by $([f], y) \mapsto f(y)$. The pair $(\mathrm{Hom}_S(Y, X), \Phi)$ has the universal property in the following sense:

> Given an S-scheme V and an S-morphism $f : V \times_S Y \to X$, there is a unique S-morphism $\phi : V \to \mathrm{Hom}_S(Y, X)$ such that $f = \Phi \circ (\phi, \mathrm{id}_Y)$.   □

As an open subset of $\mathrm{Hilb}(Y \times X)$, $\mathrm{Hom}(Y, X)$ is naturally equipped with the structure of a disjoint union of quasi-projective schemes. The morphism $\Phi : \mathrm{Hom}_S(Y, X) \times_S Y \to X$ is called the *universal morphism*.

2.15 Example Assume that Y is a connected projective curve of arithmetic genus g and that H_Y, ample divisor on Y, is of degree r, where r is the number of the irreducible components (counted with multiplicity when Y is not reduced). Then the Hilbert polynomial $h(\Gamma_f)$ is of the form $(e + r)t + 1 - g$, thus independent of the choice of H_Y, where $e = \deg f^* H_X$.

2.16 Corollary *Let Y be a flat, projective family of curves of arithmetic genus g over S, and X a projective S-scheme with relatively ample divisor H_X. Let the notation as in (2.15). Then*

$$\mathrm{Hom}_S^e(Y, X) \overset{\mathrm{def}}{=} \{f : Y \to X \mid \deg f^* H_X = e\} \subset \mathrm{Hilb}(Y \times_S X)$$

is a quasi-projective S-scheme.

Let X and Y be projective S-schemes. Fix a specific S-morphism $f_0 : Y \to X$ (a reference morphism) and a closed subscheme $B \subset Y$. Then the set of *morphisms with base subscheme* or *with base point set B* is defined by:

$$\mathrm{Hom}_S(Y, X; B) = \mathrm{Hom}(Y, X; B, f_0) \overset{\mathrm{def}}{=} \{f : Y \to X \mid f|_B = f_0|_B\},$$

which is obviously a closed subset of $\mathrm{Hom}_S(Y, X)$ and, in particular, a disjoint union of quasi-projective schemes. When Y is a flat family of curves over S, the set of morphisms of given degree

$$\mathrm{Hom}_S^e(Y, X; B) = \mathrm{Hom}_S(Y, X; B) \cap \mathrm{Hom}_S^e(Y, X)$$

is quasi-projective (and so locally of finite type) over S.

2.17 Corollary *Let R be a ring (commutative, with unity) finitely generated over $\mathbb{Z}$, with the field of fractions K of characteristic zero. Let Y be a proper smooth R-scheme of pure relative dimension one, and $B \subset Y$ a closed subscheme. Let X be a projective R-scheme. If modulo p reductions $\mathrm{Hom}^e(k \otimes_R Y, k \otimes_R X)$ [resp. $\mathrm{Hom}^e(k \otimes_R Y, k \otimes_R X; k \otimes_R B)$] are non-empty for infinitely many primes p, then $\mathrm{Hom}^e(K \otimes_R Y, K \otimes_R X)$ [resp. $\mathrm{Hom}^e(K \otimes_R Y, K \otimes_R X; K \otimes_R B)$] is non-empty, too.*

$\mathrm{Hom}_S(Y, X; B)$ is not complete in general, i.e. it has non-trivial boundary in $\mathrm{Hilb}(Y \times_S X)$. Actually, our proof of Theorem 1.2 relies on the behavior of

$\mathrm{Hom}_S(Y, X; B)$ at the boundary. To get a rough idea, let us observe what happens if $f_t \in \mathrm{Hom}(Y, X)$ tends to the boundary $\mathrm{Hilb}(Y \times X) \setminus \mathrm{Hom}(Y, X)$, when $Y = \mathbb{P}^1$.

2.18 Example $\mathrm{Hom}^0(\mathbb{P}^1, \mathbb{P}^1)$ is the set of constant maps and hence isomorphic to $\mathbb{P}^1$. $\mathrm{Hom}^1(\mathbb{P}^1, \mathbb{P}^1)$ is naturally identified with $\mathrm{Aut}(\mathbb{P}^1)$, the set of automorphisms of $\mathbb{P}^1$, thus isomorphic to $\mathbf{PGL}(2)$. The graph of $g \in \mathrm{Aut}(\mathbb{P}^1)$ is a divisor linearly equivalent to $\tilde{H} = \mathrm{pr}_1^* H + \mathrm{pr}_2^* H$, H being a divisor of degree one on $\mathbb{P}^1$ (if we identify $\mathbb{P}^1 \times \mathbb{P}^1$ with a non-singular quadric in $\mathbb{P}^3$ via the Segre embedding

$$(x_0 : x_1;\ y_0 : y_1) \mapsto (x_0 y_0 : x_0 y_1 : x_1 y_0 : x_1 y_1),$$

$\tilde{H}$ is nothing but a hyperplane section). Hence Γ_g gives an irreducible divisor $\in |\tilde{H}|$. Conversely, an irreducible member $\in |\tilde{H}|$ birationally (and hence isomorphically) projects to the both factors of $\mathbb{P}^1 \times \mathbb{P}^1$, thereby defining the graph of a certain $g \in \mathrm{Aut}(\mathbb{P}^1)$. Thus we can identify $\mathrm{Aut}(\mathbb{P}^1) = \mathrm{Hom}^1(\mathbb{P}^1, \mathbb{P}^1)$ with the open subset

$$\{D \in |\tilde{H}|;\ D \text{ is irreducible}\}$$

of the complete linear system $|\tilde{H}| \simeq \mathbb{P}^3$, which is an irreducible component of $\mathrm{Hilb}^{2t+1}(\mathbb{P}^1 \times \mathbb{P}^1)$. The complement, in this case, consists of the divisors of the form $\{p\} \times \mathbb{P}^1 + \mathbb{P}^1 \times \{q\}$, where p and q are points on $\mathbb{P}^1$, giving an isomorphism between the boundary and $\mathbb{P}^1 \times \mathbb{P}^1$.

2.19 Exercise For a given projective variety X, check that the boundary of $\mathrm{Hom}^e(\mathbb{P}^1, X)$ in $\mathrm{Hilb}^h(\mathbb{P}^1 \times X)$ $(h = (e+1)t + 1)$ consists of subschemes each of which is a union of
(a) a unique component which is isomorphic to $\mathbb{P}^1$ via the first projection $\mathrm{pr}_{\mathbb{P}^1}$,
(b) curves on $\mathbb{P}^1 \times X$ which $\mathrm{pr}_{\mathbb{P}^1}$ maps to points on $\mathbb{P}^1$, (we can eventually show that each of these curves is rational) plus, possibly,
(c) zero-dimensional embedded components.

3   Chow schemes

3.1 In general, an element of $\mathrm{Hilb}(X)$ represents a subscheme $\subset X$ with many components of various dimensions. Sometimes it will be more convenient to look at only the maximal-dimensional components. This is possible if we view a subscheme as an effective cycle, forgetting the underlying scheme structure along with the lower dimensional components. It is known that the set of effective m-cycles of degree d on a projective variety has a structure of projective scheme, called the *Chow scheme*, and denoted by $\mathrm{Chow}_m^d(X)$.

Let us quickly review the construction of the Chow scheme.

Consider an irreducible, m-dimensional subvariety $V \subset X \subset \mathbb{P}^N$. If we take $r = N - m + 1$ hyperplanes $H_1, \ldots, H_r \subset \mathbb{P}^N$ generally, then $V \cap H_1 \cap \ldots \cap H_r = \emptyset$. The hyperplanes are parametrized by $(\mathbb{P}^N)^*$, dual projective space, so the r-tuples of hyperplanes $\Pi = (H_1, \ldots H_r)$ are viewed as elements of $((\mathbb{P}^N)^*)^r$, an Nr-dimensional projective variety. Consider the Π such that $V \cap H_1 \cap H_2 \cap \ldots \cap H_r \neq \emptyset$. This condition on Π is rephrased into the vanishing of a certain resultant in the language of linear algebra, and thus it turns out that such Π's form a hypersurface of $W = ((\mathbb{P}^N)^*)^r$, defined by a single section F_V (classically called the *associated form* of V) of an ample line bundle on $((\mathbb{P}^N)^*)^r$.

Given an effective m-cycle $V = \sum a_i V_i$ on $\mathbb{P}^N$, each V_i being an irreducible subvariety of dimension m, we put $F_V = \Pi F_{V_i}^{a_i}$ and call it the *associated form* or *Chow form* of V, which is a global section of some ample line bundle L on W. F_V is uniquely determined up to non-zero constant, so that we get a mapping from the set of effective m-cycles on $\mathbb{P}^N$ to a disjoint union of the projective spaces $\mathbb{P}(H^0(W, L)^*)$. If we fix the degree of V, then we infer that its image $\mathrm{Chow}_m^d(\mathbb{P}^N)$ is contained in a single copy of these projective spaces. Thus the set of effective m-cycles on X is a closed subset $\mathrm{Chow}_m^d(X)$ of a projective scheme $\mathrm{Chow}_m^d(\mathbb{P}^N)$.

3.2 Exercise Let $V \subset \mathbb{P}^N$ be the r-dimensional linear subspace $\{(x_0 : \ldots : x_r : 0 : \ldots : 0)\}$. Let Π be the $(r+1)$-tuple of hyperplanes H_i defined by $\sum_{j=0}^N a_{ij} x_i = 0$, $i = 0, 1, \ldots, r$. Then show that $H_0 \cap H_1 \cap \ldots \cap H_r$ meets V if and only if

$$F_V = \det(a_{ij})_{0 \leq i, j \leq r} = 0.$$

Determine the line bundle L of which F_V is a global section.

3.3 Let us look at the relationship between the Hilbert schemes and Chow schemes. Take a Hilbert polynomial $h = h_Y(t)$ of a closed subscheme $Y \subset X$. The degree m of h is the highest dimension of the components of Y. By forgetting the underlying scheme structure of Y, we may think of Y as an m-cycle V. The degree d of V is given by the expression $h = \frac{d}{m!} t^m + $ (lower term in t). This correspondence gives a set-theoretically well-defined mapping $\psi : \mathrm{Hilb}^h(X) \to \mathrm{Chow}^d(X)$, and you can check this map is continuous with respect to the Zariski topology.

Let $U \subset \mathrm{Chow}(X)$ be the subset consisting of reduced effective cycles $Z = \sum_i Z_i$, where Z_i's are mutually distinct irreducible varieties of the same dimension. Then we can attach a pure-dimensional subscheme structure to Z in a natural way. There is a open subset $U' \subset U$ on which the Hilbert function is constant [Mumford 3, p.50, Corollary], which induces a morphism $\phi : U' \to \mathrm{Hilb}(X)$, such that $\psi\phi = \mathrm{id}$. If $\mathrm{Hilb}(X)$ is reduced and normal at $\phi([V])$, $[V] \in U'$, then ϕ is locally an isomorphism at $[V]$ by the Zariski Main Theorem.

If, in particular, Z is reduced and locally a complete intersection with $H^1(Z, \mathcal{N}_{Z/X}) = 0$ in a smooth projective scheme X, then $\psi : \mathrm{Hilb}(X) \to \mathrm{Chow}(X)$ is locally an isomorphism near $[Z]$, see (II.1.13.1).

3.4 Example (1) Let X be a projective variety over an algebraically closed field k. If h is the constant function 1, then $\mathrm{Hilb}^1(X)$ is nothing but the set of closed points, isomorphic to X and also to $\mathrm{Chow}_0^1(X)$.

(2) Let X be a non-singular surface over $\mathbb{C}$. Let $Y \subset X$ be a zero-dimensional subscheme, with Hilbert polynomial being the constant function 2, so that Y has degree 2. Y is reduced if and only if $Y = \{x_1, x_2\}$, for certain distinct points $x_1, x_2 \in X$. When Y is not reduced and supported by a single point x, the scheme structure of Y is uniquely determined by a one-dimensional subspace $I_Y / \mathfrak{M}_x^2 \subset \mathfrak{M}_x / \mathfrak{M}_x^2$, or equivalently a one-dimensional quotient of the two-dimensional subspaces $\mathfrak{M}_x / \mathfrak{M}_x^2$. This defines a natural identification of $\mathrm{Hilb}^2(X) \simeq Bl_\Delta(X \times X)/\iota$, the blowing up of $X \times X$ along the diagonal Δ, divided by the natural transposition ι of the two factors. On the other hand, if we forget the scheme structure, $\mathrm{Chow}_0^2(X)$ is naturally identified with the symmetric product $S^2 X = X \times X/\iota$.

(3) $\mathrm{Chow}_1^2(\mathbb{P}^3)(\mathbb{C})$ is set-theoretically a union of U, the open set of the skew lines (8-dimensional family) with Hilbert polynomial $2t + 2$, and the closed set V of the plane quadrics with Hilbert polynomial $2n + 1$. Two coplanar lines $L_1 + L_2$ represent an element in $\overline{U} \cap V$. Thus the rational map $\phi : \mathrm{Chow}(X) \dashrightarrow \mathrm{Hilb}(X)$ is not defined at $[L_1 + L_2]$.

Because we are specifically interested in rational curves, let us introduce subsets $R^d(X) \subset RC^d(X) \subset \mathrm{Chow}_1^d(X)$.

3.5 Proposition *Let X be a projective variety. Consider the subsets*

$$RC^d(X) \overset{\mathrm{def}}{=} \{[Z] \in \mathrm{Chow}_1^d(X) \mid Z \text{ is a connected chain of rational curves}\}$$

$$R^d(X) \overset{\mathrm{def}}{=} \{[Z] \in \mathrm{Chow}^d(X) \mid Z \text{ is an irreducible rational curve}\}$$

in the Chow scheme $\mathrm{Chow}_1^d(X)$. Then:

(1) *There is a natural surjective continuous mapping $\mathrm{Hom}_1^d(\mathbb{P}^1, X) \to R^d(X)$, of which each fibre is bijective to $\mathrm{Aut}(\mathbb{P}^1)$. When $\mathrm{Hom}_1^d(\mathbb{P}^1, X)$ is normal at $[f]$, and the arithmetic genus is locally constant on $R^d(X)$ near $f(\mathbb{P}^1)$, then this mapping is locally a submersion (i.e. an open morphism of maximal rank) at $[f]$. Here $\mathrm{Hom}_1^d(\mathbb{P}^1, X)$ denotes the set of morphisms $f : \mathbb{P}^1 \to X$ such that $f : \mathbb{P}^1 \to f(\mathbb{P}^1)$ is birational and $\deg f^* H = d$, which is an open subset of $\mathrm{Hom}^d(\mathbb{P}^1, X)$.*

(2) *$RC^d(X)$ is a closed subset of $\mathrm{Chow}(X)$, and hence projective.*

(3) *$R^d(X)$ is an open subset of $RC^d(X)$, and hence quasi-projective.*

Proof. (1) Given a morphism $f : \mathbb{P}^1 \to X$ birational onto the image, we attach the image $[f(\mathbb{P}^1)] \in \mathrm{Chow}(X)$, thereby defining a continuous mapping $\mathrm{Hom}_1^d(\mathbb{P}^1, X) \to R^d(X)$. Let $C \in X$ be an arbitrary irreducible rational curve. Then take the normalization $\tilde{C}$ of C. Since C is rational, there is an isomorphism between $\tilde{C}$ and $\mathbb{P}^1$. With such an isomorphism fixed, we get a morphism $f : \mathbb{P}^1 \to X$, which is clearly birational onto the image C. This shows that the above map is surjective. The fibre corresponds to the choice of isomorphisms between $\tilde{C}$ and $\mathbb{P}^1$, naturally identified with the automorphism group $\mathrm{Aut}(\mathbb{P}^1)$.

For the proof of the submersiveness, we need the infinitesimal description of Hilbert schemes and Hom-schemes, which is described in the next lecture (II.1). Once you have the description, then the assertion is clear.   □

(2) By Mumford's criterion for completeness, it suffices to show the following:

— Let A be a discrete valuation ring and K its fraction field. Then any morphism $\operatorname{Spec} K \to RC(X)$ extends to $\operatorname{Spec} A \to RC(X)$.

A heuristic interpretation of the criterion over $\mathbb{C}$ is this. Assume that there is a holomorphic mapping from a disk Δ to $\operatorname{Chow}(X)$, such that the image of the punctured disk $\Delta^* = \Delta \setminus 0$ is contained in $RC(X)$. Then the mapping: $\Delta^* \to RC(X)$ extends to one from Δ to $RC(X)$.

We check the Mumford criterion below in an abstract manner, but you should try to give a concrete proof in the geometric situation mentioned above.

Given an effective 1-cycle Z_K on $K \otimes X$ consisting of rational curves, we can take a finite extension L of K such that each irreducible component is defined over L. Let B be the integral closure of A in L. Then the morphism $\operatorname{Spec} L \to RC(X) \subset \operatorname{Chow}(X)$ induced by $\operatorname{Spec} K \to RC(X)$ extends to a morphism $\phi :$ $\operatorname{Spec} B \to \operatorname{Chow}(X)$ by the projectivity of the Chow variety. We will show that the cycle associated with the closed point of $\operatorname{Spec} B$ is a union of rational curves. Look at one irreducible component Z_0^* of $L \otimes_K Z_K$. Its normalization $W_L \subset$ $L \otimes_A X \subset B \otimes X$ is smooth over L with arithmetic genus 0. The closure $\overline{W_L}$ in $B \otimes X$ is a 2-dimensional projective scheme, equipped with a natural projection $\overline{W_L} \to \operatorname{Spec} B$ which has fibres of dimension ≤ 1. Let $W = W_B$ be the minimal resolution of the normalization of $\overline{W_L}$. Then the generic fibre of the projection $\pi : W \to \operatorname{Spec} B$ has, by construction, arithmetic genus 0. On the other hand, since the one-dimensional scheme $\operatorname{Spec} B$ is normal and hence regular, the equi-dimensional, proper projection $W \to \operatorname{Spec} B$ from the regular scheme W is flat over $\operatorname{Spec} B$. Hence by the invariance of the arithmetic genus $p_a(W_s) = -\chi(W_s) + 1$ under flat morphisms, we infer that the arithmetic genus of the closed fibre W_o is also 0. By the normality of $\operatorname{Spec} B$, $W_o = \pi^* o$ is a connected Cartier divisor of the form $\sum m_i W_{0i}$. Replacing B by a suitable finite extension if necessary, you may assume $\mathrm{G.C.D}\{m_i\} = 1$. Then you easily check that $H^0(W_o, \mathcal{O}_{W_o})$ is equal to the residue field k, so that, by $\chi(W_o) = -p_a(W_o) + 1 = 1$, you get $H^1(W_o, \mathcal{O}_{W_o}) = 0$.

Thus the special fibres of each component of $\overline{Z}_L \subset B \otimes X$ is necessarily a union of rational curves. It is clear that a limit of connected cycles is connected. Hence $[Z_o] \in RC(X_o)$, completing the proof.   □

(3) Left to the reader as an exercise.   □

The statement (3.5.2) is is paraphrased as follows:
A specialization of a connected chain of rational curves is again a connected chain of rational curves.

A specialization of an irreducible rational curve is not necessarily irreducible.

4   Essentials from Intersection Theory

4.1 Let us review some fundamentals of intersection theory, which the proof of Theorem 1.2 relies on.

Let X be a projective (perhaps singular) variety. By $\mathrm{Div}(X)$ we denote the additive group of Cartier divisors, and by $Z_1(X)$ the group of 1-cycles. There is a bilinear coupling $Z_1(X) \times \mathrm{Div}(X) \to \mathbb{Z}$ defined by the intersection number, which defines a coarse equivalence relation called the *numerical equivalence*.

A Cartier divisor D is said to be *numerically trivial* if $(C, D) = 0$ for all 1-cycles C on X (or, equivalently, for all irreducible curve C on X). Similarly a 1-cycle C is numerically trivial if $(C, D) = 0$ for all Cartier divisor D on X. Two Cartier divisors D_1, D_2 or two 1-cycles C_1, C_2 are *numerically equivalent* to each other (denoted by the symbol $\approx$) if the difference is numerically trivial.

Let $N_{\mathbb{Z}}^1(X)$ [resp. $N_1^{\mathbb{Z}}(X)$] be the Cartier divisors [resp. 1-cycles] modulo $\approx$. By definition, the intersection number induces a non-degenerate paring $N_1^{\mathbb{Z}}(X) \times N_{\mathbb{Z}}^1(X) \to \mathbb{Z}$. In particular, $N_1^{\mathbb{Z}}(X)$ and $N_{\mathbb{Z}}^1(X)$ are both torsion-free, and it is known that their rank $\rho(X)$ is finite (the Picard number of X). They are lattices of the finite dimensional $\mathbb{R}$-vector spaces $N_1(X) = \mathbb{R} \otimes N_1^{\mathbb{Z}}(X)$ and $N^1(X) = \mathbb{R} \otimes N_{\mathbb{Z}}^1$.

An element $[C]$ of $\mathbb{R} \otimes Z_1(X)$ is represented by a finite sum $\sum a_i C_i$, where $a_i \in \mathbb{R}$ and C_i is an irreducible curve. $C = \sum a_i C_i \in \mathbb{R} \otimes_{\mathbb{Z}} Z_1(X)$ is said to be an *effective* 1-cycle if all the coefficients a_i are positive or 0. An element $[C] \in N^1(X)$ is said to be effective if its numerical equivalence class contains an effective 1-cycle. The numerical equivalence classes of effective 1-cycles form a convex cone $NE(X) \subset N_1(X)$, called the *effective cone*. A real 1-cycle C is said to be *pseudo-effective* if its numerical class $[C]$ sits in the closure $\overline{NE}(X)$.

A real Cartier divisor $[D] \in \mathbb{R} \otimes \mathrm{Div}(X)$ is said to be *ample* when $[D]$ is represented as a sum $\sum a_i D_i$ of ample Cartier divisors D_i with positive coefficients. The numerical equivalence classes of real ample divisors form a convex cone $NA(X) \subset N^1(X)$. An $\mathbb{R}$-divisor D is said to be *nef* if its numerical class sits in the closure $\overline{NA}(X)$.

In terms of these symbols, Kleiman's criterion for ampleness (see [Hartshorne 1]) states the following:

4.2 Theorem   *The closed cones $\overline{NE}(X)$ and $\overline{NA}(X)$ are mutually dual in the following sense:*

$$\overline{NA}(X) = \{D \in N^1(X) \mid \forall C \in \overline{NE}(X), (C, D) \geq 0\},$$
$$\overline{NE}(X) = \{C \in N_1(X) \mid \forall D \in \overline{NA}(X), (C, D) \geq 0\}.$$

The ample cone $NA(X)$ is an open set of $N^1(X)$, so that $NA(X) = \mathrm{Int}\big(\overline{NA}(X)\big)$.
($NE(X)$ is, in general, neither open nor closed in $N_1(X)$).

4.3 Exercise   Let X be a projective scheme.

(1) From (4.2) derive the following criteria for nefness and ampleness:

A Cartier divisor D is nef $\Longleftrightarrow (C, D) \geq 0$ for any curve $C \subset X$.

A Cartier divisor D is ample
$\Leftrightarrow \exists$ an ample divisor H, $\exists \varepsilon > 0$ s.t. $(C, D) \geq \varepsilon(C, H)$ for $\forall$ irreducible curve C on X
$\Leftrightarrow D - \varepsilon H$ is nef for some ample H and some positive ε.

(2) Let n denote the dimension of X. Prove that $D^n > 0$ [resp. ≥ 0] if $D \in \mathrm{Div}(X)$ is ample [resp. nef].

(3) Show that (the numerical equivalence class of) the 1-cycle D^{n-1} (what would be a correct definition?) is effective [resp. pseudo-effective] if D is ample [resp. nef].

(4) Let $f : X \to Y$ be a morphism. Show that $f^*D \in \mathrm{Div}(X)$ is nef if $D \in \mathrm{Div}(X)$ is nef.

4.4   Consider a specific case where X is a smooth projective surface over an algebraically closed field. Then $\mathrm{Div}(X) = Z_1(X)$, $N^1(X) = N_1(X)$. Thus $N^1(X)$ is a $\mathbb{R}$-vector space of dimension $\rho(X)$ equipped with a natural non-degenerate quadratic form.

Fix an ample divisor H on X. If a divisor D is orthogonal to H (i.e. if $DH = 0$), then $D^2 \leq 0$. Indeed, if $D^2 > 0$, then the Riemann-Roch tells you that $\pm mD$, $(m \gg 0)$ must be an effective divisor, so that $\pm DH > 0$, contradicting the hypothesis. Therefore the signature of the non-degenerate quadratic form on $N^1(X)$ is $(1, \rho(X) - 1)$. This fact is called the *Hodge index theorem*; though elementary, it plays a decisive role in many crucial situations.

It is easy to see that an ample $\mathbb{R}$-divisor on X is numerically equivalent to an effective real 1-cycle. Thus we have the implications

$$
\begin{array}{ccc}
\text{ample} & \Rightarrow & \text{nef} \\
\Downarrow & & \Downarrow \\
\text{effective} & \Rightarrow & \text{pseudo-effective.}
\end{array}
$$

4.5   Let

$$
\tilde{X} = X_m \overset{\mu_m}{\to} X_{m-1} \to \ldots X_1 \overset{\mu_1}{\to} X_0 = X
$$

be a sequence of one-point blowing-ups of smooth projective surfaces. Namely, X_i is a blowing-up of X_{i-1} at a point $P_i \in X_{i-1}$. $E_i = \mu_i^{-1}(P_i) \subset X_i$ is the exceptional divisor attached to the blowing up, with self-intersection number -1.

Then $\mathrm{Div}(X_i) = \mathrm{Div}(X_{i-1}) \oplus \mathbb{Z}E_i$, where $\mathrm{Div}(X_{i-1})$ is identified as a subgroup of $\mathrm{Div}(X_i)$ via the pull-back. It follows that $N^1(X_i)$ is the direct sum of $N^1(X_{i-1})$ and $\mathbb{R}E_i$, each of which is orthogonal to the other. Let $C_i \subset X_i$ be a curve on X_i. When $P_{i+1} \in X_i$ is away from C_i, the pull-back $\mu_{i+1}^*C_i$ is clearly isomorphic to C_i. But if P_{i+1} lies on C_i, $\mu_{i+1}^*C_i$ contains E_{i+1} as an irreducible component, with multiplicity $m_{i+1} = \mathrm{mult}_{P_{i+1}} C_i$. By the *strict transform* $\mu_{i+1,\mathrm{strict}}^{-1}(C_i)$, we mean the curve $\mu_{i+1}^*C - m_{i+1}E_{i+1}$, which does not contain E_{i+1}. The *total transform* is, on the other hand, the pull-back $\mu_{i+1}^*C_i$.

Let $\nu_{ji} : X_j \to X_i$ be the morphism $\mu_{i+1} \cdots \mu_j$ $(j > i)$. The strict transform $\nu_{ji,\mathrm{strict}}^{-1}(C_i) \subset X_j$ of $C_i \subset X_i$ is inductively defined by $\mu_{j,\mathrm{strict}}^{-1}(\nu_{j-1,i,\mathrm{strict}}^{-1}(C_i))$. The total transform of C_i means the pull-back $\nu_{j,i}^* C_i$.

4.6 Let Y be a smooth projective surface and $\phi : Y \to C$ a surjective morphism onto C. If a general fibre of ϕ is a smooth rational curve, then there is a sequence of one-point blowing-ups $Y = X_m \to \ldots \to X_0 = X$ such that X is a (geometric) ruled surface $\mathbb{P}(\mathcal{E})$, where $\mathcal{E}$ is a vector bundle of rank two on C. Then $\mathrm{Pic}(X)$ is generated by the tautological divisor $\mathbf{1}_\mathcal{E}$ and a general fibre F of the projection $\mathbb{P}(\mathcal{E}) \to C$. Their intersection numbers are given by

$$F^2 = 0, \ F\mathbf{1}_\mathcal{E} = 1, \ \mathbf{1}_\mathcal{E}^2 = \deg \det \mathcal{E}.$$

The canonical divisor K_X of X is given by $-2 \cdot \mathbf{1}_\mathcal{E} + \phi^*(\det \mathcal{E} + K_C)$. Thus $N^1(Y)$ is generated by the total transforms of $F, \mathbf{1}_\mathcal{E}$ and the total transforms of the exceptional divisors $E_1, \ldots, E_m$. $\sum \mathbb{R}E_i$ and $N^1(X) = \mathbb{R}F \oplus \mathbb{R}\mathbf{1}_\mathcal{E}$ are the orthogonal complements of each other.

4.7 Exercise Prove the equalities $\mathbf{1}_\mathcal{E}^2 = \deg \det \mathcal{E}$, $K_X = -2 \cdot \mathbf{1}_\mathcal{E} + \phi^*(\det \mathcal{E} + K_C)$.

The following lemma is elementary but has important applications in what follows:

4.8 Lemma *The ground field k is assumed to be algebraically closed. Let $X = \mathbb{P}(\mathcal{E})$ be a ruled surface over a smooth projective curve C (in particular, $\rho(X) = 2$). Then,*

(1) *A curve $D \subset X$ with negative self-intersection, if any, is unique.*

(2) *Assume that the characteristic of k is zero. Then an irreducible curve $D \subset X$ satisfies $D^2 \geq 0$ unless D is a section of the projection $X \to C$.*

Proof.   (1) Suppose that there were two distinct irreducible curves D_1, D_2 with negative self-intersections. Then the intersection matrix

$$\begin{pmatrix} D_1^2 & D_1 D_2 \\ D_1 D_2 & D_2^2 \end{pmatrix}$$

is clearly negative definite, which contradicts the Hodge index theorem.

(2) Write $D \approx a\mathbf{1}_\mathcal{E} + bF$, where a is the mapping degree $\phi|_D : D \to C$, and b an integer. Then $D^2 = a(a \deg \det \mathcal{E} + 2b)$ by easy calculation. First suppose that D is smooth. Then the adjunction formula gives $K_D = (a-2)\mathbf{1}_\mathcal{E} + bF + \phi^*(\det \mathcal{E} + K_C)$. By Hurwitz formula for finite covering of algebraic curves, we have $K_D \geq \phi^* K_C$, so that

$$\begin{aligned}
0 \leq \deg_D((a-2)\mathbf{1}_\mathcal{E} + bF + \phi^* \det \mathcal{E}) &= (a\mathbf{1}_\mathcal{E} + bF)((a-2)\mathbf{1}_\mathcal{E} + bF + \phi^* \det \mathcal{E}) \\
&= a(a-2)\mathbf{1}_\mathcal{E}^2 + \mathbf{1}_\mathcal{E}(2b(a-1)F + a\phi^* \det \mathcal{E}) \\
&= (a-1)(a \deg \det \mathcal{E} + 2b) = (a-1)D^2.
\end{aligned}$$

Hence $a > 1$ implies $D^2 \geq 0$ (the equality holds only when D is étale over C). When D is not smooth, then the adjunction computes the degree of the dualizing sheaf ω_D. The pull-back of ω_D to the normalization $\tilde{D}$ is bigger than $K_{\tilde{D}}$, and the Hurwitz formula applied to the projection $\tilde{D} \to C$ gives an even better inequality $D^2 > 0$. $\square$

4.9 Exercise   Find an example of a ruled surface X in characteristic $p > 0$ that carries an irreducible curve D with negative self-intersection which projects onto C as a purely inseparable p-sheeted covering.

5   Proof of Theorem 1.2

Now we are ready to prove Theorem 1.2. For simplicity of the argument, we assume that the varieties are projective over k, an algebraically closed field.

5.1 Step I. By assumption, the quasi-projective scheme $\mathrm{Hom}(C, X; B)$ has positive dimension. Hence we can find a smooth irreducible curve (S, o) with a marked point o and a non-constant morphism $h : S \to \mathrm{Hom}(C, X; B)$ such that $h(o) = [f]$. Then we get a morphism $\tilde{f} = \{f_s\} : S \times C \to X$ by $f_s(p) = \tilde{f}(s, p) = \Phi_{h(s)}(p) \in X$, where $\Phi : \mathrm{Hom}(C, X; B) \times \mathbb{P}^1 \to X$ is the universal morphism and Φ_t is the restriction of Φ to $\{t\} \times C \simeq C$. By the definition of the base point set $B = \{p_1, \ldots, p_b\}$, we have $f_s(p_i) = f(p_i)$ for an arbitrary $s \in S$, so that $\tilde{f}|_{S \times \{p_i\}}$ is a single point. On the other hand, $\tilde{f}$ being a non-trivial deformation, $\tilde{f}|_{S \times \{p\}}$ is not a single point but one-dimensional for $p \in C$ generic.

Let $\overline{S}$ be a smooth compactification of S, i.e. $\overline{S}$ is a smooth complete curve which contains S as a Zariski open subset.

5.2 Lemma   *The morphism $\tilde{f}$ does not lift to a morphism from $\overline{S} \times C$.*

Proof.   Suppose the contrary and denote the lift by the same symbol $\tilde{f}$. Then $\tilde{f}(\overline{S} \times \{p\})$ would be a complete curve in X, for a general point $p \in C$. The intersection number $(\tilde{f}(\overline{S} \times \{p\}), H)$ would be clearly positive, where H is an ample divisor on X. On the other hand, $\{p\}$ is algebraically (or homologically) equivalent to p_i (as a cycle on C) and so is $\tilde{f}(\overline{S} \times \{p\})$ to $\tilde{f}(\overline{S} \times \{p_i\})$, which is a single point. Hence

$$0 < (\tilde{f}(\overline{S} \times \{p\}), H) = (\tilde{f}(\overline{S} \times \{p\}), H) = 0,$$

a contradiction. $\square$

5.3 Remark   (1) (5.2) is a special case of the following Mumford rigidity lemma [Mumford 3, p. 43]:

Let X be a complete variety, Y, Z varieties and $f : X \times Y \to Z$ a morphism. If $f|X \times \{y_0\} \to Z$ is a constant morphism for some closed point $y_0 \in Y$, then f comes from a morphism $Y \to Z$ via the projection $X \times Y \to Y$.

(2) In (5.2), we need once more the assumption that X is projective (or, alternatively, Kähler). Indeed, on a non-Kähler complex manifold, a non-trivial closed curve may well have trivial cohomology class. For instance, let $X = (\mathbb{C}^2 \backslash (0,0))/\langle \alpha \rangle$ be a Hopf surface, where $\alpha \in \mathbb{C}^*$ is taken to be a real number with $0 < \alpha < 1$. There is a natural fibration $X \to \mathbb{P}^1 = (\mathbb{C}^2 \backslash (0,0))/\mathbb{C}^*$, of which every fibre is isomorphic to an elliptic curve $\mathbb{C}^*/\langle \alpha \rangle \simeq \mathbb{C}/\langle 2\pi\sqrt{-1}, \log \alpha \rangle$. Let us show that the homology class of these elliptic curves are trivial. In fact, the mapping $X \to S^3 = \{z \in \mathbb{C}^2; \|z\| = 1\}$ defined by $(x, y) \mapsto \frac{z}{\|z\|}$ endows X an S^1-bundle structure over S^3, while the natural inclusion $S^3 \hookrightarrow \mathbb{C}^2 \backslash (0,0)$ induces a cross section of this projection. Thus we infer that X is diffeomorphic to $S^1 \times S^3$, and its second homology group is trivial.

(5.4) Step II. Let us return to the case where X, Y are projective.

Though $h : S \to \mathrm{Hom}(C, B; X, f(B))$ does not lift to a morphism from $\overline{S}$, we can view the natural morphim $\tilde{f} : S \times C \to X$ as a rational map: $\overline{S} \times C \dashrightarrow X$. By blowing up finitely many points of the smooth surface $\overline{S} \times C$, we can resolve the indeterminacy of the rational map to get a morphism $\overline{f} : W = Bl_{q_r}(Bl_{q_{r-1}}(\cdots (Bl_{q_1}(\overline{S} \times C)) \cdots)) \to X$, which fits into the commutative diagram

$$
\begin{array}{ccccc}
S \times C & \xrightarrow{\;\alpha\;} & \overline{S} \times C & \xleftarrow{\;\beta\;} & W \\
\tilde{f} \downarrow & & \tilde{f} \downarrow & & \overline{f} \downarrow \\
X & \xrightarrow[\text{identity}]{} & X & \xleftarrow[\text{identity}]{} & X
\end{array}
$$

Since an exceptional divisor (with respect to β) on W is a union of rational curves, so is its image unless it is a single point. The fact that $\tilde{f}$ does not lift to $\overline{S} \times C$ tells us that some exceptional curve is non-trivially mapped onto a curve on X, which is clearly a rational curve.

This shows the existence of a rational curve on X, but we need to show there is a rational curve of bounded degree.

(5.5) Step III. Let the notation be as above.

We have two cases:
1) $\overline{f}(W)$ is one-dimensional and hence coincides with $f(C)$, and
2) $\overline{f}(W)$ is a surface (i.e. two-dimensional).

Case 1. $\overline{f}(W) = f(C)$. In this case, standard theory of deformation of morphisms (see Lecture II) tells us that the image $f(C)$ must be rational, see (II,1.11).

Case 2. $\overline{f}(W)$ *is a surface.* Our proof depends completely on the intersection theory on a blown-up ruled surface W.

Let us fix the notation. Let $B = \{y_1, \ldots, y_d\}$ be the base point set. The morphism $W \to \overline{S} \times C$ is a composite of blowing-ups μ_α with exceptional divisors E_α over smooth points p_α. Since two blowing ups commute each other if the centres are mutually disjoint, we may assume that W is the fibre product of morphisms $\nu_1, \ldots, \nu_d$ and ν_0, where ν_i is a composite of blowing ups with centers over $y_i \in C$ ($i = 1, \ldots, d$) or with centers over points away from B ($i = 0$). Accordingly we have a grouping of E_α's into the $E_{i,\beta}$ over y_i and $E_{0,\beta}$ over points outside B. By abuse of notation, we let $E_{i\beta}$ also denote the total transform $\in \mathrm{Pic}(W)$. Let $\overline{S}_i \subset W$ be the strict transform of the curve $\overline{S} \times \{y_i\}$. For $s \in S$ and $y \in C$ generic, let C_s and $\overline{S}_y \subset W$ be the total ($=$ strict) transforms of $\{s\} \times C$ and $\overline{S} \times \{y\} \subset \overline{S} \times C$, respectively.

In this notation, we see that

— Since $\overline{f}(W)$ is a surface,

$$(\overline{f}^*H, \overline{f}^*H) > 0. \tag{5.6.1}$$

— Since $\overline{f}$ contracts the curves $\overline{S}_i$ into finitely many points $y_i \in X$,

$$(\overline{S}_i, \overline{f}^*H) = 0. \tag{5.6.2}$$

— By definition,

$$\deg f^*H = (C_s, \overline{f}^*H) = (\overline{f}_*C, H) > 0. \tag{5.6.3}$$

$N^1(W)$ is the direct sum of $N^1(\overline{S} \times C)$ and $\bigoplus \mathbb{R}E_{i\beta}$. $\mathbb{Z}C_s \oplus \mathbb{Z}\overline{S}_y$ is a unimodular sublattice of the first summand $N^1(\overline{S} \times C)$ so that $N^1(\overline{S} \times C)$ is a direct sum of $\mathbb{Z}C_s \oplus \mathbb{Z}\overline{S}_y$ and its orthogonal complement with respect to the intersection pairing. Thus we can find integers $a, b, c_{i\beta}$ and a divisor R orthogonal to $C_s, \overline{S}_y, E_{i\beta}$ such that

$$\overline{f}^*H \approx aC_s + b\overline{S}_y - \sum_{i,\beta} e_{i\beta}E_{i\beta} + R \tag{5.6.4}$$

Noting the pull-back $\overline{f}^*H$ of an ample divisor H is nef, we infer that

$$a = (\overline{S}_y, f^*H) \geq 0, \ e_{i\beta} = (E_{i\beta}, f^*H) \geq 0,$$

and from (5.6.3) we derive the equality $b = \deg f^*H$. On the other hand, the strict transform $\overline{S}_i$ of the smooth curve $\overline{S} \times \{y_i\}$ is of the form $\nu^*(\overline{S} \times \{b_i\}) - \sum_\beta \varepsilon_{i\beta}$, $\varepsilon_{i\beta}$ being 1 or 0 according as the center of the blowing-up sits on the strict transform of $\overline{S} \times \{b_i\}$ or away from it. Thus

$$\overline{S}_i \approx \overline{S}_y - \sum_\beta \varepsilon_{i\beta}E_{i\beta}. \tag{5.6.5}$$

Substituting $\overline{f}^*H$ and $\overline{S}_i$ in (5.6.1) and (5.6.2) by the numerical equivalences (5.6.4) and (5.6.5), we obtain

$$2ab - \sum_{i,\beta} e_{i\beta}^2 > 0, \tag{5.6.6}$$

$$a - \sum_\beta \varepsilon_{i\beta}e_{i\beta} = 0 \tag{5.6.7}$$

Replacing a in (5.6.6) by $\sum_\beta \varepsilon_{i\beta} e_{i\beta}$, we get

$$2b \sum_\beta \varepsilon_{i\beta} e_{i\beta} - \sum_{i,\beta} e_{i\beta}^2 > 0.$$

Then, by $e_{i\beta} \geq 0$ and $0 \leq \varepsilon_{i\beta} \leq 1$, we have

$$2b \sum_\beta \varepsilon_{i\beta} e_{i\beta} - \sum_{i,\beta} (\varepsilon_{i,\beta} e_{i\beta})^2 > 0.$$

Summing up in the indices $i = 1, \ldots, d$, we get

$$2d \sum_{i,\beta} \varepsilon_{i\beta} e_\varepsilon - b \sum_{i,\beta} (\varepsilon_{i\beta} e_{i\beta})^2 > 0.$$

It follows that there must be at least one pair (i, β) such that

$$0 < \varepsilon_{i\beta} e_{i\beta} = E_{i\beta} \overline{f}^* H < \frac{2d}{b}.$$

By this inequality, the image $\overline{f}(E_{i\beta})$, which is a connected union of rational curves and points, is one-dimensional, and each of its one-dimensional irreducible component has degree $< 2d/b$. By the inequality $E_{i\beta} \overline{S}_i > 0$, we see that $\overline{f}(E_{i\beta})$ passes through $b_i = \overline{f}(\overline{S}_i)$, and we can find an irreducible component $L \ni b_i$ of $\overline{f}(E_{i\beta})$ which passes through b_i. This completes the proof.   □

Historical Comments

Chow schemes are generalizations of the Grassmann varieties $\mathrm{Grass}(m, n)$, the set of n-dimensional vector subspaces of a fixed m-dimensional variety, which is embedded into $\mathbb{P}^N$, $N = \binom{m}{n} - 1$ by Plücker coordinates. For more details of the theory of Chow schemes, developed specifically by W.-L. Chow and B.L. van der Waerden, the reader is referred to [Samuel]. Hilbert schemes were first constructed by A. Grothendieck [FGA]. A similar method is applicable to construct the moduli space of stable vector bundles via Quot schemes. The description of infinitesimal structure of the Hilbert scheme around a locally complete intersection subscheme was given by [Kodaira-Spencer] and [SGA].

Analogues of the Hilbert schemes and Chow schemes in the category of compact Kähler spaces are Douady spaces [Douady] and Barlet spaces [Barlet], respectively.

There are several criteria for ampleness – due to [Nakai], [Kleiman], [Seshadri], etc. – all of which are found in [Hartshorne 1]. Nef divisors are called numerically effective in [Zariski 2], but this terminology is confusing because an effective divisor may not be numerically effective. Fundamental examples of nef divisors are D's such that $|mD|$ are free from base points for some $m > 0$; such D is said to be semiample (T. Fujita) or eventually free (M. Reid). "Nef" (from "numerically eventually free") is a neologism by Reid.

Theorem 1.2 was stated in [Mori 1] without an explicit bound of the degree of resulting rational curves. The present version is due to [Miyaoka-Mori]

Lecture II
Construction of Non-Trivial Deformations via Frobenius

Overview

In the second lecture, we discuss a technique of S. Mori to construct deformation of morphisms via reduction modulo p, and show the existence of rational curves on smooth projective varieties whose canonical divisors are not nef.

This technique, developed in the famous solution [Mori 1] of a conjecture of R. Hartshorne, was the starting point to the theory of extremal rays and minimal models. As one of its applications, we characterize a class of varieties which contain sufficiently many rational curves (a criterion for uniruledness in terms of canonical divisors).

Because of the nature of Mori's technique, we work in the category of schemes: everything in this section is algebraic.

1   Quick review on deformation theory

1.1 Let Z be a k-scheme, where k is an algebraically closed field. Pick up a k-rational point z. The *Zariski tangent space* $T_{Z,z}$ of X at z is the set of k-morphisms from the pointed Artinian scheme $(\operatorname{Spec} k[\varepsilon]/(\varepsilon^2), \operatorname{Spec} k[\varepsilon]/(\varepsilon))$ to the pointed scheme (Z, z). If there is a k-morphism $f : (Z, z) \to (V, v)$ of pointed schemes, then the composition of morphisms defines a mapping $T_{Z,z} \to T_{V,v}$, called the *differential* of f at z and denoted by $(df)_z$ or $(f_*)_z$. $T_{Z,z}$ naturally has the structure of a finite-dimensional k-vector space, which is observed as follows.

A k-morphism $j : \operatorname{Spec} k[\varepsilon]/(\varepsilon^2)$ is, rigorously speaking, a pair of the continuous map j plus the k-algebra homomorphism $j^* : \mathcal{O}_Z \to k[\varepsilon]/(\varepsilon^2)$. As a topological space, $\operatorname{Spec} k[\varepsilon]/(\varepsilon^2)$ is a one-point set $\{o\}$ corresponding to the unique maximal ideal (ε), so that the continuous map j is uniquely determined as the constant map $o \mapsto z$. The structure homomorphism j^* maps the maximal ideal $\mathfrak{M}_z$ to the maximal ideal (ε), and since $(\varepsilon^2) = 0 \in k[\varepsilon]/(\varepsilon^2)$, j^* is identified as a k-algebra homomorphism $\mathcal{O}_{Z,z}/\mathfrak{M}_z^2 \to \operatorname{Spec} k[\varepsilon]/(\varepsilon^2)$. On the other hand, the canonical injection $k \hookrightarrow \mathcal{O}_{Z,z}$, together with the canonical projection $\mathcal{O}_{Z,z} \to \mathcal{O}_{Z,z}/\mathfrak{M}_z \simeq k$, defines a natural direct sum decomposition $\mathcal{O}_{Z,z}/\mathfrak{M}_z^2 = k \oplus \mathfrak{M}_z/\mathfrak{M}_z^2$, where the ring structure is defined by the formula $(\alpha_1, \beta_1) \cdot (\alpha_2, \beta_2) = (\alpha_1 \alpha_2, \alpha_1 \beta_2 + \alpha_2 \beta_1)$, $\alpha_i \in k$, $\beta_i \in \mathfrak{M}/\mathfrak{M}^2$. (More explicitly, the decomposition is given by $\mathcal{O}_Z \ni a \mapsto (a_0, a - a_0)$, where $a_0 = a \bmod \mathfrak{M}_z \in k$.) It follows that a ring homomorphism j^* is uniquely determined by its restriction $j^*|_{\mathfrak{M}_z/\mathfrak{M}_z^2} \to \varepsilon k[\varepsilon]/(\varepsilon^2) \simeq k$, because $j^*(\alpha, 0)$ must be equal to $\alpha \in k$. Conversely, when a k-linear map $j' : \mathfrak{M}_z/\mathfrak{M}_z^2 \to k$ is given, then we can define a k-algebra homomorphism $\mathcal{O}_{Z,z}/\mathfrak{M}_z^2 \simeq k \oplus \mathfrak{M}_z/\mathfrak{M}_z^2 \to k[\varepsilon]/(\varepsilon^2)$ by $j^*(\alpha, \beta) = \alpha + \varepsilon j'(\beta)$. Thus the Zariski tangent space, or the set of j's is naturally identified with $\operatorname{Hom}_k(\mathfrak{M}_z/\mathfrak{M}_z^2, k)$, the k-vector space of the k-linear maps from $\Omega^1_{Z,z} = \mathfrak{M}_z/\mathfrak{M}_z^2$ to k.

The k-vector space $\Omega^1_{Z,z}$ is called the *cotangent space* of Z at z. Given a morphism $f : (Z, z) \to (V, v)$, f^* naturally defines a k-linear map $\Omega^1_{V,v} \to \Omega^1_{Z,z}$.

Assume that a k-scheme Z is smooth at z, a closed point. There is a local parameter system $(z_1, \ldots, z_n)$ of (Z, z), where z is locally defined by $z_1 = \cdots = z_n = 0$. An element of $\mathcal{O}_{Z,z}$ is uniquely expressed as a formal power series in the z_i. Then, to divide out $\mathcal{O}_{Z,z}$ by $\mathfrak{M}_z^2$ means that we drop the terms of order ≥ 2 from a given power series, or equivalently, we observe only the constant term and the linear terms of Taylor expansions. The projection $\mathcal{O}_{Z,z} \to \mathfrak{M}_z/\mathfrak{M}_z^2$ amounts to killing also the constant term. Thus $\Omega^1_{Z,z} = kdz_1 + \ldots + kdz_n$, where dz_i stands for the equivalence class of z_i modulo $\mathfrak{M}_z^2$. This interpretation agrees with a naive definition of the cotangent space (or the space of differentials). The apparently very abstract definition above is essentially classical infinitesimal calculus in the sense of Leibnitz.

1.2 If we let the point z move around on Z, we get the notion of the sheaf of *Kähler differentials* Ω^1_Z, which is pointwise equal to $\Omega^1_{Z,z}$. The explicit construction is this:

Let Δ_Z be the closed subscheme (the *diagonal*) of $Z \times Z$ defined by the ideal $I_{\Delta_Z} \subset \mathcal{O}_Z \otimes_k \mathcal{O}_Z$ generated by the elements of the form $\alpha \otimes 1 - 1 \otimes \alpha$. $\mathcal{O}_Z \otimes_k \mathcal{O}_Z/I^2_{\Delta_Z}$ is a direct sum of $\mathcal{O}_Z$ and $I_{\Delta_Z}/I^2_{\Delta_Z}$ by virtue of the pull-back $\mathrm{pr}_1^* : \mathcal{O}_Z \to \mathcal{O}_Z \otimes_k \mathcal{O}_Z$ via the first projection and of the quotient $\mathcal{O}_Z \otimes_k \mathcal{O}_Z \to \mathcal{O}_Z \otimes_k \mathcal{O}_Z/I_{\Delta_Z} \simeq \mathcal{O}_Z$. Then Ω^1_Z is defined as $I_{\Delta_Z}/I^2_{\Delta_Z}$, which we view as an $\mathcal{O}_Z$-module (or a coherent sheaf on Z) via the first projection. Namely, if $a \in \mathcal{O}_Z$ and $\beta \in \Omega_Z = I_{\Delta_Z}/I^2_{\Delta_Z}$, we define $a \cdot \beta$ to be $(\mathrm{pr}_1^* a)\beta = (a \otimes 1)\beta$. It is left to the reader to check that $k(z) \otimes_{\mathcal{O}_Z} \Omega^1_Z$ is naturally isomorphic to $\Omega^1_{Z,z}$.

Let Z be a k-variety, i.e. an irreducible and reduced k-scheme. The sheaf of Kähler differentials Ω^1_Z is locally free if and only if Z is smooth, i.e. the completion $\hat{\mathcal{O}}_Z$ of $\mathcal{O}_Z$ (with respect to the maximal ideal $\mathfrak{M}_z$) is isomorphic to the formal power series ring $k[[z_1, \ldots, z_n]]$ at every closed point $z \in Z$, where $n = \dim Z$. In fact, we have checked that $\Omega^1_{Z,z}$ is of rank n at a smooth point z. Conversely, if $\Omega^1_{Z,z}$ is of rank n, then $\mathfrak{M}_z/\mathfrak{M}_z^2$, and hence $\hat{\mathcal{O}}_Z = \varprojlim \mathcal{O}_Z/\mathfrak{M}_z^j$, is generated by n elements. Since $\dim \hat{\mathcal{O}}_Z = \dim \mathcal{O}_Z = n$, this implies that $\hat{\mathcal{O}}_Z$ is isomorphic to a power series ring. When Z is smooth, Ω^1_Z is usually called the *sheaf of 1-forms* of rank $n = \dim_k Z$, and the dual locally free sheaf $T_Z \overset{\mathrm{def}}{=} \mathcal{H}om_{\mathcal{O}_Z}(\Omega^1_Z, \mathcal{O}_Z)$ is said to be the *tangent bundle* of Z. Each fibre $k(z) \otimes_{\mathcal{O}_Z} T_Z$ of the tangent bundle is isomorphic to the Zariski tangent space $T_{Z,z}$, in this case. Note, however, for singular Z, $k(z) \otimes \mathcal{H}om(\Omega^1_Z, \mathcal{O}_Z)$ is usually different from $T_{Z,z}$.

If there is a k-morphism $f : Z \to V$, then we have natural sheaf homomorphisms $f^* : \Omega^1_V \to \Omega^1_Z$ and $df : T_Z \to f^*T_V$.

1.3 With these definitions in mind, we study the infinitesimal structure of the universal deformation $\Phi : (\mathrm{Hom}(Y, X), [f]) \times Y \to X$ of a given morphism $f : Y \to X$ between projective varieties. We can describe it in terms of f^*T_X, the pull-back of the tangent bundle. The target variety X is assumed to be smooth around the image $f(Y)$, but the source Y may not necessarily be smooth.

1.4 Theorem [SGA I, Exposé III] *Let X and Y be projective varieties over an algebraically closed field k of arbitrary characteristic. Assume that X is smooth around $f(X)$. Then the Zariski tangent space $T_{\mathrm{Hom}(Y,X),[f]}$ of the Hom scheme $\mathrm{Hom}(Y,X)$ at $[f]$ is naturally isomorphic to $H^0(Y, f^*T_X)$. For morphisms with base point set $B \subset Y$, we have an natural isomorphism $T_{\mathrm{Hom}(Y,X;B),[f]} \simeq H^0(Y, I_B f^*T_X)$.*

First we give an intuitive explanation to this theorem and thereafter a rigorous proof.

Explanation   Let $j : \mathrm{Spec}\, k[\varepsilon]/(\varepsilon^2) \to \mathrm{Hom}(Y,X)$ be a morphism. For each k-rational point $y \in Y$, we get a morphism $j_y : \mathrm{Spec}\, k[\varepsilon]/(\varepsilon^2) \to \mathrm{Hom}(Y,X) \times Y$ by identifying $\mathrm{Spec}\, k[\varepsilon]/(\varepsilon^2)$ with $\mathrm{Spec}\, k[\varepsilon]/(\varepsilon^2) \times \{y\}$. Then by composing j_y with $\tilde{f}$ we get a morphism $\big(\mathrm{Spec}\, k[\varepsilon]/(\varepsilon^2), \mathrm{Spec}\, k[\varepsilon]/(\varepsilon)\big) \to (X, f(y))$ and thus an element $\xi_y \in T_{X, f(y)}$. The correspondence $y \mapsto \xi_y$ gives a global section $H^0(Y, f^*T_X)$.

Proof.   For simplicity, we give a proof for the case $B = \emptyset$. Consider an arbitrary morphism $g : \mathrm{Spec}\, k[\varepsilon]/(\varepsilon^2) \times Y \to X$ such that $g|_o = f$, where o is the unique closed point of $\mathrm{Spec}\, k[\varepsilon]/(\varepsilon^2)$; or equivalently, a ring homomorphism $g^* : \mathcal{O}_X \to k[\varepsilon]/(\varepsilon^2) \otimes_k \mathcal{O}_Y$ such that $g^* \bmod(\varepsilon) = f^*$. Letting $\overline{f}$ be the trivial deformation defined by $\overline{f}^* = 1 \otimes f^*$, we look at the morphism $(\overline{f}, g) : \mathrm{Spec}\, k[\varepsilon]/(\varepsilon^2) \times Y \to X \times X$. Since $g^* \equiv \overline{f}^* \bmod \varepsilon$, it follows that $(\overline{f}, g)^*(I_{\Delta_X}) \subset (\varepsilon) \otimes \mathcal{O}_Y$, $(\overline{f}, g)^*(I_{\Delta_X}^2) \subset (\varepsilon^2) \otimes \mathcal{O}_Y = 0$, and hence

$$(\overline{f}, g)^* \in \mathrm{Hom}_{\mathcal{O}_X}(I_{\Delta_X}/I_{\Delta_Z}^2 \to \varepsilon \mathcal{O}_Y) = \varepsilon\, \mathrm{Hom}_{\mathcal{O}_X}(\Omega_X^1, \mathcal{O}_Y) \simeq \varepsilon H^0(Y, \overline{f}^* T_X).$$

Here the $\mathcal{O}_X$-module structure of $\mathcal{O}_Y$ is defined by the multiplication via $\overline{f}^* \bmod(\varepsilon) = f^*$. Thus we get an correspondence

$$\big\{ g : \mathrm{Spec}\, k[\varepsilon]/(\varepsilon^2) \times Y \to X \mid g \bmod(\varepsilon) = f \big\} \to \varepsilon H^0(Y, f^*T_X).$$

Conversely, given an $\mathcal{O}_X$-homomorphism $\delta h : \Omega_X^1 \to \mathcal{O}_Y$, we can define an $\mathcal{O}_X$-algebra homomorphism $h : \mathcal{O}_Z \otimes X \mathcal{O}_X/I_{\Delta_X}^2 = \mathcal{O}_X \oplus \Omega_X^1 \to k[\varepsilon]/(\varepsilon^2) \otimes \mathcal{O}_Y$ by $h(\alpha, \beta) = \overline{f}^*(\alpha) + \varepsilon \delta h(\beta)$. Since $h \cdot \mathrm{pr}_1 = \overline{f}^*$, we can write h in the form $(\overline{f}^*, g^*)$, where g^* is the ring homomorphism $h(\mathrm{pr}_2)^*$. This gives the inverse correspondence.

Thus the set of deformation of f parametrized by $\mathrm{Spec}\, k[\varepsilon]/(\varepsilon^2)$ is bijective to $H^0(Y, f^*T_X)$. By the definition of the universal deformation, the former set is identified with the set of morphisms: $\mathrm{Spec}\, k[\varepsilon]/(\varepsilon^2) \to (\mathrm{Hom}(Y,X), [f])$, i.e. the Zariski tangent space of the Hom scheme. $\square$

1.5 Exercise   (1) Complete the proof of Theorem 1.4 in case $B \neq \emptyset$.

(2) Let Y be a closed subscheme of a smooth projective variety X. Assume that Y is locally a complete intersection so that I_Y/I_Y^2 is a locally free sheaf on Y. Use a similar argument as above and prove that the Zariski tangent space of $\mathrm{Hilb}(X)$

at $[Y]$ is naturally identified with $\mathrm{Hom}_{\mathcal{O}_X}(I_Y/I_Y^2, \mathcal{O}_Y) \simeq H^0(Y, \mathcal{N}_{Y/X})$, where $\mathcal{N}_{Y/X} = \mathcal{H}om_{\mathcal{O}_Y}(I_Y/I_Y^2, \mathcal{O}_Y)$ is the *normal bundle*.

Hint: Take a closed subscheme $\mathcal{Y} \subset \mathrm{Spec}\, k[\varepsilon]/(\varepsilon^2) \times X$, flat over the base $\mathrm{Spec}\, k[\varepsilon]/(\varepsilon^2)$. Show that $\varepsilon\mathcal{O}_{\mathcal{Y}} \simeq \varepsilon\mathcal{O}_Y \subset k[\varepsilon]/(\varepsilon)^2 \otimes \mathcal{O}_X$. Consider the composite of the injection $I_Y \hookrightarrow \mathcal{O}_X$ and the pull-back $\mathcal{O}_X \to \mathcal{O}_{\mathcal{Y}}$. Check that the image is contained in $\varepsilon\mathcal{O}_Y$ and that the map factors through I_Y/I_Y^2.

(3) Let X be a smooth projective variety. Prove that there is a bijection between the vector space $H^1(X, T_X)$ and the set of $k[\varepsilon](\varepsilon)^2$-isomorphism classes of the proper and smooth scheme $\mathcal{X}$ over $\mathrm{Spec}\, k[\varepsilon]/(\varepsilon^2)$ with $\mathcal{X}_o \simeq X$. Here $\mathcal{X}_o$ is the closed fibre $k[\varepsilon]/(\varepsilon) \otimes_{k[\varepsilon]/(\varepsilon)^2} \mathcal{X}$.

Hint: Cover X by open affine subsets $U_\nu = \mathrm{Spec}\, A_\nu$. Show that $\mathcal{X}$ is locally $\mathrm{Spec}(A_\nu \oplus \varepsilon A_\nu)$, and that the patching data of $\mathcal{X}$ gives a system of ring automorphisms $\sigma_{\mu\nu}$ of $k[\varepsilon]/(\varepsilon^2) \otimes \mathcal{O}_{U_\mu \cap U_\nu}$ such that they are identities modulo ε (i.e. the transition functions) and subject to cocycle conditions. Prove that the $\theta_{\mu\nu} = \sigma_{\mu\nu} - \mathrm{id}$ define a cocycle with values in εT_X, and that $\mathcal{X}$ is $k[\varepsilon]/(\varepsilon)^2$-isomorphic to $\mathrm{Spec}\, k[\varepsilon]/(\varepsilon^2) \times X$ if and only if $\{\theta_{\mu\nu}\}$ is a coboundary.

1.6 Let $(R, \mathfrak{M})$ be a noetherian local k-algebra, where k is an algebraically closed field. The projective limit $\hat{R} = \varprojlim R/\mathfrak{M}^i$ is called the ($\mathfrak{M}$-adic) *completion* of R. $\mathrm{Spec}\,\hat{R} = \varinjlim \mathrm{Spec}(R/\mathfrak{M}^i)$ is the *formal neighbourhood* of the closed point $o = \mathrm{Spec}(R/\mathfrak{M})$.

Take elements $\alpha_1, \ldots, \alpha_e \in \mathfrak{M}$ such that they form a basis of the k-vector space $T^*_{\mathrm{Spec}\, R, o} = \mathfrak{M}/\mathfrak{M}^2$. Then there is a natural surjection $k[[\alpha_1, \ldots, \alpha_e]] \twoheadrightarrow \hat{R}$. In other words, the formal scheme $\mathrm{Spec}\,\hat{R}$ is embedded to $\mathrm{Spec}\, k[[\alpha_1, \ldots, \alpha_e]]$, the formal neighbourhood $\hat{\mathbb{A}}^e$ of the origin of the affine space $\mathbb{A}^e$, as a closed subscheme, thereby inducing an isomorphism between the Zariski tangent spaces. The dimension e of the Zariski tangent space is thus called the *embedding dimension* of $\mathrm{Spec}\, R$. $\mathrm{Spec}\, R$ is smooth if and only if the embedding dimension e is equal to the dimension.

Let $\mathfrak{J}$ be the defining ideal of $\mathrm{Spec}\,\hat{R}$ in $\hat{\mathbb{A}}^e$. By the isomorphism between the Zariski tangent spaces of $\mathrm{Spec}\, R$ and $\mathbb{A}^e$, we easily check that $\mathfrak{J} \subset \tilde{\mathfrak{M}}^2$, where $\tilde{\mathfrak{M}} = \mathfrak{M}k[[\alpha_1, \ldots, \alpha_e]]$ is the maximal ideal of the formal power series ring. If $\mathfrak{J} \neq 0$, then there exists a positive integer a such that

$$A_{a-1} = k[[\alpha_1, \ldots, \alpha_e]]/\tilde{\mathfrak{M}}^a \xrightarrow{\sim} R_{a-1} = R/\mathfrak{M}^a,$$
$$A_a = k[[\alpha_1, \ldots, \alpha_e]]/\tilde{\mathfrak{M}}^{a+1} \xrightarrow{\not\sim} R_a = R/\mathfrak{M}^{a+1}.$$

It follows that the isomorphism $\mathrm{Spec}\, A_{a-1} \to \mathrm{Spec}\, R/\mathfrak{M}^a$ does not lift to $\mathrm{Spec}\, A_a \to \mathrm{Spec}\, R/\mathfrak{M}^{a+1}$. Thus the smoothness of $\mathrm{Spec}\, R$ is closely related to the liftability of morphisms of artinian schemes $\mathrm{Spec}\, A_{a-1} \to \mathrm{Spec}\, R_{a-1}$.

1.7 Theorem 1.4 determines the embedding dimension of $\mathrm{Hom}(Y, X)$ at $[f]$ but not the dimension. In order to get information on the dimension of Hom, we introduce the notion of *obstruction* to extend the morphisms.

Let $(A, \mathfrak{M})$ be an artinian local k-algebra and $J \subset \mathfrak{M}$ an ideal such that $\mathfrak{M}J = 0$. Given a k-morphism $\bar{g} : \operatorname{Spec}(A/J) \to \operatorname{Hom}(Y, X)$ such that $\bar{g}|_{\operatorname{Spec}(A/\mathfrak{M}) \times Y} = f$, we want to know if we can lift $\bar{g}$ to $g : \operatorname{Spec} A \to \operatorname{Hom}(Y, X)$. Take an affine open covering U_ν of Y. $\bar{g}^*$ is a k-algebra homomorphism $\mathcal{O}_X \to (A/J) \otimes_k \mathcal{O}_{U_\nu}$. Noting that $\mathcal{O}_X$ is a regular ring, we can lift $\bar{g}^*$ to a k-algebra homomorphism $g_\nu^* : \mathcal{O}_X \to A \otimes_k \mathcal{O}_{U_\nu}$, i.e. a system of morphisms $g_\nu : \operatorname{Spec} A \times U_\nu \to X$. (g_μ, g_ν) gives a morphism $U_\mu \cap U_\nu \to X \times X$, which is the diagonal morphism when restricted to $\operatorname{Spec}(A/J) \times Y$. Namely, the ring homomorphism $(g_\mu, g_\nu)^* : \mathcal{O}_X \otimes \mathcal{O}_X \to A \otimes \mathcal{O}_{U_\mu \cap U_\nu}$ maps I_{Δ_X} to $J\mathcal{O}_{U_\mu \cap U_\nu}$, so that $(g_\mu, g_\nu)^*(I_{\Delta_X}^2) \subset J^2 \mathcal{O}_{U_\mu \cap U_\nu} \subset \mathfrak{M}J\mathcal{O}_{U_\mu \cap U_\nu} = 0$. We can thus think of $(g_\mu, g_\nu)^*$ as a map: $\Omega_X^1 \to J\mathcal{O}_{U_\mu \cap U_\nu}$, or equivalently, a 1-cochain $\in H^0(\operatorname{Spec} A \otimes U_\mu \cap U_\nu, Jg_\mu^* T_X)$. The condition that $J\mathfrak{M} = 0$ allows us to view $Jg_\mu^* T_X$ as an $A/\mathfrak{M} \otimes \mathcal{O}_{U_\mu \cap U_\nu} = \mathcal{O}_{U_\mu \cap U_\nu}$-module, so that $\gamma_{\mu\nu} = (g_\mu, g_\nu)^* \in H^0(U_\mu \cap U_\nu, Jf^* T_X)$. It is easy to check that the $\gamma_{\mu\nu}$ satisfy the cocycle condition, thereby defining a cohomology class $\in H^1(Y, Jf^* T_X)$. This cohomology class $\gamma(A, J, \bar{g})$ is determined by $A, J, \bar{g}$ and independent of the choice of g_ν's. To see this, take another system of liftings g_ν'. I_{Δ_X} is generated by the elements of the form $\delta(a) = a \otimes 1 - 1 \otimes a$, and we have $(g_\mu, g_\nu)^*(\delta(a)) = g_\mu^*(a) - g_\nu^*(a)$ for every $a \in \mathcal{O}_X$. Hence

$$\big((g_\mu, g_\nu)^* - (g_\mu', g_\nu')^*\big)(\delta(a)) = (g_\mu^* - g_\nu^* - g_\mu'^* + g_\nu'^*)(a)$$
$$= \big((g_\mu, g_\mu')^* - (g_\nu, g_\nu')^*\big)(\delta(a))$$

for arbitrary $a \in \mathcal{O}_X$, showing

$$(g_\mu, g_\nu)^* - (g_\mu', g_\nu')^* = (g_\mu, g_\mu')^* - (g_\nu, g_\nu')^*. \tag{$*$}$$

On the other hand, $(g_\nu, g_\nu')^* \in H^0(U_\nu, Jf^* T_X)$, and therefore the right hand side of $(*)$ is a coboundary.

You can globally extend $\bar{g}$ to $g : \operatorname{Spec} A \otimes Y \to X$ if and only if you can choose the g_ν such that $g_\mu^* = g_\nu^*$ on $U_\mu \cap U_\nu$ i.e. $\gamma(\bar{g}) = 0 \in H^1(Y, Jf^* T_X)$. The cohomology class $\gamma(A, J, \bar{g})$ is called the *obstruction* to extend $\bar{g}$ to $\operatorname{Spec} A \times Y$.

Provided the obstruction $\gamma(A, J, \bar{g})$ vanishes, there is at least one lift g of $\bar{g}$. If you specify one such g, then the set of the extensions of $\bar{g}$ to $\operatorname{Spec} A \times Y$ is naturally identified with $H^0(Y, Jf^* T_X)$ by the correspondence $g' \mapsto (g, g')^*$.

Let $J' \subset J$. By construction, it is obvious that the obstruction class $\gamma(A/J', J/J', \bar{g}) \in H^1(Y, (J/J')f^* T_X)$ to lift $\bar{g}$ to $\operatorname{Spec}(A/J') \times Y$ is the natural image of $\gamma(A, J; \bar{g})$ modulo J'.

1.8 Theorem [Mori 1]   *Under the same assumption as in (1.2), the formal completion of $\operatorname{Hom}(Y, X)$ [resp. $\operatorname{Hom}(Y, X; B)$] at $[f]$ can be viewed as a closed subscheme of $\operatorname{Spec} k[[V]]$, where V is the dual k-vector space of $H^0(Y, f^* T_X)$ [resp. of $H^0(Y, I_B f^* T_X)$]. The closed subscheme is defined by at most $\dim H^1(Y, f^* T_X)$ [resp. $\dim H^1(Y, I_B f^* T_X)$] equations. In particular, we have the following estimates of the local dimensions of $\operatorname{Hom}(Y, X)$, $\operatorname{Hom}(Y, X; B)$ around $[f]$:*

$$h^0(Y, f^* T_X) - h^1(Y, f^* T_X) \le \dim_{[f]} \operatorname{Hom}(Y, X) \le h^0(Y, f^* T_X)$$
$$h^0(Y, I_B f^* T_X) - h^1(Y, I_B f^* T_X) \le \dim_{[f]} \operatorname{Hom}(Y, X; B) \le h^0(Y, I_B f^* T_X).$$

Here the symbol h^i stands for $\dim_k H^i$. If $H^1(Y, f^*T_X)$ [resp. $H^1(Y, I_B f^*T_X)$] vanishes, then $\mathrm{Hom}(Y, X)$ [resp. $\mathrm{Hom}(Y, X; B)$] is smooth at $[f]$.

1.9 Remark   The vanishing of H^1 is a sufficient, but not necessary, condition for the smoothness. An easy example of smooth Hom scheme with non-trivial H^1 is given by $\mathrm{Hom}(X, X)$, where X is an abelian variety of dimension n. The identity component $(= \mathrm{Aut}(X))$ of $\mathrm{Hom}(X, X)$ consists of the translations, thus

$$\dim_{[\mathrm{id}]} \mathrm{Hom}(X, X) = d = \dim X = \dim H^0(X, \mathrm{id}^* T_X),$$

while $\dim H^1(X, \mathrm{id}^* T_X) = n^2$.

1.10 Proof of Theorem 1.8   For the sake of simplicity of notation, we show the statement when $B = \emptyset$. The proof for non-empty B is left to the reader. Put $A = k[[H^0(Y, f^*T_X)^*]]$ and let $J \subset A$ be the ideal sheaf of the completion $\mathrm{Spec}\,\hat{R}$ of $(\mathrm{Hom}(Y, X), [f])$ (namely, $\hat{R} = A/J$). By definition $J \subset \mathfrak{M}^2$. It suffices to show that J is generated by a elements, where $a = \dim H^1(X, f^*T_X)$, or, thanks to Nakayama's lemma, that $J/\mathfrak{M}J$ is generated by a elements.

Take a sufficient large integer m so that

$$J \cap \mathfrak{M}^m = \mathfrak{M}(J \cap \mathfrak{M}^{m-1})$$

(a lemma of Artin-Rees). Put $J_m = (J + \mathfrak{M}^m)/(\mathfrak{M}J + \mathfrak{M}^m) \subset B_m = A/(\mathfrak{M}J + \mathfrak{M}^m)$. The ideal J_m of B_m satisfies $\mathfrak{M}J_m = 0$. There is a natural homomorphism $R \to B_m/J_m = A/(J + \mathfrak{M}^m)$, inducing a morphism $\bar{g} : \mathrm{Spec}(B_m/J_m) \to \mathrm{Hom}(Y, X)$. The obstruction to extend $\bar{g}$ to a morphism $g : \mathrm{Spec}\,B_m \to \mathrm{Hom}(Y, X)$ is given by $\phi \in J_m H^1(Y, f^*T_X)$. By a k-basis $\{\phi_i\}_{i=1,\dots,a}$ of $H^1(X, f^*T_X)$, we can write $\phi = \sum \bar{r}_i \phi_i$, where $\bar{r}_i \in J_m$ is the equivalence class modulo $\mathfrak{M}J + \mathfrak{M}^m$ of $r_i \in J$. Since $\phi \equiv 0$ modulo $(\bar{r}_1, \dots, \bar{r}_a)$, $\bar{g}$ lifts to $\bar{g}' : \mathrm{Spec}(B_m/(\bar{r}_1, \dots, \bar{r}_a)) \to \mathrm{Hom}(Y, X)$, thus inducing a commutative diagram of ring homomorphisms:

$$
\begin{array}{ccc}
A/J & \xrightarrow{\;\pi\;} & B_m/(\bar{r}_1, \dots, \bar{r}_a) \\
\downarrow & & \downarrow \\
B_m/J_m & =\!=\!= & A/(J + \mathfrak{M}^n).
\end{array}
\qquad (*)
$$

The homomorphism π is clearly surjective. A is the formal power series ring $\simeq k[[\alpha_1, \dots, \alpha_e]]$ and we can find $\sigma \in \mathrm{Aut}(A)$ such that the following diagram $(**)$ commutes:

$$
\begin{array}{ccc}
A & \xrightarrow{\;\sigma\;} & A \\
\downarrow & & \downarrow \\
A/J & \xrightarrow{\;\pi\;} B_m/(\bar{r}_1, \dots, \bar{r}_a) & =\!=\!= \; A/(\mathfrak{M}J + \mathfrak{M}^m + (\bar{r}_1, \dots, \bar{r}_a)).
\end{array}
\qquad (**)
$$

From the diagrams (*) and (**), it follows that id $- \sigma$ maps A to $J + \mathfrak{M}^m$. In particular,

$$\sigma(J) \subset J + \mathfrak{M}^m, \quad \sigma^{-1}(J) \subset J + \mathfrak{M}^m.$$

Then by (**),

$$\sigma(J) \subset \mathfrak{M}J + \mathfrak{M}^m + (r_1, \ldots, r_a).$$

Hence the images of the a elements $r_1, \ldots, r_a$ in $J + \mathfrak{M}^m$ generate

$$J + \mathfrak{M}^m/\mathfrak{M}J + \mathfrak{M}^m = (\sigma(J) + \mathfrak{M}^m)/(\mathfrak{M}\sigma(J) + \mathfrak{M}^m) \simeq \sigma(J)/\mathfrak{M}\sigma(J). \qquad \square$$

1.11 Corollary   *Assume that the target variety X is a smooth projective curve. Let $f : Y \to X$ be a surjective k-morphism. If $\mathrm{Hom}(Y, X)$ is not the single point $[f]$ near $[f]$, then $g(X) \le 1$. If $\mathrm{Hom}(Y, X; B) \ne \{[f]\}$ near $[f]$ for $B \ne \emptyset$, then X is $\mathbb{P}^1$.*

More generally, $[f]$ is an isolated point in $\mathrm{Hom}(Y, X)$ [resp. in $\mathrm{Hom}(Y, X; B)$] if f is non-constant and Ω^1_X is ample [resp. if Ω^1_X is nef]. (A vector bundle is said to be ample or nef if so is the tautological line bundle is so.)

The proof is immediate from the fact that the line bundle T_X is negative when $g(X) \ge 2$ and trivial when $g(X) = 0$.

1.12 Corollary   *Let $f : Y \to X$ be a morphism from a smooth projective curve Y of genus g to a smooth projective variety X of dimension n. Let $B \subset Y$ be a zero-dimensional, closed, reduced subscheme of degree d. If $\deg f^*T_X > n(d + g - 1)$, then f has a non-trivial deformation with base point set B.*

This follows from the Riemann-Roch theorem for vector bundles over curves.

1.13 Exercise   (1) Let $Y \subset X$ be a subscheme which is locally a complete intersection in X. Show that the obstruction to extend an (A/J)-flat, closed subscheme $\overline{\mathcal{Y}} \in \mathrm{Spec}\, A/J \times X$ to an A-flat closed subscheme $\mathcal{Y} \subset \mathrm{Spec}\, A \times X$ is an element of $H^1(Y, J\mathcal{N}_{Y/X})$. Here $(A, \mathfrak{M})$ is an artinian local ring, $J \subset \mathfrak{M}$ satisfies $\mathfrak{M}J = 0$, and the closed fibre Y over $\mathrm{Spec}\, A/M$ is locally a complete intersection in X. When the obstruction vanishes, show that the set of liftings can be identified with $H^0(Y, J\mathcal{N}_{Y/X})$.

(2) Show that the obstruction to lift a smooth deformation over $\mathrm{Spec}\, A/J$ to $\mathrm{Spec}\, A$ lies in $H^2(X, JT_X)$, and that the set of liftings is identified with $H^1(X, JT_X)$ provided the obstruction vanishes.

2   Frobenius and deformation

2.1 Let $C \subset X$ be an irreducible curve on a smooth projective variety X. Take a surjective finite morphism $\nu : Y \to C$, where Y is a smooth projective curve, and let $f : Y \to X$ be the composite of the covering ν with the embedding $C \hookrightarrow X$. Then $\deg f^* T_X = (\deg \nu)(C, -K_X)$ and, if $(-K_X, C)$ is sufficiently bigger than the geometric genus $g = g(Y)$ of Y, then f will have a non-trivial deformation with a base point set of positive degree. Once you come across this situation, you can find a rational curve by (I.1.2).

Suppose that we are given a curve C such that $(C, -K_X) > 0$. C has possibly a very high genus, so that $\deg f^* T_X$ may not satisfy $> n(d + g - 1)$. To make the degree sufficiently big, we could take the branched cover Y of C, the normalization of C. But this does not work in characteristic zero, because $g(Y)$ would grow at least as rapidly as $[k(Y) : k(C)]$ does, when we start from a curve of geometric genus ≥ 2. In positive characteristic, say p, there are very special kind of morphisms called *Frobenius morphisms* that leaves the genus the same.

Let k be an algebraically closed field of characteristic $p > 0$ and A a noetherian k-algebra. We define a ring homomorphism (*absolute Frobenius*) $\varphi_{ab}^* : A \to A$ by $\varphi_{ab}^*(\alpha) = \alpha^p$. This operation satisfies

$$\varphi_{ab}^*(\alpha\beta) = \varphi_{ab}^*(\alpha)\varphi_{ab}^*(\beta),$$
$$\varphi_{ab}^*(\alpha + \beta) = \varphi_{ab}^*(\alpha) + \varphi_{ab}^*(\beta),$$
$$\varphi_{ab}^*(a\alpha + b\beta) = a^p \varphi_{ab}^*(\alpha) + b^p \varphi_{ab}^*(\beta), \quad a, b \in k, \alpha, \beta \in A.$$

Thus φ_{ab}^* is a ring homomorphism but not a k-algebra homomorphism. Note that the correspondence $\varphi_{ab}^{-1} : a \mapsto a^{1/p}$ is a field automorphism of the algebraically closed field k. Define $A^{(-1)}$ to be $k \otimes_k A$, where the first factor k is viewed as a k-module with a non-standard multiplication $\cdot$, given by $a \cdot b = a^p b$. $A^{(-1)}$ is isomorphic to A as a ring, by the correspondence $\iota : a \otimes b \mapsto a^p b$. The natural k-module structure of $A^{(-1)}$ is, however, different from that of A. As the tensor product $k \otimes_k A$, $A^{(-1)}$ is naturally a k-module by the multiplication in the first factor, and hence $\mu(a \otimes b) = \mu a \otimes b$, $\mu \in k$, which is mapped to $\mu^p(\iota(a \otimes b))$ by ι. By this shift of the k-algebra structure, we get a k-algebra homomorphism

$$\varphi^* = \varphi_{rel}^* : A \to A^{(-1)} = k \otimes_k A, \quad \varphi(a) = 1 \otimes \varphi_{ab}^*(a).$$

Note that $A^{(-1)}$ and A are isomorphic as abstract rings, but not necessarily so as k-algebras. The construction of φ^* is canonical, and we can globalize it into a k-algebra homomorphism $\varphi^* : \mathcal{O}_Z \to \mathcal{O}_{\tilde{Z}} = k \otimes_k \mathcal{O}_Z$ for a given k-scheme Z, and hence a finite k-morphism $\varphi : \tilde{Z} = Z^{(-1)} \to Z$. Since $\mathcal{O}_{\tilde{Z}}$ is isomorphic to $\mathcal{O}_Z$ as a ring, $\tilde{Z}$ is regular or normal if and only if Z is so.

Let Z be a regular k-variety with local parameters $z_1, \ldots, z_n$. Then the regular k-variety $\tilde{Z}$ has local parameters $\tilde{z}_1, \ldots, \tilde{z}_n$ such that $\varphi^* z_i = \tilde{z}_i^p$ (here $\tilde{z}_i$ denotes $1 \otimes z_i$). In general, $\varphi^*(\sum a_I z^I) = \sum a_I \tilde{z}^{pI}$, $a_I \in k$, under the multi-index convention. Thus $k(\tilde{Z})$ is a purely inseparable field extension of degree p^n, while it gives a bijection between the topological spaces $\tilde{Z}$ and Z, i.e. the k-morphism φ (*k-Frobenius morphism* or *geometric Frobenius*) is purely inseparable homeomorphism of degree p^n, $n = \dim_k Z$.

Given a power $q = p^m$, you can start with q-th power map $a \mapsto a^q$ instead of the p-th power map. The same construction works to obtain a purely inseparable morphism $\varphi^m : Z^{(-m)} \to Z$ of degree q^n. φ^m is also called a k-Frobenius morphism. You can easily check that φ^m is the composite $\varphi_1 \cdots \varphi_m$, where $\varphi_i : Z^{(-i)} \to Z^{(-i+1)}$ is the k-Frobenius attached to p-th powers.

2.2 Exercise Let $Z \subset \mathbb{P}^N$ be a subscheme defined by homogeneous equations $f_i(x) = \sum a_{iI} x^I$, $i = 1, \ldots, r$. Let $\varphi : \tilde{Z} \to Z$ be a geometric Frobenius. Show that $\tilde{Z}$ is a subscheme $\subset \mathbb{P}^N$ defined by $\tilde{f}_i(x) = \sum a_{iI}^{1/q} x^I$, $i = 1, \ldots, r$.

2.3 Proposition (*ch $k = p > 0$*) *Let Y be a smooth projective curve and $\varphi : \tilde{Y} \to Y$ a Frobenius k-morphism (attached to q-th powers). Then $\deg \varphi = q$, and $g(\tilde{Y}) = g(Y)$.*

Proof. The first assertion was already shown above. The second statement follows from the fact that genus of a smooth curve is a topological invariant of curves (it is the half of the first Betti number given by *étale cohomology*). An alternative proof is to look at the number of k-linearly independent 1-forms, which is left to the reader. □

2.4 Corollary (*ch $k = p > 0$*) *Suppose that ch $k = p > 0$. Let $C \subset X$ be a curve of genus ≥ 0 with $(C, -K_X) > 0$. Fix an arbitrary closed point $x \in C$. Then there exists a rational curve $\Gamma \subset X$ such that $\Gamma \ni x$ and that $(\Gamma, H) \leq \frac{2(C,H)}{(C,-K_X)}$.*

Proof. As before, let $f : Y \to X$ denote the composite of the normalization and the embedding $C \to X$. We can take iterations of Frobenius morphisms $\varphi^m : Y^{(-m)} \to Y$. Consider the composite $f_m = f \circ \varphi^m : Y^{(m)} \to X$. Then $g = g(Y^{(-m)})$ is independent of m, while $\deg f_m^*(-K_X) = p^m \deg f^* K_X$. Choose a base point set $B \subset Y^{(-m)}$. By the Riemann-Roch,

$$\dim H^0(Y^{(-m)}, I_B f_m^* T_X) - \dim H^1(Y^{(-m)}, I_B f_m^* T_X)$$
$$= p^m \deg f^*(-K_X) - n(\deg B + g - 1),$$

so that $\text{Hom}(Y^{(-m)}, X; B)$ has positive dimension at $[f_m]$ if

$$\deg B = \frac{p^m \deg f^*(-K_X)}{n} + g.$$

Hence by (1.2), for an arbitrary positive integer m, there exists a rational curve $\Gamma \subset X$ such that $\Gamma \cap f_m(B) \neq \emptyset$ and that

$$(\Gamma, H) \leq \frac{2n \deg f_m^* H}{\deg f_m^*(-K_X) + ng} = \frac{2n(C,H)}{(C,-K_X) + p^{-m} ng}.$$

Letting m sufficiently large, we get a rational curve of degree $\leq \frac{2n(C,H)}{(C,-K_X)}$ (by noticing that the degree of a curve Γ is an integer), which passes through $f_m(B)$. You can take B freely on $Y^{(m)}$ as long as the degree is fixed, and so Γ passes through a general point, if B is supported by general points. This implies that, given an arbitrary closed point, you can find Γ which passes through the point.

(*Exercise:* Prove the last statement with the aid of the properness of $RC(X)$.) $\square$

2.5 We want to apply the Frobenius technique to varieties in characteristic zero.

Let X be a smooth projective variety over a field of characteristic zero. Assume that there exists an irreducible curve $C \subset X$ such that $(C, -K_X) > 0$, and pick up a point $x \in C$. Then almost every reduction (X_p, C_p) of the pair (X, C) has the following two properties:

(a) X_p is smooth (and projective);
(b) C_p is an irreducible curve with $(C_p, -K_{X_p}) = (C, -K_X), (C_p, H_p) = (C, H)$.

Thus there exist rational curves $\Gamma_p \subset X_p$ such that $\Gamma_p \ni x_p$, $(\Gamma_p, H_p) \leq \frac{2(C,H)}{(C,-K_X)}$. Since the degree of Γ_p is bounded by a constant independent of p, the Hasse principle in (I.3.5) tells us there exist a rational curve $\Gamma \ni x$ also in characteristic 0 with the same degree estimate.

Thus we have shown the following

2.6 Theorem   *Let k be an algebraically closed field of arbitrary characteristic and X a smooth projective variety defined over k. Assume that there exists an irreducible curve $C \subset X$ of genus ≥ 0 such that $(C, -K_X) > 0$. Then, for an arbitrary closed point $x \in C$, there exists a rational curve $\Gamma \subset X$ such that $\Gamma \ni x$ and $(\Gamma, H) \leq 2(C, H)/(C, -K_X)$.*

Put in another way, a smooth projective variety X necessarily carries a rational curve whenever K_X is not nef. (Recall that a Cartier divisor D on a projective variety X is nef if $(C, D) \geq 0$ for every effective curve $C \subset X$.)

Of course there are lot of smooth projective varieties which have nef canonical divisor. Some carry rational curves and some do not.

2.7 Examples   (1) An abelian variety (with trivial canonical divisor) does not contain any rational curve, and a finite cover of an abelian variety does not, either.

(2) Any polarized K3 surface (X, H), with trivial K_X, contains at least one rational curve (Theorem of Mumford-Bogomolov-Mori-Mukai, see [Mori-Mukai 2]) A very general polarized K3 surface contains countably many rational nodal curves provided $H^2 \equiv 2 \mod 4$ (S. Nakatani (1996), unpublished).

(3) A very general Kähler K3 surface X carries no compact curve, and in particular no rational curve. The K3 surfaces that contain rational curves form a countable union of hypersurfaces in the 20 dimensional moduli space (period domain) of the complex Kähler K3 surfaces.

(4) A very general hypersurface $X \subset \mathbb{P}^3(\mathbb{C})$ of degree sufficiently large does not contain any rational curve [Brody-Green] (X is indeed a hyperbolic manifold in the sense of [Kobayashi 1]).

(5) The Fermat hypersurface in $\mathbb{P}^{n+1}, n \geq 2$ of degree d defined by the equation $X_0^d + X_1^d + \ldots + X_{n+1}^d = 0$ contains the lines defined by

$$X_1 = \zeta_d^i X_0, X_3 = \zeta_d^j X_2, X_4 = \cdots = X_{n+1} = 0, \quad \zeta_d = \exp\left(\frac{2\pi\sqrt{-1}}{d}\right), i, j \in \mathbb{Z}/d\mathbb{Z}.$$

K_X is nef [resp. ample] if $d \geq n+2$ [resp. $\geq n+3$].

In view of (4) and (5), we see that smooth projective deformation does not preserve the property that rational curves are contained in the variety in question. In particular, there *cannot* be any numerical characterization of varieties which contain a rational curve. However, there *is* a characterization of varieties that contain *sufficiently many* rational curves, which is the topic of Section 5 below.

3   Rational curves of minimum degree

3.1 We have seen above that you can construct rational curves by starting an irreducible curve C with $(C, -K_X) > 0$. Let us see what happens if the original curve C is already rational.

Let $RC(X) \subset \mathrm{Chow}(X)$ be the subset

$$\{[C] \in \mathrm{Chow}(X) \mid C_{\mathrm{red}} \text{ is a connected chain of rational curves}\},$$

as in Lecture I. $RC(X)$ is a closed subset of $\mathrm{Chow}(X)$. Given a closed zero-dimensional subset $B \subset X$, we let $RC(X, B) \subset RC(X)$ denote the closed subset consisting of the $[C]$ such that $B \subset C$.

3.2 Theorem   *Let (X, H) be a smooth, projective polarized variety of dimension n over an algebraically closed field of arbitrary characteristic. Let $C \subset X$ be an irreducible rational curve and fix two distinct closed points $x_1, x_2 \in C$. Let $S = S_0 \subset RC(X)$ be an irreducible component that contains $[C]$. Define the closed subsets $S_1, S_{12} \subset S$ to be $S \cap RC(X, \{x_1\})$ and $S \cap RC(X, \{x_1, x_2\})$, respectively. Let d_* be the dimensions of the closed subsets $V_* = \bigcup_{[C] \in S_*} C \subset X$, $* = 0, 1, 12$. Then we have $(C, -K_X) \leq d_0 + d_1 + d_{12} - n$. If $d_{12} \geq 2$, then $RC(X, \{x_1, x_2\})$ has positive dimension and there exists a reducible member $[C'] \in S_{12}$ one irreducible component C_0' of which satisfying $(C_0', H) < (C, H), (C_0', -K_X) > 0$. In particular, if C is a rational curve of minimum degree, say l, on X, then*

$$(C, -K_X) \leq d_0 + d_1 + 1 - n \leq n+1.$$

When the upper bound $(C, -K_X) \leq n+1$ is attained, then $d_0 = d_1 = n$, $d_{12} = 1$, or equivalently, for given two points (x_1, x_2) there are at least one (and only finitely many) rational curve C such that $\deg C = l$, $(C, -K_X) = n+1$, $C \supset \{x_1, x_2\}$.

Proof. Let $f : \mathbb{P}^1 \to X$ be the morphism induced by the normalization $\mathbb{P}^1 \to C \subset X$. By taking a suitable automorphism of $\mathbb{P}^1$, we may assume $f(\infty) = x_1$, $f(0) = x_2$. Let $\Phi : \text{Hom}(\mathbb{P}^1, X) \times \mathbb{P}^1 \to X$ be the universal morphism. Then the union of rational curves $C \subset X$ that contain x_1 [resp. x_1 and x_2] is nothing but the closure of $\Phi(\text{Hom}(\mathbb{P}^1, X; \infty) \times \mathbb{P}^1)$ [resp. $\Phi(\text{Hom}(\mathbb{P}^1, X; \{\infty, 0\}) \times \mathbb{P}^1)$. Thus there is an irreducible component $T \subset \text{Hom}(\mathbb{P}^1, X)$ such that $V_1 = \Phi((T \cap \text{Hom}(\mathbb{P}^1, X; \infty)) \times \mathbb{P}^1)$, $V_{12} = \Phi((T \cap \text{Hom}(\mathbb{P}^1, X; \{\infty, 0\})) \times \mathbb{P}^1)$. $\text{Hom}(\mathbb{P}^1, X; \infty)$ [resp. $\text{Hom}(\mathbb{P}^1, X; \{\infty, 0\})$] is, by definition, the set of the morphisms that send ∞ to $f(\infty) = x_1$ [resp. $(\infty, 0)$ to (x_1, x_2)], and hence, around the reference point $[f]$, we have

$$\text{Hom}(\mathbb{P}^1, X; \infty) \times \{\infty\} = \left(\Phi|_{\text{Hom}(\mathbb{P}^1, X) \times \{\infty\}}\right)^{-1}(x_1),$$

$$\text{Hom}(\mathbb{P}^1, X; \{\infty, 0\}) \times \{0\} = \left(\Phi|_{\text{Hom}(\mathbb{P}^1, X; \infty) \times \{0\}}\right)^{-1}(x_2)$$

near $[f]$. Thus we get the estimates

$$\dim_{[f]} \text{Hom}(\mathbb{P}^1, X; \infty) \geq \dim \text{Hom}(\mathbb{P}^1, X) - d_0 \geq (C, -K_X) + n - d_0,$$
$$\dim_{[f]} \text{Hom}(\mathbb{P}^1, X; \{\infty, 0\}) \geq \dim \text{Hom}(\mathbb{P}^1 \infty; X, x_1) - d_1$$
$$\geq (C, -K_X) + n - d_0 - d_1.$$

On the other hand,

$$d_{12} = \dim V_{12} = \dim\left(\text{Hom}(\mathbb{P}^1, X; \{\infty, 0\}) \times \mathbb{P}^1\right) - \dim \text{Aut}(\mathbb{P}^1, \{\infty, 0\})$$
$$= \dim(\text{Hom}(\mathbb{P}^1, X; \{\infty, 0\}).$$

Hence

$$(C, -K_X) \leq -n + d_0 + d_1 + d_{12}.$$

Suppose that $d_{12} \geq 2$. Then $\text{Hom}(\mathbb{P}^1, X; \{\infty, 0\})$ has strictly bigger dimension than $\text{Aut}(\mathbb{P}^1; \infty, 0) \simeq \mathbb{G}_m$ does, and the image $\Phi(\text{Hom}(\mathbb{P}^1, X; \{\infty, 0\}) \times \mathbb{P}^1) \subset X$ is of dimension ≥ 2. Hence we can find a morphism from a smooth curve T to $\text{Hom}(\mathbb{P}^1, X; \{\infty, 0\})$ such that the induced morphism $\Phi_T : T \times \mathbb{P}^1 \to X$ has two-dimensional image. Let $\overline{T}$ be a smooth compactification of T. Then we know that Φ_T does not extend to a morphism of $\overline{T} \times \mathbb{P}^1$. We blow up finitely many points on $\mathbb{P}^1 \times \overline{T}$ to resolve the indeterminacy of Φ_T. Let W denote the blown up surface and $\text{pr}_{\overline{T}} : W \to \overline{T}$ the naturally induced projection. Let us prove that, for some closed point $s \in \overline{T}$, the fibre $F(s) = \text{pr}_{\overline{T}}^{-1}(s)$ satisfies $F(s) \geq F_1 + F_2$, with F_i non-trivially mapped to X.

Suppose the contrary, i.e. each fibre $\text{pr}_{\overline{T}}^{-1}(s)$ contains a unique component $F_0(s)$ that are non-trivially mapped to X.

By looking at the negative definite intersection matrix of the extra components $\text{pr}_{\overline{T}}^{-1}(s) - F_0(s)$, which are mapped to finitely many points on X by assumption, we can contract these extra components to get a normal surface V with a fibre space structure over $\overline{T}$, and the morphism $\text{pr}_{\overline{T}} : W \to X$ factors through V (see [Artin]). If you note that every singular fibre $\subset W$ of $\text{pr}_{\overline{T}}$ is supported by a tree of rational curves, the fibres on V over $\overline{T}$ are just $\mathbb{P}^1$ (with multiplicity). (Note that the birational morphism $W \to V$ contracts a disjoint union of trees of smooth rational curves, whence follows that V is still projective. It is left to the reader as an exercise to construct a suitable linear system $|L|$ on W that gives the contraction $W \to V$.)

There are two cases:

Case 1. Every fibre is reduced. The equidimensional fibre space V over the smooth curve $\overline{T}$, with irreducible, reduced, smooth fibres, is a smooth surface. Every fibre is now $\mathbb{P}^1$, and V is a $\mathbb{P}^1$ bundle over $\overline{T}$. The closures σ_∞, σ_0 of $T \times \{\infty\}$, $T \times \{0\}$ are two distinct sections which are mapped to points x_1, x_2 by $\tilde{\Phi}_T$. This implies that $\sigma_\infty^2, \sigma_0^2$ are both negative by the Hodge index theorem, which is impossible by (I.4.7).

Case 2. Some fibre V_s is non-reduced of multiplicity $m \geq 2$. Then $V_s \geq mF_0(s)$, $m \geq 2$.

Thus we have shown that some fibre of $\mathrm{pr}_{\overline{T}}$ is reducible or non-reduced, which means that C deforms into a sum C' of two or more rational curves C_i'. Obviously $(C_i', H) < (C', H) = (C, H)$, for every C_i. Since $(C, -K_X) > 0$, we can choose an irreducible component C_0' so that $(C_0', -K_X) > 0$. $\square$

3.3 Examples   (1) The above bound $(C, -K_X) \leq n + 1$ for the minimum degree of rational curves is best possible. In fact the anti-canonical divisor of a projective n-space is $(n+1)H$, H being the hyperplane, and a rational curve of minimum degree in $\mathbb{P}^n$ is of course a line.

(2) Recall the lectures by Peternell. Let X be smooth projective over $\mathbb{C}$ and $R \subset N_1(X)$ an extremal ray in $NE(X)$. Assume that the linear functional $-K_X$ takes positive value on R. Then R is generated by a rational curve C. Choose C to have the smallest degree. Let $\mathrm{cont}_R : X \to Y$ be the contraction of R. In this situation, the d_i in Proposition (3.2) has the following interpretation.

Case 1. cont_R is birational. In this case, d_0 is the dimension of the cont_R-exceptional set E_R, and d_1 is the dimension of the fibre $E_{R,y} = (\mathrm{cont}_R|_{E_R})^{-1}(y)$ of the contraction. Thus $0 < (C, -K_X) \leq \dim E_R + \dim E_{R,y} - n + 1 \leq 2 \dim E_R - n + 1$, and hence $\dim E_R \geq \frac{n}{2}$. In other words, the exceptional set of a birational contraction of an extremal ray on a smooth projective variety does not have too small dimension [Ionescu]. In particular, a smooth threefold X has no small contraction [Mori 2]. (This result does not pass over to singular varieties. We can construct a birational extremal contraction which contracts a single rational curve on a singular threefold. The first example was found by P. Francia, which is discussed in Peternell's lectures in connection with flips.)

Case 2. cont_R defines a birationally non-trivial fibre space structure. In this case, $d_0 = n$, and d_1 is the fibre dimension, so that $(C, -K_X) \leq \dim X_y + 1$. This upper bounded is attained if $X \to \mathrm{cont}_R(X)$ is a certain $\mathbb{P}^r$-bundle (e.g. $X = \mathbb{P}^r \times \mathbb{P}^s$, $s \geq r$).

4   Conjectures of Hartshorne and Frankel

4.1 One of the most remarkable applications of Theorem 3.2 is a solution of the Hartshorne conjecture on a characterization of projective spaces.

R. Hartshorne [Hartshorne 1] raised a conjecture to the effect that a smooth projective variety X with ample tangent bundle would be a projective space. The conjecture was affirmatively solved by S. Mori in the following strengthened form:

4.2 Theorem [Mori 1]   *Let X be a smooth projective variety over an algebraically closed field of arbitrary characteristic. X is isomorphic to a projective space if and only if X contains at least one rational curve and the restriction of $T_X|_C$ is ample for every rational curve $C \subset X$.*

4.3 When X is a curve or a surface, the proof of (4.2) is very easy, so we assume that $n = \dim X \geq 3$ below. We give the outline of the proof, and then the detailed proof.

Let us fix the notation. Let X be as above with a fixed polarization L, and l the minimum degree of the rational curves on X with respect to L. Let $R^l \subset R(X) \subset \mathrm{Chow}(X)$ be an irreducible component of the set of the rational curves of degree l. R^l is a non-empty, projective, closed subset of $\mathrm{Chow}(X)$. Let $H(l) \subset \mathrm{Hom}(\mathbb{P}^1, X)$ be an irreducible component of $\mathrm{Hom}^l(\mathbb{P}^1, X)$, which is the set of the morphisms $f : \mathbb{P}^1 \to X$ such that $\deg f^*L = l$. $H(l)$ suitably chosen, there is $[f] \in H(l)$ with $[f(\mathbb{P}^1)] \in R^l$. For given $[f] \in H(l)$, the image $f(\mathbb{P}^1) \subset X$ is an element of $R^l \subset \mathrm{Chow}(X)$, thus defining a set-theoretical mapping $H(l) \to R^l$. Fix a general element $[f_0] \in H(l)$ such that $x = f_0(\infty) \in f_0(\mathbb{P}^1) \subset X$ is a smooth point of $f_0(\mathbb{P}^1)$. Let $H(x, l) \subset H(l)$ be the set of morphisms f such that $f(\infty) = x$. $R^l(x) = \rho(H(x, l))$ is a set of rational curves C containing x. Let $\psi : H(x, l) \times \mathbb{P}^1 \to \mathrm{Chow}(X) \times X$ be the map defined by $([f], p) \mapsto (\rho([f]), f(p))$, and let $W \subset R^l(x) \times X$ be its image.

Under this notation we prove:

(1) $\rho : H(x, l) \to R^l(X, l)$ is a surjective morphism with set-theoretical fibre $\mathrm{Aut}(\mathbb{P}^1)$.
(2) $R^l(X, x)$ is isomorphic to $\mathbb{P}^{n-1}$, and W is a $\mathbb{P}^1$-bundle over $R^l(X, x)$.
(3) W is isomorphic to the one-point blown up $Bl_x(X)$ of X.
(4) X is in fact isomorphic to $\mathbb{P}^n$.

On $\mathbb{P}^n$, the rational curves of minimum degree are lines and form the Grassmann variety $\mathrm{Grass}(n + 1, n - 1)$. The subset $G_x \subset \mathrm{Grass}(n + 1, n - 1)$ consisting of the lines passing through a given point $x \in \mathbb{P}^n$ is naturally identified with $\mathbb{P}^{n-1}$, the set of one-dimensional subspaces of $T_{\mathbb{P}^N, x}$, by associating to a given line $\in G_x$ its tangent direction at x. The incidence variety

$$W = \{([\ell], x) \in G_x \times \mathbb{P}^n \mid x \in \ell\}$$

is a $\mathbb{P}^1$-bundle over G_x, and it is indeed isomorphic to $Bl_x(\mathbb{P}^n)$. Thus we are going to prove that we are exactly in this situation.

4.4 Lemma   (1) $H(l)$ and $H(x,l)$ are smooth of dimension $2n+1$ and $n+1$, respectively. For every $[f] \in H(l)$, $f : \mathbb{P}^1 \to X$ is an immersion (i.e. the differential of f is everywhere of maximal rank).

(2) The morphism $\Phi^{(2)} : H(l) \times \mathbb{P}^1 \times \mathbb{P}^1 \to X \times X$, defined by $([f],p,q) \mapsto (f(p), f(q))$, is of maximal rank $2n$ on $H(l) \times ((\mathbb{P}^1 \times \mathbb{P}^1) \setminus \Delta_{\mathbb{P}^1})$, where Δ_* denote the diagonal in $* \times *$. Hence in particular, $\Phi : H(l) \times \mathbb{P}^1 \to X$ is surjective, and $H(x,l) \neq \emptyset$ for any $x \in X$.

(3) Assume that $x \in X$ is general. Then the subset $S = \{[C] \mid C \text{ is singular}\} \subset R^l(x)$ [resp. $S(x) = \{[C] \mid C \text{ is singular at } x\} \subset R^l(x)$] is either empty or exactly one-dimensional [resp. empty or zero-dimensional].

(4) $\rho : H(l) \to R^l$ [resp. $\rho|_{H(x,l)} : H(x,l) \to R^l(x)$] is a surjective morphism with set-theoretical fibre $\mathrm{Aut}(\mathbb{P}^1)$ [resp. $\mathrm{Aut}(\mathbb{A}^1) \subset \mathrm{Aut}(\mathbb{P}^1)$].

Proof.   (1) Let $[f] \in H(l)$ be an arbitrary point. Since $f(\mathbb{P}^1)$ is of minimum degree, we have $\deg f^*T_X = \deg f^*(-K_X) \leq n+1$ by (3.2). On the other hand, by assumption, f^*T_X is ample, so that it is a direct sum of n line bundles of degree ≥ 1. Since f is generically injective, $T_{\mathbb{P}^1} \simeq \mathcal{O}(2)$ is naturally a subsheaf of f^*T_X. This implies that $f^*T_X \simeq \mathcal{O}(2) \oplus \mathcal{O}(1)^{\oplus n-1}$, and that $T_{\mathbb{P}^1}$ is a subbundle of it, which shows in particular that $f : \mathbb{P}^1 \to X$ is an immersion. Furthermore, $H^1(\mathbb{P}^1, f^*T_X) = H^1(\mathbb{P}^1, I_\infty f^*T_X) = 0$. This proves the assertion. $\square$

(2)(3) The differential of $\Phi^{(2)}$ at $([f],p,q)$ restricted to the subspace $\mathrm{pr}_1^* T_{H(l)}$ is nothing but the evaluation map $H^0(\mathbb{P}^1, f^*T_X) \to (f^*T_X)_p \oplus (f^*T_X)_q$, which is surjective if $p \neq q$ in view of the ampleness of f^*T_X. In particular, the inverse image of $\Delta_X \subset X \times X$ [resp. (x,x)], when restricted to $H(l) \times (\mathbb{P}^1 \times \mathbb{P}^1 \setminus \Delta_{\mathbb{P}^1})$ is either empty or smooth, n-codimensional [resp. empty or purely $2n$-codimensional]. Noting that $H(l) \times \mathbb{P}^1 \times \mathbb{P}^1$ is of dimension $2n+3$ and that $\mathrm{Aut}(\mathbb{P}^1)$ is of dimension 3, this shows that the set of curves $C \ni x$ such that the inverse image of x in the normalization contains two distinct points is 0-dimensional. Since $\mathbb{P}^1 \to X$ is an immersion, so that C has no infinite near singularities, this proves that $\dim S(x) \leq 0$. $H_{\mathrm{sing}}(l) \subset H(l)$, consisting of the f with singular image in X is of dimension $n+3$, and $H_{\mathrm{sing}}(l) \times \mathbb{P}^1$ of dimension $n+4$. If x is general, we have also the dimension estimate for S; indeed, $\Phi^{-1}(x) \cap H_{\mathrm{sing}}(l)$ is exactly of dimension 4 if $\Phi : H_{\mathrm{sing}}(l) \times \mathbb{P}^1 \to X$ is dominant and empty otherwise. $\square$

(4) For a given $[C] \in R^l$, the normalization $\mathbb{P}^1 \to C$ gives a morphism $f : \mathbb{P}^1 \to X$ with $\deg f^*L = l$. By composing an automorphism of $\mathbb{P}^1$, if necessary, we can take f such that $f(\infty) = x$. This shows the surjectivity of ρ. It is easy to show that ρ is at least a rational map. Indeed, the Hilbert polynomial of $\rho(f_s(\mathbb{P}^1))$ is identically $lt + 1$ as long as $C_s = f_s(\mathbb{P}^1)$ is smooth. Hence there is a morphism from an open subset U of $H(l)$ to $\mathrm{Hilb}(X)$, which is smooth at $[C_s]$ because $\mathcal{N}_{C_s/X} \simeq f^*T_X/T_{\mathbb{P}^1}$ is ample. By locally identifying $\mathrm{Hilb}(X)$ and $\mathrm{Chow}(X)$ at a smooth rational curve C_s, we see that ρ is a morphism on U. Since $H(l)$ is smooth and hence normal, this shows that ρ is actually a morphism by the Zariski Main Theorem. The fibre

of ρ over $[C] \in R^l$ corresponds to the choice of isomorphisms between $\mathbb{P}^1$ and the normalization of C, thus naturally identified with $\text{Aut}(\mathbb{P}^1)$.   □

In what follows, $x \in X$ is always general.

4.5 Lemma   (1) *Let $\tilde{R}^l(x)$ denote the normalization of $R^l(x)$ in $k(H(x,l))$. Then there exists a natural morphism $\sigma : \tilde{R}^l(X,x) \to \mathbb{P}(T^*_{X,x}) \simeq \mathbb{P}^{n-1}$. Moreover the finite morphism $\tilde{R}^l(X) \to R^l(X)$ has connected fibres and birational.*
   (2) σ is étale and hence an isomorphism.

Proof.   (1) $\mathbb{P}(T^*_{X,x})$ is, by definition, the space of 1-dimensional subspaces of $T_{X,x}$. Define the morphism $\tilde{\sigma} : H(x,l) \to \mathbb{P}(T^*_{X,x})$ by $[f] \mapsto [df_\infty(T_{\mathbb{P}^1})]$, which makes sense because f is an immersion. Let $\tilde{\rho} : H(x,l) \to \tilde{R}^l(X,x)$ be the natural morphism. Since $\tilde{R}^l(X,x)$ is a variety, there is an open subset U contained in its smooth locus such that $\rho : H(x,l) \to \tilde{R}^I(X,x)$ has reduced, pure-dimensional fibres over U. If $[f] \in \phi^{-1}(U)$, then the fibre is set-theoretically $[f \, \text{Aut}(\mathbb{A}^1)]$, which is a reduced subscheme of $H(x,l)$, and thus the scheme-theoretical fibre. It is obvious that $\tilde{\sigma}$ takes constant value on $[f \, \text{Aut}(\mathbb{A}^1)]$ for any $[f] \in H(x,l)$, so that $\tilde{\sigma}$ defines a continuous map from $\tilde{R}^l(x)$ and a morphism $U \to \mathbb{P}(T^*_{X,x})$, i.e. a rational map $\tilde{R}^l(x) \dashrightarrow \mathbb{P}(T^*_{X,x})$. Since $\tilde{R}^l(x)$ is normal, a continuous rational map is a morphism. The connectedness of the fibres follows from the connectedness of $\text{Aut}(\mathbb{A}^1)$. Hence $\tilde{R}^l(x)$ is birational to, or purely inseparable over, $P^l(x)$. If $C = f(\mathbb{P}^1)$ is a smooth curve, then the differential of $\rho : H(l,x) \to R^l(x)$ is the natural map $H^0(I_\infty f^* T_X)/H^0(I_\infty T_{\mathbb{P}^1}) \to H^0(C, I_x N_{C/X})$ under the identification of a neibourhood of $[C]$ in $R_l(x)$ with an subset of $\text{Hilb}(X)$. Since $I_\infty f^* T_X$ is semi-positive, this differential is of maximal rank, so that $\tilde{R}^l(x)$ is separable over $R^l(x)$.   □

(2) Let us compute the differential of $\tilde{\sigma}$ at $[f]$, provided $f(\mathbb{P}^1)$ is smooth at x. $T_{H(x,l)}$ is isomorphic to $H^0(\mathbb{P}^1, I_\infty f^* T_X)$. Let $\mu : \tilde{X} = Bl_x(X) \to X$ be the blowing up at x, with $E \subset \tilde{X}$ the exceptional divisor. E is naturally isomorphic to $\mathbb{P}(T^*_{X,x})$. Then we have a canonical exact sequence

$$0 \to \mu^* T_X(-E) \to T_{\tilde{X}} \to \mathcal{N}_{E/\tilde{X}} \to 0,$$

and a natural surjection $\mu^* T_X(-E)|_E \simeq \mathcal{O}(1)^n_E \to T_E$. The universal morphism $\Phi : H(x,l) \times \mathbb{P}^1 \to X$ lifts to a rational map $\tilde{\Phi} : H(x,l) \times \mathbb{P}^1 \dashrightarrow \tilde{X}$ in an obvious manner, which is well-defined at $([f],p)$ unless $f(p) = x$ and x is a singular point of $f(\mathbb{P}^1)$. Then the differential of $\tilde{\sigma}$ is nothing but the differential of $\tilde{\Phi}|_{H(x,l)\times\infty}$, which maps $H^0(\mathbb{P}^1, I_\infty f^* T_X) \simeq H^0(\mathbb{P}^1, \tilde{f}^* \mu^* T_X(-E))$ to $T_{E,\tilde{\Phi}([f],\infty)}$ via the evaluation map to $(\mu^* T_X(-E))_{\tilde{\Phi}([f],\infty)}$. Since $I_\infty f^* T_X$ is semi-positive, the evaluation map and hence the differential $d\tilde{\sigma}$ are surjective. This shows that $d\tilde{\sigma}$ is of maximal rank, on $\tilde{R}^l(x) \setminus \tilde{S}(x)$, where $\tilde{S}(x) \subset \tilde{R}^l(x)$ is the inverse image of the finite set $S(x) \subset R^l(X,x)$ consisting of curves singular at x. $\tilde{\sigma}(\tilde{S}(x))$ is a finite set Σ in $\mathbb{P}(T^*_{X,x}) = \mathbb{P}^{n-1}$, so $\tilde{\sigma}$ gives a morphism étale over $\mathbb{P}^{n-1} \setminus \Sigma$. Since $\pi^{\text{alg}}_1(\mathbb{P}^{n-1} \setminus \Sigma) \simeq$

$\pi_1^{\text{alg}}(\mathbb{P}^{n-1}) = (1)$ (i.e. there is no étale cover of $\mathbb{P}^{n-1}$), this implies that $\tilde{\sigma}$ is birational, and finite. Hence by the Zariski Main Theorem, we get $\tilde{R}^l(x) \simeq \mathbb{P}^{n-1}$. $\square$

4.6 Lemma   *Let $\tilde{W} \subset \tilde{R}^l(x) \times X$ be the image of $H(x,l) \times \mathbb{P}^1$ via (ρ, Φ), with natural projections $\text{pr}_1 : \tilde{W} \to \tilde{R}^l(x)$ and $\text{pr}_2 : \tilde{W} \to X$.*
(1) *$\tilde{W}$ is smooth at $(\tilde{\gamma}, y)$ if y is a smooth point of C or $y \neq x$. Here $[C] = \gamma \in R^l(x)$ is the image of $\tilde{\gamma}$.*
(2) *Every member C in $R^l(x)$ is a smooth rational curve.*
(3) *$R^l(x)$ is smooth. $W = \tilde{W}$ is globally smooth.*
(4) *$\text{pr}_2 : W \to X$ is birational, and is an isomorphism over $X \setminus \{x\}$. It induces an isomorphim q from W to $Bl_x(X)$, the blown-up of X at x.*

Proof.   (1): Because the base space $\tilde{R}^l(x)$ is smooth, a smooth point y of C represents a smooth point $(\tilde{\gamma}, y)$ of the incidence variety $\tilde{W}$. Let us see that if $y \neq x$, the projection $\text{pr}_2 : \tilde{W} \to X$ is unramified. Since $\Phi : H(l,x) \times \mathbb{P}^1 \to X$ is of maximal rank on $\Phi^{-1}(y)$, $\Phi^{-1}(y)$ is a smooth, reduced scheme of pure dimension $2 = \dim \text{Aut}(\mathbb{A}^1)$. From $\Phi = \text{pr}_2(\tilde{\rho}, \Phi)$, it follows that $\text{pr}_2 : \tilde{W} \to X$ is finite, and with reduced fibres. Thus pr_2 is unramified at $(\tilde{\gamma}, y)$. Since $\tilde{W}$ is normal, we conclude that $\tilde{W}$ is smooth at $(\tilde{\gamma}, y)$. $\square$

(2): By (1), $\tilde{W}$ is smooth outside $\tilde{S}(x) \times \{x\}$. The projection pr_1, viewed as a morphism from $\tilde{W} \setminus \text{pr}_1^{-1}(\tilde{S}(x))$ to $\tilde{R}^l(x) \setminus \tilde{S}(x)$, is thus an equidimensional morphism between smooth varieties and hence flat. Every fibre of pr_1 over $\tilde{R}^l(x) \setminus \tilde{S}(x)$ is irreducible, reduced, of pure dimension one (locally a complete intersection, in fact). A general fibre is a smooth rational curve so that every fibre over $\tilde{R}^l(x) \setminus \tilde{S}(x)$ must be smooth by the invariance of arithmetic genus. Hence $\dim S \leq \dim S(x) = 0$, which was possible only when $S = \emptyset$; in other words, every element of $R^l(x)$ represents a smooth rational curve. $\square$

(3) and (4): By (2), $R^l(x) \subset \text{Chow}(X)$ can be identified with a closed subset of $\text{Hilb}(X)$. At any point $[C] \in R^l(x)$, the Zariski tangent space of $R^l(x)$ is $H^0(C, I_x \mathcal{N}_{C/X})$. $R^l(x)$ is smooth because $H^1(C, I_x \mathcal{N}_{C/X}) = 0$ for every $[C]$. It is now clear that W is smooth, too. Easy computation shows that $\text{pr}_2 : W \to X$ is étale over $X \setminus \{x\}$. Furthermore pr_2 naturally lifts to a morphism: $W \to Bl_x(X)$, by associating $([C], x)$ the tangent direction of C at x. You can also check that q is étale again. Since $R^l(x)$ is isomorphic to $\mathbb{P}^{n-1}$, and W is a $\mathbb{P}^1$-bundle on it, it follows that $\chi(W, \mathcal{O}_W) = 1$. Hence the étale morphism $q : W \to Bl_x(X)$ must be an isomorphism. $\square$

4.7 Lemma   *X is isomorphic to $\mathbb{P}^n$.*

Proof.   We identify W with $Bl_x(X)$ and $R^l(x) \times \{x\}$ with the exceptional divisor. Take a hyperplane L_0 on $R^l(x) \simeq \mathbb{P}^{n-1}$, and let $\tilde{L}$ be the divisor $\text{pr}_1^* L_0 + E$. Then, since $E|_E \simeq \mathcal{O}(-1)$, we have a natural exact sequence

$$0 \longrightarrow \mathcal{O}_W(\text{pr}_1^* L_0) \longrightarrow \mathcal{O}_W(\tilde{L}) \longrightarrow \mathcal{O}_E \longrightarrow 0.$$

Since W is a $\mathbb{P}^1$-bundle over $\mathbb{P}^{n-1}$, we get $R^i(\mathrm{pr}_1)_* = 0$, $i > 0$, so that

$$H^1(W, \mathcal{O}(\mathrm{pr}_1^* L_0)) = H^1(\mathbb{P}^{n-1}, \mathcal{O}(L_0)) = 0.$$

In particular, $\dim H^0(W, \mathcal{O}(\tilde{L})) = n + 1$. $|\mathrm{pr}^* H_0|$ is obviously free from base points, and so the base point set of $|\tilde{L}|$ is contained in E. On the other hand, since $H^0(W, \mathcal{O}(\tilde{L})) \to H^0(E, \mathcal{O}_E)$ is surjective, it follows that $\tilde{L}$ is free from base point. $\tilde{L}$ is trivial on E and hence is the pull-back of a divisor L on X.

Let us show that the morphism $\varphi_L : X \to \mathbb{P}^n$ attached to the linear system $|L|$ is finite. To see this, let us check that L satisfies $(\Gamma, L) > 0$ for every curve Γ on X. Let $\tilde{\Gamma}$ be the strict transform of Γ with respect to the blowing up $\mu : W = \tilde{X} \to X$. Clearly $\tilde{\Gamma}$ is not contained in E, and $(\tilde{\Gamma}, E) \geq 0$. Since H_0 is ample, $(C, \mathrm{pr}_1^* L_0)$ is positive whenever $\mathrm{pr}_1(\tilde{C})$ is one-dimensional. If $\mathrm{pr}_1(\tilde{C})$ is a point, then $\tilde{C}$ is a fibre of the projection, so it meets E at one point, showing that $(C, L) = (\tilde{C}, \mu^* L) \geq (\tilde{C}, E) > 0$.

Let $w \in X$ be a general point and v an arbitrary point on $X \setminus \{x\}$. We check that, if w and v are mapped to the same point by φ_L, then $v = w$, i.e. φ_L is generically one-to-one. Since $\mathrm{pr}_1^* L_0$ separates the fibres, $\varphi_L(v) = \varphi_L(w)$ implies that they sit on the same fibre $C \simeq \mathbb{P}^1$. But then $|\tilde{L}|$ restricted to C gives a non-trivial map C to a curve with mapping degree $\deg \tilde{L}|_C = 1$, i.e. an isomorphism. Hence we have $v = w$.

We check also that the morphism is of maximal rank at a general point w, which follows from:

(1) The tangent space of the fibre is non-trivially mapped by the linear system $|\tilde{L}|$, but mapped to zero by the linear system $|\mathrm{pr}_1^* L_0|$;

(2) The linear system $|\mathrm{pr}_1^* L_0|$ gives an injection $T_W/T_{W/R^l(x)}$ into $T_{\mathbb{P}^{n-1}} \subset T_{\mathbb{P}^n}$.

Hence the map φ_L is generically bijective and generically unramified, defining a birational morphism onto $\mathbb{P}^n$. Since we have checked that it is also finite, this proves that $X \simeq \mathbb{P}^n$. □

4.8 Corollary (Frankel conjecture) *Let X be a compact complex manifold with a Kähler metric $\omega = \omega_{i\bar{j}} dx^i d\bar{x}^j$. Let $\sum R^j_{ikl} \frac{\partial}{\partial x^j} dx^i dx^k d\bar{x}^l \in \mathrm{Hom}(T_X, T_X \otimes \mathcal{D}_X^{1,1})$ be the curvature form, and put $R_{i\bar{j}k\bar{l}} = \sum \omega_{\alpha i} R^\alpha_{jkl} \in \mathcal{D}^2 \otimes \mathcal{D}^2$. If $R_{i\bar{j}k\bar{l}} dx^i d\bar{x}^j dx^k d\bar{x}^l$ defines a positive definite hermitian form on $L_1 \otimes L_2$ for every $(L_1, L_2) \in \mathbb{P}(T_X^*) \times \mathbb{P}(T_X^*)$, then X is biholomorphic to a projective space. In other words, a compact Kähler manifold of positive holomorphic bisectional curvature is a projective space.*

Indeed, the positivity of holomorphic bisectional curvature implies the ampleness of T_X [Griffiths 1]. There is a purely analytic proof of the Frankel conjecture [Siu-Yau], but our algebraic proof is much easier.

5   Numerical characterization of uniruled varieties

5.1 An even more important application of Theorem 2.6 is a characterization of uniruled varieties, one of the fundamental results in classification theory.

A variety X of dimension n is said to be *uniruled* [resp. *ruled*] if there exist a variety Y of dimension $n-1$ and a dominant rational [resp. birational] map $\mathbb{P}^1 \times Y \dashrightarrow X$. X is *unirational* [resp. *rational*], by definition, if there is a dominant rational [resp. birational] map $\mathbb{P}^n \dashrightarrow X$. All the above four properties are apparently preserved under birational operations. As for the relationship between them, we have obvious implications

$$
\begin{array}{ccc}
\text{rational} & \Longrightarrow & \text{ruled} \\
\Downarrow & & \Downarrow \\
\text{unirational} & \Longrightarrow & \text{uniruled}
\end{array}
$$

In dimension one, the four notions coincide with each other. In dimension two, a unirational [resp. uniruled] complex surface is rational [resp. ruled]. In positive characteristics, there are lot of unirational [resp. uniruled] surfaces which are not rational [resp. ruled]. In higher dimension, all the four notions are different from each other even over the complex numbers.

5.2 Examples   (1) An irreducible quadric hypersurface $Q^n \subset \mathbb{P}^{n+1}$ is rational, via the birational projection $Q^n \dashrightarrow \mathbb{P}^n$ from a point $x \in Q^n$.

(2) A smooth cubic threefold $X \subset \mathbb{P}^4$ is known to always contain a line L. The hypersurfaces $H_t \subset \mathbb{P}^4$ that contain L, cutting out a quadric curve Q_t plus L out of X, form a family parametrized by $t \in \mathbb{P}^2$. The incidence variety $\bigcup_t \{t\} \times Q_t \subset \mathbb{P}^2 \times X$ turns out rational and is a two-sheeted cover of X. Thus X is unirational.

Irrationality of a cubic threefold was independently proved by [Clemens-Griffiths] and by [Artin-Mumford].

(3) Rationality of a variety is extremely delicate. There is even an irrational vrariety X such that $X \times \mathbb{P}^1$ is rational [Beauville-Colliot-Thélène-Sansuc-Swinnerton-Dyer].

(4) [Iskovskih-Manin] showed that some special quartic hypersurfaces in $\mathbb{P}^4$ are irrational and unirational, but it is an open problem whether all such varieties are unirational or not. For further example of irrational uniruled varieties, see also [Iskovskih 3].

(5.3) In the examples above, rationality or unirationality is always checked by an *ad hoc* method, depending on very specific properties of the varieties in question. The fact is that so far we do not have general principles to test rationality, unirationality or ruledness; more precisely, no characterization is known which is not tautological rephrasing of the definition but applicable to sufficiently many objects. (For some sufficient conditions for unirationality, see [Morin], [Ramero] [Ebihara].)

In the classification theory, two smooth varieties should fall into the same class if one is a smooth deformation of the other or if one is an étale cover of the other. Thus what matters is to find properties which are birational invariants, étale invariants, and deformation invariants. Rationality, unirationality and ruled-ness are invariant under birational maps and étale morphisms, but their behaviour under smooth deformation may not be very nice. The lack of good characteriza-tion of such varieties makes it difficult to tell if these properties are preserved by deformation.

We do not come across these difficulties when we deal with uniruledness.

Firstly, we have more geometric criteria for uniruledness, in terms of rational curves:

5.4 Proposition *Let X be a projective variety defined over an algebraically closed field k with uncountably many elements. The following four conditions on X are equivalent:*

(1) *X is uniruled.*

(2) *The universal morphism $\Phi : \mathrm{Hom}(\mathbb{P}^1, X) \times X \to X$ is dominant (i.e. the image contains an open subset of X).*

(3) *If x is a sufficiently general point of X (i.e. it sits in a complement of a union of countable proper subvarieties of X), then there exists a rational curve on X which passes through x.*

(4) *Given any point x on X, we can find a rational curve passing through x.*

Here we denote by $\mathrm{Hom}(\mathbb{P}^1, X)$ the set of non-constant morphisms.

Proof. The implication (1) $\Rightarrow$ (2) is straightforward. (2) is equivalent to (3) by definition. Let us show (2) $\Rightarrow$ (4). Given a point x, take a curve $S \in \mathrm{Hom}(\mathbb{P}^1, X)$ in such a way that the closure of $\Phi(S \times \mathbb{P}^1)$ in X contains x (this is possible because Φ has Zariski dense image in X). Since S can be compactified to $\overline{S}$ in $\mathrm{Hilb}(\mathbb{P}^1 \times X)$, this means that there is a point $s \in \overline{S} \subset \mathrm{Hilb}(\mathbb{P}^1 \times X)$ such that the image of the corresponding subscheme $\mathrm{pr}_X(Z_s)$ contains x. We have seen that Z_s is supported by a connected union of finitely many rational curves. The image of these curves on X cannot be a single point and hence must be a union of rational curves, one of which contains x. Since (4) is clearly stronger than (3), or equivalently (2), it suffices to show (2) $\Rightarrow$ (1) in order to complete the proof. Noting that $\mathrm{Hom}(\mathbb{P}^1, X)$ is a countable union of quasiprojective schemes $\mathrm{Hom}^d(\mathbb{P}^1, X)$, it follows that there exists an irreducible component V, which is quasiprojective, such that the restriction $\pi|_{V \times \mathbb{P}^1} : V \times \mathbb{P}^1 \to X$ is dominant. Take a general $(n-1)$-dimensional subvariety Y of V, to get a dominant morphism $Y \times \mathbb{P}^1 \to X$. $\square$

The above characterizations enable us to show the deformation invariance of uniruled varieties in characteristic zero.

5.5 Theorem [Fujiki][Levine] *If $\pi : \mathcal{X} \to S$ is a projective smooth k-morphism with $\mathrm{ch}\, k = 0$, or if π is a proper smooth analytic mapping with relative Kähler form ω, then the subset $\{s \in S; X_s \text{ uniruled}\}$ is open and closed in S. Equivalently, any smooth projective (or Kähler) deformation of a uniruled variety is uniruled (provided $\mathrm{ch}\, k = 0$.)*

Proof. For simplicity we work in the category of projective schemes over k, ch $k =$ 0. Given a proper smooth morphism $\mathcal{X} \to S$, we think of X as an S-scheme. Then $\text{Hom}_S(S \times \mathbb{P}^1, \mathcal{X})$ is a countable union of quasiprojective S-schemes. X_s is uniruled if and only if the naturally induced map $\text{Hom}(\mathbb{P}^1, X_s) \times \mathbb{P}^1 \to X_s$ is dominant. Consider the subset $\{s; X_s \text{ is uniruled}\}$ of S. It suffices to show that this set is open and closed. Because ch $k = 0$, if a rational curve passes through a sufficiently general point $x_s \in X_s$, then $f^*T_{X_s}$ is semi-positive, see (V.5.6). Since the normal bundle $\mathcal{N}_{X_s/\mathcal{X}}$ is trivial, it follows that $f^*T_{\mathcal{X}}$ is semipositive and there is no obstruction to lift f_s to a parametrized morphism. Hence the openness.

The closedness condition is proved like this. Take a curve $\Gamma \in \mathcal{X}$ such that Γ_o contains a given general point. Then for $\gamma \neq 0$, we can find a rational curve that passes through Γ_γ. Take the specialization of these curves and we get rational curves through Γ_o.

5.6 Remark We cannot drop the hypothesis that ch $k = 0$. Indeed, by [Shioda-Katsura], some K3 surfaces in positive characteristic are unirational, see (5.9). But you can deform this one to a general K3 surface, which is no more uniruled. In general, the uniruledness is always a closed condition with respect to algebraic flat deformation.

5.7 Exercise Let X be a smooth projective variety and Y an étale cover of X (the characteristic of the ground field is arbitrary.) Show that X is uniruled if and only if Y is so. (Hint: $\mathbb{P}^1$ is simply connected).

While (5.4) is viewed as a geometric characterization of uniruled varieties, the following theorem gives a characterizaion in terms of numerical data:

5.8 Theorem *Let X be a projective variety. X is uniruled if there exists a family of irreducible curves $\{C_t\}_{t \in T}$ such that (1) C_t is contained in the smooth locus of X, that (2) the family sweeps out an open dense subset of X (i.e. $\bigcup_{t \in T} C_t \subset X$ contains a non-empty open subset) and that (3) $(C, -K_X) > 0$. If the characteristic of the ground field is zero, then the converse is also true.*

Proof. Assume that there is a covering family C_t such that $(C_t, -K_X) > 0$. Then (2.6) asserts that there is a rational curve passing through an arbitrary point on C_t. Conversely, assume that there is a covering family of rational curves parametrized by an $(n-1)$-dimensional variety Y. If the things are in characteristic 0 and $y \in Y$ is general, then there is an injection $T_{Y \times \mathbb{P}^1}|_{y \times \mathbb{P}^1} \to f^*T_X|_{y \times \mathbb{P}^1}$, so that $(\deg f)(C, -K_X) \geq (\{y\} \times \mathbb{P}^1, -K_{Y \times \mathbb{P}^1}) = 2$, whence follows $(C, -K_X) > 0$. $\square$

5.9 Example In positive characteristic p, there are many smooth uniruled varieties with ample canonical divisor. For example, a Fermat surface $X_0^d + X_1^d + X_2^d + X_3^d = 0$, $d \geq 4$ in $\mathbb{P}^3$ is unirational if and only if $p^\nu \equiv -1 \bmod d$, for some integer ν [Shioda-Katsura, Theorem III and Corollary].

5.10 On a smooth variety X, the canonical divisor is defined as the highest exterior power of the cotangent sheaf Ω_X^1. It is important, however, to consider the canonical divisors of singular varieties, as the minimal model program usually introduces necessary, though mild, singularities.

Let X be a complete normal variety. The locus of smooth points $X^\circ = X_{\text{smooth}}$ is an open subset with complement $\text{Sing}(X)$ of codimension at least 2. Let $\omega_{X^\circ} = \mathcal{O}_{X^\circ}(K_{X^\circ})$ denote the dualizing sheaf of X°. The open immersion $j : X^\circ \hookrightarrow X$ gives rise to the sheaf $j_* \omega_{X^\circ}^{\otimes m}$, $m \in \mathbb{Z}$, which is known to be coherent [Serre]. We denote by $\omega_X^{(m)}$ this coherent sheaf. $\omega_X^{(m)}$ is, by definition, invertible on X° for any integer m, and, there is a natural homomorphism $(\omega_X^{(m)})^{\otimes d} \to \omega_X^{(dm)}$ for any positive integer d, making $\bigoplus_{m \geq 0} \omega_X^{(m)}$ a graded $\mathcal{O}_X$-algebra. Note, however, $\omega_X^{(m)}$ is not necessarily isomorphic to $(\omega_X^{(1)})^{\otimes m}$.

5.11 Example Let C be a smooth curve of genus $g \geq 2$ and $p \in C$ a point. The ruled surface $Y = \mathbb{P}(\mathcal{O}_C \oplus \mathcal{O}_C(-p))$ has a section σ, defined as $\mathbb{P}(\mathcal{O}_C(-p)) \simeq C$. You compute the normal bundle of σ and see that $N_{\sigma/Y} \simeq \mathcal{O}_C(-p)$, a negative line bundle. You can then contract the section σ to get a normal surface X with a unique singularity o. A local section of $\omega_X^{(m)}$ at o is identified with a section of $\omega_Y^m(*\sigma)$ defined on an open neighbourhood of σ. Here the symbol $(*\sigma)$ means that the section may have pole of arbitrarily high order along σ.

Consider an open neighbourhood U of σ and observe the exact sequence

$$0 \longrightarrow \Gamma(U, \omega_U^{\otimes m}((m+i-1)\sigma)) \longrightarrow \Gamma(U, \omega_U^{\otimes m}((m+i)\sigma)) \longrightarrow \Gamma(\sigma, \omega_U^{\otimes m}((m+i)\sigma)).$$

By the adjunction formula, the third term is the global section of $\mathcal{O}_C(mK_C - ip)$, which vanishes if $i > (2g-2)m$. Thus it follows that $\omega_X^{(m)} = \pi_* \mathcal{O}_X(mK_X + (2g-2)m\sigma)$, and $\omega_X^{(m)}$ is invertible if and only if mK_C is linearly equivalent to $(2g-2)mp$. When p is general, then the line bundle $\mathcal{O}_C(mK_C - ip)$ cannot be trivial for any pair (m, i), and hence $\omega_X^{(m)}$ is not invertible for any m. You also observe that the $\mathcal{O}_X$-algebra $\bigoplus_{m \geq 0} \omega_X^{(m)}$ is not finitely generated.

5.12 Thus the $\omega_X^{(m)}$ have pretty bad behaviour in general. However, in some cases, it happens that $\omega_X^{(m)}$ is an invertible sheaf for some $m > 0$. When this is the case, we say that X is $\mathbb{Q}$-Gorenstein, or m-Gorenstein if we want to specify m. (The terminology is a little bit confusing. Traditionally a normal variety is said to be Gorenstein when it is 1-Gorenstein and Cohen-Macauley; in our definition of $\mathbb{Q}$-Gorenstein singularities, we drop the requirement of the Cohen-Macauley property.)

When X is $\mathbb{Q}$-Gorenstein, we can define the canonical divisor K_X as an element (or a $\mathbb{Q}$-divisor) in $\text{Pic}_{\mathbb{Q}}(X) = \text{Pic}(X) \otimes_{\mathbb{Z}} \mathbb{Q}$. Indeed, take a divisor D of the invertible sheaf $\omega_X^{(m)}$ and put $K_X = \frac{1}{m} D$. The intersection number $\in \mathbb{Q}$ between curves and $\mathbb{Q}$-divisors on X is well-defined in an obvious manner. A $\mathbb{Q}$-divisor $D = \frac{1}{m} D'$, $D' \in \text{Pic}(X)$ is said to be ample if so is D.

5.13 Exercise (1) Let X be a normal variety with only quotient singularities. Thus each singularity p_i is (analytically locally) isomorphic to $(\mathbb{C}^n, 0)/G_i$, where G_i is a finite subgroup of $GL(n, \mathbb{C})$. For each element $\gamma \in G_i$, $\det \gamma$ is a root of unity, of which the order is denoted by $\mathrm{ord}(\det \gamma)$. Then show that X is m-Gorenstein, where m is the L.C.M. of the $\mathrm{ord}(\det \gamma)$. In particular, X is 1-Gorenstein (in this specific case also Gorenstein in the traditional sense) if and only if each G_i is a subgroup of $SL(n, \mathbb{C})$. (In case dimension $= 2$, rational singularities are quotient singularities. Furthermore, the quotient singularities via finite subgroups of $SL(2, \mathbb{C})$ are rational double points of type A_n, D_n, E_6, E_7, E_8.)

(2) A cone $\subset \mathbb{P}^{n+1}$ over a smooth hypersurface $\subset \mathbb{P}^n$ of degree $n + 1$, with an isolated singularity at the vertex, is 1-Gorenstein.

5.14 Corollary *Let X be a normal, projective, $\mathbb{Q}$-Gorenstein variety. If $-K_X$ is ample, then X is uniruled.*

Proof. Since X is normal, a general complete intersection curve C sits in the smooth locus. Such curves satisfies the three conditions in (5.8).

5.15 Corollary *Let X be a (possibly non-normal) projective variety such that the dualizing sheaf ω_X is well-defined as an invertible sheaf on X. If ω_X^{-1} is ample, then X is uniruled.*

Proof. Let $\nu : Z \to X$ be the normalization. Z is projective. Furthermore, on the smooth locus Z° of Z, the canonical divisor K_{Z° is of the form $\nu^* \omega_X - B$, where B is an effective divisor attached to the codimension one singularity of X. Hence, for a general complete intersection curve C, sitting on Z°, satisfies $(C, -K_{Z^\circ}) = (C, \nu^* \omega_X^{-1}) + (C, B) \geq (C, \nu^* \omega_X^{-1}) > 0$, and Theorem 5.8 applies. $\square$

A general definition of dualizing sheaves is omitted here. However, a locally complete intersection subvariety of a smooth variety has always an invertible dualizing sheaf. For a global complete intersection $Y = D_1 \cap D_2 \cap \ldots \cap D_r$ in a smooth X, we have $\omega_Y = \mathcal{O}_Y(K_X + D_1 + \ldots + D_r)$, by adjunction.

5.16 Examples (1) An n-dimensional complete intersection of type $(d_1, d_2, \ldots, d_r)$ in $\mathbb{P}^{n+r}$ is uniruled if $d_1 + \cdot + d_r \leq n + r$. For instance, any (irreducible) hypersurface of degree $\leq n + 1$ in $\mathbb{P}^{n+1}$ is uniruled.

As we will see in Lecture V, smooth complete intersections of degree in the above range are rationally connected. But singular ones are not necessarily rationally connected. For instance, a smooth cubic surface is a rational surface, but a cubic cone in $\mathbb{P}^3$ is elliptic ruled.

(2) A double cover $X \to \mathbb{P}^n$ branching over a divisor $B \subset \mathbb{P}^n$ is uniruled if $2 \deg B \leq n$.

Historical Remarks

The notion of moduli goes back to Riemann, who pointed out that curves of genus g are parametrized by $3g - 3$ essential parameters. Rigorous treatment of analytic deformation was initiated with great success by K. Kodaira and D.C. Spencer in 1950s [Kodaira-Spencer], and culminated in the construction of universal local deformation by [Kuranishi]. The functorial approach explained above is due to A. Grothendieck [SGA] and his followers, see [Schlessinger].

The idea that modulo p reduction and Frobenius can be applied to produce enough deformations and hence rational curves is due to S. Mori, who developed the technique to show the Hartshorne conjecture on a characterization of projective spaces [Mori 1], and then applied it to the analysis of extremal rays of smooth threefolds [Mori 2].

Several people tried to find analytic methods to construct rational curves (harmonic maps etc.) but this attempt was not very fruitful except under very strong conditions (positivity of holomorphic bisectional curvature etc.) [Siu-Yau] [Uhlenbeck-Yau].

It was J. Kollár who pointed out that the Mori's technique easily leads to the uniruledness of a smooth Fano variety (his proof was published in [Mori 3]). The above numerical criterion for uniruledness was obtained in [Miyaoka-Mori], enabling us to extend Kollár's result to singular $\mathbb{Q}$-Fano varieties.

The notion of $\mathbb{Q}$-Gorenstein varieties was introduced by [Reid 1], to develop the theory of canonical models of threefolds, a prototype of the minimal model program.

Lecture III
Foliations and Purely Inseparable Coverings

Overview

In Lecture II, we gave a numerical characterization of uniruled varieties in terms of a certain numerical property of anti-canonical divisors. In this lecture, we discuss a refined characterization of such varieties in terms of the tangent bundle. Namely, a smooth projective variety in characteristic zero is uniruled unless its tangent bundle is almost everywhere seminegative.

The proof of this result is made by using quotient varieties by foliations in positive characteristics. Since tangent bundles in positive characteristics, viewed as sheaves of derivations, have features quite different from those in characteristic zero, we discuss some properties of derivations and differential operators in characteristic p in detail. Then a variety in characteristic p is shown to be uniruled if it carries a foliation (a subsheaf which is closed under Lie bracket and p-th power) with certain properties.

What relates this new criterion of uniruledness in characteristic p to the one in characteristic zero is "semistability". The theory of semistable torsion free sheaves will be discussed in the second section, including numerical characterizations of semistability (in characteristic zero) and its behaviour under modulo p reductions. A renowned theorem of Mehta-Ramanathan plays an essential rôle in our argument.

There are several intriguing implications of the refined characterization of uniruledness above. One of them is the result that the 2-cocycle $c_2(X)$ of a minimal variety X looks like an effective cycle, which is derived from the famous Bogomolov inequality for semistable bundles.

1   Foliations and quotients attached to them

Let us review the essentials of the theory of foliations and associated quotients. A fully general account is found in [Ekedahl 1], but for all that we need later will be given proofs. In this section, all varieties are defined over an algebraically closed field k.

1.1 Let X be a smooth k-scheme, where k is an algebraically closed field.

A k-linear transformation ξ of $\mathcal{O}_X$ is said to be a $(k\text{-})derivation$ if $\xi(ab) = a\xi(b) + b\xi(a)$ for arbitrary $a, b \in \mathcal{O}_X$. We can think of the tangent sheaf $T_X = \mathcal{H}om_{\mathcal{O}_X}(\Omega^1_X, \mathcal{O}_X)$ as the sheaf of derivations : $\mathcal{O}_X \to \mathcal{O}_X$, by the following observation.

The second projection $\mathrm{pr}_2 : X \times X \to X$ gives rise to a ring homomorphism $\mathrm{pr}_2^* : \mathcal{O}_X \to \mathcal{O}_X \otimes \mathcal{O}_X$ and we can define the (exterior) derivative $d : \mathcal{O}_X \to \Omega^1_X = I_{\Delta_X}/I^2_{\Delta_X}$ by $d(a) = 1 \otimes a - a \otimes 1 \bmod I^2_{\Delta_X}$ (equivalently, d comes from the composite of pr_2^* and the $\mathcal{O}_X$-linear canonical projection $\delta : \mathcal{O}_X \otimes_k \mathcal{O}_X \to I_{\Delta_X}$,

$\delta(a \otimes b) = a \otimes b - ab \otimes 1$). Given $\xi \in T_X$ and $a \in \mathcal{O}_X$, define $\xi(a)$ to be $\xi(d(a))$ (recall that T_X is the dual bundle of Ω^1_X). By the identity

$$\begin{aligned}
\delta(1 \otimes ab) &= 1 \otimes ab - ab \otimes 1 \\
&= (1 \otimes a)(1 \otimes b) - ab \otimes 1 \\
&= \big(a \otimes 1 + \delta(1 \otimes a))\big)\big(b \otimes 1 + \delta(1 \otimes b)\big) - ab \otimes 1 \\
&= a\delta(1 \otimes b) + b\delta(1 \otimes a) + \delta(1 \otimes a)\delta(1 \otimes b) \\
&\equiv a\delta(1 \otimes b) + b\delta(1 \otimes a) \bmod I^2_{\Delta_X},
\end{aligned}$$

we get $\xi(ab) = \xi(d(ab)) = a\xi(b) + b\xi(a)$.

The product $\xi_1 \cdots \xi_r$ of given local derivations $\xi_1, \ldots, \xi_r$ is naturally defined as the composite of k-linear transformations of $\mathcal{O}_X$. Of course a product of derivations is no more a derivation, but is a (local) differential operator of finite order.

In general, (local) *differential operators* of order $\le r$ on X are k-linear operators acting on $\mathcal{O}_X$, which are inductively defined as follows:

(0) *A local differential operator of degree zero is an element $\in \mathcal{O}_X$ viewed as a local operator $\mathcal{O}_X \to \mathcal{O}_X$ via the multiplication.*
(i) *A local k-linear operator $D : \mathcal{O}_X \to \mathcal{O}_X$ is a differential operator of degree $\le i$ if $[D, a] = Da - aD$ is a differential operator of degree $\le i-1$ for every differential operator a of degree zero.*

The sheaf of the local differential operators of order $\le i$ is denoted by $Diff^i_X$. The left multiplication of $\mathcal{O}_X$ on $Diff^i_X$ is well-defined as the action of differential operators of degree 0 from the left, thereby making $Diff^k_X$ a left $\mathcal{O}_X$-module.

1.2 Exercise   (1) Check that αD is a differential operator of order $\le i$ when $\alpha \in \mathcal{O}_X$, $D \in Diff^i_X$.
(2)   Show that the set of differential operator of order ≤ 1 is equal to $\mathcal{O}_X \oplus T_X$, and that the left $\mathcal{O}_X$-module structures of those two sets are identical.
(3)   Prove that, for $D_1 \in Diff^i_X$, $D_2 \in Diff^j_X$, the commutator $[D_1, D_2] = D_1 D_2 - D_2 D_1$ is a differential operator of order $\le i + j - 1$. (In particular, $[T_X, T_X] \subset T_X$.)
(4)   Show that $[Diff^i_X, Diff^j_X]$ annihilates the constant functions k.

1.3 By definition, there is a natural inclusion $Diff^i_X \hookrightarrow Diff^j_X$ for $i \le j$. A differential operator of finite order D is said to *have no constant terms* if D annihilates the ground field k. By the symbol $diff^i_X$ we denote the $\mathcal{O}_X$-submodule of $Diff^i_X$ consisting of the differential operators without constant terms. The two natural inclusions $\mathcal{O}_X = Diff^0_X \hookrightarrow Diff^i_X$ and $diff^i_X \hookrightarrow Diff^i_X$ give a canonical direct sum decomposition $Diff^i_X \simeq \mathcal{O}_X \oplus diff^i_X$ (more explicitly, the isomorphism is given by $\xi \mapsto (\xi(1), \xi - \xi(1))$, where we regard $\xi(1) \in \mathcal{O}_X$ as a differential operator of order zero). By (1.2.2), $diff^1_X$ is equal to T_X, which is a Lie algebra with respect to the Lie bracket $[,]$ by (1.2.3). $Diff_X = \bigcup Diff^i_X$ is an envelope algebra of T_X.

There is another description of the sheaf $\mathit{Diff}^{\,i}_X$. Indeed, you can show that $\mathit{Diff}^{\,i}_X$ is isomorphic to $\mathcal{H}om_{\mathcal{O}_X}(\mathcal{O}_X \otimes \mathcal{O}_X / I^{i+1}_{\Delta_X}, \mathcal{O}_X)$ as a left $\mathcal{O}_X$-module. The natural injection $\mathrm{pr}^*_1 : \mathcal{O}_X \hookrightarrow \mathcal{O}_X \otimes \mathcal{O}_X$, along with the restriction map $\mathcal{O}_X \otimes \mathcal{O}_X \to \mathcal{O}_{\Delta_X} \simeq \mathcal{O}_X$ gives a canonical direct sum decomposition $\mathcal{O}_X \otimes \mathcal{O}_X = \mathcal{O}_X \oplus I_\Delta$, thus inducing $\mathcal{O}_X \otimes \mathcal{O}_X / I^{i+1}_{\Delta_X} = \mathcal{O}_X \oplus I_{\Delta_X} / I^{i+1}_{\Delta_X}$. The second factor $I_{\Delta_X} / I^{i+1}_{\Delta_X}$, the sheaf of i-th jets, has a natural decreasing filtration $\{ I^j_{\Delta_X} / I^{i+1}_{\Delta_X} \}$, $j = 1, \ldots, i$, of which the associated graded module is

$$\bigoplus_{j=1}^{i} I^j_{\Delta_X} / I^{j+1}_{\Delta_X} \simeq \bigoplus_{j=1}^{i} S^j \Omega^1_X.$$

Here the symbol S^j stands for the j-th symmetric tensor, that is the quotient module of the j-th tensor power under the natural $\mathfrak{S}_j$-action. Accordingly the dual module $\mathit{diff}^{\,i}_X$ has a natural increasing filtration, which is nothing but $\{ \mathit{diff}^{\,j}_X \}$. The associated graded module $\bigoplus \mathit{diff}^{\,j}_X / \mathit{diff}^{\,j-1}_X$ is thus naturally isomorphic to $\bigoplus \mathcal{H}om_{\mathcal{O}_X}(S^j \Omega^1_X, \mathcal{O}_X)$. Be careful: $\mathit{diff}^{\,1}_X = \mathcal{H}om_{\mathcal{O}_X}(\Omega^1_X, \mathcal{O}_X) = S^1 T_X$ by definition, but $S^i T_X$ is *not* isomorphic to $\mathcal{H}om_{\mathcal{O}_X}(S^i \Omega^1_X, \mathcal{O}_X)$, in general.

1.4 Exercise   (1) Let $ch\ k = p > 0$, $X = \mathbb{A}^1$. Show that $(I_{\Delta_X} / I^{i+1}_{\Delta_X})_o \simeq \bigoplus_{j=1}^{i} \mathcal{O}_{X,o}(\delta x)^j$, where x is the natural coordinate of $\mathbb{A}^1$ and $\delta x = \delta(1 \otimes x) = 1 \otimes x - x \otimes 1$. Determine the action of $\left(\frac{\partial}{\partial x} \right)^j$ on $I_{\Delta_X} / I^i_{\Delta_X}$, $j \leq i$ to see the action of $\left(\frac{\partial}{\partial x} \right)^p$ is trivial. From this fact, derive that $S^p T_X$ is not isomorphic to $\mathcal{H}om_{\mathcal{O}_X}(S^p \Omega^1_X, \mathcal{O}_X)$.

(2) Think of $\mathbb{A}^1$ as a scheme $\operatorname{Spec} \mathbb{Z}[x]$ defined over $\mathbb{Z}$. Show that the differential operator $(\partial / \partial x)^{[i]} = (1/i!)(\partial / \partial x)^i$ of order i defined over $\mathbb{Q} \otimes \mathbb{A}^1$ is actually a differential operator on $\mathbb{A}^1$. This operator is called the divided i-th power of $\partial / \partial x$.

(3) Define the divided powers of $\partial / \partial x_j$ for local parameters $x_1, \ldots, x_n$ of a smooth $\mathbb{Z}$-scheme. Show that the divided powers generate the sheaf of differential operators.

1.5 *Let k be of characteristic $p > 0$ in the subsequent subsections.*

For a derivation ξ, you easily show the *Leibnitz identity* $\xi(a^i) = ia^{i-1}\xi(a)$ by induction. In particular, ξ annihilates $\mathcal{O}^p_X = \{ a^p \mid a \in \mathcal{O}_X \}$. Let $\mathcal{F} \subset T_X$ be a subbundle (i.e. $\mathcal{F}$ and $T_X / \mathcal{F}$ are both locally free). Then

$$Ann(\mathcal{F}) = \{ a \in \mathcal{O}_X \mid \xi(a) = 0, \forall \xi \in \mathcal{F} \}$$

is a subring of $\mathcal{O}_X$ and contains $\mathcal{O}^p_X = \mathcal{O}_{X^{(1)}}$, where $\varphi : X \to X^{(1)}$ is the geometric Frobenius. This shows that $\xi \in T_X$ is an $\mathcal{O}^p_X$-linear transformation, so that $T_X = T_{X/X^{(1)}} := \{ \xi \in T_X \mid \xi \text{ is } \mathcal{O}_{X^{(1)}}\text{-linear} \}$. In fact, $\xi(a^p b) = b\xi(a^p) + a^p\xi(b) = b \cdot 0 + a^p\xi(b)$.

Given a morphism $f : X \to Y$, we define the sheaf of relative differential operators by

$$Diff_{X/Y} = \{\xi \in Diff_X \mid \xi(\mathcal{O}_Y) = 0\} \simeq \lim_{m \to \infty} \mathcal{H}om_{\mathcal{O}_X}(\mathcal{O}_X \otimes_{\mathcal{O}_Y} \mathcal{O}_X / I^m_{\Delta_X}, \mathcal{O}_X).$$

In particular, when $Y = X^{(1)}$, you can easily show that $Diff_{X/X^{(1)}}$ is a locally free sheaf of rank p^n, $n = \dim X$; indeed, if $(x_1, \ldots, x_n)$ is a local parameter system, $Diff_{X/X^{(1)}}$ is locally generated by

$$\prod_{j=1}^{n} \left(\frac{\partial}{\partial x_j}\right)^{i_j}, \quad 0 \leq i_j \leq p - 1,$$

and hence is identical with the left $\mathcal{O}_X$-subalgebra $\mathcal{O}_X[T_X]$ generated by T_X. Furthermore, you also verify the identity

$$Diff_{X/X^{(1)}} = \mathcal{H}om_{\mathcal{O}_X}(\mathcal{O}_X \otimes_{\mathcal{O}_X^p} \mathcal{O}_X, \mathcal{O}_X).$$

If ξ is a non-zero derivation, you easily check that ξ^i is a differential operator of order exactly i for $i = 1, \ldots, p-1$ (i.e. ξ^i is of order $\leq i$ but not of order $\leq i-1$). However, ξ^p is a derivation, i.e. a differential operator of order one. This follows from the following observation. By the Leibnitz rule, you have

$$\xi^i(ab) = \sum_{\alpha=0}^{i} \binom{i}{\alpha} \xi^\alpha(a)\xi^{i-\alpha}(b),$$

hence for $i = p$, only two terms $\alpha = 0, p$ survive, to yield $\xi^p(ab) = a\xi^p(b) + b\xi^p(a)$. Thus $\times^p : \xi \mapsto \xi^p$ is an operator: $T_X \to T_X$, called the *p-th power*.

1.6 An $\mathcal{O}_X$-submodule $\mathcal{F} \subset T_X$ is said to be *saturated* if $T_X/\mathcal{F}$ is torsion free. For a given submodule $\mathcal{F}$, we can define its saturation $\overline{\mathcal{F}}$ as the kernel of the natural surjection $T_X \to (T_X/\mathcal{F})/(torsion)$. $\overline{\mathcal{F}} \supset \mathcal{F}$ and is identical with $\mathcal{F}$ at a generic point of X. $\mathcal{F} \subset T_X$ is called a *subbundle* if $\mathcal{F}$ and the quotient $T_X/\mathcal{F}$ are both locally free. A saturated subsheaf $\mathcal{F} \subset T_X$ is always a subbundle if restricted to an open subset $X \setminus Y$, where Y is a closed subset of codimension ≥ 2 (for general properties of torsion free sheaves, see Section 2).

Let $\mathcal{F}$ be a saturated subsheaf of T_X. $\mathcal{F} \subset T_X$ is called *involutive* if it is closed under Lie bracket, i.e. $[\mathcal{F}, \mathcal{F}] \subset \mathcal{F}$. $\mathcal{F}$ is said to be *p-closed* if $\xi^p \in \mathcal{F}$ for every $\xi \in \mathcal{F}$ (denoted symbolically by $\mathcal{F}^p \subset \mathcal{F}$). By a *foliation*, we mean a saturated subsheaf $\mathcal{F} \subset T_X$ which is involutive and p-closed. A *smooth* foliation is, by definition, a foliation which is a subbundle. Any foliation is smooth in codimension one (that is, outside a closed subset of codimension ≥ 2).

1.7 Proposition *Let $\mathcal{F}$ be a subbundle of of rank r in T_X. Then $\mathcal{F}$ is a (smooth) foliation if and only if $\mathcal{F}$ generates an $\mathcal{O}_X$-algebra of rank p^r in $Diff_X = \bigcup_i Diff^i_X$.*

Proof. Let $\mathcal{O}_X[\mathcal{F}]$ be the algebra generated by $\mathcal{F}$ and let $\mathcal{O}_X[\mathcal{F}]_i$ be the submodule generated by the products of i or less elements of $\mathcal{F}$. Let $\xi_1, \ldots, \xi_r$ be a local basis of $\mathcal{F}$. Then the involutiveness shows that

$$\xi_{\iota(1)}\xi_{\iota(2)} \cdots \xi_{\iota(i)} \equiv \xi_{\iota(\sigma(1))}\xi_{\iota(\sigma(2))} \cdots \xi_{\iota(\sigma(i))} \bmod \mathcal{O}_X[\mathcal{F}]_{i-1},$$

for every permutation σ of $\{1, \ldots, i\}$ and every mapping $\iota : \{1, \ldots, i\} \to \{1, \ldots, r\}$. Thus the quotient $\mathcal{O}_X[\mathcal{F}]_i/\mathcal{O}_X[\mathcal{F}]_{i-1}$ is generated by $\xi_1^{i_1}\xi_2^{i_2} \cdots \xi_r^{i_r}$, $i_1+i_2+\ldots+i_r = i$. On the other hand, if some index, say, i_1 is p or larger, the p-closedness allows us to substitute the term $\xi_1^{i_1}$ by an element of smaller degree. Eventually we infer that a local basis of $\mathcal{O}_X[F]$ is given by $\xi_1^{i_1}\xi_2^{i_2} \cdots \xi_r^{i_r}$, $0 \le i_1, \ldots, i_r \le p-1$, hence the assertion. $\square$

1.8 Exercise Check that the above elements $\xi_1^{i_1}\xi_2^{i_2} \cdots \xi_r^{i_r}$, $0 \le i_1, \ldots, i_r \le p-1$ are independent over $\mathcal{O}_X$.

Let $Ann(\mathcal{F})$ be the elements $\in \mathcal{O}_X$ that is annihilated by $\mathcal{F} \subset T_X$, i.e.

$$Ann(\mathcal{F}) = \{a \in \mathcal{O}_X \mid \xi(a) = 0, \forall \xi \in \mathcal{F}\}.$$

Clearly $Ann(\mathcal{F})$ contains $\mathcal{O}_X^p$, and is an $\mathcal{O}_X^p$-subalgebra of $\mathcal{O}_X$ by the Leibnitz law. Furthermore it is integrally closed in $\mathcal{O}_X$. Indeed, if $a \in \mathcal{O}_X$ is in the fraction field of $Ann(\mathcal{F})$, then there is an open subset U such that $a \in Ann(\mathcal{F})(U)$, so that $\mathcal{F}(a) = 0 \in \mathcal{O}_U$, which of course implies the vanishing of $\mathcal{F}(a)$ as an element of $\mathcal{O}_X$. Hence $a \in Ann(\mathcal{F})$ by definition. Thus $\operatorname{Spec} Ann(\mathcal{F})$ gives rise to a normal variety $Y = X/\mathcal{F}$, which factors the Frobenius morphism $\varphi : X \to X^{(1)}$:

$$
\begin{array}{ccc}
X & == & X \\
{\scriptstyle \pi}\downarrow & & \downarrow{\scriptstyle \varphi} \\
Y & \longrightarrow & X^{(1)}.
\end{array}
$$

Conversely, given the commutative diagram above, we define $\mathcal{F} \subset T_X$ by

$$\mathcal{F} = Ann(\mathcal{O}_Y) \overset{def}{=} \{\xi \in T_X; \xi(\mathcal{O}_Y) = 0\}.$$

Then you easily check that $\mathcal{F}$ is saturated in T_X, involutive and p-closed. Furthermore, it is clear by definition that $Ann(Ann(\mathcal{F})) \supset \mathcal{F}$, $Ann(Ann(\mathcal{O}_Y)) \supset \mathcal{O}_Y$.

1.9 Proposition (1) *Let $\mathcal{F}$ be a foliation of rank r, and put $Y = X/\mathcal{F}$. Then $[k(X) : k(Y)] = p^r$.*

(2) *Given a foliation $\mathcal{F}$ and a normal variety between X and $X^{(1)}$, we have $Ann(Ann(\mathcal{F})) = \mathcal{F}$, $Ann(Ann(\mathcal{O}_Y)) = \mathcal{O}_Y$. Hence there is a one-to-one correspondence between the foliations on X and the normal varieties Y between X and $X^{(1)}$, by the correspondence $\mathcal{F} \mapsto Y = X/\mathcal{F}$ and the inverse correspondence $Y \mapsto \mathcal{F} = Ann(\mathcal{O}_Y)$.*

(3) *$\mathcal{O}_Y$ is a regular ring if and only if $\mathcal{F}$ is a smooth foliation.*

Proof. To show (1), it suffices to prove that $[\mathcal{O}_X : \mathcal{O}_Y] = p^r$ at a generic point of Y. In particular, we may assume that $\mathcal{F}$ is (locally) smooth. Let $\xi_1, \ldots, \xi_r$ be a local basis of $\mathcal{F}$ and $\xi_1, \ldots, \xi_n$ a local basis of T_X. In this case, $\mathcal{O}_X[\mathcal{F}]$ is locally free with basis $\xi_1^{i_1} \cdots \xi_r^{i_r}$, $0 \leq i_s < p$. On the other hand, $\mathcal{O}_X$ is a locally free $\mathcal{O}_X^p$-module of rank p^n, so that $\mathcal{O}_X \otimes_{\mathcal{O}_X^p} \mathcal{O}_X$ is naturally viewed as a locally free left $\mathcal{O}_X$-module of rank p^n. There is a natural identification $Diff_{X/X^{(1)}} = \mathcal{H}om_{\mathcal{O}_X}(\mathcal{O}_X \otimes_{\mathcal{O}_X^p} \mathcal{O}_X, \mathcal{O}_X)$. Hence the $\mathcal{O}_X$-submodule $\widetilde{Ann}(\mathcal{F}) \subset \mathcal{O}_X \otimes_{\mathcal{O}_X^p} \mathcal{O}_X$ that is annihilated by the submodule $\mathcal{O}_X[\mathcal{F}]$ of rank p^r has rank p^{n-r}. You easily check the equality $\widetilde{Ann}(\mathcal{F}) = \mathcal{O}_X \otimes_{\mathcal{O}_X^p} Ann(\mathcal{F})$, and deduce that $Ann(\mathcal{F})$ is of rank p^{n-r} as an $\mathcal{O}_X^p$-module. This proves (1).   □

(2) Suppose we are given $\mathcal{F}$ of rank r. By (1), $[\mathcal{O}_Y : \mathcal{O}_{X^{(1)}}] = p^{n-r}$. Let $x_1, \ldots, x_n$ be local parameters of X at a general point. $x_1^p, \ldots, x_n^p$ are local parameters of $X^{(1)}$ at the corresponding general point. Thus, at a general point on Y, we can choose $y_1, \ldots, y_{n-r} \in \mathcal{O}_Y$ such that $x_1^p, \ldots, x_n^p, y_1, \ldots, y_{n-r}$ are local parameters of $\mathcal{O}_Y$ (by renumbering $1, \ldots, n$ if necessary). Then $x_1, \ldots, x_r, y_1, \ldots, y_{n-r}$ is a second local parameter system of $\mathcal{O}_X$, and clearly $Ann(\mathcal{O}_Y) = \sum_{i=1}^r \mathcal{O}_X(\partial/\partial x_i)$, of rank r. Since $\mathcal{F}$ is contained in $Ann(\mathcal{O}_Y)$ and is saturated of rank r, this implies $\mathcal{F} = Ann(\mathcal{O}_Y) = Ann(Ann(\mathcal{F}))$.

Conversely, if you start with $\mathcal{O}_Y$, you repeat the same argument to conclude that $\mathcal{O}_Y = Ann(Ann(\mathcal{O}_Y))$ by using of the normality of $\mathcal{O}_Y$.   □

(3) Suppose that $Y = X/\mathcal{F}$ is smooth. Then, since $[\mathcal{O}_Y : \mathcal{O}_X^p] = p^{n-r}$, there exists a local parameter of the form $(x_1^p, \ldots, x_r^p, x_{r+1}, \ldots, x_n)$ of $\mathcal{O}_Y$. Then it is clear that $\mathcal{F} = \oplus \mathcal{O}_X \frac{\partial}{\partial x_i}$, $i = 1, \ldots, r$.

Suppose conversely that $\mathcal{F}$ is a subbundle with a local basis $\xi_1, \ldots, \xi_r$, where $\xi_1, \ldots, \xi_n$ is a local basis of T_X. Think of $\mathcal{F}$ as an locally free $\mathcal{O}_X^p$-module of rank $p^n r$ and express ξ_i as a sum of the terms of the form $x_1^{j_1} \cdots x_n^{j_n} \frac{\partial}{\partial x_m}$, $j_k \leq p-1$, $m = 1, \ldots, n$. Then your can write down the condition for $a = \sum a_J x_1^{j_1} \cdots x_n^{j_n}$, $a_J \in \mathcal{O}_X^p$ to be killed by the ξ_i. You can find solutions by inductively determine the coefficients a_J, and check that $Ann(\mathcal{F})$ contains $p-r$ solutions $y_1, \ldots, y_{n-r} \in \mathfrak{M} = (x_1, \ldots, x_n)$ that are independent in $\mathfrak{M}/\mathfrak{M}^2$. This proves that $(x_1^p, \ldots, x_r^p, y_1, \ldots, y_{n-r})$ is a local parameter system of Y (after renumbering $x_1, \ldots, x_n$, if necessary).   □

1.10 As long as $\mathcal{F}$ is a subbundle of T_X, the "quotient" variety $Y = X/\mathcal{F}$ is smooth, so that the cotangent sheaf Ω_Y^1 is well defined. In terms of a suitable local coordinate, the natural homomorphism $T_X \to \pi^* T_Y$ is clearly given by

$$\frac{\partial}{\partial z_i} \mapsto \begin{cases} 0 & i = 1, \ldots, r \\ \frac{\partial}{\partial z_i} & i = r+1, \ldots, n. \end{cases}$$

Hence, at least on the open set $U \subset X$ on which $\mathcal{F}$ is a subbundle of T_X, we have a natural exact sequence

$$0 \to \mathcal{F} \to T_X \to \pi^* T_Y \to \mathcal{F}^p \to 0,$$

where $\mathcal{F}^p$ is a locally free $\mathcal{O}_X$-module generated by the elements of the form $\frac{\partial}{\partial z_i^p}, i = 1, \ldots, r$. If the transition matrices of $\mathcal{F}$ are $A_{\mu\nu} = (a_{\mu\nu,ij}) \in \mathbf{GL}(\mathcal{O}_{U_\mu \cap U_\nu}^{\oplus r})$, then the transition matrices of $\mathcal{F}^p$ are given by $(a_{\mu\nu,ij}^p)$. In particular, $-q^* K_Y = p(\det \mathcal{F}) + (\det T_X / \mathcal{F}) = -K_X + (p-1)(\det \mathcal{F})$.

1.11 Corollary   *Given a foliation $\mathcal{F}$ on a smooth projective variety X over an algebraically closed field k of characteristic $p > 0$, we have the identity*

$$-\pi^* K_Y = K_X + (p-1)(\det \mathcal{F})$$

on the open subset U on which $\mathcal{F}$ is a subbundle of T_X.

Here det denotes the divisor attached to the highest exterior product of a locally free sheaf.

1.12 Corollary   *Let (X, H) be a smooth, polarized projective variety over a field of characteristic $p > 0$. Assume that the tangent sheaf T_X contains a saturated subsheaf $\mathcal{F}$ such that*
(1) $\mathcal{F}$ is involutive and p-closed, and that
(2) $\big(-K_X + (p-1)(\det \mathcal{F})\big) H^{n-1} > 0$.

Then X contains a rational curve C passing through a general point of X such that

$$(C, H) \leq \frac{2p H^n}{\big(-K_X + (p-1)(\det \mathcal{F})\big) H^{n-1}}.$$

In particular X is uniruled.

Proof.   Let $\pi : X \to Y$ be the natural map. Let $H^{(1)}$ be an ample divisor on $X^{(1)}$ such that $\varphi^* H^{(1)} = pH$. Take a suitable positive integer m such that $mH^{(1)}$ is very ample. Let $\Gamma^{(1)} \subset X^{(1)}$ be a general complete intersection curve cut out by divisors $\in |mH^{(1)}|$, and $\Gamma^* \subset Y$, $\Gamma \subset X$ its inverse images (with reduced structure). The natural projection: $\Gamma \to \Gamma^{(1)}$ is nothing but the Frobenius of mapping degree p, and Γ is numerically equivalent to $m^{n-1} H^{n-1}$ as a 1-cycle on X. Let d be the mapping degree of $\pi|_\Gamma : \Gamma \to \Gamma^*$, which is either 1 or p. Then we have:

$$\begin{aligned}
d(\Gamma^*, K_Y) &= (\Gamma, -\pi^* K_Y) \\
&= (\Gamma, -K_X + (p-1)(\det \mathcal{F})) \\
&= \big(m^{n-1} H^{n-1}, -K_X + (p-1)(\det \mathcal{F})\big) > 0.
\end{aligned}$$

Hence you can apply (II.2.6) to Y, to infer that Y is covered by rational curves. The geometric Frobenius does not affect the property of being uniruled (the only difference between Z and $Z^{(-1)}$ lies in k-scheme structure), and so $Y^{(-1)}$ is also uniruled. In view of the natural finite surjection $Y^{(-1)} \to X$, we conclude that X is also uniruled. $\square$

The above new criterion for uniruledness was given by N. Shepherd-Barron (see his contribution in [Kollár *et al.*]), but a similar argument had been used in [Rudakov-Shafarevich] in order to prove the non-existence of global vector fields on a K3 surface.

1.13 Corollary [Rudakov-Shafarevich]   *Assume that a K3 surface X admits a non-zero vector field. Then X would be unirational, and the Picard number $\rho(X)$ would attain the possible maximum $b_2(X)$, the second Betti number (i.e. X is "supersingular").*

Proof. By a K3 surface over an algebraically closed field of arbitrary characteristic, we mean a smooth projective surface with $q(X) = \dim H^1(X, \mathcal{O}_X) = 0, K_X \sim 0$. The second condition gives an isomorphism $T_X \simeq \Omega^1_X$, and if X carries a non-zero vector field, we have $H^0(X, \Omega^1_X) \neq 0$. This is impossible in characteristic zero because of the Hodge symmetry $\dim H^1(X, \mathcal{O}_X) = \dim H^0(X, \Omega^1_X)$, but here we work in characteristic p. By Noether's formula and the Serre duality, we calculate the Euler number as $c_2(X) = c_2(T_X) = 24$.

Take a non-zero vector field ξ, which generates the subsheaf $\mathcal{O}_X \xi \subset T_X$. Let $\mathcal{F}$ be its saturation in T_X. It is well known that an invertible saturated subsheaf of the tangent bundle is always involutive. Let us check that $\mathcal{F}$ is p-closed. First note that ξ has non-trivial zeros by $c_2(X) > 0$ (essentially by the Lefschetz point formula, or alternatively, by the very definition of c_2; in general, if a global section ξ of a vector bundle $\mathcal{E}$ of rank r vanishes on an r-cocycle Z, $c_r(\mathcal{E}) = [Z]$). Suppose that $\mathcal{F}$ were not p-closed. Then ξ^p would be a global vector field not contained in $\mathcal{F}$. Hence $\xi \wedge \xi^p$ would define a global section of $\bigwedge^2 T_X \simeq \mathcal{O}_X$ with non-trivial zero, which is clearly impossible, proving the p-closedness of $\mathcal{F}$.

Consider the following two cases separately:

Case 1. $c_1(\mathcal{F}) = \det \mathcal{F} > 0$. In this case, Corollary 1.11 applies to show that $-\pi^* K_Y = (p-1) \det \mathcal{F}$ is effective $\neq 0$. Hence the Enriques classification of surfaces in characteristic p [Bombieri-Mumford] tells that Y is a ruled surface. On the other hand, since $q(X) = 0$, there is no nontrivial morphism from X to a curve of genus ≥ 1, thereby ruling out the possibility of Y being a non-rational ruled surface. Thus Y is rational and X unirational.

Case 2. $c_1(\mathcal{F}) = 0$. It suffices to show that $Y = X/\mathcal{F}$ is ruled or rational. Suppose otherwise. Let Y^* be a minimal resolution of Y. The canonical bundle formula (1.11) shows that $K_Y = 0$ on the smooth locus of Y. Then $-K_{Y^*}$ is an effective divisor supported by the exceptional set with respect to the projection $Y^* \to Y$. Thus $-K_{Y^*}$ is effective.

Since Y^* is neither rational nor ruled by assumption, it has non-negative Kodaira dimension by the Enriques classification, which is possible only when the effective divisor $-K_{Y^*}$ is numerically trivial and Y has only rational double points. Since Y is dominated by X, we have $q(Y^*) = q(Y) = 0$, which means that Y^* is a minimal K3 surface with $b_2(Y^*) = 22$ or an Enriques surface with $b_2(Y^*) = 10$.

Hence the second Betti number $b_2(Y)$ of Y is bounded by $b_2(Y^*) = 22$, the equality being attained if only if Y is a smooth K3 surface. In the meantime, the purely inseparable morphism with connected fibres $X \to Y$ is bijective, and hence a homeomorphism, so that $b_2(Y) = b_2(X) = 22$, yielding the smoothness of Y and hence of $\mathcal{F}$. But $\mathcal{F} = \mathcal{O}_X \xi \subset T_X$ is a subbundle with trivial c_1 if and only if ξ is nowhere vanishing, which is impossible because $c_2(X) > 0$.   □

1.14 Corollary   *Let X be a smooth projective variety with $K_X \approx 0$. If X admits a non-zero global vector field ξ which vanishes along a non-zero effective divisor D, then X is uniruled.*

Proof. $\xi^i, i = 1, 2, \ldots$, generates a subring $R \subset \textit{Diff}_X$, and put $\mathcal{F} = R \cap T_X$. It is easy to check that $\mathcal{F}$ is a foliation and that $c_1(\mathcal{F}) \geq D > 0$.   □

Assume that a smooth surface X has negative Euler characteristic $\chi(X, \mathcal{O}_X)$. In characteristic zero, such X is necessarily ruled and its Kodaira dimension is $-\infty$. In positive characteristics, on the other hand, X can be of general type, but still we can show that such X must be uniruled (a conjecture of M. Raynaud):

1.15 Corollary [Shepherd-Barron]   *Let X be a smooth projective surface of general type over an algebraically closed field of characteristic $p \geq 3$. If $\chi(X, \mathcal{O}_X) < 0$, then X is uniruled.*

The outline of the proof is this. We may assume that X is a minimal surface with $K_X^2 > 0$. From the negativity of the Euler characteristic, we derive that X is a fibre space over a curve C of genus ≥ 1. Let $\mathcal{L} \subset T_X$ be the saturation of the relative tangent sheaf $T_{X/C}$. Put $c_1(\mathcal{L})K_X = tK_X^2$, $t \in \mathbb{Q}$. From the exact sequence

$$0 \to \mathcal{L} \to T_X \to T_X/\mathcal{L} \to 0,$$

we get

$$c_2(T_X) = c_2(X) \geq -K_X c_1(\mathcal{L}) - c_1(\mathcal{L})^2 \geq -t(1+t)K_X^2.$$

From the inclusion $T_X/\mathcal{L} \to \pi^* T_C$, we infer that $t \geq -1$. In view of the Noether formula $12\chi(X, \mathcal{O}_X) = K_X^2 + c_2(X)$, we get $c_2(X) < -K_X^2$, and hence

$$t > \frac{\sqrt{5} - 1}{2} > \frac{1}{2}.$$

In view of this inequality, we infer that $\mathcal{L}$ is a foliation and that the pull-back of $K_{X/\mathcal{L}}$, which is linearly equivalent to $-(p-1)c_1(\mathcal{L}) + K_X$, negatively intersects K_X. This shows that $X/\mathcal{L}$ is ruled and hence X is uniruled.   □

2   Semistable sheaves and generically semipositive sheaves

2.1 In order to apply the uniruledness criterion (1.12) to the geometry in characteristic zero, we need the notion of semistability of tangent bundles.

Let X be a variety over an algebraically closed field k. A coherent sheaf $\mathcal{E}$ is said to be *torsion-free* if it is a torsion-free $\mathcal{O}_X$-module, or equivalently, if the restriction map $H^0(U, \mathcal{E}) \to H^0(V, \mathcal{E})$ is always injective for arbitrary open subsets U, V with $\emptyset \neq V \subset U \subset X$. Any coherent sheaf is locally free at a general point. By the rank of a torsion free sheaf, we mean the one at a generic point. If X is smooth, then it is locally free in codimension one i.e. outside a closed subset of codimension ≥ 2.

For a given coherent sheaf $\mathcal{E}$, the dual sheaf $\mathcal{E}^* = \mathcal{H}om_{\mathcal{O}_X}(\mathcal{E}, \mathcal{O}_X)$ is torsion-free. A torsion free sheaf is naturally a subsheaf of its double dual $\mathcal{E}^{**} = (\mathcal{E}^*)^*$. If $\mathcal{E} = \mathcal{E}^{**}$, $\mathcal{E}$ is called a *reflexive sheaf*. A reflexive sheaf on a smooth variety is locally free in codimension two (i.e. outside a subset of codimension ≥ 3).

2.2 Exercise   (1) Show that a coherent sheaf on a variety is torsion free if and only if it is a subsheaf of a locally free sheaf.

(2) A torsion free sheaf $\mathcal{E}$ is reflexive if and only if there exists a locally free sheaf $\mathcal{F}$ such that $\mathcal{E} \subset \mathcal{F}$ and that $\mathcal{F}/\mathcal{E}$ is torsion free.

(3) Show that a torsion free sheaf on a smooth curve is locally free and that a reflexive sheaf on a smooth surface is locally free.

2.3 Let X be a smooth projective variety of dimension n with ample divisor H. Since a torsion free sheaf $\mathcal{E}$ is locally free in codimension one, the determinant $\det \mathcal{E}$ is a line bundle in codimension one, and so the intersection number $c_1(\mathcal{E})H^{n-1}$ is well-defined. (For example, take m so that mH is very ample. Take a general complete intersection curve C cut out by hyperplane sections $\sim mH$ such that $\mathcal{E}$ is locally free around C. Then define $c_1(\mathcal{E})H^{n-1}$ to be $(1/m)^{n-1} \deg(\mathcal{E}|C)$. This does not depend on the choice of m or C.) Given a torsion free sheaf $\mathcal{E} \neq 0$ on X, we define the *normalized degree* $\mu(\mathcal{E})$ with respect to H by

$$\mu(\mathcal{E}) = \frac{c_1(\mathcal{E})H^{n-1}}{\operatorname{rank}\mathcal{E}} \in \frac{1}{\operatorname{rank}\mathcal{E}}\mathbb{Z} \subset \mathbb{Q}.$$

$\mathcal{E}$ is said to be *H-semistable* [resp. *H-stable*] if

$$\mu(\mathcal{F}) \leq \mu(\mathcal{E}) \ [\text{resp.} < \mu(\mathcal{E})]$$

for every non-zero subsheaf $\mathcal{F} \subset \mathcal{E}$ [resp. for every saturated subsheaf $\mathcal{F} \neq 0, \mathcal{E}$]. When $\mathcal{E}$ is not H-semistable, $\mathcal{E}$ is called *H-unstable*. If $\mathbb{R}_+[H_1] = \mathbb{R}_+[H_2]$ in $N^1(X)$, then H_1-(semi)stability coincides with H_2-(semi)stability. Thus the notion of H-(semi)stability depends only on the ray $\mathbb{R}_+[H]$ generated by the numerical equivalence class $[H]$. Furthermore, tensor products with line bundles do not affect H-(semi)stability (i.e. $\mathcal{E}$ is H-(semi)stable if and only if so is $\mathcal{E} \otimes L$, where L is a line bundle). In this sense semistability is a property of $\mathbb{P}(\mathcal{E})$ rather than a property of $\mathcal{E}$ itself.

2.4 Proposition   *Fix a torsion free sheaf $\mathcal{E}$ of rank r on a projective polarized variety (X, H). Then the set of normalized degrees $\{\mu(\mathcal{F}) \mid 0 \neq \mathcal{F} \subset \mathcal{E}\} \subset \frac{1}{r!}\mathbb{Z} \subset \mathbb{Q}$ is bounded from above by a constant depending only on $(X, H, \mathcal{E})$. Let μ_1 be the maximum of the values. Then the set of subsheaves $\{\mathcal{F} \subset \mathcal{E} \mid \mu(\mathcal{F}) = \mu_1\}$ contains the largest element with respect to the inclusion relation. (Call this largest element the "maximal destabilizer".)*

Proof.   Let $\mathcal{E}^*$ denote $\mathcal{H}om(\mathcal{E}, \mathcal{O}_X)$ (the dual bundle when $\mathcal{E}$ is a vector bundle). Take a positive integer m such that $\mathcal{E}^*(mH)$ is generated by global sections. Thus there is a surjection $H^0(X, \mathcal{E}^*(mH)) \otimes_k \mathcal{O}_X \to \mathcal{E}^*(mH)$. Take the dual to get an injection $\mathcal{E}^{**} \to \mathcal{O}_X(mH)^N$, where $N = \dim H^0(X, \mathcal{E}^*(mH))$. Since $\mathcal{E}$ is torsion free, the natural homomorphism $\mathcal{E} \to \mathcal{E}^{**}$ is an injection, being an isomorphism on the open subset on which $\mathcal{E}$ is locally free. Thus we can view $\mathcal{E}$ as a subsheaf of $\mathcal{O}_X(mH)^N$. Pick up an arbitrary non-zero subsheaf $\mathcal{F} \subset \mathcal{E}$ of rank, say s. Then by composing the injection $\mathcal{F} \to \mathcal{O}_X(mH)^N$ with the projection $\mathcal{O}_X(mH)^N \to \mathcal{O}_X(mH)^s$ onto general s factors, we get an injection $\mathcal{F} \hookrightarrow \mathcal{O}_X(mH)^s$. It is now clear that $\det \mathcal{F} \leq msH$, so that $\mu(\mathcal{F}) \leq mH^n$. Thus the $\mu(\mathcal{F})$ is bounded from above by mH^n, independent of the choice of $\mathcal{F}$.

In order to prove the second statement, it suffices to show that $\mu(\mathcal{F}' + \mathcal{F}'') = \mu_1$ whenever $\mu(\mathcal{F}') = \mu(\mathcal{F}'') = \mu_1$. From the exact sequence

$$0 \longrightarrow \mathcal{F}' \cap \mathcal{F}'' \longrightarrow \mathcal{F}' \oplus \mathcal{F}'' \longrightarrow \mathcal{F}' + \mathcal{F}'' \longrightarrow 0,$$

we get

$$c_1(\mathcal{F}' + \mathcal{F}'') = c_1(\mathcal{F}') + c_1(F'') - c_1(\mathcal{F}' \cap \mathcal{F}'').$$

By taking the intersection number with H^{n-1}, we have

$$r\mu(\mathcal{F}' + \mathcal{F}'') = r'\mu(\mathcal{F}') + r''\mu(\mathcal{F}'') - r^*\mu(\mathcal{F}' \cap \mathcal{F}'')$$
$$= (r' + r'')\mu_1 - r^*\mu(\mathcal{F}' \cap \mathcal{F}''),$$

where r, r', r'' and r^* stand for the ranks of $\mathcal{F}' + \mathcal{F}'', \mathcal{F}', \mathcal{F}''$ and $\mathcal{F}' \cap \mathcal{F}''$, respectively, so that $r^* = r' + r'' - r$. By the definition of μ_1, we have a trivial inequality $\mu(\mathcal{F}' \cap \mathcal{F}'') \leq \mu_1$, whence follows that $r\mu(\mathcal{F}' + \mathcal{F}'') \geq r\mu_1$. This implies that $\mu(\mathcal{F}' + \mathcal{F}'') = \mu(\mathcal{F}' \cap \mathcal{F}'') = \mu_1$ (unless $\mathcal{F}' \cap \mathcal{F}'' = 0$). $\qquad\square$

2.5   Thus $\mu_{\max}(\mathcal{E}) = \mu_1$ is an invariant of $(X, H, \mathcal{E})$. $\mathcal{E}$ is H-semistable if and only if its maximal destabilizer coincides with $\mathcal{E}$. The maximal destabilizer of a torsion free sheaf is saturated because of its maximality. If you divide out $\mathcal{E}$ by the maximal destabilizer $\mathcal{E}_1$, you easily check that $\mu_{\max}(\mathcal{E}/\mathcal{E}_1) < \mu_{\max}(\mathcal{E})$ as long as $\mathcal{E}$ is H-unstable. Reiterating this process, we get a unique filtration (the *Harder-Narasimhan filtration*)

$$0 = \mathbf{F}^0(\mathcal{E}) \subset \mathbf{F}^1(\mathcal{E}) \subset \cdots \subset \mathbf{F}^s(\mathcal{E}) = \mathcal{E}$$

in such a way that $\mathbf{F}^i(\mathcal{E})/\mathbf{F}^{i-1}(\mathcal{E})$ is the maximal destabilizer of $\mathcal{E}/\mathbf{F}^{i-1}(\mathcal{E})$. By construction, $\mathbf{Gr}^i = \mathbf{F}^i(\mathcal{E})/\mathbf{F}^{i-1}(\mathcal{E})$ is H-semistable. If the Harder-Narasimhan filtration $\mathbf{F}^{\cdot}(\mathcal{E})$ is non-trivial (i.e. $s \geq 2$), then

$$\mu_{\max}(\mathcal{E}) = \mu(\mathbf{Gr}^1) > \mu(\mathbf{Gr}^2) > \ldots > \mu(\mathbf{Gr}^s) = \mu_{\min}(\mathcal{E}),$$

where $\mu_{\min}(\mathcal{E}) = -\mu_{\max}(\mathcal{E}^*)$ is another invariant of $(X, H, \mathcal{E})$. $\mu_{\min}(\mathcal{E})$ is equal to the minimum of the normalized degrees of the non-trivial torsion free quotients of $\mathcal{E}$.

When the variety X is a curve, we use the terminology μ-(semi)stability instead of H-(semi)stability because the notion is independent of the choice of H in this case.

2.6 Exercise (1) Assume that $(X, H, \mathcal{E})$ is defined over a subfield $k_0 \subset k$. Assume that $k \cap \overline{k}_0$ is separable over k_0 (which is the case when k_0 is a finite field). Show that the Harder-Narasimhan filtration is defined over k_0.

(2) Let $\mathcal{E}_i$ be H-stable reflexive sheaves ($i = 1, 2$) with $\mu(\mathcal{E}_1) = \mu(\mathcal{E}_2)$. Prove that an $\mathcal{O}_X$-homomorphism $\mathcal{E}_1 \to \mathcal{E}_2$ is either an isomorphism or 0.

(3) Let $\mathcal{E}$ be an H-semistable sheaf. Show that there exists a filtration $0 = \mathcal{E}_0 \subset \mathcal{E}_1 \subset \ldots \subset \mathcal{E}_t = \mathcal{E}$ such that (a) $\mathcal{E}_i$ is saturated in $\mathcal{E}$, (b) $\mu(\mathcal{E}_i) = \mu(\mathcal{E})$, and (c) $\mathrm{Gr}_i = \mathcal{E}_i/\mathcal{E}_{i-1}$ is H-stable. When $\mathcal{E}$ is a vector bundle, show that $\sum_i \mathrm{Gr}_i$ is uniquely determined by $\mathcal{E}$.

Let H be a very ample divisor on a variety X of dimension ≥ 2, and $Y \in |H|$ a generic member. It is easy to see that, if $\mathcal{E}$ is an H-unstable sheaf, then $\mathcal{E}|Y$ is $(H|_Y)$-unstable. Equivalently, $\mathcal{E}$ is semistable if $\mathcal{E}|_Y$ is $H|_Y$-semistable. The converse is false in general.

2.7 Exercise (1) Show that the tangent bundle T_X on $X = \mathbb{P}^n$ is H-stable with $\mu(T_X) = \frac{n+1}{n}$, where H is the hyperplane. Check that $T_X|_H$ is an extension of $\mathcal{O}_H(1)$ by T_H, and hence unstable ($n \geq 2$).

However, if you replace H by a sufficiently high multiple, then the restriction of a semistable sheaf is semistable.

2.8 Theorem [Mehta-Ramanathan]   *Let X be a normal projective variety of dimension ≥ 2 and H an ample divisor. Let $\mathcal{E}$ be an H-semistable sheaf. Then, for a sufficiently large positive integer m, the restriction of $\mathcal{E}$ to a general member $Y \in |mH|$ is $H|_Y$-semistable. In particular, $\mathcal{E}$ is H-semistable if and only if so is its restriction to a general complete intersection curve cut out by sufficiently large multiples of H.*

We cannot give the proof of this extremely useful theorem in this lecture, but we would like to mention that this is another remarkable application of the theory of Hilbert schemes.

Thus H-semistability of $\mathcal{E}$ is checked by restricting $\mathcal{E}$ to complete intersection curves of large degree. Over a field of characteristic zero, the μ-semistability of a locally free sheaf on a smooth curve is rephrased by a certain numerical property.

To see this, let us introduce a new symbol. Let $\mathcal{E}$ be a vector bundle (locally free sheaf) of rank r on a projective variety and $\pi : \mathbb{P}(\mathcal{E}) = \operatorname{Proj} S^{\cdot}\mathcal{E} \to X$ the associated projective bundle with tautological divisor $\mathbf{1}_{\mathcal{E}}$. We denote by $\tilde{\mathbf{1}}_{\mathcal{E}}$ the *normalized tautological divisor* $\mathbf{1}_{\mathcal{E}} - (1/r)\pi^* \det \mathcal{E} \in \operatorname{Pic}(\mathbb{P}(\mathcal{E}))\otimes\mathbb{Q}$. The normalized tautological divisor stays the same if we replace $\mathcal{E}$ by $\mathcal{E}' = \mathcal{E}\otimes L$ for an arbitrary line bundle L. Indeed, $r\tilde{\mathbf{1}}_{\mathcal{E}}$ is nothing but the relative anti-canonical divisor $-K_{\mathbb{P}(\mathcal{E})/X}$, canonically attached to the projective bundle $\mathbb{P}(\mathcal{E})$. In this notation, we have:

2.9 Proposition   *Let C be a smooth projective curve over an algebraically closed field of characteristic zero. Let $\mathcal{E}$ be a locally free sheaf of rank r and $\pi : \mathbb{P}(\mathcal{E}) \to C$ the associated projective bundle, with $\tilde{\mathbf{1}}_{\mathcal{E}}$ denoting the normalized tautological divisor. Then the following six conditions are equivalent:*

(1)   *For any finite morphism $f : \overline{C} \to C$, $\overline{\mathcal{E}} = f^*\mathcal{E}$ is μ-semistable.*
(2) $\tilde{\mathbf{1}}_{\mathcal{E}}$ *is nef;*
(3) $\tilde{\mathbf{1}}_{\mathcal{E}} - \pi^*D$ *is not pseudo-effective for any $\mathbb{Q}$-divisor D of positive degree.*
(4) $\tilde{\mathbf{1}}_{\mathcal{E}} + \pi^*D$ *is ample, where D is some $\mathbb{Q}$-divisor of degree d, $0 < d < 1/r!$.*
(5) $\tilde{\mathbf{1}}_{\mathcal{E}} - \pi^*D$ *is not pseudo-effective, where D is some $\mathbb{Q}$-divisor of degree d,*
      $0 < d < 1/r!$.
(6) $\mathcal{E}$ *is μ-semistable;*

In positive characteristic, we have the implications (1) $\Leftrightarrow$ (2) $\Leftrightarrow$ (3) $\Rightarrow$ (4) $\Leftrightarrow$ (5) $\Rightarrow$ (6).

Proof.   (2) $\Leftrightarrow$ (3), as well as (4) $\Leftrightarrow$ (5), is immediate by the equalities between the intersection numbers:

$$\tilde{\mathbf{1}}_{\mathcal{E}}^2 = (\tilde{\mathbf{1}}_{\mathcal{E}} + \pi^*D, \tilde{\mathbf{1}}_{\mathcal{E}} - \pi^*D) = 0.$$

(1) $\Leftrightarrow$ (3)($\Leftrightarrow$ (2)). Assume that the condition (3) fails. Then there exists a curve $\Gamma \subset \mathbb{P}(\mathcal{E})$ such that $(\Gamma, \tilde{\mathbf{1}}_{\mathcal{E}}) < 0$. Γ cannot be a fibre of π because $\tilde{\mathbf{1}}_{\mathcal{E}}$ is relatively ample. Consider the base change $\overline{C} \times_C \mathbb{P}(\mathcal{E})$, which is identical with $\mathbb{P}(f^*\mathcal{E})$. Here $\overline{C}$ is the normalization of Γ and $f : \overline{C} \to C$ is induced by the projection $\Gamma \to C$. Then we get a section $\sigma : \overline{C} \to \mathbb{P}(f^*\mathcal{E})$ such that $\overline{\Gamma} = \sigma(\overline{C})$ is birationally mapped to $\Gamma \subset \mathbb{P}(\mathcal{E})$ via the natural projection $\operatorname{pr} : \mathbb{P}(f^*\mathcal{E}) \to \mathbb{P}(\mathcal{E})$. By construction we have $\tilde{\mathbf{1}}_{f^*\mathcal{E}} = \operatorname{pr}^* \tilde{\mathbf{1}}_{\mathcal{E}}$, and $(\overline{\Gamma}, \tilde{\mathbf{1}}_{f^*\mathcal{E}}) = (\Gamma, \tilde{\mathbf{1}}_{\mathcal{E}}) < 0$. Consider the exact sequence

$$0 \to I_{\overline{\Gamma}}(\mathbf{1}_{f^*\mathcal{E}}) \to \mathcal{O}(\mathbf{1}_{f^*\mathcal{E}}) \to \mathcal{O}_{\overline{\Gamma}}(\mathbf{1}_{f^*\mathcal{E}}) \to 0,$$

and the associated exact sequence of direct image sheaves

$$0 \to \mathcal{F} \to \mathcal{E} \to \mathcal{O}_{\overline{\Gamma}}(\mathbf{1}_{\mathcal{E}}) \to 0.$$

The third term is a line bundle of degree

$$(\Gamma, \tilde{\mathbf{1}}_{\mathcal{E}}) + (\Gamma, (1/r)\pi^* \det \mathcal{E}) < (\Gamma, (1/r)\pi^* \det \mathcal{E}) = (1/r)\deg f^* \det \mathcal{E} = \mu(f^*\mathcal{E}).$$

This implies that $\mu(\mathcal{F}) > \mu(f^*\mathcal{E})$, i.e. $f^*\mathcal{E}$ is μ-unstable.

Conversely assume that $f^*\mathcal{E}$ is μ-unstable for some $f : \overline{C} \to C$. Let $\mathcal{F}$ be the maximal destabilizer of $\overline{\mathcal{E}} = f^*\mathcal{E}$. Let D be a $\mathbb{Q}$-divisor on C such that $\mu(\overline{\mathcal{E}}) <$ $\deg f^*D < \mu(\mathcal{F})$. Then by Riemann-Roch, $\chi(\overline{C}, S^m(\overline{\mathcal{F}}(-f^*D))) > 0$, meaning that $H^0(\overline{C}, S^m(\overline{\mathcal{F}}(-f^*D))) \neq 0$. Therefore $S^m(\overline{\mathcal{E}}(-f^*D))$ has non-zero global sections for sufficiently large and divisible m (to make mD integral). In other words,

$$m\big(\tilde{\mathbf{1}}_{f^*\mathcal{E}} - \tilde{\pi}^*(f^*D - (1/r)\det f^*\mathcal{E})\big) = m\mathrm{pr}^*\big(\tilde{\mathbf{1}}_{\mathcal{E}} - \pi^*(D - \frac{1}{r}\det\mathcal{E})\big)$$

is effective. Since $\deg D > (1/r)\deg\det\mathcal{E} = \mu(E)$, the condition (3) fails.

(2) $\Rightarrow$ (4): Let D be a $\mathbb{Q}$-divisor of positive degree on C. Since $\tilde{\mathbf{1}}_{\mathcal{E}}$ is relatively ample, $\tilde{\mathbf{1}}_{\mathcal{E}} + N\pi^*D$ is ample for sufficiently large N. Then $N(\tilde{\mathbf{1}}_{\mathcal{E}} + \pi^*D) = (N - 1)\tilde{\mathbf{1}}_{\mathcal{E}} + (\tilde{\mathbf{1}}_{\mathcal{E}} + N\pi^*D)$, being a sum of nef divisor and an ample divisor, is ample.

(5) $\Rightarrow$ (6). Assume that $\mathcal{E}$ is μ-unstable. Let $\mathcal{F} \subset \mathcal{E}$ be the maximal destabilizer of rank s. Then, for any $\mathbb{Q}$ divisor D on C of degree $< \mu(\mathcal{F})$, we have $\chi(C, (S^m\mathcal{F})(-mD)) > 0$ for sufficiently large and divisible m by Riemann-Roch. Hence $H^0(\mathbb{P}(\mathcal{E}), \mathcal{O}(m(\mathbf{1}_{\mathcal{E}} - \pi^*D)) \neq 0$, and $\mathbf{1}_{\mathcal{E}} - \pi^*D$ is effective for any $D <$ $(1/s)\det\mathcal{F}$. Thus $\mathbf{1}_{\mathcal{E}} - (1/s)\det\mathcal{F}$ is pseudo-effective. Since $\mathcal{E}$ is unstable, we have $\mu(\mathcal{F}) \geq \mu(\mathcal{E}) + \frac{1}{r!}$ and so $\tilde{\mathbf{1}}_{\mathcal{E}} - \pi^*E$ is pseudo-effective, where $E = (1/s)\det\mathcal{F} -$ $(1/r)\det\mathcal{E}$ of degree $\mu(\mathcal{F}) - \mu(\mathcal{E}) \geq 1/r!$, thus (5) fails.

Up to now, we never used the assumption on the characteristic. But we need the assumption to prove (6) $\Rightarrow$ (1).

(6) $\Rightarrow$ (1) (*ch* $k = 0$). Suppose that there exists a finite morphism $f : \overline{C} \to C$ such that $f^*\mathcal{E}$ is μ-unstable. f is of course separable. Let $\overline{C}'$ be the Galois closure of the extension f. It is trivial to see that the pull-back of $f^*\mathcal{E}$ is unstable on $\overline{C}'$. Thus we may assume that $\overline{C}$ is Galois over C without loss of generality. Under this assumption, $G = \mathrm{Gal}(\overline{C}/C)$ acts on the pull-back $f^*\mathcal{E}$ in an obvious manner. By the uniqueness of the Harder-Narasimhan filtration, the maximal destabilizer of $f^*\mathcal{E}$ is G-stable, and comes from a subbundle $\mathcal{F} \subset \mathcal{E}$. Now it is clear that $\mathcal{E}$ is unstable.

2.10 Corollary *Let (X, H) be a normal, projective, polarized scheme over a ring R of characteristic zero, finitely generated over $\mathbb{Z}$. Let $\mathcal{E}$ be a torsion free sheaf on X. Let K be the algebraic closure of the field of fractions of R. If $K \otimes \mathcal{E}$ is H-semistable on $K \otimes X$, then $\mathcal{E}$ stays H-semistable on reduction modulo p, for almost every prime p.*

Proof. Let $C \approx mH^{n-1}$ be a general complete intersection curve on X. Then by the theorem of Mehta-Ramanathan, we may assume that $\mathcal{E}|_C$ is μ-semistable over $K\otimes C$. By (2.9.4), $\tilde{\mathbf{1}}_{\mathcal{E}} + \frac{1}{2(r!)mH^n}\pi^*H$ is ample on $K\otimes\mathbb{P}(\mathcal{E}|_C)$. Noting that ampleness is an open condition for projective morphisms, the condition (2.9.4) is satisfied for reductions of $\mathcal{E}|_C$ modulo almost every p, which implies the H-semistability on $\mathcal{E}$ modulo p. $\square$

2.11 Corollary   *Let m be a positive integer. Over a field of characteristic zero, $\mathcal{E}^{\otimes m}$ is H-semistable if $\mathcal{E}$ is so. Hence the direct summands $S^m\mathcal{E}$, $\wedge^m\mathcal{E}$, etc. are also H-semistable. More generally the tensor product of two H-semistable sheaves is H-semistable (in characteristic zero).*

2.12 Exercise   (1) Prove (2.11), in the following two steps. (a) $\mathcal{E}|_C$ is μ-semistable if and only if $\mathcal{E}(-\frac{\det\mathcal{E}}{\operatorname{rank}\mathcal{E}})|_C$ is semipositive, with semipositivity suitably defined. (b) The tensor product of semipositive bundles are semipositive.

(2) Show that $\mathcal{E}$ is H-semistable if and only if $\mathcal{E}^* = \mathcal{H}om(\mathcal{E}, \mathcal{O}_X)$ is H-semistable, and that $\mathcal{E}$ is H-semistable if and only if $\mathcal{E}(-\frac{\det\mathcal{E}}{\operatorname{rank}\mathcal{E}})|_C$ is seminegative.

Remark. The semi-positivity and the semi-negativity of the same bundle is not contradictory. Actually, if $\mathcal{E}$ a μ-stable bundle on a complex projective curve C, $\mathcal{E}(-\frac{\det\mathcal{E}}{\operatorname{rank}\mathcal{E}})$ is flat (with respect to some hermitian connection) over a suitable étale cover $\tilde{C} \to C$, and is obtained from a unitary representation of the fundamental group $\pi_1(\tilde{C}) \to SU(r)$ [Narasimhan-Seshadri]. A flat bundle is always semi-positive and semi-negative.

Let us see that the modulo p reduction of the maximal destabilizer $\mathcal{F}$ of T_X defines a foliation on $X \bmod p$, provided $\mu(\mathcal{F}) > 0$. Note that, in this case, there is a $\mathbb{Q}$-divisor D of degree $DH^{n-1} > 0$ such that $\mathcal{F}(-D)$ is ample and that $(T_X/\mathcal{F})(-D)$ is negative when restricted to a general complete intersection curve.

2.13 Proposition   *Let X be a smooth projective variety over a field of characteristic $p > 0$. Let $\mathcal{F} \subset T_X$ be the maximal destabilzer with respect to a fixed polarization H. Assume that there exists a $\mathbb{Q}$-divisor D such that $\deg D > 0$, $\mathcal{F}(-D)$ is ample and $(T_X/\mathcal{F})(-D)$ is negative when restricted to a general complete intersection curve. Then, on the open subset U on which $\mathcal{F}$ is a subbundle of T_X, $\mathcal{F}$ is a foliation, i.e. involutive and closed under p-th powers.*

Proof.   The Lie bracket $[,] : \mathcal{F} \times \mathcal{F} \to T_X$ is not $\mathcal{O}_X$-bilinear. However, if you consider $[,]$ as a map from $\mathcal{F} \times \mathcal{F}$ to $T_X/\mathcal{F}$, it is $\mathcal{O}_X$-bilinear, and anti-commutative. Thus you get a global $\mathcal{O}_X$-homomorphism $\wedge^2\mathcal{F} \to T_X/\mathcal{F}$. This must be zero, because $(\wedge^2\mathcal{F})(-2D)$ is ample while

$$(T_X/\mathcal{F})(-2D) = (T_X/\mathcal{F})(-D) \otimes \mathcal{O}(-D)$$

is negative if restricted to a general curve. Hence $\mathcal{F}$ is involutive.

The p-th power map $\times^p : T_X \to T_X$ induces the map $\overline{\times}^p : \mathcal{F} \to T_X/\mathcal{F}$. Though not $\mathcal{O}_X$-linear, this is additive and p-linear, i.e. $(a\xi)^p \equiv a^p\xi^p \bmod \mathcal{F}$. In particular, $\mathcal{K} = \operatorname{Ker}\overline{\times}^p$ is an $\mathcal{O}_X$-submodule of $\mathcal{F}$, and $\operatorname{Im}\overline{\times}^p$ is an $\mathcal{O}_X^p$-submodule of $T_X/\mathcal{F}$. $\mathcal{O}_X \otimes_{\mathcal{O}_X^p} \operatorname{Im}\overline{\times}^p$ is then an $\mathcal{O}_X$-module isomorphic to $(\mathcal{F}/\mathcal{K})^p$, the sheaf obtained by replacing the entries of the transition matrices of $\mathcal{F}/\mathcal{K}$ by their p-th powers. Since $\mathcal{F}(-D)$ is ample, so are $(\mathcal{F}/\mathcal{K})(-D)$ and $(\mathcal{F}/\mathcal{K})^p(-pD)$ on a generic curve. This proves that $\operatorname{Hom}_{\mathcal{O}_X}(\mathcal{O}_X \otimes \operatorname{Im}\overline{\times}^p, T_X/\mathcal{F}) = 0$, by a similar argument as above. $\square$

A torsion free sheaf $\mathcal{E}$ on a polarized normal variety (X, H) is called *generically seminegative* or *generically H-seminegative* if $\mu_{\max}(\mathcal{E}) \leq 0$, i.e. the maximal destabilizer $\mathcal{E}_1$ satisfies $\mu(\mathcal{E}_1) \leq 0$. $\mathcal{E}$ is *generically (H-)semipositive*, by definition, if $\mu_{\min} \geq 0$, or equivalently, if $\mathcal{H}om(\mathcal{E}, \mathcal{O})$ is generically seminegative.

2.14 Theorem   *Let X be a normal projective variety defined over an algebraically closed field of characteristic zero. If T_X is not generically semi-negative (or, equivalently, if $\mu_{\max}(T_X) > 0$), then X is uniruled.*

Proof.   Let $\mathcal{F} \subset T_X$ be the maximal destabilizer. Then $\mathcal{F}$ defines a foliation on almost every reduction modulo p by (2.13), and you can apply (1.12). The degree of rational curves is estimated asymptotically by $2H^n/(\det \mathcal{F}, H^{n-1})$, and hence uniformly bounded for almost every p, enabling you to lift the curves back to characteristic zero. $\square$

Put in another way, this implies that the cotangent bundle Ω^1_X of a non-uniruled variety X is semi-positive when restricted to a general complete intersection curve of sufficiently high degree:

2.15 Theorem (generic semi-positivity theorem)   *Let X denote a normal projective variety over k, an algebraically closed field of characteristic zero, and let H be an ample divisor. Let $j : X_{\mathrm{smooth}} \hookrightarrow X$ be the open embedding of the smooth locus. If X is not uniruled, then the restriction of $j_*\Omega^1_X$ to a general smooth complete intersection curve cut out by elements $\in |mH|$, $m \gg 0$, is semi-positive.*

In the above proof of (2.14), we followed an idea of N. Shepherd-Barron to use purely inseparable quotient $X/\mathcal{F}$ in order to find rational curves. This method has a merit that it considerably simplifies the proof. However, it does not explicitly give direct connection between the resulting rational curves and the original $\mathcal{F}$.

An involutive subbundle $\mathcal{F} \subset T_X$ defined on a complex manifold X has unique analytic integral submanifold passing through a given point $x \in X$ by a theorem of Frobenius. Here a complex submanifold $Y \subset X$ is called an integral submanifold if $T_{Y,y} = \mathcal{F}|_y$ for every point $y \in Y$. There is a second proof of (2.14), more complicated but more geometric, which shows that the integral submanifolds of $\mathcal{F}$ is algebraizable and covered by rational curves.

2.16 Theorem (algebraic integrability theorem)   *Let the notation and assumption be as in (2.14). Let $U \subset X$ be the open subset on which $\mathcal{F}$ is a subbundle. Then any analytic leaf (i.e. integral submanifold) of the involutive subbundle $\mathcal{F}|_U$ can be compactified to algebraic subvarieties of X. Furthermore, two general points on a compactified leaf can be connected by an irreducible rational curve (i.e. each leaf is "rationally connected" in the terminology defined in Lecture V).*

For the proof, we refer to [Miyaoka 2].

3   Second Chern class of minimal varieties

The generic semipositivity of Ω_X^1 involves certain information on the topology of minimal varieties. Recall that a *minimal* variety X over an algebraically closed field of characteristic zero is, by definition, a normal, projective, $\mathbb{Q}$-factorial variety with at worst terminal singularities of which the canonical divisor is nef.

The objective of this section is the following:

3.1 Theorem *("pseudo-effectivity" of c_2)   Let (X, H) be a normal polarized variety, smooth in codimension two over a field of characteristic zero. Let $\mathcal{E}$ be a torsion free sheaf which is locally free in codimension two. If $\det \mathcal{E}$ is nef and $\mathcal{E}$ is generically semipositive, then $c_2(\mathcal{E})H^{n-2} \geq 0$. In particular, if X is a minimal variety, then $c_2(j_*\Omega_{X_{\mathrm{smooth}}}^1)H^{n-2} \geq 0$. Here $j : X_{\mathrm{smooth}} \hookrightarrow X$ is the open embedding of the smooth locus.*

A heuristic interpretation is that the second Chern class of a generically semipositive sheaf, which is a two-cocycle in X, looks like an effective cycle. For a minimal variety, we can strengthen the inequality as follows:

3.2 Theorem   *For a minimal n-fold X over a field of characteristic zero, we have the following inequality*

$$3c_2(j_*\Omega_{X_{\mathrm{smooth}}}^1)H^{n-2} \geq K_X^2 H^{n-2}.$$

Given a resolution $\rho : \tilde{X} \to X$, the above inequality is equivalent to the following

$$3c_2(\Omega_{\tilde{X}}^1)(\rho^*H)^{n-2} \geq K_X^2 H^{n-2}.$$

3.3   You do not have to bother to think about what Chern classes exactly are. What you need to memorize in this lecture is minimal:

(a) Given a coherent sheaf $\mathcal{F}$ on X of dimension n, $c_i(\mathcal{F}) \in Z_{n-i}(X)$ ($\in Z^i(X)$ if X is smooth).

(b) If $\mathcal{F}$ is locally free, then $c_i(\mathcal{F}) \in Z^i(X)$, and c_i is compatible with the pull-back: $c_i(f^*\mathcal{F}) = f^*c_i(\mathcal{F})$.

(c) If $\mathcal{F}$ is a line bundle, then $c_1(\mathcal{F})$ is the Cartier divisor of a meromorphic section of $\mathcal{F}$ and $c_i(\mathcal{F}) = 0$ for $i > 1$.

(d) Let $c(\mathcal{F}) = 1 + c_1(\mathcal{F}) + c_2(\mathcal{F}) + \ldots + c_n(\mathcal{F}) \in \sum Z^j(X)$ be the total Chern class of $\mathcal{F}$, a coherent sheaf on a smooth complete variety X of dimension n. Given an exact sequence

$$0 \to \mathcal{F}' \to \mathcal{F} \to \mathcal{F}'' \to 0$$

the equality $c(\mathcal{F}) = c(\mathcal{F}')c(\mathcal{F}'')$ holds.

(e) If $Y \subset X$ is an r-codimensional closed subscheme on a smooth variety X, then

$$c_1(I_Y) = \ldots = c_{r-1}(I_Y) = 0, \ c_r(I_Y) = (-1)^r[Y] \in Z^r(X),$$
$$c_1(\mathcal{O}_Y) = \ldots = c_{r-1}(\mathcal{O}_Y) = 0, \ c_r(\mathcal{O}_Y) = -(-1)^r[Y] \in Z^r(X).$$

(You do not need to know higher Chern classes of such sheaves in this lecture.)

The proof of (3.1) relies on a fundamental result due to F. Bogomolov:

3.4 Theorem (Bogomolov inequality)   *Let X be a smooth algebraic surface over an algebraically closed field of characteristic zero and $\mathcal{E}$ a vector bundle of rank r. If $\mathcal{E}$ is H-semistable for some ample divisor H, then the inequality*

$$c_2(\mathcal{E}) \geq \frac{r-1}{2r} c_1(\mathcal{E})^2 \tag{$*$}$$

holds. Alternatively, if the inequality ($*$) fails, then there exists a positive rational number ε such that $\hat{1}_{\mathcal{E}} - \varepsilon H$ is pseudo-effective in $N^1(\mathbb{P}(\mathcal{E}))$.

In view of the theorem of Mehta-Ramanathan, from (3.4) follows:

3.5 Corollary   *Let k be an algebraically closed field of characteristic zero. Let X be a normal projective k-variety smooth in codimension two. If $\mathcal{E}$ is an H-semistable torsion free sheaf of rank r, then*

$$c_2(\mathcal{E})H^{n-2} \geq \frac{r-1}{2r} c_1(\mathcal{E})^2 H^{n-2}.$$

Before entering the proof of the Bogomolov inequality, recall once more elementary facts on projective bundles and Chern classes.

3.6   Let $S^i\mathcal{E}$ denote the i-th symmetric tensor product of a vector bundle $\mathcal{E}$. The projective bundle $\mathbb{P}(\mathcal{E})$ associated with $\mathcal{E}$ was, by definition, $\mathrm{Proj}(\sum_{i=0}^{\infty} S^i\mathcal{E})$. When the ground field is $\mathbb{C}$, we have a geometrically concrete description:

$$\mathbb{P}(\mathcal{E}) = (\mathcal{E}^* \setminus 0\text{-section})/\mathbb{C}^{\times},$$

where $\mathcal{E}^*$ is the dual bundle $\mathcal{H}om(\mathcal{E}, \mathcal{O}_X)$ and the the action of $\mathbb{C}^{\times}$ is the fibrewise multiplication.

Any non-zero element $s \in \mathcal{E}$, fibrewise a linear function on $\mathcal{E}^*$, defines a family of hyperplanes on the fibres of $\pi : \mathbb{P}(\mathcal{E}) \to X$, thus defining a relatively very ample divisor $1 = 1_{\mathcal{E}}$ on $\mathbb{P}(\mathcal{E})$, (tautological divisor). From the definition we derive a canonical isomorphism

$$\pi_* \mathcal{O}_{\mathbb{P}(\mathcal{E})}(m1) \simeq S^m\mathcal{E},$$
$$R^i\pi_* \mathcal{O}_{\mathbb{P}(\mathcal{E})}(m1) = 0, \ i \geq 1,$$

for an arbitrary non-negative integer m. The first Chern class of $\mathbf{1} = \mathbf{1}_{\mathcal{E}}$ computes the Chern classes of $\mathcal{E}$ thanks to the following formula of [Grothendieck 1]:

$$\sum_{i=0}^{r}(-1)^i(\pi^*c_i(\mathcal{E}))\mathbf{1}^{r-i} = 0 \in Z^r(\mathbb{P}(\mathcal{E})). \tag{3.5.1}$$

In particular, we have the same identity when all terms are viewed as cohomology classes. For example, given an $(n-1)$-cocycle (1-cycle) Γ on a smooth projective variety X of dimension n, and a vector bundle $\mathcal{E}$ of rank r, we get

$$(\pi^*\Gamma)\mathbf{1}^r = (\pi^*\Gamma\pi^*c_1(\mathcal{E}))\mathbf{1}^{r-1} = \Gamma c_1(\mathcal{E}) = (\Gamma, c_1(\mathcal{E})),$$

because $\mathbf{1}^{r-1}$ is a single point in $\mathbb{P}^{r-1}$ over each point of the zero-cycle $\Gamma c_1(\mathcal{E})$. For an $(n-2)$-cocycle Δ on X, we have

$$(\pi^*\Delta)\mathbf{1}^r = (\pi^*(\Delta c_1(\mathcal{E})))\mathbf{1}^{r-1} - (\pi^*(\Delta c_2(\mathcal{E})))\mathbf{1}^{r-2}$$

By multiplying the both hand sides by $\mathbf{1}$, we obtain

$$(\pi^*\Delta)\mathbf{1}^{r+1} = (\pi^*(\Delta c_1(\mathcal{E})))\mathbf{1}^r - (\pi^*(\Delta c_2(\mathcal{E})))\mathbf{1}^{r-1}$$
$$= (\Delta, c_1(\mathcal{E})^2) - (\Delta, c_2(\mathcal{E})).$$

Assume that X is smooth projective of dimension two. Let $\tilde{\mathbf{1}} = \mathbf{1} - (1/r)\pi^*\det\mathcal{E}$ be the normalized tautological divisor. Compute $\tilde{\mathbf{1}}^3$ in a similar manner as above to get

$$\tilde{\mathbf{1}}^3 = -c_2(\mathcal{E}) + \frac{r-1}{2r}c_1(\mathcal{E})^2.$$

Thus the Bogomolov inequality is equivalent to the inequality $\tilde{\mathbf{1}}^3 \leq 0$, which has an apparent relation between the *non-bigness* of $\tilde{\mathbf{1}}$.

3.7 Proposition   *Let (X, H) be a polarized variety of dimension n over an algebraically closed field. Given a Cartier divisor D, the following five conditions are equivalent:*

(1)   *There is a Cartier divisor E such that $h^0(X, \mathcal{O}(mD + E))$ grows like m^n (i.e. there are two polynomials q_1, q_2 of degree n with positive leading coefficients such that $q_1(m) \leq h^0(mD + E) \leq q_2(m)$.)*
(2) $h^0(X, \mathcal{O}(mD))$ *grows like m^n.*
(3)   *For any fixed divisor E, $h^0(X, \mathcal{O}(mD + E))$ grows like m^n.*
(4)   *For a given $\mathbb{Q}$-divisor M, there exists a positive rational number ε such that $D - \varepsilon M$ is effective.*
(5) $[D]$ *sits in the interior of the cone of effective divisors;*

When one of these equivalent conditions is satisfied, D is said to be "big".

3.8 Exercise   Let X be smooth projective over k, and D a divisor on X.

(1) Show that $h^i(X, mD + E) \leq O(m^n)$, $n = \dim X$, $i = 0, 1, \ldots, n$ by induction on n. ($A(m) \leq O(m^n)$ means that $A(m)/m^n$ is bounded from above as a function in m.)

(2) Verify (3.7) by using (1) above.

(3) Let D be nef and assume that $ch\ k = 0$. Show that $h^i(X, \mathcal{O}(mD)) \leq O(m^{n-1})$, $i \geq 1$. (*Hint:* Take an ample divisor H such that $|H - K_X|$ contains a smooth member Y. Consider the exact sequence $0 \to \mathcal{O}(mD) \to \mathcal{O}(mD+Y) \to \mathcal{O}_Y(mD - Y) \to 0$, and apply Kodaira vanishing to the middle column of the long exact sequence.)

(4) Assume that $ch\ k = 0$ and D nef. Show that D is big if and only if $D^n > 0$. (*Hint:* "If" part is by Riemann-Roch and (3). To show the "only if" part, choose $\varepsilon > 0$ such that $D - \varepsilon H$ is effective. Show the inequality $D^i H^{n-i} \geq \varepsilon D^{i-1} H^{n-i+1}$ to conclude $D^n \geq \varepsilon^n H^n$.)

(5) Prove that $D^2 > 0$ implies that D or $-D$ is big, when $n = 2$.

(6) If D is not nef, $D^n > 0$ does not necessarily yield the bigness of $\pm D$. Construct an example of such D on a threefold (say, on $\mathbb{P}^1 \times \mathbb{P}^2$).

3.9 Proof of the Bogomolov inequality   Let the notation be as above. The Riemann-Roch tells us that

$$\chi(\mathbb{P}(\mathcal{E}), \mathcal{O}(m\tilde{1})) = \frac{m^{r+1}}{(r+1)!}\tilde{1}^{r+1} + \text{ terms of order } \leq r \text{ in } m.$$

By the functor π_*, this is equivalent to the equality

$$\chi(X, (S^m\mathcal{E})(-\frac{m}{r}\det\mathcal{E})) = \frac{m^{r+1}}{(r+1)!}\tilde{1}^{r+1} + O(m^r).$$

Suppose that $\tilde{1}^{r+1}$ is positive. Then it follows that either

(a) $h^0(X, (S^m\mathcal{E})(-\frac{m}{r}\det\mathcal{E})) = h^0(\mathbb{P}(\mathcal{E}), \mathcal{O}(m\tilde{1}))$ grows like m^{r+1} or

(b) $h^2(X, (S^m\mathcal{E})(-\frac{m}{r}\det\mathcal{E}))$ does so.

Assume that (a) occurs. Then $\tilde{1}$ is big and there exists $\varepsilon > 0$ such that $\tilde{1} - \varepsilon\pi^* H$ is effective. In particular $\mathcal{E}(-\frac{1}{r}\det\mathcal{E})$ is not seminegative on a generic curve C, whence follows that $\mathcal{E}$ is H-unstable. When (b) occurs, then the Serre duality gives

$$h^0(X, (S^m\mathcal{E}^*)(-\frac{m}{r}\det\mathcal{E}^* + K_X)) = h^0(\mathbb{P}(\mathcal{E}^*), \mathcal{O}(m\tilde{1}_{\mathcal{E}^*} + K_X))$$

grows like m^{r+1}. Thus $\tilde{1}_{\mathcal{E}^*}$ is big, and hence $\mathcal{E}^*$ is H-unstable. This implies $\mathcal{E}$ is unstable, too. $\square$

The Bogomolov inequality depends on (2.6) and (2.8), and so does not necessarily hold in characteristic p. For a counter example, see [Gieseker 1].

3.10 *Proof of (3.1).* By the theorem of Mehta-Ramanathan, we may assume that X is two-dimensional without any essential loss of generality.

(1) Assume that $c_1(\mathcal{E}) = D$ is big, i.e. $D^2 > 0$. Let $\mathbf{F}^i\mathcal{E}$ be the Harder-Narasimhan filtration and let $\mathrm{Gr}\mathbf{F} = \bigoplus \mathcal{G}_i = \bigoplus \mathbf{F}^i\mathcal{E}/\mathbf{F}^{i-1}\mathcal{E}$ the associated graded module. Each piece $\mathcal{G}_i$ is torsion free and H-semistable, but not necessarily locally free. Since X is a surface, the double dual $\mathcal{G}_i^{**}$ is locally free of rank r_i with $c_1(\mathcal{G}_i^{**}) = c_1(\mathcal{G}_i)$, $c_2(\mathcal{G}_i^{**}) \leq c_2(\mathcal{G}_i)$. Put $c_1(\mathcal{G}_i) \approx r_i(a_iD + \Delta_i)$, where Δ_i is a $\mathbb{Q}$-divisor orthogonal to D (with respect to the intersection pairing). By the multiplicativity of the total Chern class, we have:

$$
\begin{aligned}
c_2(\mathcal{E}) &= \sum_i c_2(\mathcal{G}_i) + \sum_{i<j} c_1(\mathcal{G}_i)c_1(\mathcal{G}_j) \\
&= \sum_i c_2(\mathcal{G}_i) + \frac{1}{2}\Big\{\Big(\sum_i c_1(\mathcal{G}_i)\Big)^2 - \sum_i (c_1(\mathcal{G}_i))^2\Big\} \qquad \text{by (3.4)} \\
&\geq \sum_i \Big(\frac{r_i - 1}{2r_i} - \frac{1}{2}\Big)c_1(\mathcal{G}_i)^2 + \frac{1}{2}c_1(\mathcal{E})^2 \\
&= -\sum_i \frac{r_i}{2}(a_i^2 D^2 + \Delta_i^2) + \frac{1}{2}D^2 \\
&\geq \frac{1}{2}\Big\{1 - \sum_i (a_i^2 r_i)\Big\}D^2 \qquad \text{by the Hodge index theorem.}
\end{aligned}
$$

Noting that $r_i \geq 1$, $a_i \geq 0$ and $\sum_i r_i a_i = 1$, we infer that $\sum_i a_i^2 r_i \leq 1$, proving the inequality.

(2) In case D is not big, consider $\mathcal{E}(\varepsilon H)$ for sufficiently small positive rational number ε. Then $\mathcal{E}_\varepsilon = \mathcal{E}(\varepsilon H)$ is of course generically semi-positive with nef and big determinant. Hence $c_2(\mathcal{E}_\varepsilon) \geq 0$ for every ε, and the limit $\varepsilon \to 0$ fulfills the desired inequality. □

We omit the proof of (3.2), which is essentially similar, except that it uses the more or less well-known inequality $c_2(\mathcal{E}) \geq 3c_1(\mathcal{E})^2$ for vector bundles of rank two on a surface Y subject to the following two conditions:

(1) $\mathcal{E}$ is a subsheaf of Ω_Y^1.
(2) $\det \mathcal{E}$ is pseudo-effective.

For the proof of the inequality mentioned, see [Miyaoka 1], and for the detail of the proof of (3.2), see [Miyaoka 3]. The perturbation technique (replace a divisor or a vector bundle by a small perturbation of it) used in (3.10) is very useful, yielding a generalization (3.12) below.

3.11 Let X be a minimal projective variety of dimension n over an algebraically closed field of characteristic zero. Recall that X is smooth outside a closed subset of codimension at least 3 (smooth in codimension 2, for short).

Let $D_1, \ldots, D_{n-1}$ be arbitrary Cartier divisors. We can write D as a differ-ence of two very ample divisors so that the cycles $D_1 \cdots D_{n-2}$ and $D_1 \cdots D_{n-1}$ are numerically equivalent to a 1-cycle and a 2-cycle which are away from the singularities of X. Since Ω_X^1 is a locally free sheaf on the smooth locus of X, the intersection numbers $c_1^2(\Omega_X)D_1 \cdots D_{n-2}$, $c_2(\Omega_X^1)D_1 \cdots D_{n-2}$ are well-defined, depending only on X and the numerical classes of D_i's. The situation is quite similar if you replace D_i by $\mathbb{Q}$-Cartier divisors $\in \mathrm{Pic}_{\mathbb{Q}}(X)$. In particular the inter-section number $c_1(X)^2 D_1 \cdots D_{n-2} = K_X^2 D_1 \cdots D_{n-2}$ makes sense, for arbitrary $D_i \in N_{\mathbb{Q}}^1(X)$.

3.12 Theorem   *Let X be a minimal projective variety of dimension n defined over an algebraically closed field of characteristic zero. Let $D_1, \ldots, D_{n-2}$ be nef $\mathbb{Q}$-Cartier divisors on X. Then the following inequality holds:*

$$3c_2(\Omega_X^1)D_1 \cdots D_{n-2} \geq K_X^2 D_1 \cdots D_{n-2} \geq 0.$$

Proof.   Since a nef $\mathbb{Q}$-divisor is a limit of ample $\mathbb{Q}$-divisors, it suffices to show the inequality in case D_i's are ample. By replacing D_i by its high multiple, we may further assume that D_i is a very ample integral Cartier divisor and $S = D_1 \cdots D_{n-2}$ is smooth surface such that $\Omega_X^1|_S$ is generically semi-positive. Then exactly the same argument as above works.

The pseudo-effectivity of c_2 yields the non-negativity of the Kodaira dimen-sion for smooth minimal threefolds, as will be seen below.

3.13 Corollary   *Let X be a smooth projective threefold over the complex numbers. If X is minimal, i.e. if K_X is nef, then the Kodaira dimension $\kappa(X)$ is non-negative.*

Proof.   The Noether formula for threefolds tells us

$$\chi(X, \mathcal{O}_X) = \frac{1}{24}c_1(X)c_2(X) = -\frac{1}{24}K_X c_2(\Omega_X^1),$$

which is non-positive by Theorem 3.13. Since

$$\chi(X, \mathcal{O}_X) = h^0(X, \mathcal{O}_X) + h^2(X, \mathcal{O}_X) - h^1(X, \mathcal{O}_X) - h^3(X, \mathcal{O}_X)$$
$$\geq 1 - h^1(X, \mathcal{O}_X) - h^3(X, \mathcal{O}_X),$$

it follows that $H^1(X, \mathcal{O}_X)$ or $H^3(X, \mathcal{O}_X)$ is non-trivial. If $H^3(X, \mathcal{O}_X) \neq 0$, then the Serre duality shows $H^0(X, \mathcal{O}_X(K_X)) \neq 0$, so that $\kappa(X) \geq 0$. If the irregularity $q(X) = \dim H^1(X, \mathcal{O}_X) > 0$, or, by the Hodge symmetry, $H^0(X, \Omega_X^1) > 0$, then we have non-trivial Albanese mapping $\alpha : X \to \mathrm{Alb}(X)$, see (V.1). If $\alpha(X)$ is 3-dimensional, then the pull-back of a general 3-form on $\mathrm{Alb}(X)$ gives a non-zero holomorphic 3-form on X, so that $H^0(X, \mathcal{O}_X(K_X)) \neq 0$. (In general, a variety which is generically finite over a subvariety of an abelian variety has non-negative Kodaira dimension.) Suppose that $\alpha(X)$ has dimension one or two. Then by taking the Stein factorization of α, we can define a fibre space structure on X over a curve or a surface, with general fibre being accordingly a surface or a curve. In these cases, the Kodaira dimension of the total space X is subject to the subadditivity law:

$$\kappa(X) \geq \kappa(\text{general fibre}) + \kappa(\text{the base space})$$

(theorems due to [Ueno 2], [Viehweg 2] and [Kawamata 4]; see (IV.1) for a brief introduction to Kodaira dimension and subadditivity principle in classification theory. Also compare [Mori 4][Kollár 1] for a general survey on the current state of the subadditivity or the Iitaka conjecture). The base space, being a finite cover of a subvariety of an abelian variety, has non-negative Kodaira dimension. The fibres are not uniruled because X is not; a non-uniruled curve or surface has non-negative Kodaira dimension by classical classification theory. $\square$

In the next lecture, we generalize this result to singular minimal threefolds.

Historical Comments

Let $\mathcal{F} \subset T_X$ be an involutive subbundle. A smooth subvariety $Y \subset X$ is said to be an *integral submanifold* if $T_{Y,y} = (\mathcal{F}) \otimes k(y)$ at each closed point $y \in Y$. Over complex numbers, given $\mathcal{F}$ and $x \in X$, there exists a unique germ of analytic integral submanifolds passing through x (Theorem of Frobenius), though it is not algebraic in general. (For instance, a global vector field of an abelian variety X generates an involutive invertible subsheaf of T_X, but their integrable submanifolds are almost always non-algebraic.) Our theorem (2.15) states something very non-trivial, in this sense.

The positive characteristic case is quite different. As an analogue of the Frobenius theorem, we would like to know whether there exists a unique smooth formal scheme Y in $\hat{X}$ such that $T_Y = \mathcal{F} \otimes_{\mathcal{O}_X} \hat{\mathcal{O}}_X \otimes_{\hat{\mathcal{O}}_X} \mathcal{O}_Y$. Here $\hat{X}$ is the formal neighbourhood of x. The answer is definitely No; in some case, there does not exist such Y, and there are infinitely many Y's if there is one. However, Proposition (1.9) means that there exists a unique maximal closed subscheme $Y \simeq \mathrm{Spec}\, k[z_1,...,z_r]/(z_1^p,...,z_r^p)$ in the Artinian scheme $\mathrm{Spec}\, \mathcal{O}_X/\mathfrak{M}_X^p$ such that $T_Y = \mathcal{F}|_Y$.

As was mentioned in the lecture, the theory of foliations was implicitly used by [Rudakov-Shafarevich]. It was, however, systematically studied first by [Miyaoka 2], and was developed in full generality by [Ekedahl 1]. The theory was effectively exploited to the study of pluricanonical maps of surfaces of general type [Ekedahl 2] along with the one we explained above [Kollár *et al.*] [Shepherd-Barron].

The original proof of the Bogomolov inequality [Bogomolov] was based on geometric invariant theory of D. Mumford [Mumford-Fogarty]. The simpler proof cited here follows [Miyaoka 3]. An alternative proof by a differential geometric approach was given by Donaldson [Donaldson] whose result involves, however, stronger implication. Namely, if a vector bundle is H-stable with $c_1 \approx 0, c_2 \approx 0$, then it is flat with respect to some hermitian metric. In particular, it comes from a hermitian local system, or equivalently from an irreducible representation $\pi_1(X) \to U(r)$.

Lecture IV
Abundance for Minimal 3-Folds

Overview

In the first section, we give a quick introduction to the birational classification theory due to Enriques-Kodaira-Shafarevich-Iitaka, which divides the n-dimensional algebraic varieties into $n+2$ disjoint classes according to a single invariant called the "Kodaira dimension" κ. A nice thing of the theory is that a variety of intermediate Kodaira dimension has a canonical structure of a fibre space unique up to birational equivalence ("Iitaka fibration").

This classification would be incorporated to the Minimal Model Program through the "abundance conjecture", which asserts that the canonical divisor of a minimal variety should be semiample, meaning in particular that a numerical invariant $\nu(X)$ computes the Kodaira dimension $\kappa(X)$. Some easy cases where the abundance conjecture is actually verified are discussed in Section 2. Along with the minimal model theory, the existence of Iitaka fibration is also essential in the argument.

In Section 3, the non-negativity of the Kodaira dimension of a minimal threefold is proved. The key to the proof is the pseudo-effectivity of c_2 proved in Lecture III. We are exceptionally lucky in this case, because the Todd classes involve only c_1 and c_2 in dimension three.

By the result in Section 3, we can find an effective divisor S in a pluricanonical linear system on a minimal threefold. The proof of the abundance conjecture relies on the analysis of linear systems on S. Principal difficulties are caused from the fact that S may be highly reducible and non-reduced. The complete proof of the three-dimensional abundance conjecture requires fairly technical materials (log-minimal models etc.), which would be beyond the scope of these lectures intended for non-specialists. Instead, Section 4 gives a proof in easier cases where S is smooth. Still the discussion involves the heart of the idea and, hopefully, will illustrate guiding principles toward the proof in general cases.

1   Quick introduction to the Enriques-Kodaira-Shafarevich-Iitaka classification

1.1 Several times we have mentioned birational classification theory of algebraic varieties in very vague ways. Here we would like to explain in clearer language what we mean by birational classification, following an idea of S. Iitaka.

As suggested in Lecture I, a good classification theory would satisfy the following two requirements:

A. Each class in the classification should be stable under birational equivalence and finite étale morphisms as well as under smooth (projective) deformation;

B. There should be a reasonable (preferably, effective) criterion to determine which class a given variety falls under. (In this sense, rationality, unirationality etc. are, as of today, not nice classes.)

Furthermore, at least for some of the classes in the classification table, there should be non-trivial statements about the geometric structure of varieties in there.

In dimension one (i.e. for curves), a classification theory which satisfies the requirement **A** is uniquely determined and it satisfies **B** as well. In fact we have three classes:

a)   the projective line $\mathbb{P}^1$, the unique curve of genus 0;
b)   the elliptic curves (or curves of genus 1);
c)   the curves of genus ≥ 2.

For each class, there are simple characterizations, natural from the points of view of algebraic geometry, topology, and function theory. In fact, for a given smooth projective curve C and its universal cover $\tilde{C}$, we have:

$$C \simeq \mathbb{P}^1 \Leftrightarrow \deg K_C < 0 \Leftrightarrow H^0(C, \mathcal{O}(K_C)) = 0$$
$$\Leftrightarrow b_1(C) = 0 \Leftrightarrow \pi_1(C) = (1) \Leftrightarrow \tilde{C} \simeq \mathbb{P}^1$$
$$\Leftrightarrow \dim H^0(C, T_C) = 3 \Leftrightarrow \operatorname{Aut}(C)^\circ \text{ is an non-abelian Lie group,}$$
$$g(C) = 1 \Leftrightarrow \deg K_C = 0 \Leftrightarrow K_C \sim 0 \Leftrightarrow \dim H^0(C, \mathcal{O}(K_C)) = 1$$
$$\Leftrightarrow b_1(C) = 2 \Leftrightarrow \pi_1(C) = \mathbb{Z}^2 \Leftrightarrow \tilde{C} \simeq \mathbb{C}$$
$$\Leftrightarrow \dim H^0(C, T_X) = 1 \Leftrightarrow \operatorname{Aut}(C)^\circ \text{ is non-trivial, abelian}$$
$$g(C) \geq 2 \Leftrightarrow \deg K_C > 0 \Leftrightarrow \dim H^0(C, \mathcal{O}_X(K_C)) \geq 2$$
$$\Leftrightarrow b_1(C) \geq 4 \Leftrightarrow \pi_1(C) \text{ non abelian} \Leftrightarrow \tilde{C} \simeq \mathcal{H}$$
$$\Leftrightarrow H^0(C, T_C) = 0 \Leftrightarrow \operatorname{Aut}(C) \text{ finite.}$$

1.2 Exercise Show that this classification is the finest among the ones that satisfy the requirement **A**. Check the equivalences given above (you may assume that the moduli space of curves of genus g is irreducible).

1.3 In dimension two, the situation is not that clear. Blowing-ups, which is a fundamental birational morphism, changes topological data and kills automorphism groups. Thus many of the characterizations which were useful for curves cannot be adopted in birational classification in dimension two or higher.

The fundamental group π_1 is invariant under birational maps and smooth deformation. Finite étale morphisms affect it little. In this sense the fundamental group is a good invariant. A problem to this invariant is that, in higher dimension, information on the fundamental group is usually too small. For example, every smooth hypersurface in $\mathbb{P}^{n+1}$ is simply connected for $n \geq 2$, but a hypersurface of high degree is quite different from a hyperplane $\simeq \mathbb{P}^n$.

The irregularity $q(X) = \dim H^1(X, \mathcal{O}_X) = \frac{1}{2} b_1(X)$ is a birational and topological invariant for smooth varieties X of arbitrary dimension and gives useful geometric information via the Albanese map. The geometric genus $p_g(X) = \dim H^0(X, \mathcal{O}_X(K_X))$ is a third birational invariant. Either of the two invariants

q and p_g is not stable under étale morphisms. Furthermore, they do not separate quite different type of varieties. For example, Enriques found that there are smooth non-rational surfaces X such that $q(X) = p_g(X) = 0$. He discovered, nevertheless, that if you introduce *plurigenera* $P_m = \dim H^0(X, \mathcal{O}_X(mK_X))$, $m = 1, 2, \ldots$, you have a more reasonable classification of surfaces. His classification of the algebraic surfaces X via the $P_m(X)$ and $q(X)$ is restated by [Kodaira 1,2] and [Shafarevich *et al.*] in the following form:

1.4 Theorem (Enriques classification of surfaces) *Smooth complex projective surfaces fall into the following four classes, two of which are divided into disjoint subclasses. Every class or subclass is stable under birational maps, finite étale morphisms and smooth deformation.*

Class I. $\forall m > 0, P_m(X) = 0$.
  Subclass Ia: $q(X) = 0$: X *is a rational surface.*
  Subclass Ib: $q(X) = 1$: X *is birational to* (*elliptic curve*) $\times \mathbb{P}^1$.
  Subclass Ic: $q(X) \geq 2$: X *is birational to* (*curve of genus* ≥ 2) $\times \mathbb{P}^1$.

Class II. $\forall m > 0, P_m(X) \leq 1$ *and* $\exists m > 0, P_m(X) = 1$.
  Subclass IIa: $q(X) = 0$: X *is birational to a K3 or Enriques surface* (*étale* $\mathbb{Z}/(2)$-*quotient of a K3 surface*).
  Subclass IIb: $q(X) \geq 1$: X *is birational to an étale quotient of an abelian surface* (*abelian or bielliptic surface*).

Class III. $\forall m \gg 0, P_m(X) \sim O(m)$. *For large* m, *the complete linear system* $|mK_X|$ *provides* X *with a canonical structure of a fibre space* $\pi : X \to C$, *of which a general fibre is an elliptic curve* (*properly elliptic surface*).

Class IV. $\forall m \gg 0, P_m(X) \sim O(m^2)$. *The complete linear system* $|mK_X|$ *gives a birational morphism of* X *onto its image in a projective space when* $m \gg 0$ (*surface of general type*).

There are almost complete descriptions of the surfaces in the classes I, II, III.

A most accessible reference to the Enriques classification will be [Beauville]. [Barth-Peters-Van de Ven] extends the table to compact complex surfaces including non-Kähler ones (Kodaira's classification of compact complex surfaces [Kodaira 1][Kodaira 2]). A similar classification in positive characteristics is found in [Mumford 2][Bombieri-Mumford].

Inspired by this classification table, [Iitaka 2] introduced the notion of *Kodaira dimension* $\kappa(X)$ for a given algebraic variety (or compact complex variety) of arbitrary dimension and gave a classification of varieties according to the value of κ. Kodaira dimension is defined by the asymptotic behaviour of the plurigenera P_m, and hence a birational invariant.

1.5 Exercise   (1) Let $\rho : \mathbf{GL}(\mathbb{C}^n) \to \mathbf{GL}(\mathbb{C}^m)$ be a rational representation. Let X be a smooth variety of dimension n. Show that ρ and the vector bundle Ω_X^1 naturally define a vector bundle Ω^ρ of rank m (e.g. Ω_X^r corresponds to the exterior product representation $\mathbf{GL}(\mathbb{C}^n) \to \mathbf{GL}(\wedge^r \mathbb{C}^n)$, while T_X corresponds to the adjoint representation $A \mapsto {}^t A^{-1}$).

(2) Assume that ρ in (1) is a polynomial representation. Then, given a dominant, generically finite morphism $f : Y \to X$, define the pull-back $f^* : \Omega_X^\rho \to \Omega_Y^\rho$.

(3) Let $f : X \dashrightarrow X'$ be a birational (or bimeromorphic) map between smooth, complete varieties over k. Then for any polynomial representation ρ as above, f induces an isomorphism $H^0(X', \Omega^\rho) \simeq H^0(X, \Omega^\rho)$.

Hint: Given a birational morphism $X \to X'$, you can take $U \subset X$ such that U is a complement of a subset of codimension ≥ 2 and $f : U \to X'$ is a morphism. This shows that

$$H^0(X', \Omega^\rho) \hookrightarrow H^0(U, \Omega^\rho) \simeq H^0(X, \Omega^\rho).$$

(4) Show that either $\dim h^0(X, T_X)$ or $\dim h^0(X, \mathcal{O}_X(-K_X))$ is not a birational invariant.

Let X be a smooth projective variety, and K_X its canonical divisor. [Iitaka] proved the following

1.6 Theorem *Let X be a smooth complex projective variety (or simply a compact complex manifold).*

(1) *The Kodaira dimension $\kappa(X)$, defined by $\kappa(X) = \limsup\limits_{m \to \infty} \dfrac{\log P_m}{\log m}$, takes value in $\{-\infty, 0, 1, \ldots, \dim X\}$, where we put $\log 0 = -\infty$ as convention.*

(2) *Assume that $\kappa(X) \geq 0$, so that $P_m(X) \neq 0$ for some $m > 0$. For such m, let φ_m be the rational map from X to $\mathbb{P}^{P_m - 1}$ given by the complete linear system $|mK_X|$. Then $\kappa(X) = \max \dim \varphi_m(X)$.*

(3) *If $0 < \kappa(X) < \dim X$, there exists a positive integer m such that φ_m gives a dominant rational map $X \dashrightarrow Y$, $\dim Y = \kappa(X)$ by (2). Resolve the singularities and the indeterminacy to get a surjective morphism $\pi : X^\# \to Y^\#$ between smooth projective varieties. Then a general fibre of π is connected and has Kodaira dimension zero (for sufficiently large and divisible m).*

The reader is referred to [Ueno 1] for further properties of the Kodaira dimension. The proof of the theorem is not very difficult; the reader should try to give a complete proof.

An important implication of this theorem is that birational classification of complex algebraic varieties reduces to the following three (apparently) independent problems:

(a) Classification of varieties of Kodaira dimension $-\infty$. (Conjecturally these varieties will be uniruled.)

(b) Classification of varieties of Kodaira dimension zero and theory of their moduli. (We have so far no idea how the classification looks like in higher dimension.)

(c) Theory of varieties of maximal Kodaira dimension (varieties of general type).

1.7 Exercise Let X_1 and X_2 be two smooth projective varieties over $\mathbb{C}$.

(1) Show that $P_m(X_1) \geq P_m(X_2)$ and hence $\kappa(X_1) \geq \kappa(X_2)$ if there is a dominant, generically finite morphism $X_1 \to X_2$.

(2) Show that $P_m(X_1) = P_m(X_2)$ and hence $\kappa(X_1) = \kappa(X_2)$ if X_1 is birationally equivalent to X_2.

(3) Show that $\kappa(X_1) = \kappa(X_2)$ if X_1 is an étale cover of X_2.

(4) Show that $\kappa(X \times Y) = \kappa(X) + \kappa(Y)$ (product formula).

(5) Given a fibre space $\pi : X \to Y$ with smooth general fibre F, show that $\kappa(X) \leq \kappa(F) + \dim Y$ (easy addition). In particular, $\kappa(F) \geq 0$ if $\kappa(X) \geq 0$. (*Hint:* Take an ample divisor H such that $\pi_* \mathcal{O}_X(mK_X + m\pi^* H)$ is generated by global sections, and show that $h^0(X, \mathcal{O}_X(mK_X + m\pi^* H))$ grows like $m^{\kappa(F) + \dim Y}$.)

Given a (possibly singular) variety X, the Kodaira dimension $\kappa(X)$ is defined as that of a smooth, projective model of X. Of course this is independent of the choice of smooth models.

1.8 Theorem 1.6 gives a birational classification of n-dimensional varieties into $n + 2$ classes, and is an obvious generalization of the (coarser) Enriques classification of surfaces into four large classes. $\kappa(X)$ is invariant under finite étale morphisms (1.7.3). The deformation invariance of the Kodaira dimension is, however, much more delicate. In the category of compact complex manifolds, Kodaira dimension is not constant under proper smooth deformation, but, conjecturally it is expected to be invariant under projective (or Kähler), smooth deformation (suggesting that the key to the invariance might be Hodge theory or something like that). [Nakayama] showed that, if you assume the existence of minimal models and the abundance conjecture, the plurigenera are invariant under projective smooth deformation. If the deformation invariance is indeed the case, then the coarse Enriques classification smoothly passes over to arbitrary dimension, the varieties of dimension n being divided into $n + 2$ disjoint classes according to their Kodaira dimension, which behaves nicely under standard operations.

Finer Enriques classification (1.4) involves subdivisions of varieties of Kodaira dimension $-\infty$ or 0. Hopefully, there might be higher-dimensional analogues.

Natural subdivision of the varieties of Kodaira dimension $-\infty$ will be discussed in Lecture V.

As for varieties of Kodaira dimension zero, an idea of Iitaka was to use Albanese varieties to get new information (for the definition and properties of Albanese varieties, see (V.1)). (When X is smooth and K_X is numerically trivial, then we know that X is, up to finite étale covering, a product of an abelian variety and a simply connected manifold – Berger-Bogomolov decomposition, see [Beauville 2] – and Iitaka's observation below is easily verified.)

Suppose that a smooth projective variety X has Kodaira dimension zero and that $q(X) > 0$. Then, by Hodge symmetry $q(X) = \dim H^1(X, \mathcal{O}_X) = \dim H^0(X, \Omega^1_X)$, the Albanese variety $\mathrm{Alb}(X)$ is non-trivial, and the Stein factorization of the Albanese morphism $\alpha_X : X \to \mathrm{Alb}(X)$ gives a fibre space structure $\psi : X \to Z$ over

Z, a finite cover of $\alpha_X(X) \subset \mathrm{Alb}(X)$. Suppose that $\alpha_X(X)$ of of dimension r. Noting that there are $q(X)$ independent 1-forms on $\mathrm{Alb}(X)$, we easily see that the wedge product of general r elements define a global r form on (a smooth model of) $\alpha_X(X)$. Thus $\kappa(Z) \geq \kappa(\alpha_X(X)) \geq 0$. Furthermore, the equality $\kappa(\alpha_X(X)) = 0$ is attained if and only if $\alpha_X(X)$ coincides with the total space $\mathrm{Alb}(X)$, see [Ueno 1]. On the other hand, the fibre F of $\psi : X \to Z$ also has Kodaira dimension ≥ 0 by (1.7.5). Hence, if we know the "subadditivity property" for fibre spaces, then we will be able to deduce that X is a quotient of an étale fibre bundle over an abelian variety with fibres of Kodaira dimension zero. Here the "subadditivity" means the following conjecture.

1.9 Subadditivity Conjecture   *Let $\psi : X \to Y$ be a surjective morphism between smooth projective varieties with smooth connected general fibre F. Then $\kappa(X) \geq \kappa(Y) + \kappa(F)$. The equality holds only if two general fibres F_1 and F_2 are mutually birational (i.e. the fibre space is birationally "isotrivial").*

Though still open in general, the conjecture is verified in considerably many cases. Specifically (1.9) is verified when $\dim X = 3$.

1.10   The formulation of the Enriques classification of surfaces stated above was purely of birational nature. Iitaka's original idea also was that you would not need any specific "good models" to get birational classification. However, you would notice that the proof of (1.4) heavily depends on the theory of minimal models once you have a closer look at it. In other words, the Enriques classification is essentially of biholomorphic nature.

In fact, the original Castelnuovo-Enriques theorem states something stronger than what is explicitly stated in the table (1.4).

1.11 Theorem (Castelnuovo-Enriques)   *Let X be a smooth projective surface over an algebraically closed field of characteristic zero.*

(1) *(Minimal models). When X is not ruled (i.e. not in class I in (1.4)), there exists a smooth projective model $X_{\min}$ with the following property:*

— *Any birational map $\tilde{X} \dashrightarrow X_{\min}$ from a smooth $\tilde{X}$ is a morphism.*

By this property, it follows that $X_{\min}$ is unique (up to isomorphisms), and is called the minimal model of X. $X_{\min}$ is explicitly constructed by successively contracting (-1)-curves from a given smooth model X. Furthermore, the minimal model is characterized by a completely numerical property: X is minimal (i.e. X is identical with the minimal model) if and only if K_X is nef.

(2) *(Abundance for surfaces). If X is non-ruled and minimal, then there exists a positive integer m such that the linear system $|mK_X|$ is free from base points; in other words, K_X is semiample (or eventually free). The minimum m_0 of such m is explicitly bounded ($m_0 = 1, 2, 3, 4, 6$ for X in class II (Enriques); $m_0 \leq 84$ for class III [Iitaka 1], $m_0 \leq 4$ for class IV [Bombieri]). In particular, the canonical ring $R = \bigoplus_{i \geq 0} H^0(X, \mathcal{O}(iK_X))$ is a finitely generated graded k-algebra by a theorem of [Mumford 1].*

The minimal model program (MMP) is an attempt to establish a generalization of the statement (1) above, while the abundance conjecture concerns a higher dimensional analogue of (2), connecting minimal models directly to Iitaka's extended classification table in (1.6).

2 Abundance conjecture in general

Assume that we have a minimal variety X. Namely X is a normal, $\mathbb{Q}$-factorial, projective variety with only terminal singularities and its canonical divisor K_X is nef. Then the abundance conjecture is stated as follows:

2.1 Conjecture   *There exists a positive integer m such that mK_X is Cartier and that the linear system $|mK_X|$ is free from base points. Put in another way, K_X is semiample.*

In this section, we study related general facts known to hold in arbitrary dimension.

2.2 Proposition   *Let X be a minimal variety of dimension n. Then X is of general type (i.e. $\kappa(X) = n$) if and only if $K_X^n > 0$. Furthermore, if $\kappa(X) = n$, then K_X is semiample.*

Proof.   By definition, $\kappa(X) = n$ amounts to the bigness of the nef divisor K_X; then the first statement was an exercise (III.3.9.4). The second statement is a direct consequence of the Kawamata's base-point-free theorem [Kawamata 3]. □

Let us define the *numerical Kodaira dimension $\nu(X)$* of a minimal variety X. Fix an ample divisor H, and put

$$\nu(X) = \max\{i|\ K_X^i H^{n-i} \neq 0\} = \max\{i|\ [K_X]^i \neq 0 \in H^{2i}(X, \mathbb{Q})\}.$$

Thus $\nu(X) = 0$ if and only if K_X is numerically trivial and $\nu(X) = n$ if and only if K_X is big.

2.3 Proposition   *Let X be a minimal variety of dimension n. Then we have the inequality $\nu(X) \geq \kappa(X)$.*

Proof.   If $\kappa(X) = 0, -\infty$, there is nothing to prove, and (2.2) meanwhile asserts that $\kappa(X) = n \Leftrightarrow \nu(X) = n$. Assume that $\kappa(X) = r > 0$. Then we have a rational map $\varphi : X \dashrightarrow Y$ to an r-dimensional variety Y via the linear system $|mK_X|$. Take a smooth model $\mu : \tilde{X} \to X$ to resolve the indeterminacy of φ. Let $\tilde{\varphi} : \tilde{X} \to Y$ the resulting morphism. Then there exists an effective divisor (possibly zero) E such that $\mu^* mK_X - E = \tilde{\varphi}^* L$, or equivalently $\mu^* mK_X = \tilde{\varphi}^* L + E$, where L is a very ample divisor on Y. Since both $\tilde{\varphi}^* L$ and $\mu^* mK_X$ are nef, we get

$$(\mu^* mK_X)^i (\tilde{\varphi}^* L)^{r-i} \mu^* H^{n-r} = (\mu^* mK_X)^{i-1} (\tilde{\varphi}^* L + E)(\tilde{\varphi}^* L)^{r-i} \mu^* H^{n-r}$$
$$\geq (\mu^* mK_X)^{i-1} (\tilde{\varphi}^* L)^{r-i+1} \mu^* H^{n-r},$$

to conclude

$$\mu^*((mK_X)^r H^{n-r}) \geq (\tilde{\varphi}^* L)^r (\mu^* H)^{n-r}. \tag{2.3.1}$$

Since L is very ample on Y, $(\varphi^* L)^r$ is the class of r-codimensional, freely moving subvarieties F on $\tilde{X}$. In particular, $\mu(F)$ is $n-r$-dimensional, so that $\mu(F) H^{n-r} > 0$. Now the assertion is clear. $\qquad\square$

2.4 Proposition (*cf.* [Kawamata 2, Theorem 6.1]) *Assume that the minimal model conjecture and the abundance conjecture are true for varieties of dimension $\leq n-1$. Then for an n-dimensional minimal variety X, the Kodaira dimension $\kappa(X)$ is $-\infty, 0$ or $\nu(X)$.*

Proof.   Assume that $\kappa(X) > 0$. Then we have a diagram

$$
\begin{array}{ccc}
X^\# & \xrightarrow{\ \mu\ } & X \\
\downarrow{\scriptstyle\varphi} & & \\
Y & &
\end{array}
$$

where μ is birational and φ gives the Iitaka fibration, i.e. $X^\#$, Y are smooth and projective, and a general fibre W of φ has Kodaira dimension 0. Since X has only terminal singularities, $K_{X^\#} = \mu^* K_X + \Sigma$, where Σ is effective and exceptional with respect to μ. By adjunction, we have $K_W = K_{X^\#}|_W = (\mu^* K_X + \Sigma)|_W$. By changing the model $X^\#$ if necessary, we may assume that there is a morphism $\psi : W \to W_{\min}$, where $W_{\min}$ is a minimal model. Then the abundance conjecture for $W_{\min}$ tells you that $K_{W_{\min}}$ is numerically trivial. Hence K_W is an effective divisor F which is exceptional with respect to ψ. Thus we have $\mu^* K_X|_W = F - E$, where $E = \Sigma|_W$. If there are common components in F and E, cancel them out, to get $\mu^* K_X|_W = F_0 - E_0$, effective divisors F_0, E_0 having no common component. We derive a contradiction assuming that F_0 is non-zero.

Let d be the dimension of $\psi(F_0) \subset W_{\min}$. Fix very ample divisors H' on $W_{\min}$ and H on $X^\#$. Cut out W by a general complete intersection surface S by d hyperplanes on $W_{\min}$ and $(m - d - 2)$ hyperplanes on $X^\#$, where $m = \dim W < n$. Let F_0, E_0 denote also their restrictions to S, by abuse of notation. By construction $F_0 \subset S$ is a non-empty effective divisor, which is contracted to finitely many points on $W_{\min}$. Hence we get an inequality on the intersection number F_0^2 on S:

$$F_0^2 < 0. \tag{2.4.1}$$

Since K_X is nef,

$$F_0^2 - F_0 E_0 = F_0 \mu^* K_X \geq 0. \tag{2.4.2}$$

Comparing (2.4.1) and (2.4.2), you get $F_0 E_0 \leq F_0^2 < 0$. But since F_0 and E_0 has no common component, this is absurd; therefore $F_0 = 0$ on W.

Thus $E_0 = -\mu^* K_X$ is also 0 on W, and $\mu^* K_X|W \sim_{\mathbb{Q}} 0$. In particular $\mu^* K_X$ comes from a divisor on Y (up to torsion), and hence $\nu(X) \leq \dim(Y) = \kappa(X)$. Looking at the converse inequality (2.2), you get the assertion. □

2.5 Theorem *If $\kappa(X) = \nu(X)$ for a minimal variety X, then K_X is semiample, i.e. the abundance conjecture is true for X.*

The proof is omitted here because it requires several technical lemmas. For details, see [Kawamata-Matsuda-Matsuki, Corollary 6.1.13].

3 Non-negativity of the Kodaira dimension of minimal threefolds

In this section, varieties will be defined over the complex numbers.

The main result of this section is the following

3.1 Theorem *Let X be a minimal threefold. Then there exists a positive integer m such that $mK_X \in \mathrm{Pic}(X)$ and that $H^0(X, \mathcal{O}_X(mK_X)) \neq 0$. Equivalently, $\kappa(X) \geq 0$.*

There are four cases to check: (1) $q(X) = \dim H^1(X, \mathcal{O}_X) > 0$; (2) $K_X \approx 0$, $q(X) = 0$; (3) $K_X^2 \not\approx 0$, $q(X) = 0$; and finally (4) $K_X \not\approx 0$, $K_X^2 \approx 0$, $q(X) = 0$. Let us check case by case.

3.2 *Case 1. $q(X) > 0$.*
It is known that terminal singularities are rational singularities (i.e. given a resolution $\mu : \tilde{X} \to X$, $R^i \mu_* \mathcal{O}_{\tilde{X}} = 0$, $i \geq 1$ [Elkik]), which implies among others that the irregularity of a smooth model is equal to $q(X)$. Hence we get non-trivial Albanese map of a smooth model, and the proof of (III.3.14) also applies to this case. □

3.3 *Case 2. $K_X \approx 0$, $q(X) = 0$.*
Since $q(X) = 0$, we see that $\mathrm{Pic}(X) \hookrightarrow H^2(X, \mathbb{Z})$ and that an integral divisor which is numerically equivalent to zero is a torsion in $\mathrm{Pic}(X)$. In other words, there is a positive integer m such that $mK_X \sim 0$, $H^0(X, \mathcal{O}_X(mK_X)) \simeq \mathbb{C}$. □

3.4 *Case 3. $K_X^2 \not\approx 0$.*
Let $\rho : Y \to X$ be a resolution of singularities. By the rationality of terminal singularities, we have $H^i(Y, \mathcal{O}_Y) \simeq H^i(X, \mathcal{O}_X)$, $i = 0, 1, 2, 3$. In case $\chi(X, \mathcal{O}_X) = \chi(Y, \mathcal{O}_Y) \leq 0$, we have $q(Y) \neq 0$ or $p_g(Y) \neq 0$. So we may assume $\chi(X, \mathcal{O}_X) > 0$, without loss of generality. Fix a positive integer r such that rK_X is integral. Then by virtue of:

(a) the Riemann-Roch theorem on Y,

(b) the rationality of the singularities and

(c) the equality $(\rho^* K_X)(K_Y)^2 = (\rho^* K_X)^2 K_Y = (\rho^* K_X)^3 = K_X^3$, (this follows from the fact that a 3-dimensional terminal singularity is isolated, so that we can choose the cohomology class K_X to avoid the singularities), we get

$$\chi(X, \mathcal{O}_X(mrK_X)) = \chi(Y, \mathcal{O}_Y(\rho^* mrK_X))$$
$$= \frac{mr}{12}(2m^2r^2 - 3mr + 1)K_X^3 + \frac{mr}{12}c_2(Y)\rho^* K_X + \chi(Y, \mathcal{O}_Y).$$

Since $K_X^3 \geq 0$, $c_2(Y)\rho^* K_X \geq 0$ and $\chi(Y, \mathcal{O}_Y) > 0$, we have

$$\dim H^0(X, \mathcal{O}_X(mrK_X)) + \dim H^2(X, \mathcal{O}_X(mrK_X)) \geq \chi(X, \mathcal{O}_X(mrK_X)) > 0.$$

On the other hand, we show:

3.5 Lemma   *Under the condition $K_X^2 \not\approx 0$, the second cohomology group $H^2(X, \mathcal{O}_X(mrK_X))$ vanishes, so that $H^0(X, \mathcal{O}_X(mrK_X)) = H^0(Y, \mathcal{O}_Y(mrK_Y)) \neq 0$.*

Proof.   Let $D \subset X$ be a smooth, sufficiently ample effective divisor away from the singularity of X. Consider the short exact sequence

$$0 \longrightarrow \mathcal{O}_X(mrK_X) \longrightarrow \mathcal{O}_X(mrK_X + D) \longrightarrow \mathcal{O}_D(mrK_X + D) \longrightarrow 0$$

and the associated long exact sequence

$$\cdots \longrightarrow H^1(D, \mathcal{O}_D(mrK_X + D)) \longrightarrow H^2(X, \mathcal{O}_X(mrK_X))$$
$$\longrightarrow H^2(X, \mathcal{O}_X(mrK_X + D)).$$

The restriction of K_X on D is a nef and big Cartier divisor. Since D lies on the smooth locus of X, the adjunction formula gives $K_D = (K_X + D)|_D$, so that

$$\dim H^1(D, \mathcal{O}_D(mrK_X + D)) = \dim H^1\big(D, \mathcal{O}_D(-(mr - 1)(K_X|_D))\big) = 0,$$

by the vanishing theorem of Kawamata-Viehweg (see the lecture by Peternell). The third term $H^2(X, \mathcal{O}_X(mrK_X + D))$ also vanishes whenever D is sufficiently ample, showing the vanishing of the middle term as wanted.   □

3.6 *Case 4. $K_X \not\approx 0$, $K_X^2 \approx 0$, $q(X) = 0$.*
     This is the hardest part. We may assume that $H^1(X, \mathcal{O}_X) = H^3(X, \mathcal{O}_X) = 0$, $\chi(X, \mathcal{O}_X) > 0$. Let X° denote the smooth locus of X.

(3.6.A) *Subcase A. The algebraic fundamental group $\pi_1^{\mathrm{alg}}(X^\circ)$ is infinite.*
     A threefold terminal singularity is a finite cyclic quotient of an isolated hypersurface singularity called a compound Du Val singularity [Reid 2, Main Theorem]. Any isolated normal hypersurface singularity of dimension ≥ 3 is known to have trivial local fundamental group [Milnor]. Hence, for each singular point $x \in Sing(X)$, there exists an analytic neighbourhood U such that $\pi_1^{\mathrm{top}}(U \setminus \{x\})$ is a finite cyclic group.

By our assumption, there is an infinite tower of normal, finite, Galois coverings

$$\cdots \to X_2 \to X_1 \to X_0 = X$$

étale over X°. It is easily shown that X_i is a minimal threefold for every i, with K_{X_i} identical with the pull-back of K_X. By the finiteness of local fundamental groups around the singular points, we can find an index m such that $X_i \to X_j$ is étale and hence $\chi(X_i, \mathcal{O}_{X_i}) = [k(X_k) : k(X_j)]\chi(X_j, \mathcal{O}_{X_j})$ for arbitrary i, j with $i > j \geq m$. If $\chi(X_m, \mathcal{O}_{X_m}) \leq 0$, we derive $\kappa(X_m) = \kappa(X) \geq 0$, and thus we may assume that $\chi(X_m, \mathcal{O}_{X_m}) > 0$. In particular for large i, $\chi(X_i, \mathcal{O}_{X_i}) \geq 4$. Since $\kappa(X) = \kappa(X_i)$, we may substitute X by X_i and assume that $m = i = 0$ without loss of generality. By the trivial inequality $1 + \dim H^2(X, \mathcal{O}_X) \geq \chi(X, \mathcal{O}_X)$, we thus get $\dim H^2(X, \mathcal{O}_X) \geq 3$. Take a resolution $\rho : Y \to X$. Then $\dim H^0(Y, \Omega_Y^2) = \dim H^2(Y, \mathcal{O}_Y) = \dim H^2(X, \mathcal{O}_X) \geq 3$, by the Hodge symmetry and the rationality of the singularities. Three independent global sections of Ω_Y^2 give a $\mathcal{O}_X$-homomorphism $\mathcal{O}_Y^{\oplus 3} \to \Omega_Y^2$ in an obvious manner. Take the saturation $\mathcal{E} \subset \Omega_Y^2$ of the image of this homomorphism.

When $\operatorname{rank} \mathcal{E} = 3$ (and hence $\mathcal{E} = \Omega_Y^2$), it readily follows that

$$0 = c_1(\det(\mathcal{O}_Y^{\oplus 3})) \leq c_1(\det \Omega_Y^2) = 2K_Y,$$

deriving $H^0(Y, \mathcal{O}_Y(2K_Y)) \neq 0$. When $\operatorname{rank} \mathcal{E} = 1$ or 2, you can check that the linear system $|\det(\mathcal{E})|$ has dimension at least 1, with non-trivial moving part $|M|$. We want to show that $\rho_* M$ is numerically equivalent to aK_X for some positive rational number a. Once this is checked, we can find a positive integers m, m', such that $mK_X \in \operatorname{Pic}(X)$, $mK_X \approx m'\rho_* M > 0$. Since $q(X) = 0$, the numerical equivalence coincides with the linear equivalence up to torsion and we are done.

Let us check that $\mathbb{Q}\rho_* M = \mathbb{Q}K_X$. Take an ample divisor $H \in \operatorname{Pic}(X)$ and an ample $\mathbb{Q}$-divisor $L \in \operatorname{Pic}_{\mathbb{Q}}(X)$. Since Ω_X^1 is generically semi-positive, so is Ω_X^2. Hence $c_1(\Omega_Y^2/\mathcal{E})\rho^* H\rho^* L \geq 0$, or equivalently $c_1(\mathcal{E})\rho^* H\rho^* \leq 2K_Y\rho^* H\rho^* L$ for any H, L. Letting L tend to the nef divisor K_X, we thus get $c_1(\mathcal{E})\rho^* K_X\rho^* H \leq 2\rho^* K_X^2\rho^* H$. Therefore

$$0 \leq M\rho^* K_X\rho^* H \leq c_1(\mathcal{E})\rho^* K_X\rho^* H \leq 2K_Y\rho^* K_X\rho^* H = 2(K_X)^2 H = 0,$$

or equivalently,

$$(\rho_* M)K_X H = K_X^2 H = 0. \tag{3.6.1}$$

On the other hand, since $|M|$ has no fixed component, M^2 is an effective 1-cycle (intersection of two divisors without common irreducible component is an effective 2-cocycle) and so

$$(\rho_* M)^2 H \geq 0. \tag{3.6.2}$$

The intersection matrix with respect to the divisors (= 1-cycles) $\rho_* M|_H \cdot K_X|_H$ on the surface H is thus

$$\begin{pmatrix} (\rho_* M)^2 H & 0 \\ 0 & 0 \end{pmatrix}$$

with determinant 0. The Hodge index theorem on H then tells us that $\mathbb{Q}(\rho_* M|_H) = \mathbb{Q}(K_X|_H) \subset N^1(H)$, from which we derive $\mathbb{Q}(\rho_* M) = \mathbb{Q}(K_X) \subset N^1(X)$ thanks to the Hard Lefschetz theorem.

(3.6.B) *Subcase B.* $\pi_1^{\mathrm{alg}}(X^\circ)$ *is finite.*

Let $\rho : Y \to X$ be a resolution and take $r \geq 2$ such that rK_X is Cartier. Then by Riemann-Roch, we get $H^0(X, \mathcal{O}_X(rK_X)) \neq 0$ or $H^2(X, \mathcal{O}_X(rK_X) \neq 0$. We may assume that H^2 is non-trivial. Take a smooth, sufficiently ample effective divisor $H \subset X$. Then you can easily check that

$$0 \neq H^2(X, \mathcal{O}_X(rK_X)) \simeq H^1(H, \mathcal{O}_H(rK_X + H)) \simeq H^1(H, \mathcal{O}_H(-(r-1)K_X)).$$

Suppose that there exists a positive constant m, depending only on X and independent of H, such that

$$H^0(H, \mathcal{O}_H(mr(r-1)K_X) \neq 0.$$

Then, by the Serre vanishing theorem, you can show that

$$H^0(X, \mathcal{O}_X(mr(r-1)K_X)) \simeq H^0(H, \mathcal{O}_H(mr(r-1)K_X)) \neq 0$$

for sufficiently ample H.

$\pi_1^{\mathrm{alg}}(H) \simeq \pi_1^{\mathrm{alg}}(X^\circ)$ by [SGA 2, Exposé X, Exemple 2.2]. Furthermore you can show that, if the restriction of a Cartier divisor $D \in \mathrm{Pic}(X)$ to general H is divisible by ι in $\mathrm{Pic}(H)$, then $D|_{X^\circ}$ is also divisible by ι in $\mathrm{Pic}(X^\circ)$, by using the descent technique [Grothendieck 2, Théorème 2.5, Lemme 2.6]. (For the technical details for the proof of the two isomorphisms, see [Miyaoka 3].)

Thus the proof reduces to the following *compensation theorem*:

3.7 Theorem *Let D be a nef divisor with $D^2 = 0$ on a smooth projective surface Z with finite algebraic fundamental group $\pi_1^{\mathrm{alg}}(Z)$ of order g. Put $\iota = [N_{\mathbb{Z}}(Z) \cap \mathbb{Q}D : \mathbb{Z}D]$. Then either*

$$H^1(Z, \mathcal{O}_Z(-D)) = 0, \quad or \tag{a}$$
$$H^0(Z, \mathcal{O}_Z(2\iota! g D)) \neq 0. \tag{b}$$

Proof. The assertion is easily reduced to the case where Z is (algebraically) simply connected. Assume that $H^1(Z, \mathcal{O}(-D)) = \mathrm{Ext}_{\mathcal{O}}^1(\mathcal{O}(D), \mathcal{O}) \neq 0$, and choose an non-trivial extension

$$0 \longrightarrow \mathcal{O} \longrightarrow \mathcal{E} \longrightarrow \mathcal{O}(D) \longrightarrow 0.$$

The vector bundle $\mathcal{E}$ satisfies $c_1(\mathcal{E}) = D$, $c_2(\mathcal{E}) = D^2 = 0$. Fix an ample divisor H.

Case 1. $\mathcal{E}$ is not H-stable.

By definition, there exists an invertible, saturated subsheaf $\mathcal{O}(E) \subset \mathcal{C}$, with

$$EH \geq (1/2)DH. \tag{3.7.1}$$

The natural homomorphism $\mathcal{O}(E) \to \mathcal{O}(D)$ is clearly non-trivial, and is not an isomorphism (if so, $\mathcal{E} \simeq \mathcal{O} \oplus \mathcal{O}(D)$, contradicting our hypothesis). Thus

$$E = D - M, \tag{3.7.2}$$

where M is a non-zero effective divisor with $MH \leq (1/2)DH$.

The injection $\mathcal{O}(E) \hookrightarrow \mathcal{E}$ gives the exact sequence

$$0 \longrightarrow \mathcal{O}(E) \longrightarrow \mathcal{E} \longrightarrow \mathcal{E}/\mathcal{O}(E) \longrightarrow 0.$$

$\mathcal{E}/\mathcal{O}(E)$ is isomorphic to $\mathcal{O}(D - E)$ in codimension one, so that its double dual is the line bundle $\mathcal{O}(D - E)$. From this follows that $\mathcal{E}/\mathcal{O}(E) = I_V(D - E)$, where I_V is the ideal sheaf of a zero-dimensional subscheme $V \subset Z$. Hence

$$0 = c_2(\mathcal{E}) = c_1(\mathcal{O}(E))c_1(I_V(D - E)) + c_2(I_V(D - E)) = E(D - E) + \deg V.$$

Thus we have

$$-E(D - E) = \deg V \geq 0. \tag{3.7.3}$$

For simplicity of the calculation, put $L = E - (1/2)D$. Then we get

$$LH \geq 0,$$

$$L = \frac{1}{2}D - M,$$

$$L^2 = L^2 - \frac{1}{4}D^2 = (L + \frac{1}{2}D)(L - \frac{1}{2}D) = E(E - D) \geq 0,$$

from (3.7.1),(3.7.2),(3.7.3), respectively.

If $L^2 > 0$, then L or $-L$ would be big. Since $LH \geq 0$, the latter is impossible, so L would be big and so must be $(1/2)D = L + M$ since M is effective. But this is impossible because D is nef with $D^2 = 0$, see (III,3.9.4). We thus have $L^2 = 0$. Since D is nef and M is effective, $MD \geq 0$, or equivalently $LD \leq (1/2)D^2 \leq 0$. Consider the intersection matrix A with respect to the three divisors H, D, L.

$$\begin{pmatrix} H^2 & HD & HL \\ HD & D^2 & DL \\ HL & DL & L^2 \end{pmatrix} = \begin{pmatrix} H^2 & HD & HL \\ HD & 0 & DL \\ HL & DL & 0 \end{pmatrix}.$$

If H, D, L are mutually independent then the signature of this symmetric matrix is $(1,2)$ by Hodge index theorem, so that $\det A > 0$ ($H^2 > 0$ rules out the possibility that the signature is $(0,3)$). $\det A = 0$ if the three divisors are dependent. In any case, we have the inequality

$$\det A = 2(HD)(HL)(DL) - (H^2)(DL)^2 \geq 0.$$

Noting that $H^2 > 0$, $HD \geq 0$, $HL \geq 0$, $DL \leq 0$, this implies that $DL = 0$. Thus D, L generates an isotropic subspace in $N^1(Z)$ of signature $(1, \rho - 1)$, whence follows that $\mathbb{R}[L] = \mathbb{R}[D] = \mathbb{R}[M]$ in $N^1(Z)$. Since $\pi_1(Z) = (1)$, $\mathrm{Pic}(Z)$ is a free $\mathbb{Z}$-module, so that there are an integral divisor G and positive integers ι, a such that $D = \iota G$, $M = aG$, $a \leq (1/2)\iota$. Then we see that $cG = (c/\iota)D = (c/a)M$ is effective, where c is the least common multiple of (ι, a). We can easily check that c/ι divides $\iota!$. $\square$

Case 2. $\mathcal{E}$ is H-stable.

We first assume that D is divisible by 2. Then $\mathcal{F} = \mathcal{E}(-\frac{1}{2}D)$ is an H-stable bundle with trivial c_1 and $c_2 = 0$. Then a theorem of S.K. Donaldson (the solution of a conjecture by S. Kobayashi) tells you that $\mathcal{F}$ is a hermitian flat bundle, i.e. it comes from an irreducible representation of $\pi_1^{\mathrm{top}}(X) \to SU(2)$ [Donaldson]. On the other hand, since π_1^{alg} is trivial, we can show that $\mathcal{F}$ is actually isomorphic to $\mathcal{O}^{\oplus 2}$. Hence we get an non-trivial map $\mathcal{O} \to \mathcal{O}(\frac{1}{2}D)$, which proves that D is effective.

If D is not divisible by 2, then we construct a finite $(\mathbb{Z}/2\mathbb{Z})^{\oplus 2}$ covering $f : \tilde{Z} \to Z$ such that
(a) f^*D is divisible by 2;
(b) $\pi_1^{\mathrm{alg}}(\tilde{Z}) = \pi_1^{\mathrm{alg}}(Z)$.
Once such a covering is constructed, the above proof says that $|\frac{1}{2}f^*D|$ is non-empty on $\tilde{Z}$, and then the trace map gives an effective member of $|2D|$ on Z.

Let us construct $\tilde{Z}$. Take an ample divisor L such that $|2L + D|$ is very ample. By choosing three general sections, you get a finite morphism $g : Z \to \mathbb{P}^2$. The pull-back of the hyperplane is of course $2L + D$. Consider the 4:1 morphism $\mathbb{P}^2 \to \mathbb{P}^2$ defined by $(z_0 : z_1 : z_2) \mapsto (z_0^2 : z_1^2 : z_3^2)$. The pull-back of a hyperplane divisor via this map is clearly a quadric hypersurface, and hence divisible by 2. Define $\tilde{Z}$ to be the fibre product $\mathbb{P}^2 \times_{\mathbb{P}^2} Z$ via this morphism. f^*D on $\tilde{Z}$ has clearly Property (a). You can also check that it has Property (b), too, by using the connectedness theorem of [Fulton-Hansen]. For details, see [Miyaoka 3]. $\square$

4   Abundance conjecture for threefolds

In this section, we verify the abundance conjecture for threefolds in some special cases, and give an outline of the proof in general cases. All varieties and schemes in this section are defined over complex numbers.

4.1 Theorem (Abundance in dimension three)   *Let X be a minimal threefold and $K_X \in \mathrm{Pic}_{\mathbb{Q}}(X)$ its canonical divisor. Then there exists a positive integer m such that mK_X is an integral Cartier divisor and that $|mK_X|$ is free from base points, i.e. K_X is semiample.*

By the existence theorem for minimal models of threefolds, this theorem yields the following *finer Enriques classification* in dimension three.

4.2 Corollary   *Complex algebraic threefolds fall into the following five classes according to its Kodaira dimension:*

$\kappa = -\infty$: *the class of the uniruled threefolds.*

$\kappa = 0$: *varieties which have minimal models X with $mK_X \sim 0$ for some positive integer $m = m(X)$.*

$\kappa = 1$: *a minimal model X of a variety in this class has a canonical fibre space structure $\pi : X \to C$ onto a smooth projective curve C, a general fibre of which is a smooth surface isomorphic an abelian surface, a bielliptic surface, a K3 surface, or an Enriques surface. K_X is the pull-back of an ample $\mathbb{Q}$-divisor on C (up to torsion).*

$\kappa = 2$: *a minimal model X in this class has a canonical fibre space structure $\pi : X \to S$ over a normal surface S, a general fibre being a smooth curve of genus 0 (an elliptic threefold structure). K_X is the pull-back of an ample $\mathbb{Q}$-Cartier divisor on S.*

$\kappa = 3$: *There is a positive integer such that $|mK_X|$ defines a birational morphism of X into a subvariety $X_{\mathrm{can}} \subset \mathbb{P}^{\dim |mK_X|}$. X_{can} is uniquely determined (up to isomorphisms) by the birational class of X and called the canonical model.*

For the proof of (4.1), we may assume that $\nu(X)$, the numerical Kodaira dimension, is one or two by (3.8). In particular, we assume that $K_X^3 = 0$, and $H^3(X, \mathcal{O}_X(mK_X)) = 0$ for large, divisible m. (The second assumption amounts to $H^0(Y, \mathcal{O}_Y(K_Y - \rho^* mK_X)) = 0$ by Serre duality, where $\rho : Y \to X$ is a resolution.)

4.3 Lemma   *Let X be a minimal threefold with $\nu(X) = 1$ or 2. Then*

$$\chi(X, \mathcal{O}_X(mK_X)) \geq \chi(X, \mathcal{O}_X),$$

for positive integer m such that mK_X is Cartier. When $\nu(X) = 2$, we have

$$h^0(X, \mathcal{O}_X(mK_X)) \geq h^1(X, \mathcal{O}_X(mK_X)) + \chi(X, \mathcal{O}_X).$$

Proof. Let $\rho : Y \to X$ be a resolution of singularities. Then by the Riemann-Roch, we get

$$\chi(X, \mathcal{O}_X(mK_X)) = \frac{1}{6}(\rho^* mK_X)^3 + \frac{1}{4}(\rho^* mK_X)^2 c_1(Y)$$
$$+ \frac{1}{12}(\rho^* mK_X)(c_1(Y)^2 + c_2(Y)) + \chi(Y, \mathcal{O}_Y).$$

By the rationality of terminal (or, more generally, canonical) singularities, we have $\chi(Y, \mathcal{O}_Y) = \chi(X, \mathcal{O}_X)$. On the other hand, since the cohomology class of mK_X can be taken to avoid the isolated singularities of X, we have

$$(\rho^* mK_X)^2 K_Y = (\rho^* mK_X)^2(\rho^* K_X) = 0,$$
$$(\rho^* mK_X)K_Y^2 = (\rho^* mK_X)(\rho^* K_X)^2 = 0.$$

Noting that $c_1(Y) = c_1(T_Y) = -K_Y$, we get thus

$$\chi(X, \mathcal{O}_X(mK_X)) = \frac{1}{12}(\rho^* mK_X)c_2(Y) + \chi(X, \mathcal{O}_X) \geq \chi(X, \mathcal{O}_X)$$

by (III.3.13). When $\nu(X) = 2$, the term $H^2(X, \mathcal{O}_X(mK_X)) = 0$ by (3.5), whence immediately follows the second assertion.   □

4.4 Corollary *Assume that $\nu(X) = 2$. Let a be a positive integer such that aK_X is Cartier and $|aK_X| \neq \emptyset$. Pick up $S \in |aK_X|$. If $h^1(S, \mathcal{O}_S(maK_X))$ grows like m, then K_X is semiample.*

Proof. Note first that $h^0(S, \mathcal{O}_S(maK_X))$ does not have quadratic growth order in m. Indeed $aK_X|_S$ is clearly nef with $(aK_X|_S)^2 = (aK_X)^2 S = (aK_X)^3 = 0$. Thus $h^0(S, \mathcal{O}_S(maK_X))$ does not grow like m^2.

Consider the exact sequence

$$0 \to \mathcal{O}_X((m-1)aK_X) \to \mathcal{O}_X(maK_X) \to \mathcal{O}_S(maK_X) \to 0$$

and the associated long exact sequence. Then since

$$H^2(X, \mathcal{O}_X((m-1)aK_X)) = H^2(X, \mathcal{O}_X(maK_X)) = 0$$

by (3.5), we get a surjection

$$H^1(X, \mathcal{O}_X(maK_X)) \to H^1(S, \mathcal{O}_X(maK_X)).$$

Therefore $h^1(X, \mathcal{O}_X(maK_X))$ grows at least like m, and so does $h^0(X, \mathcal{O}_X(maK_X))$ by (4.3). This shows that $\kappa(X) \geq 1$, and by (2.4),(2.5), you get the semiampleness of K_X.

4.5 *Proof of Theorem 4.1 when $\nu(X) = 2$: Special case*
    Assume that X is smooth minimal, and that there exists $S \in |aK_X|$ such that S is a smooth surface. Then let us see that $h^1(S, \mathcal{O}_S(maK_X))$ grows like m.
    The adjunction formula says that $K_S = (a+1)K_X|_S$, which is nef with self-intersection zero. Furthermore, for an ample divisor H on X, we have $(K_S, H|_S) = (a+1)K_X SH = a(a+1)K_X^2 H > 0$. Namely the smooth surface S has nef canonical divisor with $\nu(S) = 1$. By the Enriques classification, we have that $\kappa(S) = 1$ (i.e. S is a minimal, properly elliptic surface). On the other hand, applying the Riemann-Roch for the surface S, you get

$$h^0(S, \mathcal{O}_S(mK_X)) - h^1(S, \mathcal{O}_S(mK_X)) = \chi(S, \mathcal{O}_S(mK_S))$$
$$= \frac{1}{2}(mK_X|_S)^2 - \frac{1}{2}(mK_X|_S)K_S + \chi(S, \mathcal{O}_S)$$
$$= \chi(S, \mathcal{O}_S) = (\text{a constant independent of } m).$$

Thus $\kappa(S) = 1$ implies that $h^1(S, \mathcal{O}_S(mK_X))$ grows like m, which proves the assertion by (4.4).   □

4.6 *Proof of Theorem 4.1 when $\nu(X) = 2$ - Outline*

It suffices to show that $h^1(S, \mathcal{O}_S(maK_X))$ grows like m, where $S \in |aK_X|$. There are three obstacles to imitating (4.5).

(1) S may hit the singular locus of X;
(2) Components of S may be very singular;
(3) S may be reducible and non-reduced.

To overcome the difficulties (1) and (2), we can take a resolution $\rho : Y \to X$ such that the pull-back $\rho^*S \in |\rho^*mK_X|$ is supported by a divisor with simple normal crossings. But, then, the multiplicities of the components of ρ^*S would be even higher.

Y. Kawamata's idea was to consider a log-minimal model of the pair $(Y, (\rho^*S)_{\text{red}})$. Roughly speaking, he was able to construct a new model (Z, E) with only log-terminal singularities, with E being reduced and nef on Z. Then E looks like a degeneration of properly elliptic surfaces, and he proves that $h^0(E, \mathcal{O}_E(mK_Z))$ and $h^1(E, \mathcal{O}_E(mK_Z))$ actually grow linearly in m, to show the desired result. For details, the reader is referred to his original paper [Kawamata 6].   □

Now consider the case $\nu(X) = 1$.

4.7 Lemma   *Assume that $\nu(X) = 1$ and pick up $S \in |aK_X|$. Let $mS \subset X$ be the non-reduced closed subscheme defined by the ideal sheaf $\mathcal{O}_X(-mS)$. If $h^0(mS, \mathcal{O}_{mS}(mS))$ grows like m, then K_X is semiample.*

Proof.   Consider the exact sequence

$$0 \to \mathcal{O}_X \to \mathcal{O}_X(mS) \to \mathcal{O}_{mS}(mS) \to 0.$$

Since the terms $H^i(X, \mathcal{O}_X)$ are independent of m, our condition on $h^0(mS, \mathcal{O}_{mS}(mS))$ implies that $h^0(X, \mathcal{O}_X(mS))$ also grows like m. In particular, there exists m such that $h^0(X, \mathcal{O}_X(mS)) > h^0(X, \mathcal{O}_X((m-1)S))$, which induces a non-trivial homomorphism from $H^0(X, \mathcal{O}_X(mS))$ to $H^0(S, \mathcal{O}_S(mS))$. Since $S|_S$ is a numerically trivial divisor (thanks to $S^2 \approx 0$), this means that $\mathcal{O}_S(mS) \simeq \mathcal{O}_S$. In particular, a global section $\alpha \in H^0(X, \mathcal{O}_X(mS))$ which is non-trivially mapped to $H^0(S, \mathcal{O}_S(mS))$ has no zero on S. By definition, $|mS|$ has no base point outside S, and we have now directly checked that $|mS| = |maK_X|$ is free from base points, without appealing to (2.4) or (2.5).   □

In view of this lemma, the proof of the three-dimensional abundance conjecture in case $\nu = 1$ is reduced to the following

4.8 Theorem   *Let X be a minimal threefold with $\nu(X) = 1$. Take an effective divisor $S \in |aK_X|$, $a > 0$. Then $h^0(mS, \mathcal{O}_{mS}(mS))$ grows like m.*

4.9 *Proof of (4.8): Special case.*

Assume that $S \in |aK_X|$, $a > 0$ is smooth (in particular, S is away from the singular locus of X) with trivial normal bundle $\mathcal{O}_S(S) \simeq \mathcal{O}_S$ and trivial canonical bundle $K_S \sim 0$ (hence S is a K3 surface or an abelian surface).

It is known that there is a universal smooth deformation of S, a K3 or abelian surface. More precisely, there is a smooth proper holomorphic map $\mathcal{S} \to T$, where T is a small disk in $\mathbb{C}^n$, $n = \dim H^1(S, T_S)$, which has the following two properties:

(a) The central fibre S_0 over the origin $0 \in T$ is isomorphic to S, and the relative canonical divisor $K_{\mathcal{S}/T}$ is trivial.

(b) Given a smooth morphism $\pi : V \to \operatorname{Spec} A$ over the spectrum of an Artinian local $\mathbb{C}$-algebra such that $V_o = (A/\mathfrak{M}) \otimes_A V = S$, there exists a unique $\mathbb{C}$-morphism $\phi : \operatorname{Spec} A \to T$ such that (i) $\phi(o) = 0$, and that (ii) $V \simeq \operatorname{Spec} A \times_T S$. Here $o \in \operatorname{Spec} A$ is the unique closed point. (Namely, V is the pull-back of the family $\mathcal{S}$ via the morphism ϕ.)

The base space T is called (a small neighbourhood in) the *Kuranishi space*.

Let us show the following

4.10 Proposition   *Let the notation and the assumption as in (4.9). Then, for any positive integer m, we have:*

$(a)_m$ $\mathcal{O}_{mS}(S) \simeq \mathcal{O}_{mS}$.

$(b)_m$ *There exists a k-morphism $\phi_m : \operatorname{Spec} \mathbb{C}[\varepsilon]/(\varepsilon^m) \to T$ which induces an isomorphism*

$$mS \simeq \operatorname{Spec} \mathbb{C}[\varepsilon]/(\varepsilon^m) \times_T \mathcal{S}.$$

$(c)_m$ *The restriction map $H^0(mS, \mathcal{O}_{mS}(bS)) \to H^0(cS, \mathcal{O}_{cS}(bS))$ is surjective for all $c \le m$ and $b \in \mathbb{Z}$.*

In particular, $h^0(mS, \mathcal{O}_{mS}(mS)) = h^0(mS, \mathcal{O}_{mS}) = \dim_{\mathbb{C}} \mathbb{C}[\varepsilon]/(\varepsilon^m) = m$.

Proof.   We begin with fixing the notation. Choose a small analytic neighbourhood M of S, which is an open complex manifold. Cover M by Stein open subsets U_i, and let $f_i \in H^0(U_i, \mathcal{O}_M)$ be a defining equation of $S \cap U_i$. On $U_i \cap U_j$, there is a non-vanishing function $\sigma_{ij} \in H^0(U_i \cap U_j, \mathcal{O}_M^*)$ such that

$$f_i = \sigma_{ij} f_j.$$

In particular $\{f_i\}$ is a global section of the line bundle $\mathcal{O}_M(S)$ defined by the transition functions $\{\sigma_{ij}\}$.

The proof is by induction on m. $(a)_1$ and $(c)_1$ are both trivial by assumption and definition. The morphism ϕ_1 in $(b)_1$ is just the constant map $\operatorname{Spec} \mathbb{C} \to \{0\} \subset T$.

Let us prove the case $m = 2$. By the isomorphism $\mathcal{O}_S(S) \simeq \mathcal{O}_S$, we can choose a everywhere non-vanishing section $s = \{s_i\} \in H^0(S, \mathcal{O}_S(S))$, the s_i being subject to the relation

$$s_i = \sigma_{ij} s_j, \quad s_i \in H^0(U_i \cap S, \mathcal{O}_S^*).$$

Take local liftings $\tilde{s}_i \in H^0(U_i, \mathcal{O}_M)$, and define the divisor $\tilde{S}_i \subset \operatorname{Spec}\mathbb{C}[\varepsilon]/(\varepsilon^2) \times U_i$ by the equation

$$f_i - \varepsilon \tilde{s}_i = 0.$$

The system of divisors $\{\tilde{S}_i\}$ determines a global divisor $\tilde{S}$ on $\operatorname{Spec}\mathbb{C}[\varepsilon]/(\varepsilon)^2 \times M$. Indeed, on $U_i \cap U_j$,

$$
\begin{aligned}
(f_j - &\varepsilon\tilde{s}_j)\big(\mathbb{C}[\varepsilon]/(\varepsilon)^2 \otimes \mathcal{O}_M\big) \\
&= \sigma_{ij}(f_j - \varepsilon\tilde{s}_j)\big(\mathbb{C}[\varepsilon]/(\varepsilon)^2 \otimes \mathcal{O}_M\big) \qquad \text{by } \sigma_{ij} \in \mathcal{O}_M^* \\
&= (f_i - \varepsilon\sigma_{ij}\tilde{s}_j)\big(\mathbb{C}[\varepsilon]/(\varepsilon)^2 \otimes \mathcal{O}_M\big) \\
&= \big\{(f_i - \varepsilon\tilde{s}_i + \varepsilon(\tilde{s}_i - \sigma_{ij}\tilde{s}_j))\big\}\big(\mathbb{C}[\varepsilon]/(\varepsilon)^2 \otimes \mathcal{O}_M\big) \\
&\subset (f_i - \varepsilon\tilde{s}_i)\big(\mathbb{C}[\varepsilon]/(\varepsilon)^2 \otimes \mathcal{O}_M\big) + \varepsilon(\tilde{s}_i - \sigma_{ij}\tilde{s}_j)\mathcal{O}_M \qquad \text{by } \varepsilon^2 = 0 \\
&\subset (f_i - \varepsilon\tilde{s}_i)\big(\mathbb{C}[\varepsilon]/(\varepsilon)^2 \otimes \mathcal{O}_M\big) + \varepsilon f_i \mathcal{O}_M \qquad \text{by } \tilde{s}_i - \sigma_{ij}\tilde{s}_j \in f_i \mathcal{O}_M \\
&= (f_i - \varepsilon\tilde{s}_i)\big(\mathbb{C}[\varepsilon]/(\varepsilon)^2 \otimes \mathcal{O}_M\big) + \varepsilon(f_i - \varepsilon\tilde{s}_i)\mathcal{O}_M \qquad \text{by } \varepsilon^2 = 0 \\
&= (f_i - \varepsilon\tilde{s}_i)\big(\mathbb{C}[\varepsilon]/(\varepsilon)^2 \otimes \mathcal{O}_M\big).
\end{aligned}
$$

By symmetry, you also get the converse inclusion, and hence the ideal generated by $f_i - \varepsilon\tilde{s}_i$ does not depend on i. Consider the natural projections $\operatorname{pr}_1 : \tilde{S} \to \operatorname{Spec}\mathbb{C}[\varepsilon]/(\varepsilon^2)$ and $\operatorname{pr}_M : \tilde{S} \to M$. The first projection pr_1 gives a natural exact sequence

$$0 \to \varepsilon\mathcal{O}_S \to \mathcal{O}_{\tilde{S}} \to \mathcal{O}_S \to 0,$$

showing that $\tilde{S}$ is flat and smooth over $\operatorname{Spec}\mathbb{C}[\varepsilon]/(\varepsilon)^2$. Hence by the universal property of the family $\mathcal{S} \to T$, we get a unique morphism $\phi_2 : \operatorname{Spec}\mathbb{C}[\varepsilon]/(\varepsilon)^2 \to T$ such that $\tilde{S}$ is the pull-back of $\mathcal{S}$.

Let us prove that the ring homomorphism $\operatorname{pr}_M^* : \mathcal{O}_M \to \mathcal{O}_{\tilde{S}}$ is surjective (i.e. $\operatorname{pr}_M : \tilde{S} \to M \subset X$ is a closed immersion). To see this, it suffices to show that ε is contained in the image, which follows from the identity $f_i = \varepsilon\tilde{s}_i$ on $\mathcal{O}_{\tilde{S}}$ and the fact that $\tilde{s}_i$ is invertible (near S). Let us determine the kernel of pr_M^*. By definition and easy calculation

$$
\begin{aligned}
\operatorname{Ker}\operatorname{pr}_M^* &= \mathcal{O}_M \cap \big((f_i - \varepsilon\tilde{s}_i)(\mathbb{C}[\varepsilon]/(\varepsilon)^2 \otimes \mathcal{O}_M)\big) \\
&= f_i^2 \mathcal{O}_M = \mathcal{O}_M(-2S).
\end{aligned}
$$

In particular, pr_M gives a natural isomorphism $\tilde{S} \simeq 2S$, which verifies $(b)_2$. Furthermore, $\mathcal{O}_{2S} \to \mathcal{O}_S$ is trivially surjective. To see $(a)_2$, we compute the dualizing sheaf $\omega_{2S} \simeq \omega_{\tilde{S}}$ in two different ways. Since $2S$ is a Cartier divisor on M,

$$\omega_{2S} = \mathcal{O}_{2S}(K_M + 2S) = \mathcal{O}_S((2a+1)K_X).$$

On the other hand, since $\tilde{S}$ is the pull-back of $\mathcal{S}$,

$$
\begin{aligned}
\omega_{\tilde{S}} &= \operatorname{pr}_1^* \omega_{\operatorname{Spec}\mathbb{C}[\varepsilon]/(\varepsilon)^2} \otimes_{\mathcal{O}_S} \omega_{\mathcal{S}/T} \\
&= \mathcal{O}_{\tilde{S}} \otimes_{\mathcal{O}_S} \otimes \mathcal{O}_S = \mathcal{O}_{\tilde{S}}.
\end{aligned}
$$

Thus $\mathcal{O}_{2S}((2a+1)K_X) \simeq \mathcal{O}_{2S}$, and hence K_X is a torsion element in $\mathrm{Pic}(2S)$ since $2a+1 > 0$. By our assumptions $\mathcal{O}_S(S) = \mathcal{O}_S(aK_X) \simeq \mathcal{O}_S(K_S) \simeq \mathcal{O}_S$, we have

$$K_X \in \mathrm{Ker}\big(\mathrm{Pic}(2S) \to \mathrm{Pic}(S)\big),$$

so that $K_X|_{2S}$ is a torsion element in $\mathrm{Ker}\big(\mathrm{Pic}(2S) \to \mathrm{Pic}(S)\big)$. Let us see that K_X is trivial on $2S$, by looking at the commutative diagram consisting of the exponential sequences:

$$
\begin{array}{ccccccc}
H^1(2S,\mathbb{Z}) & \xrightarrow{\quad i \quad} & H^1(2S,\mathcal{O}) & \longrightarrow & \mathrm{Pic}(2S) & \longrightarrow & H^2(2S,\mathbb{Z}) \\
\Big\downarrow{\simeq} & & \Big\downarrow & & \Big\downarrow & & \Big\downarrow{\simeq} \\
H^1(S,\mathbb{Z}) & \xrightarrow{\quad j \quad} & H^1(S,\mathcal{O}) & \longrightarrow & \mathrm{Pic}(S) & \longrightarrow & H^2(2S,\mathbb{Z}).
\end{array}
$$

Since S is a compact Kähler manifold, j is an injection, and hence so is i. Consequently,

$$\mathrm{Ker}\big(\mathrm{Pic}(2S) \to \mathrm{Pic}(S)\big) = \mathrm{Ker}\big(H^1(2S,\mathcal{O}) \to H^1(S,\mathcal{O})\big)$$

is a vector group, with no torsion. Thus $\mathcal{O}_{2S}(S) = \mathcal{O}_{2S}(aK_X) \simeq \mathcal{O}_{2S}$, showing $(a)_2$, and $(c)_2$.

Let us prove $(a)_m, (b)_m, (c)_m$ by assuming $(a)_{m-1}, (b)_{m-1}, (c)_{m-1}$, where $m \geq 3$.

Identify $\mathcal{O}_{(m-1)S}$ with the flat $\mathbb{C}[\varepsilon]/(\varepsilon)^{m-1}$ algebra $\mathbb{C}[\varepsilon]/(\varepsilon)^{m-1} \otimes_{\mathcal{O}_T} \mathcal{O}_S$ via the morphism ϕ_{m-1}. Since

$$S \subset (m-1)S \simeq \mathbb{C}[\varepsilon]/(\varepsilon)^{m-1} \otimes_{\mathcal{O}_T} S$$

is defined by f_i in M and by ε on $\mathbb{C}[\varepsilon]/(\varepsilon)^{m-1} \otimes_{\mathcal{O}_T} S$, we get

$$\varepsilon \mathcal{O}_{(m-1)S} = f_i \mathcal{O}_{(m-1)S},$$

or equivalently,

$$\varepsilon \equiv f_i \alpha_i \bmod f_i^{m-2} \mathcal{O}_M, \quad \exists \alpha_i \in H^0(U_i, \mathcal{O}_M^*).$$

Then

$$f_i(\alpha_i - \sigma_{ij}^{-1}\alpha_j) = f_i\alpha_i - f_j\alpha_j \equiv \varepsilon - \varepsilon = 0 \bmod f_i^{m-1}\mathcal{O}_M,$$
$$\alpha_i \equiv \sigma_{ij}^{-1}\alpha_j \bmod f_i^{m-2}\mathcal{O}_M,$$

showing that $\{\alpha_i\}$ is a nowhere vanishing global section α of $\mathcal{O}_{(m-2)S}(-S)$. By $(c)_{m-1}$, we can lift α to a global section $\tilde{\alpha} = \{\tilde{\alpha}_i\}$ of $\mathcal{O}_{(m-1)S}(-S)$, which satisfies

$$\tilde{\alpha}_i \equiv \sigma_{ij}^{-1}\tilde{\alpha}_j \bmod f_i^{m-1}\mathcal{O}_M.$$

We define a $\mathbb{C}[\varepsilon]/(\varepsilon)^m$-algebra structure on $\mathcal{O}_{mS}$ by

$$\varepsilon g = (f_i \tilde{\alpha}_i)g, \quad g \in \mathcal{O}_{mS}.$$

This multiplication is well-defined by

$$f_i \tilde{\alpha}_i - f_j \tilde{\alpha}_j = (\sigma_{ij} f_j)(\sigma_{ij}^{-1} \tilde{\alpha}_j + \tilde{\alpha}_i - \sigma_{ij}^{-1} \tilde{\alpha}_j) - f_j \tilde{\alpha}_j$$
$$= \sigma_{ij} f_j (\tilde{\alpha}_i - \sigma_{ij}^{-1} \tilde{\alpha}_j) \in \sigma_{ij} f_j f_j^{m-1} \mathcal{O}_M = f_j^m \mathcal{O}_M,$$

and hence lifts the $\mathbb{C}[\varepsilon]/(\varepsilon)^m$-algebra structure $(= \mathbb{C}[\varepsilon]/(\varepsilon)^{m-1}$-algebra structure) of $\mathcal{O}_{(m-1)S}$. By the natural exact sequence

$$0 \to \varepsilon^m \mathcal{O}_S \to \mathcal{O}_{mS} \to \mathcal{O}_{(m-1)S} \to 0,$$

we infer that $\mathrm{pr}_1 : mS \to \operatorname{Spec} \mathbb{C}[\varepsilon]/(\varepsilon)^m$ is flat and smooth. By the universality, we get $\phi_m : \operatorname{Spec} \mathbb{C}[\varepsilon]/(\varepsilon)^m \to T$, which induces the isomorphism

$$mS \simeq \operatorname{Spec} \mathbb{C}[\varepsilon]/(\varepsilon)^m \times_T \mathcal{S},$$

verifying $(b)_m$. We compute the dualizing sheaf again in two ways to get

$$\mathcal{O}_{mS}((am+1)K_X) \simeq \omega_{mS} \simeq \mathcal{O}_{mS}.$$

Hence we see that K_X is a torsion element in $\operatorname{Ker}(\operatorname{Pic}(mS) \to \operatorname{Pic}(S))$, which is a vector group. Thus we have an isomorphism $\mathcal{O}_{mS}(K_X) \simeq \mathcal{O}_{mS}$, to show $(a)_m$, and now $(c)_m$ is clear in view of the natural surjection $\mathcal{O}_{mS} \to \mathcal{O}_{cS}$. $\quad\rlap{\sqsupset}$

4.11 *Proof of (4.8) in the general case – Outline.* Again in general, we have difficulties because S might be singular and reducible, non-reduced. Furthermore, the normal divisor $S|_S$ and the canonical divisor K_S might not be trivial. The second difficulty is not very serious, if you replace the formal neighbourhood of S by an a finite cover of it (unramified outside the singular locus of S). So take a suitable resolution $\rho : Y \to X$ and look at the pull-back $\tilde{S} = \rho^* S$, which is supported by a divisor with only simple normal crossings, but highly non-reduced.

There are two ways to reduce the problem to a statement on a reduced divisor. The first method [Kawamata 6] is to consider the log-minimal model of the pair $(Y, \tilde{S}_{\mathrm{red}})$, (theory of three-dimensional log-minimal models was initiated by [Shokurov]). The second is to take a semi-stable reduction [Miyaoka 4], following the techniques of [Kulikov] and [Persson-Pinkham]. By either method, you have only to look at a formal neighbourhood of a reduced nef divisor, and a similar argument as in (4.10) works. $\quad\square$

Historical Comments

The classification of algebraic surfaces was essentially due to the great Italian algebraic geometers G. Castelnuovo and F. Enriques. Castelnuovo's theory of minimal models took rigorous form (in any characteristic) by the work of O. Zariski [Zariski 1]. K. Kodaira extended the classification to the compact complex surfaces, and specifically gave decisive results on the structure of elliptic surfaces [Kodaira 1][Kodaira 2]. Enriques classification was restated in a lecture note [Shafarevich *et al.*], in which the symbol κ was introduced. The analogous classification in positive characteristics was given by [Mumford 1][Bombieri-Mumford].

As was mentioned before, the overall classification for higher dimensional varieties via κ was first formulated by S. Iitaka, who also proposed many intriguing questions, including the subadditivity conjecture. The conjecture was extensively studied by K.Ueno, E. Viehweg, T. Fujita, Y. Kawamata, J. Kollár and many others. Results on subadditivity obtained so far are derived from "semipositivity" of the direct image sheaves of relative pluricanonical bundles. The semipositivity was first discovered by [Fujita] for direct images of relative canonical sheaves and was generalized to relative pluricanonical systems by many authors. The proof heavily depends on Hodge theory, especially results of [Griffiths 2] and [Schmid]. For the current state of the conjecture and semipositivity, [Viehweg 4] is the most extensive reference available.

The abundance conjecture was first formulated by M. Reid, and then by Y. Kawamata. Essentially the proof of the abundance in dimension three depends on the following three results:

(1) Subadditivity for fibre spaces of relative dimension one or two;
(2) Analysis of pluricanonical systems on singular surfaces with nef canonical divisor;
(3) Non-negativity of the Kodaira dimension via the pseudo-effectivity of c_2 for minimal varieties.

The importance of (1) to the classification theory was well known from the 1970's. It was perhaps M. Reid who first realized the critical importance of (2) and (3) for the study of threefolds (private discussion with the author at Warwick in 1983 and correspondence in 1985). The first two results are essentially statements on low-dimensional varieties; when we have nice understanding of threefolds, then it may well happen that analogues of (1) and (2) are available for the four-dimensional abundance conjecture. However, we have so far no idea how to find a member of pluricanonical linear system in dimension four or higher.

Lecture V
Rationally Connected Fibrations and Applications

Overview

In this lecture, we study the finer structure of uniruled varieties. We show that a uniruled variety has a canonical structure (*MRC-fibration*) with *maximally rationally connected* fibres. This structure provides us a splitting of a uniruled variety into rationally connected varieties and a non-uniruled variety. *Rational connectedness* is a natural generalization of unirationality, and in dimension two or three, we can completely characterize rationally connected varieties in terms of global holomorphic differential forms. Thanks to the MRC-fibrations, we get a classification of the complex uniruled threefolds into three clearly distinguished classes.

Except in Section 5, all varieties in this section are defined over the complex numbers, and are often viewed as complex manifolds.

1   Structure of ruled surfaces

1.1 Before going to general theory, it would be worthwhile to review birational theory of ruled surfaces. By ruled surfaces, we mean smooth projective surfaces which are birationally equivalent to $\mathbb{P}^1 \times C$, where C is a smooth projective curve. A theorem of Enriques [Beuville, VI,18], [Barth-Peters-Van de Ven, VI.1.1] states that a smooth projective surface X is ruled if and only if its 12-genus $P_{12}(X) = \dim H^0(X, \mathcal{O}_X(12K_X))$ vanishes. A ruled surface X birational to $\mathbb{P}^1 \times C$ is rational if and only if C is rational, or equivalently, the irregularity $q(X)$ vanishes.

There is a remarkable difference between rational surfaces and irrational ruled surfaces, the most important one being that an irrational ruled surface has non-trivial Albanese variety.

1.2 Given an algebraic variety over an algebraically closed field k (or, alternatively, a compact Kähler manifold) X, the *Albanese variety* $\mathrm{Alb}(X)$ is an abelian variety over k (or a complex torus) together with a morphism (the *Albanese mapping*) α_X characterized by the following universal property [Weil 1][Weil 2]:

If $f : X \to A$ is a k-morphism (or a holomorphic mapping) to an abelian variety (or a complex torus), then there exists a morphism g, unique up to translations on $\mathrm{Alb}(X)$, such that the diagram

$$
\begin{array}{ccc}
X & \xrightarrow{\ id\ } & X \\
\alpha_X \downarrow & & \downarrow f \\
\mathrm{Alb}(X) & \xrightarrow{\ g\ } & A
\end{array}
$$

is commutative.

In particular, $\mathrm{Alb}(X)$ and an Albanese mapping α_X are unique up to isomorphisms and translations, respectively.

When X is a complex smooth algebraic variety or a compact Kähler manifold, we have a very explicit description of $\mathrm{Alb}(X)$ and α_X. $\mathrm{Alb}(X)$ is defined to be the complex torus $H^0(X, \Omega_X^1)^*/H_1(X, \mathbb{Z})$, where the symbol $*$ denotes the dual $\mathbb{C}$-vector space. By Hodge theory, there is a canonical injection $H^0(X, \Omega_X^1) \hookrightarrow H^1(X, \mathbb{C})$, which induces a surjection $H_1(X, \mathbb{C}) \to H^0(X, \Omega_X^1)^*$, with kernel $H^1(X, \mathcal{O}_X)^*$. The image of the integral homology group $H_1(X, \mathbb{Z})$ forms a lattice in $H^0(X, \Omega_X)^*$ because the kernel $H^1(X, \mathcal{O}_X)^*$ satisfies the totally imaginary condition

$$H^1(X, \mathcal{O}_X)^* \cap \overline{H^1(X, \mathcal{O}_X)^*} = 0,$$

$$H^1(X, \mathcal{O}_X)^* + \overline{H^1(X, \mathcal{O}_X)^*} = H_1(X, \mathbb{C}),$$

where $\overline{}$ denotes the complex conjugation with respect to the natural $\mathbb{C}$-vector space structure of $H_1(X, \mathbb{C})$.

An Albanese map α_X is defined in terms of *period integrals*. Fix a basis $\{\omega_1, \ldots, \omega_g\}$ of $H^0(X, \Omega_X^1)$ and a base point $o \in X$. Given a point $x \in X$, choose a path γ from o to x on X. Then the integral

$$\alpha_{X,\gamma}(x) = \left(\int_\gamma \omega_1, \ldots, \int_\gamma \omega_g \right) \in \mathbb{C}^g$$

along γ is well-defined; in other words, the path γ from o to x gives an element $\in H^0(X, \Omega_X^1)^*$. The dependence on the choice of γ takes value in the integrals of $(\omega_1, \ldots, \omega_g)$ on the cycles in $H_1(X, \mathbb{Z})$. Thus $\alpha_X(x) = \alpha_{X,\gamma}(x)$ modulo the integrals on $H_1(X, \mathbb{Z}) \subset H^0(X, \Omega_X)^*$ defines a point $\in \mathrm{Alb}(X)$ uniquely determined by the triple (X, x, o). It is easy to see that a different choice of o gives a translation in $\mathrm{Alb}(X)$.

If X is a smooth projective curve C, $\mathrm{Alb}(X)$ is nothing but the Jacobian variety $J(C)$. The Albanese map embeds C of genus $g \geq 1$ into $\mathrm{Alb}(C) = J(C)$.

1.3 Exercise (1) Show that the pair $(\mathrm{Alb}(X), \alpha_X)$ constructed above has the universal property for the morphisms of X to complex tori. (*Hint:* Given a morphism $g : X \to A$, there are natural homomorphisms $g_* : H_1(X, \mathbb{Z}) \to H_1(A, \mathbb{Z})$, ${}^tg^* : H^0(X, \Omega_X^1)^* \to H^0(A, \Omega_A^1)^*$, and A is identified with $\mathrm{Lie}(A)/H_1(A, \mathbb{Z}) = H^0(A, \Omega_A^1)^*/H_1(A, \mathbb{Z})$.)

(2) Show that $\mathrm{Alb}(X)$ is projective if X is complex projective, by checking the Riemann bilinear relation, cited as a theorem of Lefschetz in [Mumford 3, p. 29]. (*Hint:* Represent the Chern class of a very ample line bundle H by $\eta \in H^2(X, \mathbb{Z})$. Define a $\mathbb{Z}$-valued alternating form E^* on $H^1(X, \mathbb{Z})$ by $E^*(a, b) = (\eta^{n-1} \wedge a \wedge b)([X])$, where n is the dimension of X and $([X])$ denotes the evaluation on the fundamental class $[X] \in H_{2n}(X)$. E^* naturally extends to an alternating form on $H^1(X, \mathbb{C})$, and $H^*(a, b) = \sqrt{-1}E^*(a, Jb)$ is a hermitian form. Check that for $a, b \in H^0(X, \Omega_X^1) \subset H^1(X, \mathbb{C})$,

$$H^*(a, b) = \int_X \sqrt{-1}a \wedge \bar{b} \wedge \eta^{n-1} = \int_\Gamma \sqrt{-1}a \wedge \bar{b},$$

where η is regarded as a real (1,1)-form on X and Γ is a complete intersection curve cut out by $n - 1$ hyperplanes linearly equivalent to H. It follows that H^* is positive definite.

The coupling E^* induces a $\mathbb{Z}$-valued alternating form E on the dual lattice $H_1(X, \mathbb{Z})$ in an obvious manner, and $H(\cdot, \cdot) = \sqrt{-1}E(\cdot, J\cdot)$ is positive definite.)

1.4 For an irrational ruled surface X birationally equivalent to $\mathbb{P}^1 \times C$, we have a natural isomorphism $\mathrm{pr}_2^* : H^0(C, \Omega_C^1) \xrightarrow{\sim} H^0(X, \Omega_X^1)$ for a ruled surface $X \sim \mathbb{P}^1 \times C$, and hence the image of the Albanese map α_X is identical with the that of α_C, naturally identified with C. The Albanese mapping imposes strong conditions on the birational structure of X.

Firstly, if X is an irrational surface, there is a unique morphism $\alpha : X - C$ such that a general fibre is $\mathbb{P}^1$, and any birational map σ between X and $\mathbb{P}^1 \times C$ is a *relative* birational map over C. More precisely, the following diagram is always commutative:

$$
\begin{array}{ccc}
X & \xrightarrow{\ \sigma\ } & \mathbb{P}^1 \times C \\
\alpha_X \downarrow & & \downarrow \mathrm{pr}_2 \\
C & \xrightarrow{\ \tau\ } & C \, ,
\end{array}
$$

where τ is an isomorphism. The morphism α is thus canonical and is identical with the Albanese mapping. In particular, the group of birational automorphism (k-automorphism group of $k(X)$) is very simple; i.e. there is an exact sequence

$$1 \to \mathbf{PGL}(2, k(C)) \to \mathrm{Aut}(k(X)) \to \mathrm{Aut}(C) \to 1.$$

On the other hand, the birational automorphism group of a rational surface, called the *Cremona group*, is much more complicated.

Secondly, there are far fewer rational curves on irrational ruled surfaces than on rational surfaces. Indeed, since an abelian variety (or a complex torus) does not carry a rational curve, any rational curve on X is contained in a fibre of α_X. In particular, given a point x on X, there are only finitely many rational curves that pass through x (in case X is geometrically ruled, there is exactly one rational curve), each being smooth. This means that every rational curve C on X is smooth and has non-positive self intersection, i.e. the normal bundle is not ample. In particular, $T_X|_C$ is never ample.

On the other hand, the rational surface carries lot of rational curves. Indeed any two points on a rational surface can be joined by a rational curve on it, and a smooth projective surface is rational if and only there exists an irreducible smooth rational curve C with $C^2 > 0$. (Smoothness is important. In fact, a general algebraic K3 surface has Picard number one, and hence every effective curve on it is an ample divisor. But every algebraic K3 surface is known to carry a (possibly singular) rational curve [Mori-Mukai 2].)

Our goal in this lecture is to extend these elementary observations on ruled surfaces to higher dimension.

2   Rationally connected varieties

2.1 A (not necessarily smooth or complete) variety X defined over an algebraically closed field (of any characteristic) is said to be *rationally connected* if given two general points x_1, x_2 are joined by a single rational curve on X (i.e. there is an open dense subset $U \subset X \times X$ such that any two points x_1, x_2 are joined by a single (possibly singular and non-complete) rational curve on X whenever $(x_1, x_2) \in U$). By definition, the property being rationally connected is a birational invariant, because two birationally equivalent varieties are mutually isomorphic on open dense subsets.

2.2 Examples   (1) Any two points on $\mathbb{P}^n$ are joined by a line, which is of course a rational curve. Hence an arbitrary rational variety is rationally connected.

(2) An image of a rationally connected variety is again rationally connected. In fact, let $f : X \to Y$ be a surjective morphism, with X being rationally connected. Take y_1, y_2, two general points on Y. Then we can choose two general points x_1, x_2 on X such that $f(x_i) = y_i$. Take a rational curve C that connects x_1 and x_2. Then $f(C)$ is a rational curve that connects y_1 and y_2.

(3) Since rational connectedness is invariant under birational equivalence, the image of rationally connected variety via a dominant rational map is again rationally connected. In particular, unirational varieties (= images of rational varieties) are rationally connected.

(4) A surface is rationally connected if and only if it is rational.

2.3 By the observations in (4.2.2), we have implications

$$(\text{rational}) \Rightarrow (\text{unirational}) \Rightarrow (\text{rationally connected}) \Rightarrow (\text{uniruled})$$

for varieties of positive dimension. (A single point is, by convention, rational, unirational and rationally connected, but *not* uniruled.)

   We do not know so far if there exists a rationally connected variety which is not unirational. However, as we will see below, rationally connected varieties form a good subclass of uniruled varieties, closed under smooth deformation. Up to dimension three, we have a nice characterization for this subclass. As mentioned in Lecture II, unirationality is very hard to check in general.

2.4 Theorem [Kollár-Miyaoka-Mori 2]   *A variety X over complex numbers is rationally connected if and only if the following equivalent two conditions (a) and (b) are satisfied:*

(a) *For an arbitrary finite set B of points $x_1, \ldots, x_m$ on X, there exists an irreducible rational curve $C \subset X$ that contains B plus an extra general point $x \in X$.*

(b) *For a sufficiently general point (x_1, x_2) in $X \times X$ (i.e. $(x_1, x_2) \in X \times X$ is contained in some countable intersection of Zariski open dense subset of $X \times X$), there exists an irreducible rational curve $C \subset X$ which contains x_1 and x_2.*

If X is smooth and projective, then the rational connectedness of X is also equivalent to the following three conditions:

(c) Every two points x_1, x_2 are joined by a connected chain of finitely many rational curves $C_1, \ldots, C_r \subset X$, r depending on the choice of the pair (x_1, x_2);

(d) Two general points x_1, x_2 are joined by a connected chain of finitely many rational curves $C_1, \ldots, C_r \subset X$;

(e) X contains a complete rational curve C such that $f^* T_X$ is ample on $\tilde{C} \simeq \mathbb{P}^1$, where $\tilde{C}$ is the normalization of C and $f : \tilde{C} \to X$ is the composite of the normalization and the embedding $C \hookrightarrow X$.

2.5 Remark In the latter half of the theorem, we cannot drop the two assumptions that (i) the ground field is of characteristic 0 and (ii) X is non-singular.

For example, in characteristic $p > 0$, there are many unirational (hence rationally connected) varieties X with ample canonical divisor K_X. In this case, (e) fails for X because for any rational curve C, $\det f^* T_X = -K_X$ is negative on C so that $f^* T_X|_C$ cannot be ample.

If you allow singularities, then there are varieties X which satisfies (c) but not rationally connected. Indeed, consider a hypersurface X defined by $z_1^n + z_2^n + z_3^n = 0$ in $\mathbb{P}^3$ with homogeneous coordinate (z_0, z_1, z_2, z_3). This is the cone over the plane curve

$$C = \{z_1^n + z_2^n + z_3^n = 0\}$$

in $\mathbb{P}^2$, birationally equivalent to $\mathbb{P}^1 \times C$, and hence not rationally connected provided $n \geq 3$. On the other hand, there is a 1-parameter family of lines through the vertex $(1, 0, 0, 0)$ which sweeps out the whole X. Thus every two points are joined by two lines in this family.

Before going into the proof of (2.4), we prepare several lemmas, which turn out to be very useful by themselves.

2.6 Lemma *Over the complex number field (or over a field of characteristic zero with uncountalble elements), we have:*

(1) Let x be a general closed point of X, a smooth projective variety. X is rationally connected if and only if the universal morphism $\Phi : \mathrm{Hom}(\mathbb{P}^1, \infty; X, x) \times \mathbb{P}^1 \to X$ is dominant. If $f^* T_X$ is ample for some morphism $f : \mathbb{P}^1 \to X$, then $\mathrm{Hom}(\mathbb{P}^1, \infty; X, f(\infty))$ is smooth at $[f]$ and the universal morphism is of maximal rank $(= \dim X)$ at $([f], p)$, $p \neq \infty$, whence follows the rational connectedness of X. Conversely, in characteristic zero, if $\Phi : \mathrm{Hom}(\mathbb{P}^1, \infty; X, x) \times \mathbb{P}^1 \to X$ is dominant, we can find $[f] \in \mathrm{Hom}(\mathbb{P}^1, \infty; X, x)$ and $p \in \mathbb{P}^1$ such that

$$d\Phi : T_{\mathrm{Hom}(\mathbb{P}^1, \infty; X, x), [f]} \oplus T_{\mathbb{P}^1, p} = H^0(\mathbb{P}^1, I_\infty f^* T_X) \oplus T_{\mathbb{P}^1, p} \to T_{X, f(p)}$$

is of maximal rank at $([f], p)$, from which it follows that $f^* T_X$ is ample.

(2) Let $X' \to X$ be a surjective morphism. If one of the conditions (a) -(d) in (2.4) holds on X', then the same condition holds on X.

The proof is obvious in view of the definition. Thus we may assume that X is smooth projective for the proof of (2.4).

An irreducible complete rational curve C on a smooth variety X is said to be *free* if $T_X|_C$ is semi-positive, i.e. the pull back of T_X to the normalization $\simeq \mathbb{P}^1$ of C is a direct sum of line bundles of non-negative degrees. The following lemma guarantees that a rational curve passing through a sufficiently general point is always free (valid only in characteristic zero).

2.7 Lemma *Let X be a smooth, projective, uniruled variety with ample divisor H over a field of characteristic zero. For an arbitrary positive integer d, there exists an open dense subset U_d such that an arbitrary rational curve $C \subset X$ with $C \cap U_d \neq \emptyset$ and with $(C, H) \leq d$ is free (i.e. the pull-back of T_X to the normalization $\tilde{C} \simeq \mathbb{P}^1$ is semi-positive). Hence when d tends to infinity, there exists an intersection U of countably many Zariski open subsets such that any rational curve C that meets U is free.*

Proof. An arbitrary C of degree $(C, H) \leq d$ is of the form $f(\mathbb{P}^1)$ for some point $[f]$ on the quasi-projective subset $S_d = \{[f] \mid \deg f^*H \leq d\} \subset \mathrm{Hom}(\mathbb{P}^1, X)$. By replacing S_d by a resolution of singularities, we may assume that S_d is a finite disjoint union of finite-dimensional complex manifolds $S_{d,i}$. Then the classical Sard theorem asserts that the analytic map $\Phi|_{S_{d,i} \times \mathbb{P}^1} \to X$ between manifolds is smooth over an open dense subset $U_{d,i} \subset X$ whenever it is dominant. The uniruledness of X amounts to saying that $\Phi|_{S_{d,i}} : S_{d,i} \times \mathbb{P}^1 \to X$ is dominant for some d and i. Putting

$$\Sigma_{d,i} = \begin{cases} \overline{X \setminus U_{d,i}} & \text{if} \quad \Phi|_{S_{d,i} \times \mathbb{P}^1} \text{ is dominant} \\ \overline{\pi(\mathbb{P}^1 \times S_{d,i})} & \text{otherwise,} \end{cases}$$

you easily check that $\Phi|_{S_d \times \mathbb{P}^1}$ is of maximal rank over $U_d = X \setminus \left(\bigcup_i \Sigma_{d,i} \right)$.

Suppose $C = \Phi([f], \mathbb{P}^1) = f(\mathbb{P}^1)$, $([f] \in S_d)$ meets U_d, or equivalently, that $x = f(p) \in U_d$, for some $p \in \mathbb{P}^1$. Then by construction, $d\Phi$ is of maximal rank at $([f], p)$. We have an isomorphism $T_{S_d \times \mathbb{P}^1}|_{[f] \times \mathbb{P}^1} \simeq H^0(\mathbb{P}^1, f^*T_X) \otimes_k \mathcal{O}_{\mathbb{P}^1} \oplus T_{\mathbb{P}^1}$, and $d\Phi$ is the natural map

$$H^0(\mathbb{P}^1, f^*T_X) \otimes_k \mathcal{O}_{\mathbb{P}^1} \oplus T_{\mathbb{P}^1} \simeq \mathcal{O}_{\mathbb{P}^1}^{\oplus N} \oplus \mathcal{O}_{\mathbb{P}^1}(2) \to f^*T_X,$$

which is of maximal rank at p. Hence $d\Phi$ is a generically surjective homomorphism from a semi-positive bundle to f^*T_X, showing that f^*T_X is semipositive.

2.8 Lemma *Let X be a smooth complex projective variety.*

(1) (Moving Lemma) *Let C be a free rational curve and C' an irreducible curve. Assume that C meets C' at x_0, and fix an arbitrary finite subset $B \subset C'$. Then there exists a one-parameter family $\{C_t\}$ of irreducible, free rational curves parametrized by a pointed curve $(T, 0)$ which has the following three properties:*

(a) $C_0 = C$,
(b) C_t meets C' at a point $g(t)$, where $g : T \to C'$ is a non-trivial morphism with
     $g(0) = x_0$,
(c) C_t is away from B for general t.

(2) (Weak gluing lemma) Let C_1, C_2 be irreducible free rational curves on X meeting at a point. Pick up an arbitrary point $y \in C_1 \cup C_2$. Then there is a family of
curves C_s, parametrized by a pointed curve $(S,0)$ such that
(a) $C_0 = C_1 \cup C_2$,
(b) C_s is irreducible, rational and free for $s \neq 0$,
(c) $C_s \ni y$.

Proof. (1) Let $f : \mathbb{P}^1 \to X$ be the morphism induced by the normalization of C.
Choose a point $p_0 \in \mathbb{P}^1$ such that $f(p_0) = x_0$. Since C is free, $\mathrm{Hom}(\mathbb{P}^1, X)$ is smooth
at $[f]$. The differential of the universal morphism $\Phi : \mathrm{Hom}(\mathbb{P}^1, X) \times \mathbb{P}^1 \to X$ is
nothing but the natural evaluation map $H^0(\mathbb{P}^1, f^*T_X) \oplus T_{\mathbb{P}^1, p} \to (f^*T_X) \otimes k(p)$ at
$([f], p)$, and hence of maximal rank for arbitrary $p \in \mathbb{P}^1$. In particular, Φ is locally
a submersion, so that $\dim \Phi^{-1}(C') = \dim \Phi^{-1}(x_0) + 1 \geq 1$. Around $\{[f]\} \times \mathbb{P}^1$,
$\Phi^{-1}(B)$ is either empty or has dimension one less than $\dim \Phi^{-1}(C')$. Choose a
general curve $\tilde{S}$ in $\Phi^{-1}(C')$ passing through $([f], p_0)$, and let S be its image via the
first projection $\mathrm{Hom}(\mathbb{P}^1, X) \times \mathbb{P}^1 \to \mathrm{Hom}(\mathbb{P}^1, X)$. Then after taking a suitable base
change $\phi : T \to S$, you get a section $T \to T \times \mathbb{P}^1$, $t \mapsto (t, h(t))$ which is induced by
the finite cover $\tilde{S} \subset S \times \mathbb{P}^1$ of S. By construction, $C_t = \Phi(\phi, \mathrm{id})(\{t\} \times \mathbb{P}^1)$ contains
$\Phi(\phi, \mathrm{id})(t, h(t)) \in C'$, and C_t is away from B for general t.   □

(2) Consider a family $\mathcal{Q}$ of quadrics $x_1 x_2 = t x_0^2$ in $\mathbb{P}^2$, parametrized by $t \in \mathbb{A}^1$.
$\mathcal{Q} \subset \mathbb{A}^1 \times \mathbb{P}^2$ is smooth (and is naturally compactified in $\mathbb{P}^1 \times \mathbb{P}^2$). Let $f_i : \mathbb{P}^1 \to C_i$
be the normalization and choose $p_i \in \mathbb{P}^1$ such that $f_i(p_i) = x \in C_i$. Think of $\mathbb{P}^1$ as
a line in $\mathbb{P}^2$. Then the union of two copies of $\mathbb{P}^1$ with two points p_1, p_2 identified
is viewed as the reducible quadric $Q_0 \subset \mathcal{Q}$ consisting of two lines meeting at one
point $p = p_1 = p_2$. Thus we get a morphism $f : Q_0 \to C_1 \cup C_2 \subset X$ in an obvious
way.

   Let $\Gamma_0 \subset \mathcal{Q} \times X$ denote the graph f. $\Gamma_0 \simeq Q_0$ is locally a complete intersection
in the smooth quasi-projective variety $\mathcal{Q} \times X$, with normal bundle $\mathcal{N}$ isomorphic
$f^*T_X \oplus \mathcal{O}$. Looking at the natural exact sequence

$$0 \to f^*T_X \to f_1^*T_X \oplus f_2^*T_X \to f^*T_X \otimes k(p) \to 1,$$

and the semi-positivity of $f_i^*T_X$, we infer that f^*T_X is generated by global sections
and that $H^1(\Gamma_0, \mathcal{N}) = 0$. This shows that there is a unique irreducible component
W of $\mathrm{Hilb}(\overline{\mathcal{Q}} \times X)$ which is smooth at the reference point $[\Gamma_0]$, of dimension
$\dim H^0(Q_0, f^*T_X) + 1$. It is clear that any $\Gamma \subset \mathcal{Q} \times X$ associated with an element
of W is contained in $Q_s \times X$ for some $s = s(w) \in \mathbb{A}^1$, giving rise to a morphism
$W \to \mathbb{A}^1$, $w \mapsto s(w)$.

   A small displacement of Γ_0 in $Q_0 \times X$ is nothing but an element $\mathrm{Hom}(Q_0, X)$,
of dimension $\dim H^0(Q_0, f^*T_X)$. Hence $\dim_{\Gamma_0} W = \dim_{[f]} \mathrm{Hom}(\mathbb{P}^1, X) + 1 =$

$\dim_{\Gamma_0} \text{Hilb}(Q_0 \times X) + 1$, so that the morphism $W \ni w \mapsto s(w) \in \mathbb{A}^1$ is dominant. Replacing W by a small neighbourhood of $[\Gamma_0]$, we assume that each element $[\Gamma_w] \in W \subset \text{Hilb}(Q \times X)$ induces a morphism $f_w : Q_{s(w)} \to X$.

First we prove the assertion when $y \in C_1 \setminus C_2$. Pick up a smooth point $q \in Q_0$ such that $f(q) = y$. Then, after replacing $\mathbb{A}^1$ by a suitable quasi-finite cover S which is unramified over 0, we can find a section $\sigma : S \to Q$ such that $\sigma(0) = q$. Given an element $[\Gamma_w] \in W$, or equivalently, the induced morphism $f_w : Q_{s(w)} \to X$, we associate a point $f_w(\sigma(s(w))) \in X$. This defines a morphism $g_\sigma : W \to X$. g_σ is of maximal rank at $[f]$ (or at $[\Gamma_0]$). Indeed, $\text{Hom}(Q_0, X)$ is a closed subscheme of W in an obvious manner, and the differential of $g_\sigma|_{\text{Hom}(Q_0,X)}$ is the evaluation map $H^0(Q_0, f^*T_X) \to f^*T_{X,q}$, which is surjective as was shown above. Hence we have a trivial dimension count:

$$\dim g_\sigma^{-1}(y) = \dim W - n = \dim_{[f]} \text{Hom}(Q_0, X) - n + 1,$$
$$\dim g_\sigma^{-1}(y) \cap \text{Hom}(Q_0, X) = \dim_{[f]} \text{Hom}(Q_0, X) - n \geq 0.$$

This means that there is a curve $T \subset g_\sigma^{-1}(y)$ such that $T \ni [f]$, $T \not\subset \text{Hom}(Q_0, X)$. Hence for a generic $t \in T$, we get a morphism $f_t : Q_{s(t)} \to X$ with $f_t(\sigma(s(t))) = y$, i.e. $f_t(Q_{s(t)}) \subset X$ is an irreducible rational curve which contains y.

In case $y \in C_1 \cap C_2$, we can deform C_2 to C_2' in such a way that C_2' still meets C_1 but is away from y by (2.8.1). Then we deform this new chain $C_1 \cup C_2' \ni y$ to an irreducible rational curve that contains y. □

The lemma above is a toy case of the following general

2.9 Theorem (Gluing lemma)   *Let X be a smooth projective variety. Let $C \simeq \mathbb{P}^1$ be a smooth rational curve and $h : C \to X$ a non-constant morphism. Take two finite subsets $B = \{b_1, \ldots, b_m\}$ and $D = \{p_1, \ldots, p_N\}$ of C, such that $h(B)$ and $h(D)$ are mutually disjoint. Assume that there exist non-constant morphisms $f_i : \mathbb{P}^1 \to X$, $i = 1, \ldots N$ such that $f_i(\mathbb{P}^1)$ is a free rational curve, contains $h(p_i)$, and is disjoint from $h(B)$. If N is sufficiently large, then there exist a subset $I \subset \{1, \ldots N\}$, a flat family of curves $\mathcal{C} \to S$, parametrized by a pointed curve (S, o), and a morphism $\{h_s\} : \mathcal{C} \to X$ which have the following properties:*
(a) *C_s is an irreducible smooth projective curve of genus zero for $s \neq o$ and C_o is a union of C and $|I|$ copies of mutually disjoint $\mathbb{P}^1$'s (call them F_j, $j \in I$), F_j transversally meeting C at p_j. (C_o is thus a tree of $N + 1$ copy of $\mathbb{P}^1$'s whose dual graph looks like a bouquet.)*
(b) *$h_o|_C = h$, $h_o|_{F_j} = f_i$.*
(c) *$h_s(C_s) \supset h(B)$.*

Proof.   First consider the case where B is empty. It is easy to construct a flat family of curves $\mathcal{M} \to \mathbb{A}^N$ with sections β_i, such that

(0) The total space $\mathcal{M}$ is regular,
(1) The central fibre M_0 is C_o,

(2) For $s \neq 0$, the number of irreducible components of M_s is strictly less than $N+1$,

(3) A general fibre M_s is a smooth $\mathbb{P}^1$.

The morphisms $h, f_1, \ldots, f_N$ define a morphism $h_o : M_0 = C_o \to X$. Its graph $\Gamma_0 \simeq C_o$ is a closed curve in the non-singular quasi-projective variety $\mathcal{M} \times X$, which is locally a complete intersection. Hence its normal bundle $\mathcal{N}$ is well defined and isomorphic to $h_o^* T_X \oplus \mathcal{O}^{\oplus N}$.

Thus the local dimension of the $\mathrm{Hilb}(\mathcal{M} \times X)$ at $[\Gamma_o]$ is at least

$$\dim H^0(\Gamma_o, \mathcal{N}) - \dim H^1(\Gamma_o, \mathcal{N}) = \deg h^*(-K_X) + \sum_i \deg f_i^*(-K_X) + (n + N).$$

On the other hand, the set of $\Gamma_s \subset \mathcal{M} \times X$ which is isomorphic to Γ_o is nothing but $\mathrm{Hom}(\Gamma_o, X)$ of dimension $\leq \dim H^0(C_o, h_o^* T_X)$. Look at the natural exact sequence

$$0 \to h_o^* T_X \to h^* T_X \oplus \sum_i f_i^* T_X \to \sum_i h_o^* T_X \otimes k(p_i) \to 0.$$

Since $f_i^* T_X$ is semi-positive, the evaluation map $H^0(F_i, f_i^* T_X) \to h_o^* T_X \otimes k(P_i)$ is surjective, so that

$$\dim H^0(C_o, h_o^* T_X) = \dim H^0(C, h^* T_X) + \sum_i (\dim H^0(F_i, f_i^* T_X) - n)$$

$$= \dim H^0(C, h^* T_X) + \sum_i \deg f_i^*(-K_X).$$

Hence if $N > \dim H^0(C, h^* T_X) - \deg h^*(-K_X) - n$, then there exists Γ_s which is not isomorphic to C_o, i.e. the number of irreducible component of Γ_s is strictly smaller than $N+1$. Take a local irreducible component T of $\mathrm{Hilb}(\mathcal{M} \times X)$. A generic member Γ_t in T may not be irreducible. So choose an irreducible component Γ_t' of which the specialization Γ_o' contains the component over C. Then a one-parameter subfamily connecting Γ_o' and Γ_t' gives a family as desired.

When B is not empty, replace X by the blowing-up $\mu : \tilde{X} \to X$ at $h(B)$. The morphism $h : C \to X$ naturally lifts to a morphism $\tilde{h} : C \to \tilde{X}$. Apply the above argument to this new morphism, viewed as a curve in $\mathcal{M} \times \tilde{X}$. Then a (small) displacement of $\Gamma_o \subset \mathcal{M} \times \tilde{X}$ defines a morphism $\tilde{h}_t$ with image non-trivially intersecting the exceptional divisors, so that $\mu \tilde{h}_t$ satisfies the base condition. $\quad\square$

2.10 Remark The above estimate of N, the number of free rational curves needed to deform C_o, can be improved. In fact, $N > \dim_{[h]} \mathrm{Hom}(\mathbb{P}^1, X) - \deg h^*(-K_X) - n$ is all right (when B is empty). If h is birational and $C = h(\mathbb{P}^1)$ is incompressible, i.e. C cannot be deformed to a reducible rational curve, it is known that $\dim \mathrm{Hom}(\mathbb{P}^1, X) \leq 2n + 1$, see (II,3.2). Thus, with the aid of (2.8.1), we have:

2.11 Proposition   *Assume that an incompressible rational curve $C \subset X$ meets a free rational curve F of degree d at a point p. If $N > n + 1 - (C, -K_X)$, then there exists N free rational curves F_i, $i = 1, \ldots, N$ meeting C at distinct points p_i such that $C \cup F_1 \cup \ldots \cup F_N$ deforms to an irreducible free rational curve of degree $\deg C + Nd$.*

2.12 *Proof of Theorem (2.4):* As noted above, we may assume that X is smooth, projective. We have obvious or easy implications

$$\begin{array}{ccccc} \text{(a)} & \Rightarrow & \text{(b)} & \Leftrightarrow & \text{(e)} \\ \Downarrow & & \Downarrow & & \\ \text{(c)} & \Rightarrow & \text{(d)} & & \end{array}$$

Thus it suffices to check the following two implications: (d) $\Rightarrow$ (c), (c) $\Rightarrow$ (a).

(d) $\Rightarrow$ (c): This follows from the fact that a specialization of a union of rational curves is again a union of rational curves.

(c) $\Rightarrow$ (a): Fix a sufficiently general point x. By (c), we find a connected chain of rational curves $\bigcup C_i$ which contains a prescribed finite set B and x. Let $R(B, x)$ be the set of connected chain of rational curves which contain $B \cup \{x\}$. Given $[C] \in R(B, x)$, let $r(C)$ be the number of irreducible components of C, and define $r(B, x)$ and $r(B)$ to be the minimum of $\{r(C); [C] \in R(B, x)\}$ and $\min_x\{r(B, x)\}$, where x runs through the set of sufficiently general points on X (i.e. a point of U in (2.7)).

   We derive a contradiction by assuming that $r = r(B) \geq 2$. Take $[C] \in R(B, x)$ with $r(C) = r \geq 2$. Fix a component C_1 which contains a general point x. Let $B_1 \subset B$ denote the subset of the points which sit on $B \setminus B \cap (\bigcup_{i=2}^r C_i)$. Let $\mu : \tilde{X} \to X$ be the blowing up at B_1 and $\tilde{C}_i \subset \tilde{X}$ the strict transform of C_i (actually C_i, $i \geq 2$ is not affected by the blowing up). An irreducible component $\tilde{C}_1$ which contains a very general point $x \in \tilde{X}$ is free. Pick up a second irreducible component $\tilde{C}_2$ which meets $\tilde{C}_1$. Let $f : \mathbb{P}^1 \to \tilde{C}_1 \hookrightarrow X$ be the normalization. $\mathrm{Hom}(\mathbb{P}^1, X)$ is smooth around $[f]$ and the universal morphism $\Phi : \mathrm{Hom}(\mathbb{P}^1, X) \times \mathbb{P}^1 \to X$ is of maximal rank at $([f], p)$, where $p \in \mathbb{P}^1$ is a point which is mapped to $\tilde{C}_1 \cap \tilde{C}_2$. This implies that there exists a one-parameter family of morphisms $f_t : \mathbb{P}^1 \to X$ such that $f_t(p) \subset \tilde{C}_2$ with $\bigcup_t f_t(p)$ containing an open subset of $\tilde{C}_2$ (Moving Lemma). Since a small deformation of a free rational curve is again free, we can choose arbitrarily many free rational curves F_k which meets $\tilde{C}_2$ at distinct smooth points of $\tilde{C}_2 \setminus B$. Then the gluing lemma tells us that we can deform $\tilde{C}_2 \cup F_1 \cup \ldots$ to an irreducible rational curve $\tilde{C}'_{12}$ which contains (the pull-back of) $B \cap C_2$. Since F_i is a small deformation of $\tilde{C}_1$, it contains a sufficiently general point x_i, so that we may assume $\tilde{C}'_{12}$ contains a sufficiently general point x'. By construction F_i passes through the exceptional divisors on $\tilde{X}$ over B_1. Let $C'_{12} \subset X$ be the natural image of $\tilde{C}'_{12}$. Then C'_{12} contains B_1 and $B \cap C_2$. Hence by replacing $C_1 \cup C_2 \cup C_3 \cup \ldots C_r$ by $C'_{12} \cup C_3 \cup \ldots \cup C_r$, we get a member $C' \in R(B, x')$, with $r(C) = r - 1$, contradicting the definition of $r(B)$. Thus there must exist an irreducible rational curve C which contains B and a sufficiently general point x. $\square$

Rational connectedness, which is by definition a birational invariant, is a deformation invariant, too, and hence nicely fits into the classification theory of algebraic varieties.

2.13 Proposition   *Any smooth projective deformation of a rationally connected variety is again rationally connected.*

Proof.   Let $\pi : \mathcal{X} \to S$ be a proper smooth morphism, with connected S. Take two points o and s on S. We prove that X_s is rationally connected provided X_o is so. By taking a suitable base change if necessary, we may assume that S is a smooth curve. Under this condition, let us prove that the $s \in S$ such that X_s is rationally connected form an open and closed subset of S.

(1) Openness: Assume that a central fibre $X = X_o$ is rationally connected. Take a morphism $f : \mathbb{P}^1 \to X$ such that f^*T_X is ample. We can view f as a morphism: $\mathbb{P}^1 \to \mathcal{X}$. Then, since $T_{\mathcal{X}}|_X = \mathcal{O}_X \oplus T_X$, we get $H^0(\mathbb{P}^1, f^*T_{\mathcal{X}}) = H^0(\mathbb{P}^1, f^*T_X) \oplus k$, $H^1(\mathbb{P}^1, f^*T_{\mathcal{X}}) = 0$. Hence there exists a one parameter family of morphisms $f_t : \mathbb{P}^1 \to \mathcal{X}$ such that $f_t(\mathbb{P}^1) \not\subset X_o = X$. Since $f_0(\mathbb{P}^1)$ is contained in a fibre, $f_t(\mathbb{P}^1)$ is contained in some fibre $X_{s(t)}$ for every t, and the morphism $t \mapsto s(t)$ is unramified near 0. Furthermore, the ampleness of $f_t^*T_{\mathcal{X}/S}$ is an open condition on t, and hence $X_{s(t)}$ contains a rational curve $f_t(\mathbb{P}^1)$ such that $f_t^*T_{X_{s(t)}}$ is ample if t is near o. Thus there is an open neighbourhood of o over which X_s is rationally connected. Since o is arbitrary, the openness is proved.   □

(2) Closedness: Let $\Delta \subset S$ be a small open subset. Take u such that X_s is rationally connected for $s \in \Delta \setminus u$. After taking a suitable base change, choose two general sections $\sigma_1, \sigma_2 : S \to \mathcal{X}$. Let $R(X_s; \sigma_1, \sigma_2) \subset \text{Chow}(X_s)$ denote the set of chains of rational curves that join $\sigma_1(s)$ and $\sigma_2(s)$. Then $\mathcal{R}(\sigma_1, \sigma_2) = \bigcup_s R(X_s; \sigma_1, \sigma_2)$ is a closed subscheme of $\text{Chow}(\mathcal{X}/\Delta)$, a countable union of projective Δ-schemes. For $s \neq u$, $R(X_s; \sigma_1, \sigma_2)$ is non-empty by assumption. Hence by the properness of each component of $\mathcal{R}(\sigma_1, \sigma_2)$ over Δ, $R(X_u, x_1, x_2)$ is not empty, either; i.e. the rational connectedness of the limit fibre X_u.   □

As you will guess, rationally connected varieties are subject to various geometric constraints:

2.14 Proposition   *Let X be a smooth, projective, rationally connected variety over $\mathbb{C}$. Then*
(1) $\pi_1^{top}(X) = (1)$.
(2) $\chi(X, \mathcal{O}_X) = 1$.
(3) $H^i(X, \mathcal{O}_X) = 0$ *for* $i > 0$.
(4) $H^0(X, (\Omega_X^1)^{\otimes m}) = 0$ *for* $m > 0$.
(5) $H^0(X, (\Omega_X^j)^{\otimes m}) = 0$ *for* $j > 0, m > 0$.

Proof.   (4) Assume that $(\Omega_X^1)^{\otimes m}$ admits a non-zero global section s. Then $s|_x \neq 0$ at a general point $x \in X$. By the rational connectedness of X, there is a morphism

$f : \mathbb{P}^1 \to X$ such that $f(\mathbb{P}^1) \ni x$ and f^*T_X ample. If $f(\mathbb{P}^1)$ passes through x, then $H^0(\mathbb{P}^1, f^*(\Omega_X^1)^{\otimes m}) \neq 0$, but this is impossible because $f^*\Omega_X^1$ is negative.

(5) $(\Omega_X^j)^{\otimes m}$ is a direct summand of $(\Omega_X^1)^{\otimes mj}$ (recall that the ground field is of characteristic zero).

(3) Hodge theory tells us that $\dim H^i(X, \mathcal{O}_X) = \dim H^0(X, \Omega_X^i)$, and the right hand side is zero by (5).

(2) Clear by (3).

(1) By the condition (e) in (4.2.4), rational connectedness is preserved by taking finite étale covers. On the other hand, by (2), there is no non-trivial finite étale cover of a rationally connected variety X. This shows that $\pi_1^{alg}(X) = (1)$, i.e. the profinite completion of π_1^{top} is trivial. For the proof of the triviality of the topological fundamental group, we need the cycle space of non-compact Kähler manifolds, and the reader is referred to [Campana] or [Kollár 1] for details. $\square$

2.15 Example   It is known that there are smooth quartic hypersurfaces in $\mathbb{P}^4$ that are unirational. Hence every smooth quartic hypersurface in $\mathbb{P}^4$ is rationally connected. (This result will be strengthened in the next section to a statement on the rational connectedness of Fano manifolds.)

3   Maximal rationally connected fibrations

3.1 Theorem (maximal rationally connected fibration)   *Let X be a smooth, uniruled, projective variety over $\mathbb{C}$. Then there exists a dominant rational map $\pi : X \dashrightarrow \pi(X)$ with the following properties:*

(a) *There exist open subsets $U \subset X$ and $V \subset \pi(X)$ such that $\pi : U \to V$ is a proper morphism;*

(b) *A general fibre (which makes sense because of (a)) of π is irreducible and rationally connected;*

(c) *Given a sufficiently general $y \in \pi(X)$, all rational curves on X that meet $X_y = \pi^{-1}(y)$ are contained in X_y.*

This rational fibre space structure has a universal property in the following sense:

(d) *If $f : X \to Z$ is a dominant rational map with rationally connected general fibre, then there exists a dominant rational map $\sigma : Z \to \pi(X)$, unique up to birational automorphism of $\pi(X)$, such that $\pi = \sigma f$.*

In particular, π is uniquely determined by X up to birational equivalence.

Call this dominant rational map π the *maximal rationally connected fibration*, or *MRC fibration* for short, of X. By convention, the MRC fibration of a non-uniruled variety X is the identity map $\mathrm{id} : X \to X$. X is rationally connected [resp. non-uniruled] if and only if the image of MRC fibration is a single point [resp. birational to X].

In case dimension $= 2$, the MRC fibration of a ruled surface is nothing but the Albanese map.

3.2 *Proof of (3.1), Step I:* Fix an ample divisor H on X, and pick up a general point $x \in X$. Given a positive integer d, define $H(x, d) \subset \operatorname{Hom}(\mathbb{P}^1, X)$ by the locally closed subset

$$\{f : \mathbb{P}^1 \to X \mid f(\mathbb{P}^1) \ni x, \ \deg f^*H \le d, \ f^*T_X \text{ is semipositive}\}.$$

The restriction of the universal morphism $\Phi : \operatorname{Hom}(\mathbb{P}^1, X) \times \mathbb{P}^1 \to X$ defines a morphism $\Phi : H(x, d) \times \mathbb{P}^1 \to X$. We let $V_x^d \subset X$ denote its image:

$$V_x^d = \Phi(H(x, d) \times \mathbb{P}^1) = \bigcup_{f \in H(x,d)} f(\mathbb{P}^1),$$

which is a constructible subset of X. When x is general, f^*T_X is automatically semipositive for the f such that $f(\mathbb{P}^1) \ni x$, so that the semi-positivity condition is redundant, and $H(x, d) \subset \operatorname{Hom}(\mathbb{P}^1, X)$ is closed in this case. The correspondence $x \mapsto [\overline{V_x^d}]$ defines a rational map $\pi^d : X \dashrightarrow \operatorname{Chow}(X)$. Our MRC fibration is nothing but π^d for sufficiently large d.

We have to check this rational map enjoys the properties required in (3.1).

3.3 Claim   *If x is sufficiently general and d sufficiently large, $V_x = V_x^d$ is irreducible, closed, and independent of d.*

Proof.   Note the obvious inclusions $H(x, d) \subset H(x, d')$, $V_x^d \subset V_x^{d'}$ for $d \le d'$. By definition, every point of V_x^d is connected to x by an irreducible free rational curve. In particular, by (2.8), two general points of the closure $\overline{V_x^d}$ can be connected by a rational curve, hence by the equivalence (a) $\Leftrightarrow$ (b) in (2.5), an arbitrary point $\in \overline{V_x^d}$ can be joined to x by a rational curve of sufficiently large degree, i.e. $\overline{V_x^d} \subset V_x^e$, $e \gg d$.

Assume that V_x^d is reducible. Take two irreducible components A and B in V_x^d. $C_a \cup C_b$, $([C_a], [C_b]) \in A \times B$ is a chain of two free rational curves meeting at x. Then the weak gluing lemma tells us that $C_a \cup C_b$ will deform to a single free rational curve $C \ni x$ of degree $\le 2d$. This implies that $\dim V_x^{2d}$ is strictly bigger than $\max\{\dim A, \dim B\}$. Hence if we choose d sufficiently large so that $\dim V_x^e = \dim V_x^d$ for all $e \ge d$, we conclude that V_x^d is irreducible.

Thus if d is sufficiently large, the increasing sequence $V_x^d \subset V_x^{d+1} \subset \cdots$ satisfies $V_x^e \subset \overline{V_x^d}$, $\bigcup V_x^e = \overline{V_x^d}$. Since each V_x^e is a constructible set, this means that $V_x^e = \overline{V_x^d}$ for sufficiently large e, completing the proof of Claim. $\square$

3.4 *Proof of (3.1), Step II:* Let x_1 and x_2 be two general points. Then the closed subvarieties $V_{x_1}^d$ and $V_{x_2}^d$ are either mutually disjoint or identical. Indeed, if there is a point y contained in V_{x_i}, then we get a connected chain of free rational curves $C_1 \cup C_2$ joining x_1 and x_2, which deforms to a single rational curve $\ni x_1$. In other words, $C_1 \cup C_2$ is a limit of C_t, where $C_t \in H(x_1, 2d)$. Thus $x_2 \in \overline{V_{x_1}^{2d}} = V_{x_1}^d$, yielding $V_{x_1} = V_{x_2}$. In particular, the rational map $\pi : X \dashrightarrow \mathrm{Chow}(X)$ defines a proper morphism from an open subset $U \subset X$ onto a locally closed subset $W \subset \mathrm{Chow}(X)$. We denote by Z the closure of W.

Let $z \in W \subset Z$ be a sufficiently general point. We derive contradiction if there is a rational curve $C \in X$ which meets $V_x = \pi^{-1}(z)$ at finitely many points. Since z is sufficiently general, we may assume that so is x. Choose $x_0 \in C \cap V_x$. There is a free rational curve A_0 passing through x and x_0. Fix a general point $\eta \in C \setminus V_x$. We can find a one-parameter deformation of rational curves A_t, $t \in (T, 0)$ of A_0 such that A_t meets C at $p(t) \in C$, $p(t)$ being a non-trivial morphism from T to C. Take sufficiently general members A_{t_i}, $i = 1, \ldots, N$. Then gluing lemma tells us that $C \cup \bigcup A_{t_i}$ deforms to an irreducible rational curve $C' \supset \{\eta, x_0\}$. A specialization of C' is $C \cup N A_0$ which contains the sufficiently general point x, and hence C' also contains a sufficiently general point x'. This shows that $\{\eta, x_0\} \subset V_{x'}$, i.e. $V_{x'} \cap V_x \ni x_0$, which is possible only if $V_x = V_{x'}$, and hence $\eta \in V_x$, which contradicts the choice of η. This completes the proof of Theorem 3.1. $\qquad \square$

You can conduct the construction of the MRC fibration in a relative situation. Thus given a smooth family of uniruled varieties, the MRC fibration also deforms.

3.5 Proposition   *Let $\mathcal{X} \to S$ be a proper smooth morphism. Then there exist a Zariski dense subset $U \subset \mathcal{X}$ such that $U \to S$ is surjective, a quasi-projective scheme W over S, and an S-morphism $U \to W$ such that $U_s \to W_s$ induces the MRC fibrations of X_s.*

3.6   It is not known if the image $\pi(X)$ of the rationally connected fibration π can remain uniruled. If it does happen, then we could take the second MRC fibration $\pi_2 : \pi(X) \to \pi_2(\pi(X))$. Reiterating this process, we have a tower of rational maps

$$\pi = \pi_1 : X = X_0 \dashrightarrow X_1 = \pi_1(X_0) \dashrightarrow \ldots \dashrightarrow X_r = \pi_r(X_{r-1})$$

such that a general fibre of each step is rationally connected and the bottom variety $Y = X_r$ is no more a uniruled variety (possibly a point). The naturally induced rational map $\alpha : X \dashrightarrow Y = \alpha(X)$ is uniquely determined by X up to birational equivalence, and has the following universal property:

If $f : X \dashrightarrow Z$ is a dominant rational map onto a non-uniruled variety Z, then we have a rational map $g : Y \dashrightarrow Z$ which completes the commutative diagram

$$
\begin{array}{ccc}
X & =\!=\!= & X \\
\big\downarrow{\scriptstyle \alpha} & & \big\downarrow{\scriptstyle f} \\
Y & \xrightarrow[\;g\;]{} & Z\,.
\end{array}
$$

We can view this tower as a sort of splitting of a uniruled variety into rationally connected pieces and a non-uniruled piece $Y = \alpha(X)$. For a given uniruled variety

X, you can think of $\alpha : X \dashrightarrow Y$ as a high-dimensional analogue of the Albanese mapping of ruled surfaces.

Because we well understand the structure of rationally connected varieties of dimension ≤ 2 and know the abundance for minimal 3-folds, we have the following

3.7 Theorem   *Let X be a smooth projective threefold. Then the image $\pi(X)$ of the maximal rationally connected fibration is not uniruled. Furthermore, we can determine $\dim \pi(X)$ in terms of global differential forms:*

$\dim \pi(X) = 0 \Leftrightarrow H^0(X, (\Omega_X^1)^{\otimes m}) = 0$ *for* $\forall m > 0$.

$\dim \pi(X) = 1 \Leftrightarrow H^0(X, \Omega_X^1) \neq 0, H^0(X, (\Omega_X^2)^{\otimes m}) = 0$ *for* $\forall m > 0$.

$\dim \pi(X) = 2 \Leftrightarrow H^0(X, (\Omega^2)^{\otimes 12}) \neq 0, H^0(X, (\Omega_X^3)^{\otimes m}) = 0$ *for* $\forall m > 0$.

$\dim \pi(X) = 3 \Leftrightarrow H^0(X, (\Omega^3)^{\otimes m}) \neq 0$ *for* $\exists m > 0$.

Proof.   In order to prove the first assertion, assume the contrary. Hence $\pi(X)$ is either $\mathbb{P}^1$ or a ruled surface. In the second case, a general fibre X_y is $\mathbb{P}^1$. Take a rational curve C on Y that passes through y. Then $\pi^{-1}(C)$ is a rational surface and this violates the property (b) in (4.3.1). Hence it suffices to show that a family of rational surfaces parametrized by $\mathbb{P}^1$ is rationally connected. View X as a surface over the function field $k(\mathbb{P}^1)$. Then a theorem of Colliot-Thélène [Colliot-Thélène] tells us that X has a $k(\mathbb{P}^1)$ rational point so that X is a rational surface over $k(\mathbb{P}^1)$, thereby proving that X is birational to $\mathbb{P}^2 \times \mathbb{P}^1$. Thus X is a rational threefold and, in particular, rationally connected. $\square$

4   Fano manifolds

4.1 A smooth projective variety X is said to be a *Fano manifold* if the anticanonical divisor $-K_X$ is ample. Fano manifolds, or rather a broader class of varieties called $\mathbb{Q}$-*Fano varieties*, play a fundamental role in classification theory.

The Minimal Model Program (MMP) asserts that, starting from a given variety, you reiterate contraction of extremal rays and flips to eventually reach either

(a) a minimal model birational to the original variety, or

(b) a non-birational contraction of which a general fibre is a $\mathbb{Q}$-Fano variety.

Here by a $\mathbb{Q}$-Fano variety, we mean a normal, $\mathbb{Q}$-factorial variety with only terminal singularities and with ample anti-canonical divisor $-K_X \in \text{Pic}(X) \otimes \mathbb{Q}$. Thus, MMP being assumed, birational geometry would reduce to the study of minimal varieties and that of $\mathbb{Q}$-Fano fibre spaces. In particular, the geometry of varieties without minimal models, which are presumably uniruled varieties, would be essentially equivalent to the structure theory of $\mathbb{Q}$-Fano varieties. In this sense, the study of the moduli of $\mathbb{Q}$-Fano varieties will be crucial for better understanding of uniruled varieties.

4.2 The best way to understand the moduli of a given class of varieties is to completely classify and explicitly describe them. However, in view of the classification tables of Del Pezzo surfaces and Fano threefolds, which is explained in (4.3) below, you will realize that the complete classification of Fano n-folds won't be very fruitful, because the number of the non-equivalent deformation classes of Fano n-folds grows very rapidly as n increases. Still it would be better if you knew the $\mathbb{Q}$-Fano n-folds are bounded; otherwise we would have no means to control the structure of uniruled varieties, and thus all the theory would be a complete mess. Since the singularities of $\mathbb{Q}$-Fano varieties can be very complicated, we should better start the program with smooth objects, i.e. Fano manifolds, to get insight to the general case.

4.3 Examples   (1) A projective space $\mathbb{P}^n$ is a Fano manifold, with $-K \simeq \mathcal{O}(n+1)$.

(2) A smooth hypersurface $X_d^n \subset \mathbb{P}^{n+1}$ of degree d is a Fano manifold if and only if $d \leq n+1$ (quadric hypersurfaces; cubic surfaces,3-folds, ...; quartic 3-folds, 4-folds, ...; etc.).

(3) A smooth complete intersection variety $X_{d_1,\ldots,d_r}^n \subset \mathbb{P}^{n+r}$ of type $(d_1,\ldots,d_r)$ is a Fano n-fold if and only if $d_1 + \ldots + d_r \leq n+r$. Thus if $X = X_{d_1,\ldots,d_r}^n$ is full (i.e. X is not contained in any hyperplane on $\mathbb{P}^{n+r}$), then $d_i \geq 2$, whence follows $r \leq n$. Therefore, if you fix the dimension n of a full, complete intersection variety X, there are only finitely many choices of d_i's to make X a Fano manifold. In other words, the set of complete intersection Fano n-folds is (perhaps non-effectively) parametrized by a finite dimensional algebraic variety S, which is a disjoint union of open subsets of varieties of the form $(\mathbb{P}^m)^r$.

In general, a set of varieties $W = \{X\}$ is said to be *bounded* if there exists $\mathcal{X} \to S$, an algebraic family of finite type, such that every member $X \in W$ is isomorphic to some fibre $\mathcal{X}_s$ of the family. The set of complete intersection Fano n-folds is bounded, in this sense.

(4) The terminology "Del Pezzo surfaces" is conventionally used in place of two-dimensional Fano manifolds. The complete list of the Del Pezzo surfaces (up to isomorphism) are classically known:
(a) $S_9 = \mathbb{P}^2$,
(b) $\mathbb{P}^1 \times \mathbb{P}^1 = \Sigma_0$,
(c) S_d, $d = 1,\ldots,8$, where S_d is a blown-up of $\mathbb{P}^2$ at $9 - d$ distinct points $p_1,\ldots,p_{9-d}$ subject to the following condition:
   (i)   no condition for $d = 8, 7$;
   (ii)   When $d = 6, 5, 4$, no three points $p_{i_1}, p_{i_2}, p_{i_3}$ lie on a single line $\subset \mathbb{P}^2$;
   (iii) When $d = 3, 2, 1$, no three points lie on a single line and no six points lie on a single conic.

You easily compute the *degree* K_X^2 of X by $= d$ for $X = S_d$, and $= 8$ for $X = \mathbb{P}^1 \times \mathbb{P}^1$. You also check that the S_d have $\max\{0, 5 - d\}$ essential parameters, or number of moduli. Thus the set of Del Pezzo surfaces is again bounded.

(5) [Iskovskih 1,2] and [Mori-Mukai 1] extended the classification of Fano manifolds to dimension three. There are 108 deformation classes of Fano 3-folds, and in particular, the set of Fano 3-folds are bounded.

(6) There are scores of interesting examples of Fano manifolds which arise from group theory. Let G be a reductive algebraic group and P a parabolic subgroup. Then the quotient G/P is a Fano manifold. For example, the Grassmann variety $Gras(N, m)$ as well as the flag variety is a Fano manifold.

4.4 Exercise   Show that the quotient G/P is a Fano manifold. (Hint: It suffices to find sufficiently many vector fields on $X = G/P$ out of the Lie algebra of G.)

4.5   The proof of the boundedness of the n-dimensional Fano manifolds is done by showing their rational connectedness.

In Lecture 1, I noted that the boundedness of the Fano n-folds will follow by bounding K_X^n from above. An idea of G. Fano, the initiator of the theory of Fano manifolds, further reduces the boundedness to the existence of rational curves of bounded degree, which is within the reach of our technique:

4.6 Proposition   *Let X be a Fano manifold of dimension n, and x_1, x_2 two sufficiently general points on X. Assume that there exists an irreducible curve C such that $x_1, x_2 \in C$, $(C, -K_X) \leq d$, then $(-K_X)^n \leq d^n$.*

Proof.   Let $\rho : \tilde{X} \to X$ be the blowing up at x_1 and $E \subset \tilde{X}$ the exceptional divisor. Consider the linear system $|\rho^*(-mK_X) - rE|$. By easy dimension count, we can take $r \sim m\sqrt[n]{(-K_X)^n}$ as $m \gg 0$, keeping the linear system non-empty. A member of this linear system defines an effective divisor D on X with multiplicity at least r at x_1. Since C contains another general point x_2, we may assume that C is not contained in D. Hence we get $m\sqrt[n]{(-K_X)^n} \sim r \leq (C, D)_{x_1} \leq (C, D) = m(C, -K_X) = md$. $\square$

5   Rational connectedness of Fano n-folds

In this section, all varieties and schemes are defined over a field k of characteristic ≥ 0. Indeed, we need reductions modulo p in crucial steps.

The rational connectedness of a Fano manifold is equivalent to the triviality of the MRC fibration, which will follow from:

5.1 Theorem   *Let X be a Fano manifold. Assume that there exist a non-empty open subset $U \subset X$, a smooth quasi-projective variety V of positive dimension, and a proper surjective morphism $\pi : U \to V$. Let z be a general point on V. Then there exists a horizontal rational curve $C \subset X$ passing through $\pi^{-1}(z)$; in other words C meets $\pi^{-1}(z)$ but is not contained in $\pi^{-1}(z)$. Furthermore we can choose C such that $(C, -K_X) \leq \dim X + 1$.*

This result is proved via the theory of *relative deformations*.

5.2 Definition   Let X, Y and Z be irreducible schemes. Let $U \subset X$ be an open dense subset and $\pi : U \to Z$ a (possibly non-proper) morphism. Let $f : Y \to X$ be a morphism such that $f(Y)$ meets U. A *relative deformation* over Z of f, parametrized by a connected pointed scheme (S, o), and equipped with a base subscheme $B \subset Y$ is simply a morphism

$$F = \{f_s\} : S \times Y \to X$$

which satisfies the following three conditions:
(1) $f_o = f$;
(2) $F|S \times B = f \circ \mathrm{pr}_Y |_{S \times B}$ (i.e. f_s is a trivial deformation if restricted to B);
(3) $\pi \circ f_s = \pi \circ f$, for every $s \in S$, where we think of the maps as rational maps.

5.3   Let $\mathcal{H}om_Z(Y, X; f, B)(S, o)$ denote the set of relative deformations of f parametrized by (S, o). Then $\mathcal{H}om_Z(Y, X; f, B)$ is a contravariant functor from the category of connected pointed schemes to the category of sets. When X and Y are projective, it is a locally closed subfunctor of $\mathcal{H}ilb(Y \times X)$, and hence represented by a quasi-projective scheme $\mathrm{Hom}_Z(Y, X; f, B)$, the *universal relative deformation*.

The differential $d\pi : T_U \to \pi^* T_Z$ defines a subsheaf $f^* T_{X/Z} \subset f^* T$ by

$$\Gamma(V, f^* T_{X/Z}) = \{\eta \in \Gamma(V, f^* T_X); d\pi(\eta|_{f^{-1}(U) \cap V}) = 0\},$$

for an open set $V \subset Y$.
It is clear that the Zariski tangent space of $\mathrm{Hom}_Z(Y, X; f, B)$ is a vector subspace of $H^0(Y, I_B f^* T_{X/Z})$. When $U \supset f(Y)$ and π is smooth near $f(Y)$, then $T_{\mathrm{Hom}_Z(f, B), [f]}$ is actually identical with $H^0(Y, I_B f^* T_{X/Z})$, and $\dim_{[f]} \mathrm{Hom}(Y, X; f, B) \geq \dim H^0(Y, I_B f^* T_{X/Z}) - \dim H^1(Y, I_B f^* T_{X/Z})$.

5.4 Lemma   *Let X and Z be smooth projective variety, $U \subset X$ a non-empty open subset and $\pi : U \to Z$ a morphism. Let Y be a smooth projective curve of genus $g \geq 1$ with base point set $B \subset Y$, and $f : Y \to X$ a morphism such that $f(B) \subset U$. Assume that $\mathrm{Hom}_Z(Y, X; f, B)$ is of positive dimension, and take a curve in it, with the smooth projective model Δ. Then the naturally induced rational map $F : Y \times \Delta \dashrightarrow X$ is not a morphism unless B is empty. Let $\overline{F} : W \to X$ be a resolution of the indeterminacy of F, W being a blown-up of $Y \times \Delta$. Then every exceptional curve on W with respect to the blowing up $W \to Y \times \Delta$ is mapped by $\overline{F}$ to a fibre of $\pi : X \dashrightarrow Z$. Furthermore, if we fix an ample divisor H on X, we can find an irreducible component E of the exceptional divisor which satisfies the following two conditions:*
(a) $\overline{F}(E) \subset X$ *is not a point and contains some point of $f(B)$;*
(b) $(E, \overline{F}^* H) \leq \frac{2 \deg f^* H}{\deg B}$.

Proof.   The proof is essentially the same as (II.2.6). We have, however, to check that the image of each exceptional curve lies on a fibre of π. Replace W by a suitable model such that $\pi \overline{F} : W \to Z$ is a morphism. Choose a very ample divisor D on Z. Then $(\pi \overline{F})^* D$ is a nef divisor on W. Pick up and fix $y \in Y$, a general point. The blown up surface W is isomorphic to $Y \times \Delta$ around $\Delta_y = \{y\} \times \Delta$. Thus we may think of Δ_y as a closed subset of W. Since $\pi F : \Delta \times Y \to Z$ is a trivial deformation of f, $\pi \overline{F}$ maps Δ_y to the single point $\pi f(y) \in Z$. In particular, the nef effective divisor $\overline{F}^* D$ is disjoint from Δ_y, and you can easily check that $\overline{F}^* D$ is algebraically equivalent to a multiple of $\Delta_y \in \mathrm{Pic}(W)$. Hence any exceptional curve on W intersects trivially with $\overline{F}^* D$, so that its image in Z is a point. $\square$

5.5 In the same situation as in the lemma above, pick up $\delta \in \Delta$ such that the total transform of $\{\delta\} \times Y$ in W contains an exceptional curve E which satisfies the conditions (a) and (b). Let $Y_\delta \subset W$ be the strict transform of $\{\delta\} \times Y$, and f' the restriction of $\overline{F}$ to $Y_\delta \simeq Y$. Then we have a new morphism $f' : Y \to X$ such that

(1) $\pi f = \pi f'$ unless $f'(Y)$ is contained in the locus of indeterminacy of π.
(2) $\deg f'_*(Y) < \deg f_*(Y)$.
(3) $\pi f'(Y)$ contains a point in $\pi f(B)$.

In case $U \subset X$ is of the form $\pi^{-1}(V)$, where $V \subset Z$ is an open subset, then the property (3) rules out the possibility that $f'(Y)$ is contained in the locus of indeterminacy of π. Thus we find a morphism $f' : Y \to X$ such that the degree of the image $f'(Y)$ is strictly smaller than that of $f(Y)$ with $\pi f'(Y) = \pi f(Y)$. If f' has non-trivial relative deformation with a non-empty base subscheme B', reiterate the same procedure to get a third morphism $f'' : Y \to X$ with image of even smaller degree. Eventually we come across a morphism $g : Y \to X$ with no relative deformation with base points.

5.6 Corollary   *In the notation as in (4.5.4), assume that there exists an open subset $V \subset Z$ such that $U = \pi^{-1}(V) \subset X$ (hence $\pi|_U : U \to V \subset Z$ is proper). Let $f : Y \to X$ be a non-constant morphism from a smooth projective curve of genus $g \geq 1$. If Z is of positive dimension and the base subscheme $B \subset Y$ is non-empty with $f(B) \subset U$. then there exists a morphism $f' : Y \to X$ such that*
(1) $\pi f' = \pi f$,
(2) $\mathrm{Hom}_Z(Y, X; f', B)$ is zero-dimensional.

In other words, we can deform f into a morphism relatively rigid over Z. On the other hand, when X is a Fano manifold, f often has non-trivial (absolute) deformation:

5.7 Theorem   *Let $\pi : X \dashrightarrow Z$ be a dominant rational map from a Fano manifold X to a smooth projective variety Z. Assume that there exist open subsets $U \subset X$, $V \subset Z$ such that $\pi : U \to V$ is proper. Let Y be a smooth projective curve of genus $g \geq 1$ with base subscheme $B \subset Y$ of degree $b > 0$, and let $f : Y \to X$ be a morphism such that $f(B) \subset U$. Then there exists a constant α depending only on (X, Z, π) such that $\dim_{[f]} \mathrm{Hom}(Y, Z; f, B) > 0$ whenever $\deg(\pi f)_*(Y) > \alpha(b + g)$.*

Proof. f has non-trivial deformation with base points B if

$$\chi(Y, I_B f^* T_X) = \deg f^*(-K_X) - (\dim X)(g + b - 1) > 0.$$

Noting that $-K_X$ is ample, the assertion reduces the following lemma.   □

5.8 Lemma *Let $\pi : Z \dashrightarrow Y$ be a dominant rational map from a smooth projective variety X to a projective variety Z. Let H and D be ample divisors on X and Z, respectively. Then there is a constant α, which depends only on (π, X, Z, H, D), such that*

$$\alpha \deg f^* H \geq \deg(\pi f)^* D$$

for any smooth projective curve Y and any morphism $f : Y \to X$ of which image meets $U \subset X$, the domain where π is defined.

Proof. We may assume that π is defined by a linear system, since X is smooth projective and Z is projective. By taking the normalization of the monoidal transformation along the ideal sheaf of the base locus, we get a resolution $\rho : \overline{X} \to Z$ of indeterminacy of π. Then there exist a positive constant a and an effective divisor E supported by the exceptional locus of ρ such that $a\rho^* H - E$ is ample on $\overline{X}$. Hence there is a constant b such that $b(a\rho^* H - E) - (\pi\rho)^* D$ is ample. Since $f(Y)$ is not contained in the locus of indeterminacy of π, the morphism $f : Y \to X$ canonically lifts to $\overline{f} : Y \to \overline{X}$ with $\deg \overline{f}^* E \geq 0$. Thus we have:

$$\begin{aligned}
ab \, \deg f^* H - \deg(\pi f)^* D &= ab \, \deg \overline{f}^* \rho^* H - \deg \overline{f}^* (\pi\rho)^* D \\
&\geq b\overline{f}^* (\deg a\rho^* H - E) - \deg \overline{f}^* (\pi\rho)^* D \\
&= \deg \overline{f}^* \big(b(a\rho^* H - E) - (\pi\rho)^* D\big) \geq 0.
\end{aligned}$$

We have now the assertion, putting $\alpha = ab$.   □

5.9 *Proof of Theorem 5.1.* We show the assertion in case the ground field k is algebraically closed in characteristic $p > 0$. The characteristic zero case is proved by (1) reductions modulo p, and then (2) lifting to characteristic zero; this is possible because the degree $(C, -K_X)$ of the rational curve in question is uniformly bounded by $n + 1$, independent of the characteristic.

Take a general point z on Z and a smooth complete intersection curve $Y_0 \subset X$ which meets $\pi^{-1}(z)$. Fix a point $P_0 \in Y_0 \cap \pi^{-1}(z)$. Choose a Frobenius k-morphism $f : Y \to Y_0 \hookrightarrow X$ such that

$$\deg \pi f > \alpha(g + 1),$$

where g is the genus of Y_0 and Y, and α is the constant in (5.7). By (5.6), we can replace f by f' such that
(1) $\pi f' = \pi f$, and that
(2) f' has no relative deformation with base point $P = (f^{-1}(P_0))_{\mathrm{red}}$ over Z.

By (5.7), the condition (1) guarantees the existence of non-trivial absolute defor-mation $\{f_s'\}$ of f' with base point P, parametrized by a smooth affine curve Δ. Then you can find rational curves $\subset X$ passing through P_0; they arise from the exceptional divisors on the resolution of indeterminacy $\mu : W \to Y \times \overline{\Delta}$, where $\overline{\Delta}$ is the smooth compactification of Δ. By the relative rigidity of f', the one-parameter family $\pi f_s'$ is necessarily a non-trivial deformation of $\pi f' = \pi f$. Hence some of the exceptional divisors are also mapped to rational curves $\subset Z$ passing through $z = \pi(P_0)$. Thus we get a horizontal rational curve passing through P_0.

Given a horizontal rational curve that meets $\pi^{-1}(z)$ at x, the similar argu-ment as in (II.3.2) shows that it deforms to a union of (horizontal and vertical) rational curves of degree $\leq n+1$ such that the union still contains x. Then at least one of the irreducible component is horizontal and contains $x \in \pi^{-1}(z)$.   □

5.10 Exercise Elaborate the final step of the proof above.

5.11 By carefully checking how many free rational curves are needed to make a horizontal rational curve deform to an irreducible rational curve, you can explicitly estimate the minimum of the $d = d(n)$ such that $V_x^d = X$. (Recall that V_x^d is the set of points joined to given general point x by a single rational curve C of degree $(C, -K_X) \leq d$.) For details, the reader is referred to [Kollár-Miyaoka-Mori 3] or [Campana]. The estimate available so far is of exponential order in n, which will not be optimal. The case where the Picard number is one, we have a better estimate [Kollár-Miyaoka-Mori 1]. Thus we can show:

5.12 Theorem *There exists a function $c(n)$ in $n \in \mathbb{N}$ with the following property:*

For an arbitrary n-dimensional complex Fano manifold X and two sufficiently general points x_1, x_2 on X, there exists an irreducible rational curve $C \subset X$ with $(C, -K_X) \leq c(n)$ that contains x_1, x_2.

In particular, an n-dimensional Fano manifold satisfies $c_1(X)^n \leq c(n)^n$, and the set of n-dimensional complex Fano manifolds is bounded.

Finally we give an easy application of the theory of relative deformation.

5.13 Corollary *Let $\pi : X \to Z$ be a surjective smooth morphism between smooth projective varieties of positive dimensions. Then*
(1) The relative anti-canonical divisor $-K_{X/Z} = -K_X + \pi^* K_Z$ is not ample.
(2) If X is a Fano manifold, then so is Z.

Proof. (1) It suffices to show the statement in case the ground field is of char-acteristic $p > 0$, since the ampleness is inherited by almost every reduction. The case where π is étale is trivial, so we assume that $n = \dim X \geq \dim Z + 1 \geq 2$. Let H and D be ample divisors on X and Z, respectively. Then $aH - \pi^* D$ is ample on X, for a suitable positive integer a. Let $Y_0 \subset X$ be a smooth curve of genus $g > 0$ such that $m = \deg \pi^* D|_{Y_0} > 0$ and let $f : Y \to X$ be the morphism induced by a Frobenius morphism $Y \to Y_0$ of sufficiently high degree q. Pick $P \in Y$ as a base point. Then we find a morphism f' such that $\pi f' = \pi f$ and that f' has no non-trivial deformation with base point P. By standard relative deformation theory, this implies that

$$\chi(Y, I_P f'^* T_{X/Z}) = \deg f'^*(-K_{X/Z}) - ng \leq 0.$$

Hence for any fixed positive integer a,

$$\deg f'^*(\pi^* D) = \deg(\pi f')^* D = q \deg(\pi f)^* D = qm > ang \geq a \deg f'^*(-K_{X/Z})$$

whenever q is sufficiently large compared with a. This means that, for any positive integer a, the divisor $a(-K_{X/Z}) - \pi^* D$ cannot be nef, so that $-K_{X/Z}$ is not ample.

(2) Fix an ample divisor D on Z, and choose a positive rational number α such that $-K_X - \alpha \pi^* D$ is ample. Let C_0 be an arbitrary curve on Z and C the normalization. Consider the fibre product $X_C = C \times_Z X$, with the natural smooth projection $\pi_C : X_C \to C$ and a finite, generically injective morphism $h : X_C \to X$. $h^*(-K_X)$ is ample, while the relative anti-canonical divisor $-K_{X_C/C} = h^* K_{X/Z}$ is not. In particular, for any positive rational number ε, $h^*(-K_{X/Z} - \varepsilon(-K_X))$ is not nef, i.e. there is an irreducible curve $C' \subset X_C$ such that $(C', h^*(-K_{X/Z})) < \varepsilon(C', h^* - (K_X))$. Noting that $-K_{X/Z} = -K_X + \pi^* K_Z$, we have

$$(C', h^* \pi^*(-K_Z)) > (1 - \varepsilon)(C', h^*(-K_X)) \geq (1 - \varepsilon)(C', \alpha h^* \pi^* D) \geq 0.$$

This inequality implies, in particular, $\pi_C : C' \to C$ is of positive degree, say d. Hence, dividing out by d the right-hand and left-hand sides of the inequality, we get

$$(C_0, -K_Z) > \alpha(1 - \varepsilon)(C_0, D).$$

Since C_0 is an arbitrary curve on Z and ε is an arbitrary positive rational number, this shows that $-K_Z - \alpha D$ is nef on Z and, in particular, that $-K_Z$ is ample. $\square$

5.14 Example   Assume that an $(n + r - 1)$-dimensional Fano manifold X is of the form $\mathbb{P}(\mathcal{E})$, where $\mathcal{E}$ is a vector bundle of rank r on a smooth projective variety Z. Then Z is again a Fano manifold of dimension n. Since the anti-canonical divisor $-K_X$ is linearly equivalent to $\mathcal{O}(r) - \pi^* \det \mathcal{E} + \pi^*(-K_Z)$, the classification of such X's reduces to the classification of n-dimensional Fano manifolds Z and that of the vector bundles $\mathcal{E}$ such that $Sym^r \mathcal{E}(-\det \mathcal{E} - K_Z)$ is ample.

5.15 Exercise   (1) Assume that an r-dimensional Fano manifold X has a projective bundle structure $\mathbb{P}(\mathcal{E})$ over a curve C. Then show that X is isomorphic to either $\mathbb{P}^1 \times \mathbb{P}^{r-1}$ or the blown-up of $\mathbb{P}^r$ along a linear $\mathbb{P}^{r-2}$.

(2) Assume that X is a projective bundle $\mathbb{P}(\mathcal{E})$ over $\mathbb{P}^2$, with rank $\mathcal{E} = r$, $\det \mathcal{E} = \mathcal{O}(d)$. Given a line ℓ on $\mathbb{P}^2$, define $m(\ell)$ to be the minimum of $\deg L$ where L runs through the set of invertible quotients of $\mathcal{E}|_\ell$. Show that

$$\inf_\ell m(\ell) \geq \frac{d - 3}{r},$$

if X is Fano. For instance, when $d = 0$ and $r \geq 4$, this implies that $\mathcal{E}$ is trivial on every line ℓ, whence follows that $\mathcal{E}$ is trivial, i.e. $X = \mathbb{P}^2 \times \mathbb{P}^{r-1}$.

Historical Comments

The notion of rational connectedness, as well as its application to Fano manifolds, goes back to Giuseppe Fano. Miyaoka (1988) proposed a modified version of the notion, which is essentially the condition (d) in (4.2.4), and gave an outline of the construction of the MRC fibration. [Campana] uses a similar idea to work out the construction. The Gluing Lemma, which yields far stronger statements and significantly simplifies the argument as well, is due to S. Mori.

The classification of smooth Fano 3-folds is partly extended to normal Gorenstein Fano 3-folds by S. Mukai [Mukai]. He found beautiful descriptions of *principal series* of Fano 3-folds. In his theory, you find beautiful interplay among canonical curves, K3 surfaces and Fano 3-folds.

A. Nadel [Nadel] showed the boundedness of smooth Fano n-folds of Picard number one under the assumption of their rational connectedness. J. Kollár pointed out that the rational connectedness of a Fano manifold follows from the existence of horizontal rational curves with respect to a given fibration. The existence of such curves on a Fano manifold was independently verified by F. Campana and Y. Miyaoka.

The complete classification of Fano n-folds is out of reach as was pointed out above, but it would be interesting to classify restricted classes of Fano manifolds. For instance: Fano n-folds (1) with Picard number one; (2) with nef tangent bundles; (3) with high indices; (4) with big lengths.

A vector bundle $\mathcal{E}$ on a projective variety X is said to be *nef* if any quotient line bundle of $f^*\mathcal{E}$ has non-negative degree for an arbitrary morphism $f : C \to X$ from a smooth projective curve (in other words, the tautological line bundle $1_{\mathcal{E}}$ on $\mathbb{P}(\mathcal{E})$ is nef). The index of a Fano manifold X is defined to be the maximum integer $i = i(X)$ such that $-K_X = iH$ for some ample line bundle, while by the length $l = l(X)$ we mean the minimum degree $(C, -K_X)$ of the curves C on X.

The index of a smooth Fano n-fold is always bounded by $n + 1$; indeed there is always a curve of degree $\leq n + 1$ on X. Fano n-folds with index $n + 1$ and n are projective spaces and hyperquadrics, respectively (a theorem of S. Kobayashi and T. Ochiai [Kobayashi-Ochiai]). The length is also bounded by $n + 1$, but the classification of Fano n-folds with length $n + 1$ and n is not available as of 1995 (conjecturally the answer will be the same as in the case of the index).

A smooth projective variety with ample tangent bundle is a projective space $\mathbb{P}^n$ in any characteristic (Hartshorne conjecture, see (II.4.2)). If a smooth complex projective variety X has nef tangent bundle, there should be an étale cover $\tilde{X}$ which is isomorphic to a product of an abelian variety and a Fano manifold (with nef tangent bundle). As for Fano manifolds with nef tangent bundles, a naïve guess is that such X would be a homogeneous variety G/P, where G is an algebraic group and P is a parabolic subgroup. This has been checked up to dimension three by using the classification table [Peternell-Schneider]. In dimension two, the Del Pezzo surfaces with nef tangent bundle are: $\mathbb{P}^2 = \mathbf{SL}(3)/P_3$, and $\mathbb{P}^1 \times \mathbb{P}^1 =$

$(\mathbf{SL}(2) \times \mathbf{SL}(2))/(P_2 \times P_2)$, where

$$P_n = \left\{ \begin{pmatrix} * & * & * & \cdots & * \\ 0 & * & * & \cdots & * \\ 0 & * & * & \cdots & * \\ \vdots & \vdots & \vdots & \ddots & \vdots \\ 0 & * & * & \cdots & * \end{pmatrix} \right\} \subset \mathbf{SL}(n).$$

If you replace the nef condition to the semi-positivity of the holomorphic bisectional curvature, then X must be a hermitian symmetric space [Mok]. To be more precise, if (X, g) is a compact Kähler manifold (i.e. $\omega(*, *) = \sqrt{-1}g(*, J*)$ is a Kähler form) and if its holomorphic bisectional curvature $R_{i\bar{i}j\bar{j}}$, hermitian form on $T_X \otimes T_X$ is positive semi-definite, X is biholomorphic to a hermitian symmetric space. There are a lot of homogeneous projective varieties which are not biholomorphic to hermitian symmetric spaces. (There is a complete list of hermitian symmetric spaces due to E. Cartan; the reader is referred to [Helgason]. For instance, flag varieties are not symmetric in general.) Thus the "semi-positivity" in the differential geometric sense is much stronger than the nefness of the tangent bundle.

Frankly speaking, Fano manifolds with Picard number one (classically called *of the first species*) are actually hard nuts in the classification. If the Picard number is two or more, you will have a non-trivial contraction morphism attached to an extremal ray, which will give you important extra information. Fano 3-folds with Picard number one have been studied in connection with K3 surfaces and canonical curves. By a theorem of V.V. Shokurov [Shokurov], a smooth Fano 3-fold X contains a smooth surface $S \in |-K_X|$. S is a polarized K3 surface with ample divisor $-K_X|_S$, from which you can recover much information on X (for details, you could consult an excellent exposition by S. Mukai [Mukai] on Fano 3-folds and articles in the reference thereof). Because of absolute lack of knowledge on Calabi-Yau n-folds ($n \geq 3$), such technique is not applicable to Fano $(n + 1)$-folds.

References

M. Artin, *Some numerical criteria for for contractibility of curves on algebraic surfaces*, Amer. J. Math. **84** (1962) 485–496.

M. Artin and D. Mumford, *Some elementary examples of unirational varieties which are not rational*, Proc. London Math. Soc. **25** (1972) 75–95.

D. Barlet, *Espace analytique réduit des cycles analytiques complexes*, Séminaire Norguet, Vol. 1, Lect. Notes in Math. 482, Springer, Heidelberg, (74) 1–158.

W. Barth, C. Peters and A. Van de Ven, Compact complex surfaces, Springer. Berlin-Heidelberg-New York, (1984)

A. Beauville 1, Surfaces algébriques complexes, Société Mathématique de France. Paris, 1978,

Beauville 2, *Variétés kähleriennes dont la première classe de Chern est nulle*, J. Diff. Geom. **18** (1983) 755–782.

A. Beauville, J.-L. Colliot-Thélène, J.J. Sansuc and P. Swinnerton-Dyer, *Variétés stablement rationnel non rationnelles*, Ann. of Math. **121** (1985) 283–318.

S. Bloch and D. Gieseker, *The positivity of the Chern classes of an ample vector bundle*, Invent. Math. **12** (1971) 112–117.

F. Bogomolov, *Holomorphic tensors and vector bundles on projective varieties*, Math. USSR Izv. **13** (1979) 499–555.

E. Bombieri, *Canonical models of surfaces of general type*, Publ. Math. I.H.E.S. **42** (1973) 447–495.

E. Bombieri and D. Mumford, *Enriques classification of surfaces in characteristic p*, Complex Analysis and Algebraic Geometry (W.L. Bailey and T. Shioda, eds.), Iwanami-Cambridge Univ. Press, Tokyo/Cambridge, (1977) 23–42.

A. Borel and J.-P. Serre, *Le théorème de Riemann-Roch*, Bull. Soc. Math. de France **86** (1958) 97–136.

E. Brieskorn, *Ein Satz über die komplexen Quadriken*, Math. Ann. **155** (1964) 184–193.

R. Brody and M. Green, *A family of smooth hyperbolic hypersurfaces in P_3*, Duke Math. J. **44** (1977) 873–874.

F. Campana, *Connexité rationnelle des variétés de Fano*, Ann. Sci. E.N.S. **25** (1992) 539–545.

F. Campana and H. Flenner, *Projective threefolds containing a smooth rational surface with ample normal bundle*, J. Reine u. Angew. Math. **440** (1993) 77–98.

F. Campana and T. Peternelle, *Projective manifolds whose tangent bundles are numerically effective*, Math. Ann. **289** (1991) 169–187.

K. Cho and E. Sato, *Manifolds with ample vector bundle $\wedge^2 T_X$*, preprint, Kyushu University,

H. Clemens and P.A. Griffiths, *The intermediate Jacobian of the cubic threefold*, Ann. of Math. **95** (1972) 281–356.

H. Clemens, J. Kollár and S. Mori, *Higher dimensional complex geometry*, Astérisque **166** (1988)

J.-L. Colliot-Thélène, *Arithmétique des variétés rationelles et problèmes biratio-nelles*, Proc. ICM 1986, Berkeley, 641–653.

S.K. Donaldson, *Anti self-dual Yang-Mills connections over complex algebraic surfaces and stable bundles*, Proc. Lond. Math. Soc. **50** (1985) 1–26.

A. Douady, *Le problème des modules locaux pour les espaces **C**-analytiques compactes*, Ann. Sci. Ecole Norm. Sup. **4** (1974) 569–602.

M. Ebihara, *Formal neighborhoods of a toric variety and unirationality of algebraic varieties*, J. Math. Soc. Japan **46** (1994) 385–462.

[EGA IV] A. Grothendieck *et al.* Éléments de Géométrie Algébrique, IV. Études locales des schémas et des morphismes de schémas, Pub. Math. I.H.E.S. **32** (1967)

T. Ekedahl 1, *Foliations and inseparable morphisms*, Algebraic Geometry Bowdoin, Proc. Symp. Pure Math. **46** (1986) 139–149.

Ekedahl 2, *Canonical models of surfaces of general type in positive characteristic*, Publ. Math. I.H.E.S. **67** (1988) 97–144.

R. Elkik, *Rationalité des singularités canonique*, Inv. Math. **64** (1981) 1–6.

[FGA] A. Grothendieck, *Fondements de Géométrie Algébrique*, Secrétariat Math., Paris, 1962,

R. Friedman, *Global smoothings of varieties with normal crossings*, Ann. Math **118** (1983) 75–114.

W. Fulton and J. Hansen, *A connectedness theorem for projective varieties, with applications to intersections and singularities of mappings*, Ann. Math. **110** (1979) 159–166.

A. Fujiki, *Deformation of uni-ruled manifolds*, Publ. RIMS Kyoto Univ. **17** (1981) 687–702.

T. Fujita, *On Kähler fibre space over curves*, J. Math. Soc. Japan **30** (1978) 779–794.

D. Gieseker 1, *Stable vector bundles and the Frobenius morphism*, Ann. École Norm. Sup. **6** (1973) 95–101.

Gieseker 2, *Global oduli of surfaces of general type*, Invent. Math. **43** (1977) 233–282.

P.A. Griffiths 1, *Hermitian differential geometry, Chern classes, and positive vector bundles*, Global Analysis: papers dedicated to K. Kodaira, Iwanami-Princeton Univ. Press, Tokyo/Princeton, (1969) 184–251.

Griffiths 2, *Periods of integrals on algebraic manifolds, III*, Bulletin A.M.S. **76** (1970) 229–296.

P.A. Griffiths and J. Harris, Principles of algebraic geometry, John Wiley, New York, (1978)

A. Grothendieck 1, *La théorie des classes de Chern*, Bull. Soc. Math. France **86** (1958) 137–154.

Grothendieck 2, *Technique de descente et théorème d'existence en géometrie algébrique, VI: Les schémas de Picard, Propriétés génerales.*, In: Séminaire Bourbaki (1961/62), Benjamin, New York, (1966)

G. Harder and M.S. Narasimhan, *On the cohomology groups of moduli space of vector bundles on curves*, Math. Ann. **212** (1975) 215–248.

R. Hartshorne 1, Ample vector bundles, Lecture Notes in Math. vol. 156, Springer, Berlin-Heidelberg-New York, (1970)

Hartshorne 2, Algebraic Geometry, Springer, Berlin-Heidelberg-New York, (1977)

S. Helgason, Differential geometry, Lie groups and symmetric spaces, Academic Press, New York, (1978)

F. Hirzebruch, Topological method in algebraic geometry, Springer, Berlin-Heidelberg-New York, (1966)

F. Hirzebruch and K. Kodaira, *On the complex projective spaces*, J. Math. Pures Appl. **36** (1956) 201–216.

S. Iitaka 1, *Deformation of compact complex surfaces, II*, J. Math. Soc. Japan **22** (1970) 247–261.

Iitaka 2, *On D-dimension of algebraic varieties*, J. Math. Soc. Japan **23** (1971) 356–373.

P. Ionescu, *Generalized adjunction and applications*, Math. Proc. Cambridge Phil. Soc. **99** (1986) 457–472.

V.A. Iskovskih 1, *Fano 3-folds, I*, Math. USSR Izv. **11** (1977) 485–527.

Iskovskih 2, *Fano 3-folds, II*, Math. USSR Izv. **12** (1978) 469–506.

Iskovskih 3, *On the rationality problem for conic bundles*, Duke Math. J. **54** (1987) 271–294.

V.A. Iskovskih and Yu. I. Manin, *Three dimensional quartics and counter examples to the Lüroth problem*, Math. USSR-Sb. **15** (1971) 141–166.

Y. Kawamata 1, *Characterization of abelian varieties*, Compositio Math. **43** (1980) 253–276.

Kawamata 2, *Pluricanonical systems on minimal algebraic varieties*, Inv. Math. **79** (1985) 567– 588.

Kawamata 3, *Elementary contractions of algebraic 3-folds*, Ann. of Math. **119** (1984) 95–110.

Kawamata 4, *Minimal models and the Kodaira dimension of algebraic fibre spaces*, J. Reine Angew. Math. **363** (1985) 1–46.

Kawamata 5, *Boundedness of $\mathbb{Q}$-Fano threefolds*, Contemp. Math. **131** (1992) 439–445.

Kawamata 6, *Abundance theorem for minimal threefolds*, Inv. Math. **108** (1992) 229–246.

Y. Kawamata, K. Matsuda and K. Matsuki, *Introduction to the minimal model problem*, Algebraic Geometry, Sendai; Adv. Stud. Pure Math. **10** (1987) 283–360.

S.L. Kleiman, *Toward a numerical theory of ampleness*, Ann. of Math. **84** (1966) 293–344.

S. Kobayashi 1, *Hyperbolic Manifolds and Holomorphic Mappings*, Marcel-Dekker, New York, (1970)

Kobayashi 2, *Curvature and Stability of vector bundles*, Proc. Japan Acad. **58** (1982) 156–162.

S. Kobayashi and K. Nomizu, Foundations of differential geometry, I, II, Interscience, New York-London-Sydney, (1969)

S. Kobayashi and T. Ochiai, *Characterizations of complex projective spaces and hyperquadrics*, J. Math. Kyoto Univ. **13** (1973) 31–47.

K. Kodaira 1, *On compact complex surfaces I, II, III*, Collected Papers, vol. 3, Iwanami, Tokyo, (1975)

Kodaira 2, *On the structure of compact complex surfaces I, II, III, IV*, Collected Papers, vol. 3, Iwanami, Tokyo, (1975)

K. Kodaira and D.C. Spencer, *On deformations of complex analytic structures, I,II*, Ann. Math. **67** (1958) 328–466.

J. Kollár 1, *Subadditivity of the Kodaira dimension: Fibres of general type*, Algebraic Geometry, Sendai, Adv. Stud. in Pure Math. 10, Kinokuniya-Northholland, Tokyo/Amsterdam, (1987)

Kollár 2, Shafarevich maps and automorphic forms, Princeton Univ. Press, Princeton, (1995)

Kollár 3, *Rational curves on algebraic varieties*, Springer, Berlin-Heidelberg-New York-Tokyo, (1996)

J. Kollár *et al.*, *Flips and abundance for algebraic threefolds*, Astérisque **211** (1993)

J. Kollár and T. Matsusaka, *Riemann-Roch type inequalities*, Amer. J. Math. **105** (1983) 229–252.

J. Kollár, Y. Miyaoka and S. Mori 1, *Rational curves on Fano varieties*, Lecture Notes in Math. **1515** (1992) 100–105.

Kollár-Miyaoka-Mori 2, *Rationally connected varieties*, J. Alg. Geom. **1** (1992) 429–448.

Kollár-Miyaoka-Mori 3, *Rational connectedness and boundedness of Fano manifolds*, J. Diff. Geom. **36** (1992) 765–779.

V.S. Kulikov, *Degeneration of K3 and Enriques surfaces*, Math. USSR Izvestija **11** (1977) 957–989.

M. Kuranishi, *On the locally complete families of complex analytic structures*, Ann. of Math. **75** (1962) 536–577.

M. Levine, *Deformations of uni-ruled varieties*, Duke Math. J. **48** (1981) 467–473.

V.B. Mehta and A. Ramanathan, *Semi-stable sheaves on projective varieties and their restriction to curves*, Math. Ann. **258** (1982) 213–224.

J. Milnor, Singular points of complex hypersurfaces, Ann. of Math. Studies, vol. 61, Princeton Univ. Press, Princeton, (1968)

Y. Miyaoka 1, *The maximal number of qotient singularities on surfaces with given numerical invariants*, Math. Ann **268** (1984) 225–237.

Miyaoka 2, *Deformation of a morphism along a foliation*, Algebraic Geometry Bowdoin, 1985; Symp. Pure Math. **46** (1987) 245–268.

Miyaoka 3, *The Chern classes and Kodaira dimension of a minimal variety*, Advanced Study in Pure Math. **10** (1987) 449–476.

Miyaoka 4, *On the Kodaira dimension of minimal threefolds*, Math. Ann. **281** (1988) 325–332.

Miyaoka 5, *Abundance conjeture for 3-folds – case* $\nu = 1$, Comp. Math. **68** (1988) 203–220.

Miyaoka 6, *Relative deformations of morphisms and applications to fibre spaces,* Comment. Math. Univ. St. Pauli **42** (1993) 1–7.

Miyaoka 7, *Vector fields on Calabi-Yau manifolds in characteristic p,* Preprint, RIMS, Kyoto Univ., (1995)

Y. Miyaoka and S. Mori, *Numerical criterion for uniruledness,* Ann. Math. **124** (1986) 65–69.

N.-M. Mok, *The uniformization theorem for compact Kähler manifolds of non negative holomorphic bisectional curvature,* J. Diff. Geom. **27** (1988) 179–214.

S. Mori 1, *Projective manifolds with ample tangent bundles,* Ann. Math. **110** (1979) 593–606.

Mori 2, *Threefolds whose canonical bundles are not numerically effective,* Ann. Math. **116** (1982) 133–176.

Mori 3, *Cone of curves and Fano 3-folds,* Proc. ICM, Warszawa, 1983, 747–752.

Mori 4, *Classification of higher-dimensional varieties,* Proc. of Symposia in Pure Math. **46** (1987) 269–331.

Mori 5, *Flip theorem and the existence of minimal models for 3-folds,* J. of AMS **1** (1988) 117–253.

S. Mori and S. Mukai 1, *Classification of Fano threefolds with* $B_2 \geq 2$, Manuscripta Math. **36** (1981) 147–162.

Mori-Mukai 2, *Mumford's theorem on curves on K3 surfaces* (Appendix to *The uniruledness of the moduli space of curves of genus 11*), Springer Lect. Notes in Math. **1016** (1983) 334–353.

U. Morin, *Sull' unirationalità dell' ipersuperficie algebrica di qualunque ordine e dimensione sufficientemente alta,* Atti dell II. Congresso Unione Math. Italiana (1940) 298–302.

S. Mukai, *New development in the theory of Fano manifolds – vector bundle method and moduli problem, (in Japanese),* Sûgaku **47** (1995) 125–144.

D. Mumford 1, *Canonical ring of an algebraic surface, Appendix to [Zariski 2],*

Mumford 2, *Enriques classification in characteristic p, I,* Global Analysis: Papers in honor of K. Kodaira (S. Iyanaga, ed.), Univ. Tokyo Press-Princeton Univ. Press, Tokyo-Princeton, (1969) 324–339.

Mumford 3, Abelian Varieties, Oxford Univ. Press, Oxford, (1974)

D. Mumford and J. Fogarty, Geometric Invariant Theory, 2nd Edition, Springer, Berlin-Heidelberg-New York, (1982)

A. Nadel, *The boundedness of degree of Fano varieties with Picard number one,* J. of AMS **4** (1991) 681–692.

Y. Nakai, *A criterion of an ample sheaf on a projective scheme,* Amer. J. Math. **85** (1963) 14–26.

N. Nakayama , *Invariance of the plurigenera of algebraic varieties under minimal model conjectures,* Topology **25** (1986) 237–251.

M.S. Narasimhan and C.S. Seshadri, *Stable and unitary vector bundles on a compact Riemann surface,* Nagoya Math. J. **43** (1971) 540 –567.

U. Persson and H. Pinkham , *Degeneration of surfaces with trivial canonical bundle*, Ann. Math. **113** (1981) 45–66.

H. Popp, *On moduli of algebraic varieties, II*, Compositio Math. **28** (1974) 51–81.

L. Ramero, *Effective estimates for unirationality*, Manuscripta Math. **68** (1990) 435–445.

M. Reid 1, *Canonical threefolds*, Géométrie Algébrique d'Anger (A. Beauville, ed.), Sijthoff & Nordhoff, Alphen aan den Rijn, (1980) 273–310.

Reid 2, *Minimal models of canonical threefolds*, Algebraic Varieties and Analytic Varieties; Adv. Stud. Pure Math. **1** (1983) 131–180.

Reid 3, *Young person's guide to canonical singularities*, Proc. in Symposia Pure Math. **46** (1987) 345–414.

A.N. Rudakov and I.R. Shafarevich, *Inseparable morphisms of algebraic surfaces* Math. USSR Izvestiya, **10** (1976) 1205–1237.

J.-P. Serre, *Prolongement de faisceaux analytiques cohérents*, Ann. Inst. Fourier **16** (1966) 363–374.

P. Samuel, Méthodes d'algèbre abstraite en géométrie algébrique, Springer, Berlin-Heidelberg-New York, (1955)

M. Schlessinger, *Functors of Artin rings*, Trans. Amer. Math. Soc. **130** (1968) 208–222.

W. Schmid, *Variation of Hodge structure: the singularities of the period mapping*, Invent. Math. **22** (1973) 211–319.

[SGA 2] A. Grothendieck, Cohomologie locales des faisceaux cohérents et théorème de Lefschetz locaux et globaux, Masson, North-Holland, Paris-Amsterdam, (1962)

I.R. Shafarevich et al., Algebraic Surfaces, Moscow, (1965)

T. Shioda and T. Katsura, *On Fermat varieties*, Tôhoku Math. J. **31** (1979) 97–115.

V.V. Shokurov, *Smoothness of the general anticanonical divisor on a Fano 3-fold*, Math. USSR Izvestia **14** (1980) 395–405.

Y.-T. Siu, *Curvature characterization of hyperquadrics*, Duke Math. J. **47** (1980) 641–654.

Y.-T. Siu and S.-T. Yau, *Compact Kähler manifolds of positive bisectional curvature*, Inv. Math. **59** (1980) 189–204.

M. Szurek and J.A. Wiśniewski, *Fano bundles over $\mathbb{P}^3$ and Q_3*, Pacific J. Math. **141** (1990) 197–208.

K. Ueno 1, Classification theory of algebraic varieties and compact complex spaces, Lecture Notes in Math., Nr. 439, Springer, Berlin-Heidelberg-New York, (1975)

Ueno 2, *Bimeromorphic geometry of algebraic and analytic threefolds*, Algebraic Threefolds (A. Conte, ed.), Lecture Notes in Math. vol. 947, Springer, Berlin-Heidelberg-New York, (1982) 1–34.

K. Uhlenbeck and S.T. Yau, *On the existence of Hermitian-Yang-Mills connections on stable bundles over compact Kähler manifolds*, Comm. Pure Appl. Math. **39** (1986) 257–293.

E. Viehweg 1, *Canonical divisors and the additivity of the Kodaira dimension for morphisms of relative dimension one*, Compositio Math. **35** (1977) 197–223.

Viehweg 2, *Die Additivität der Kodaira Dimension für projektive Faserräume über Varietäten des allgemeinen Typs*, J. Reine Angew. Math. **330** (1982) 132–142.

Viehweg 3, *Weak positivity and the additivity of the Kodaira dimension for certain fibre spaces*, Algebraic Varieties and Analytic Varieties, Adv. Stud. in Pure Math. 1, Kinokuniya-Northholland, Tokyo/Amsterdam, (1983) 329–353.

Viehweg 4, Quasi-projective moduli for polarized manifolds, Springer, Berlin-Heidelberg, (1995)

A. Weil 1, Variétés abéliennes et courbes algébriques, Hermann, Paris, (1948)

Weil 2, Variétés kähleriennes, Hermann, Paris, (1958)

O. Zariski 1, Introduction to the problem of minimal models in the theory of algebraic surfaces, Math. Soc. Japan, Tokyo, (1958)

Zariski 2, *The theorem of Riemann-Roch for high multiples of an effective divisor on a surface*, Ann. of Math. **76** (1962) 550–612.

Part II
An Introduction to the Classification
of Higherdimensional Complex Varieties

Thomas Peternell

Math. Institut
Universität Bayreuth
D-95440 Bayreuth
Germany

Preface

These notes are intended to be an *elementary* introduction to (parts of) the classification theory of complex algebraic varieties of dimension at least 3. They are based on lectures the author gave at a joint seminar with Y. Miyaoka which took place in Oberwolfach in April 1995 under the auspices of the Deutsche Mathematiker-Vereinigung.

From a differential geometric point of view, classification of algebraic manifolds means to distinguish manifolds by their curvature. The broadest classes are manifolds of positive curvature, curvature 0 and negative curvature. Of course a general manifold will fall in none of these classes, it will be somewhere "in between". Curvature should mean here Ricci curvature. However, manifolds with positive, negative or vanishing curvature can serve as model manifolds for the general case. From an algebraic point of view – which we will focus on throughout this article – this means that we distinguish between manifolds with negative canonical bundle, trivial canonical bundle, and positive canonical bundle. The canonical bundle is by definition the determinant of the cotangent bundle; note that the change of the sign comparing to differential geometry comes from the change from the tangent to the cotangent bundle. Again the truth is in between, and this "in between" can be well measured by algebraic methods. To explain the position of an algebraic manifold between these classes, how it is build up from negative and positive parts, will be the aim of these notes. In particular we want to single out the part coming from negativity properties of the canonical bundle.

Manifolds with negative canonical bundles (Fano manifolds) are much more concrete than manifolds with positive canonical bundle; they have in some sense a richer geometry. Most prominent examples of projective manifolds are in this class (or have trivial canonical bundle), while the study of negatively curved manifolds is often the study of moduli spaces rather than of individuals. This becomes already clear in the case of Riemann surfaces: there is only one with positive curvature (the projective line), a "few" with curvature 0 (elliptic curves) and large families with negative curvature (Riemann surfaces of genus at least 2). Of course in the case of Riemann surfaces there is nothing else, an interplay between the classes will happen only from dimension 2 on. In summary it is possible to select the positively curved out of a given variety; the "rest" will then be semi-negative.

A more concrete and less vague-philosophical introduction to the subject is provided by section 0 where we look at the classical classification theory of surfaces and work out the questions to be asked in higher dimensions.

We have almost completely ignored two important subjects in classification theory: the theory of varieties with "semipositive" (nef) canonical bundles K (the so-called abundance, i.e. the problem whether some multiple of K is generated by global sections) and the theory of rational curves, insofar as existence questions and constructions via characteristic p are concerned. Both topics are instead treated intensively in Miyaoka's notes.

As already said, these notes are intended to be elementary. This means that I found it more important to explain the ideas rather than to give always full proofs; however the reader will always find references where to obtain all details. On the other hand I also included discussions of very recent developments so that the reader will be able to go over to more advanced studies.

For other, mostly more advanced, introductions to the subject we refer to [KMM87], [CKM88], [Ko87a], [Mo87], [Wi87]. The book [Ko92] builds up on these and covers very recent developments. For connections to more differential geometric questions, see also [Pe96a].

Prerequisites

Only basic knowledge from algebraic geometry and complex analysis is required, as covered in [Ha77, chapI-III] and [GH78, chap.0 and I] plus some rudimentary knowlegde on spectral sequences as found in [GH78]. There are two more advanced tools not covered by these two books which will be used over and over: the theorem of Riemann-Roch on projective manifolds (see [Ha77]) and Hironaka's desingularisation (see Hironaka's original paper or, for references and statements, [GPR94]).

N Notations

N.1 A variety X is an integral separated scheme or algebraic space of finite type over $\mathbb{C}$. Alternatively one can think of X as a Zariski open set of a reduced compact complex space which is projective (i.e. a subspace of some projective space) or which is Moishezon, (this means that for every irreducible X_i the transcendence degree of the field of meromorphic functions on X_i is maximal, i.e. $= \dim X_i$). Usually X will be normal and quasi-projective.

N.2 Let X be a normal variety. A *Weil* divisor on X is a (formal) linear combination $D = \sum a_i D_i$, where a_i are integers and the D_i are irreducible reduced hypersurfaces in X, i.e they have codimension 1 but are not necessarily defined by one single equation. We let $\operatorname{Div}(X)$ be the group of *Cartier* divisors; then $\operatorname{Div}(X) \otimes \mathbb{Q}$ is the group of $\mathbb{Q}$-Cartier divisors. An element of $\operatorname{Div}(X) \otimes \mathbb{R}$ is consequently called $\mathbb{R}$-Cartier. In the same way, $\mathbb{Q}$-Weil divisors are defined.

If D_1, D_2 are $\mathbb{Q}$-Weil divisors, then D_1 and D_2 are linearly equivalent if rD_1 and rD_2 are linear equivalent in the usual sense, where r is chosen such that rD_i are Weil divisors.

As usual we denote by $\operatorname{Pic}(X)$ the group of line bundles modulo isomorphy.

It is also very common in the literature to confuse line bundles, locally free sheaves of rank 1 and divisors. We will find this attitude very convenient and use it from time to time without any further comment.

N.3 If D is a Cartier divisor, then one can associate to D the line bundle (= locally free sheaf of rank 1) $\mathcal{O}_X(D)$. This yields an isomorphism

$$\operatorname{Div}(X)/\sim \; \longrightarrow \operatorname{Pic}(X).$$

This correspondence can be extended to Weil divisors in the following way. First of all a coherent sheaf $\mathcal{F}$ is *reflexive*, if the canonical map

$$\mathcal{F} \longrightarrow \mathcal{F}^{**}$$

is an isomorphism. The rank of $\mathcal{F}$ is by definition the rank of the free part (= generic rank). Now we associate to any Weil divisor D a reflexive sheaf $\mathcal{O}_X(D)$ in the following way. Let X_0 be the regular part of X. Then $D|X_0$ is Cartier, hence $\mathcal{O}_{X_0}(D)$ is locally free. Now let $i : X_0 \longrightarrow X$ be the inclusion map and define

$$\mathcal{O}_X(D) = i_*(\mathcal{O}_{X_0}(D)).$$

Then it is a basic fact that $\mathcal{O}_X(D)$ is reflexive. Conversely every reflexive sheaf of rank 1 comes from a Weil divisor: first define a Weil divisor D_0 on X_0 such that $\mathcal{O}_{X_0}(D_0) \simeq \mathcal{F}|D_0$ and then extend D_0 to a Weil divisor on X by taking closure. Therefore we obtain an isomorphism

$$\{\text{Weil divisors }\}/ \sim \longrightarrow \{\text{reflexive sheaves of rank 1}\}/ \simeq .$$

N.4 Let X be a normal projective variety and D a Weil divisor on X. Then either

$$H^0(X, \mathcal{O}_X(mD)) = 0$$

in which case we let the Kodaira(-Iitaka) dimension $\kappa(D) = -\infty$. Or

$$h^0(X, \mathcal{O}_X(mD)) \sim m^q,$$

for $m \longrightarrow \infty$, then we set
$$\kappa(D) = q.$$

By (N3) we have also the notion of the Kodaira dimension of a reflexive sheaf of rank 1.

N.5 In (N4) we have already used the following notation

$$h^q(X, \mathcal{F}) = \dim H^q(X, \mathcal{F}).$$

N.6 Let X be a normal n-dimensional projective variety and $\mathcal{F}$ a coherent sheaf on X. We let
$$\chi(X, \mathcal{F}) = \sum_i (-1)^i h^i(X, \mathcal{F})$$

be the holomorphic Euler characteristic of $\mathcal{F}$. Now let D be a Cartier divisor on X. Then by Riemann-Roch applied to a desingularisation of X, the function

$$t \longrightarrow \chi(X, \mathcal{O}_X(tD) \otimes \mathcal{F})$$

is a polynomial of degree at most n.

Now let $Y \subset X$ be an irreducible algebraic set of dimension s. Then we define the intersection number $D^s \cdot Y$ to be the coefficient in degree s in the polynomial $\chi(Y, \mathcal{O}_Y(tD))$.

Alternatively $D^s \cdot Y$ can be defined in a topological way as follows. By possibly passing to a desingularisation we may assume X smooth. The divisor D determines the first Chern class $c_1(\mathcal{O}_X(D)) \in H^2(X, \mathbb{Z})$ via the exponential sequence which gives a map

$$c_1 : \mathrm{Pic}(X) = H^1(X, \mathcal{O}_X^*) \longrightarrow H^2(X, \mathbb{Z}).$$

On $H^2(X, \mathbb{Z})$ we have the usual topological intersection product, therefore we set

$$D^s \cdot Y = c_1(\mathcal{O}_X(D))^s.$$

Choosing a metric on the line bundle associated to $\mathcal{O}_X(D)$ with curvature ω we furthermore have

$$D^s \cdot Y = (\frac{i}{2\pi})^s \int_Y \omega^s.$$

Of course the intersection product is then also well-defined for $\mathbb{Q}$-Cartier divisors.

N.7 Two $\mathbb{Q}$-Cartier divisors D_1, D_2 are numerically equivalent, in signs $D_1 \equiv D_2$, if $D_1 \cdot C = D_2 \cdot C$ for every irreducible curve $C \subset X$.

N.8 Let X be a smooth variety and $D = \sum D_i$ a reduced divisor. D is said to have only normal crossings if every $x \in X$ has a coordinate neighborhood U such that $D|U$ is given by the equation $z_1 \cdot \dots \cdot z_k = 0$. Note that nevertheless some D_i can be singular!

N.9 Let $x \in \mathbb{R}$. Then $[x]$ denotes the largest integer $r \leq x$. Let $\lceil x \rceil$ be the round up and $\{x\}$ be the fractional part.

0 Surfaces and a First View to Higher Dimension

In this section we review those parts of the Kodaira classicification of smooth projective surfaces (= surfaces) which are relevant for the higher dimensional theory to be developed later. As general classical reference we recommend [BPV84] or [Be78]. Proofs will be given in sect.7.

0.1 Definition *Let X be a surface.*

(1) A (-1)-curve in X is a smooth rational curve C with self-intersection $C^2 = -1$, i.e the normal bundle $N_{C|X} = \mathcal{O}(-1)$.

(2) X is minimal if it does not contain any (-1)-curve.

We note that the notion of a minimal variety (sect.8) will be different from the notion of minimality in (0.1).

0.2 Theorem *Let X be a surface, C be a (-1)-curve on X. Then there exists a (smooth) surface Y and a birational morphism $\phi : X \longrightarrow Y$ such that $\phi(C)$ is a point $y \in Y$ and $\phi|X \setminus C \longrightarrow Y \setminus \{y\}$ is an isomorphism. ϕ is nothing than the blow-up of y in Y.*

By blowing down successively (-1)-curves in X one finally obtains a surface X' without any (-1)-curves. The reason why the procedure "blowing down (-1)-curves" stops is just that the Betti number b_2 always drops by 1. So the study of the structure of surface is reduced to the study of minimal surfaces, at least for our purposes.

0.3 Theorem *Let X be a minimal surface, $K_X = \det T_X^*$ its canonical bundle. Then either K_X is nef, i.e. $K_X \cdot C \geq 0$ for every curve $C \subset X$, or $X = \mathbb{P}_2$, or X is a ruled surface over a compact Riemann surface C, i.e. X is a $\mathbb{P}_1$-bundle over C.*

The generalisation of these two theorems to higher dimensions is roughly speaking the aim of these notes. Both theorems explain the structure of surfaces X with K_X not nef, i.e. there is a curve C with $K_X \cdot C < 0$. Actually we have very special curves $C \subset X$ with $K_X \cdot C < 0$. Either we have a (-1)-curve C (then $K_X \cdot C = -1$ or X has a $\mathbb{P}_1$-bundle structure in what case we can take C to be a ruling line and $K_X \cdot C = -2$ or X is the projective plane, then we take C to be a line and $K_X \cdot C = -3$. If however X is the blow-up of a point p in the plane and C the strict transform of a line not passing through p, then C is not such a distinctive curve although $K_X \cdot C = -3$. The reason is that some deformation of C splits into two components which is not allowed in the three types above. But if we let the line pass through p, then C is a ruling line as described above. Observe that in the beginning we only required the existence of **some** curve C with $K_X \cdot C < 0$ but we end up with a very special **rational** curve C with the same property. These C are called extremal rational curves; the terminus extremal will be explained in sect.6. Note also that these rational curves are connected with maps: in the first case the blow-down of the (-1)-curve, in the second the projection to the base curve, in the third the constant map. This seems silly in the last case but there is a concrete meaning behind: the above maps have the remarkable property that they contract exactly those curves to points which are homologous to a (rational) multiple of the given extremal rational curves. In the third case we just have $b_2(X) = 1$, i.e. all curves are homologous to the given rational curve, so the map is constant. These maps are called contractions of extremal rays resp. extremal contractions; the most significant property is that the anticanonical bundle $-K_X$ is relatively ample with respect to the map. The general method to construct these contractions in higher dimensions will be explained in sect.4. Once one has a map between two surfaces such that the anticanonical bundle is relatively ample which cannot be factored in some maps of the same type, then this map is one of those described above. This is rather easy to see. Therefore the maps are to some extend more important than the extremal curves. On the other hand extremal rational curves determine completely a unique contraction.

In higher dimensions it is not at all clear that a manifold with K_X not nef has any rational curve C (with $K_X \cdot C < 0$). In fact, this is rather delicate and requires a reduction to characteristic p. There is no complex-algebraic proof of the existence of rational curve up to now. This topic is explained in Miyaoka's article in this volume. In these notes we will see that if K_X is not nef, and if X carries a contraction, then sometimes one can detect directly rational curves contracted by this map (e.g. in the case of surfaces or threefolds).

There is another important theorem in surface theory: if X is a surface such that K_X is nef, then mK_X is generated by global sections for large m. The corresponding theorem is known in dimension three as the abundance theorem of Kawamata and Miyaoka), but it is open in higher dimensions. Also this topic is treated in Miyaoka's article.

Having in mind the picture for surfaces we can now ask the following questions in higher dimensions:
(1) Let K_X be a projective manifold such that K_X is not nef. Is there a contraction $f : X \longrightarrow Y$ such that $-K_X$ is f-ample?
(2) What is the structure of Y (and of f)?
(3) Can we repeat the process in case $\dim X = \dim Y$: i.e. if K_Y is not nef, will we have another contraction? (The difficulty will be that Y is in general no longer smooth.)
(4) Will this process terminate? This would mean that after finitely many steps we either a have fiber space structure or the resulting space Z has K_Z nef. Such a Z is then called a *minimal model*.

It is most important that already after the first step Y can be singular. However the singularities are very mild: they still allow to define the canonical "bundle" and hence the Kodaira dimension directly on Y. Moreover the Kodaira dimension will be a birational invariant in this new category of slightly singular spaces. The singularities are so mild that they occur only in codimension 3 or higher, so they do not appear on surfaces. This is the deeper reason why birational geometry on surfaces is so easy.

1 Singularities

1.1 Why are singular spaces necessary in our contexts? In the surface case we have seen several types of morphisms $\varphi : X \longrightarrow Y$ with the property that $-K_X$ is φ-ample. We want to construct such morphisms also in higher dimensions. Now notice

(a) If X is a smooth surface, Y a normal surface and $\varphi : X \longrightarrow Y$ a morphism with connected fibers such that $-K_X$ is φ-ample, then Y is actually smooth.

The proof of this fact is easy and left to the reader.

(b) The analogous statement in higher dimensions is false. We here give the following example. Let Z be the total space of the line bundle $\mathcal{O}(-2)$ on the projective plane $\mathbb{P}_2$. Let X be the compact threefold arising by compactifying every fiber of $Z \longrightarrow \mathbb{P}_2$. Note that $X = \mathbb{P}(\mathcal{O} \oplus \mathcal{O}(-2))$. The zero section $E \subset Z \subset X$ can be blown down because its normal bundle N_E is negative, namely $\mathcal{O}(-2)$, see [Gr62]. We obtain a birational morphism $f : X \longrightarrow Y$ to a singular(!) normal complex space Y. Then Y is projective, in fact the image of the section at infinity $(= X \setminus Z)$ gives an ample divisor on Y. Clearly $-K_X$ is φ-ample.

1.2 What type of singularities should we allow? We certainly want to have the following

(a) the canonical "bundle" (or sheaf) K_X should exist on X and should have reasonable properties: some multiple at least should be locally free and we want to have vanishing theorems;

(b) the Kodaira dimension, defined via K_X and a priori not by the canonical bundle of a desingularisation, should be a birational invariant (in the category of singular spaces we are looking for);

(c) the category of singular spaces we have in mind should be as small as possible;

(d) very vaguely, the "smooth theory" should work in the singular case, too.

We first address to the point (a):

1.3 Definition Let X be a normal variety.

(a) X is Cohen-Macaulay, if every stalk $\mathcal{O}_{X,x}$ is a Cohen-Macaulay ring in the sense of commutative algebra.

(b) The canonical sheaf ω_X on X is defined in the following way. Let X_0 be the regular part of X with inclusion map $i : X_0 \longrightarrow X$. Recall that $\omega_{X_0} = \Omega^n_{X_0}$, the sheaf of regular n-forms, where $n = \dim X$. Then we let

$$\omega_X = i_*(\omega_{X_0}).$$

Hence ω_X is a (coherent) reflexive sheaf and it is nothing than the dualising sheaf of duality theory. The canonical bundle K_X is the Weil divisor associated to ω_X (see N.4). More generally we define

$$\omega_X^{[m]} = i_*(\omega_{X_0}^{\otimes m})$$

and let mK_X be the Weil divisor associated to $\omega_X^{[m]}$.

(c) X is said to be $\mathbb{Q}$-Gorenstein, if X is Cohen-Macaulay and if some multiple mK_X is a Cartier divisor. The least m such that mK_X is Cartier is called the index of X. If $m = 1$ and X is Cohen-Macaulay, then X is called Gorenstein.

Every 2-dimensional normal variety is Cohen-Macaulay; every (locally) complete intersection is Gorenstein.

The example in (1.1)(b) gives a $\mathbb{Q}$-Gorenstein singularity of index 2. In fact. if $\varphi : X \longrightarrow Y$ denotes the blow-down of $E \simeq \mathbb{P}_2$ with normal bundle $\mathcal{O}(-2)$, then by adjunction $K_X = f^*(K_Y) + \frac{1}{2}E$.

1.4 Let X be a normal projective $\mathbb{Q}$-Gorenstein variety. As already mentioned, the canonical sheaf is just the dualising sheaf of duality theory (see e.g. [Ha77]). Since X is Cohen-Macaulay, we therefore have the Serre duality

$$H^q(X, \mathcal{L} \otimes \omega_X) \simeq H^{n-q}(X, \mathcal{L}^*)^*$$

for every locally free sheaf $\mathcal{L}$ on X.

Now assume that X is a normal projective $\mathbb{Q}$-Gorenstein variety. Let $\pi : \hat{X} \longrightarrow X$ be a desingularisation. Let E_i denote the components of the exceptional set π of codimension 1. (If π is a blow-up, then all components of the exceptional set have codimension 1). Then there is an equation of $\mathbb{Q}$-divisors

$$K_{\hat{X}} = \pi^*(K_X) + \sum \mu_i E_i$$

with rational numbers μ_i. This equation makes perfectly sense as an equation of Cartier divisors after multiplication with the index m of X. We are looking for conditions which guarantee that

$$\kappa(\hat{X}) = \kappa(X),$$

where $\kappa(X) = \kappa(K_X) = \kappa(mK_X)$. If some μ_i is negative, then in general we will have $\kappa(\hat{X}) < \kappa(X)$. Therefore we make the following

1.5 Definition Let X be a normal $\mathbb{Q}$-Gorenstein variety. X has only terminal (canonical) singularities, if the following holds. Let $\pi : \hat{X} \longrightarrow X$ be a desingularisation and write as above

$$K_{\hat{X}} = \pi^*(K_X) + \sum \mu_i E_i.$$

Then $\mu_i > 0 (\mu_i \geq 0.)$ The number μ_i is called the *discrepancy* at E_i.

1.6 The names "terminal" and "canonical" will be explained later. It is an easy exercise to show that the definition is independent of the choice of the desingularisation.

Note that X has only canonical singularities if and only if

$$\pi_*(\omega_{\hat{X}}^r) = \omega_X^{[r]}$$

for some positive integer r.

1.7 Proposition *(Elkik, Flenner) Canonical singularities are rational. More precisely: if X has only canonical singularities and if $f : \hat{X} \longrightarrow X$ is a desingularisation, then*

$$R^q f_*(\mathcal{O}_{\hat{X}}) = 0, q \geq 1.$$

Proposition 1.7 allows us to compare cohomology on $\hat{X}$ and on X by the Leray spectral sequence.

Proof. Let us assume $\dim X = 2$ for simplicity. A standard technique to be discussed in the next topic (1.8) says that we may assume X to be Gorenstein (of course our claim is local in X). So ω_Y is locally free and the condition "canonical" tells us that there is an effective divisor A (possibly 0) such that

$$\omega_{\hat{X}} = f^*(\omega_X) \otimes \mathcal{O}_{\hat{X}}(A). \tag{$*$}$$

The essential point in the proof is the vanishing theorem of Grauert-Riemenschneider, to be discussed in (2.15), which says that

$$R^1 f_*(\omega_{\hat{X}}) = 0.$$

The projection formula therefore gives via $(*)$ that

$$R^1 f_*(\mathcal{O}_{\hat{X}}(A)) = 0. \tag{$**$}$$

In order to get from $\mathcal{O}_{\hat{X}}(A)$ to $\mathcal{O}_{\hat{X}}$ we use the exact sequence

$$0 \longrightarrow \mathcal{O}_{\hat{X}} \longrightarrow \mathcal{O}_{\hat{X}}(A) \longrightarrow \mathcal{O}_A(A) \longrightarrow 0.$$

Taking direct images we obtain the exact sequence

$$f_*(\mathcal{O}_A(A)) \longrightarrow R^1 f_*(\mathcal{O}_{\hat{X}}) \longrightarrow R^1 f_*(\mathcal{O}_{\hat{X}}(A)).$$

Hence it suffices to show $f_*(\mathcal{O}_A(A)) = 0$, i.e.

$$H^0(A, \mathcal{O}_A(A)) = 0.$$

By the adjunction formula and $(*)$ we have

$$K_A = \mathcal{O}_A(2A).$$

Hence Serre duality gives

$$H^0(A, \mathcal{O}_A(A)) = H^1(A, \mathcal{O}_A(A)) = 0,$$

the last vanishing coming from $(**)$ by the general machinery of higher direct images, see e.g. [Ha77, chap.3, sect.11]. Since similar arguments occur several times, we give here some details. Consider the formal completion $\hat{A} \subset \hat{X}$ of $\hat{X}$ along A. Now $(**)$ can be restated as

$$R^1 f_*(\mathcal{O}_{\hat{X}}(A))_{\hat{x}} = 0,$$

where $x = f(A)$ and the hat means formal completion as $\mathcal{O}_{X,x}$-module. Hence the Grauert comparison theorem yields

$$H^1(\hat{A}, \mathcal{O}_{\hat{A}}(A)) = 0.$$

Now the obstruction to extend a cohomology class from one infinitesimal neighborhood (defined by $\mathcal{I}_A^k$ to the next (defined by $\mathcal{I}_A^{k+1}$) vanishes since it lies in

$$H^2(A, \mathcal{O}_A(A) \otimes N^{*\mu}) = 0.$$

Therefore our last vanishing implies $H^1(A, \mathcal{O}_A(A)) = 0$.

Of course the above arguments work mutatis mutandis in every dimension. In dimension 3 for example we however need another vanishing: $R^2 f_*(\mathcal{O}_{\hat{X}}) = 0$. This follows from "Grothendieck duality" which yields an isomorphism

$$R^2 f_*(\mathcal{O}_{\hat{X}}) \simeq \omega_X / f_*(\omega_{\hat{X}}),$$

the last sheaf being 0 by $(*)$ (take direct images).

Another proof of (1.7), avoiding Grothendieck duality, goes as follows. We immediately treat the n-dimensional case. By compactifying suitably, we may assume X and $\hat{X}$ projective (this means we are working strictly in the algebraic category, while the above arguments also work in the analytic case). Choose a sufficiently ample line bundle L on X such that

$$H^p(X, R^q f_*(\mathcal{O}_{\hat{X}}) \otimes L) = 0, p \geq 1, q \geq 0. \tag{1}$$

Let $\hat{L} = f^*(L)$. From Serre duality, the Grauert-Riemenschneider vanishing theorem and the assumption that X has only canonical singularities we immediately deduce via the Leray spectral sequence

$$H^j(\hat{X}, \hat{L}) \simeq H^j(X, L) = 0. \tag{2}$$

We now compute $H^*(\hat{X}, \hat{L})$ by the Leray spectral sequence $(E_2^{p,q})$. From (1) we have $E_2^{p,q} = 0$ for $p \geq 1, q \geq 0$. Thus the general machinery of spectral sequences yields

$$H^r(\hat{X}, \hat{L}) \simeq E_2^{0,r} = H^0(X, R^r f_*(\mathcal{O}_{\hat{X}}) \otimes L).$$

Hence in total

$$H^0(X, R^r f_*(\mathcal{O}_{\hat{X}}) \otimes L) = 0.$$

But this is absurd (possibly substitute L by a multiple) unless $R^r f_*(\mathcal{O}_{\hat{X}}) = 0$.

1.8 Proposition *Let X be normal $\mathbb{Q}$-Gorenstein. Let m be an integer such that mK_X is Cartier. Assume $mK_X \simeq \mathcal{O}_X$ (this can always be achieved by passing to small affine parts of X). Then there exists a covering $f : Y \longrightarrow X$ such that Y has index 1.*

Y is constructed as follows. Let V be the line bundle associated with mK_X, and W the trivial line bundle. Via the isomorphism $mK_X \simeq \mathcal{O}_X$, the m-th tensor power gives a map $g : V \longrightarrow W$. Now let Y be the g-preimage of the section 1 in W.

$f : Y \longrightarrow X$ is called the *canonical cover* of X. It is a Galois covering and has the following important property:

$$f_*(\mathcal{O}_Y(K_Y)) = \mathcal{O}_X \oplus \mathcal{O}_X(K_X) \oplus \cdots \oplus \mathcal{O}_X((m-1)K_X).$$

Note that f is étale in codimension 1. For details we refer e.g. to [Re87].

1.9 Examples (1) We first show that terminal singularities of surfaces are necessarily smooth points. In fact, let X be a surface having only terminal singularities. Let $f : \hat{X} \longrightarrow X$ be a minimal desingularisation, meaning that no component of the exceptional set E is a (-1)-curve. Write $K_{\hat{X}} = f^*(K_X) + \sum a_i E_i$. From the adjunction formula it follows

$$K_{\hat{X}} \cdot E_j = -2 - E_j^2 \geq 0,$$

since $E_j^2 \leq -2$. Hence $\sum a_i E_i \cdot E_j \geq 0$. Since all $a_j > 0$, this contradicts the classical fact that the intersection matrix $(E_i \cdot E_j)$ is negative definite.

(2) By the fact that local general hyperplane sections of varieties with at most terminal singularities have the same properties, it follows that terminal singularities occur at most in codimension 3.

(3) A similar argument as in (1) shows that in the case of canonical singularities we have $K_{\hat{X}} = f^*(K_X)$. It is now a well-known result in the theory of surface singularities that the canonical singularities are exactly the rational double points (or smooth points), see e.g. [La86]. Hence the canonical surface singularities are the singularities of type A_n (equation $x^2 + y^2 + z^{n+1}$) resp. D_n (equation $x^2 + y^2 z + z^{n-1}, n \geq 4,$) resp. E_6 (with $x^2 + y^3 + z^4$) resp. E_7 (with $x^2 + y^3 + yz^3$) or finally E_8 (with $x^2 + y^3 + z^5$).

(4) By (2) all 3-dimensional terminal singularities are isolated. Here are two examples by equations:

$$x^2 + y^2 + z^2 + w^2 = 0,$$

$$x^2 + y^2 + z^2 + w^3 = 0.$$

Blowing up the singular point, one sees by calculation in coordinates that the exceptional divisor E is $\mathbb{P}_1 \times \mathbb{P}_1$ in the first case with normal bundle $\mathcal{O}(-1)$ and the quadric cone in the second with the same normal bundle. Hence the adjunction formula implies that both singularities are terminal.

1.10 Structure of terminal threefold singularities

Here we say something more on the structure of terminal singularities in dimension 3. We refer to [Re87] for details.

(a) By (1.8) the study of terminal singularities is reduced to the Gorenstein case.

(b) A threefold singularity x is terminal Gorenstein iff it is isolated and the general hyperplane section through x is a rational double point.

(c) Terminal Gorenstein threefold singularities x are hypersurface singularities: x can locally (in the euclidean topology) be given by $\{f = 0\}$ with f a local holomorphic function near 0 in $\mathbb{C}^4$.

(d) The finer classification of terminal singularities is given by Mori [Mo85].

We finally introduce log-terminal singularities. The reason why to introduce the log category will become clear later. A *log pair* consists of a normal variety X and an effective $\mathbb{Q}$-Weil divisor $D = \sum d_i D_i$ such $K_X + D$ is $\mathbb{Q}$-Cartier. A *log resolution* is a desingularisation $f : \hat{X} \longrightarrow X$ such that the union of the exceptional locus of f and $f^{-1}(\text{Supp} D)$ is a divisor with normal crossings. In this situation we have an equation

$$K_{\hat{X}} = f^*(K_X + D) + \sum a_i E_i \qquad (*)$$

where the E_i are prime divisors on $\hat{X}$ and a_i are rational numbers.

1.11 Definition Let (X, D) be a log pair with log resolution $f : \hat{X} \longrightarrow X$.

(1) We say that (X, D) is *log terminal* (or has only log terminal singularities) if the following holds.

(a) $d_i < 1$,

(b) in the equation the log discrepancies satisfy $a_i > -1$ for all those E_i which are exceptional for f.

(2) (X, D) is *Kawamata log terminal (klt)* if (b) holds and moreover

(c) $d_i \leq 1$,

(d) there exists an f-ample divisor whose support is exactly the exceptional set for f.

One should be aware that the notions of log terminality vary in the literature. Also notice that it is not required that X is $\mathbb{Q}$-Gorenstein. An important special case is $D = 0$. This means that in (1.6) we have $\mu_i > -1$ instead of $\mu_i > 0$.

1.12 Examples and Remarks (1) Let X be a normal surface. Then $(X, 0)$ is log terminal iff X has only quotient singularities, i.e. locally $X = \mathbb{C}^2/G$ with a finite subgroup of $Gl(2, \mathbb{C})$. The difference to canonical singularities is that these are quotient singularities by a finite subgroup of $Sl(2, \mathbb{C})$.

(2) Let us look at a concrete example. Let X be the Hirzebruch surface

$$\mathbb{P}(\mathcal{O} \oplus \mathcal{O}(-a))$$

over $\mathbb{P}_1$, where $a \geq 2$. Let C_0 be the section with $C_0^2 = -a$. Let $f : X \longrightarrow Y$ be the blow-down of C_0. Then we compute easily:

$$K_X = f^*(K_Y) + (\frac{2}{a} - 1)F,$$

where F is a ruling line. Hence $(Y, 0)$ is log terminal for all a but not canonical unless $a = 0$.

(3) Let X be smooth and D an effective divisor whose support has normal crossings. Then we can take $f = id$ to verify the condition $a_i > -1$ in $(*)$, which is an empty condition. So (X, D) is log terminal if $d_i < 1$.

(4) If (X, D) is log terminal then we can take any log resolution to verify (b) in (1.10). This relies on the so-called logarithmic ramification formula and is not obvious, see [KMM87]. If (X, D) is klt, then the corresponding statement is false; the reason is that $d_i = 1$ is allowed here. As an example one can take the projective plane with D a rational curve of degree 3 having one node. As explained in (3), we can take $f = id$ but if we take f to be the blow up of the node, then the resulting discrepancy of the exceptional divisor will be 0.

There is the following analogue to (1.7), see [KMM87, 1-3-6]. The proof relies somehow on a refined version of the Grauert-Riemenschneider vanishing theorem.

1.13 Theorem *Assume that (X, D) is klt. Then X has only rational singularities.*

We end this section by mentioning the following result due to Reid [Re80], [Re83].

1.14 Theorem *Let X be $\mathbb{Q}$-Gorenstein and assume that X has a canonical cover V (which is always true locally in the Zariski topology). Then $(X, 0)$ is log terminal if and only if V has only rational (Gorenstein) singularities.*

2 Vanishing Theorems

Given a projective manifold X such that K_X is not nef, i.e. there is a curve $C \subset X$ such that $K_X \cdot C < 0$, we want to construct certain maps from X to other projective varieties. These maps should be defined by sections of holomorphic line bundles which are attached in a natural way to K_X. So we need methods to construct sections in line bundles. Usually one tries to construct sections by using vanishing theorems for H^1-cohomology groups and therefore the aim of this section is to discuss the vanishing theorems needed for these notes. We start by quoting the most famous and basic vanishing theorem.

2.1 Kodaira vanishing theorem *Let X be a projective manifold and L an ample line bundle on X. Then*

$$H^q(X, \mathcal{O}(K_X + L)) = 0, q \geq 1.$$

This is usually proved by harmonic theory, see e.g. [We80], [GH78], [SS85]. For another approach in connection with topology and Hodge decomposition see [EV92], [Ko87], [CKM88]. We shall see later a proof using L^2-methods (the Hörmander approach).

The ampleness of L is in the harmonic approach rephrased in the following way: there exists a metric on L whose curvature is positive. In [GR70] this condition was weakened: one needs only the curvature to be semipositive and positive almost everywhere (actually positivity at some point is sufficient). Grauert and Riemenschneider called these line bundles "almost positive". We need however more general vanishing theorems; almost positive line bundles do not have enough flexibility. The first comes by extracting the algebraic essence from almost positivity.

2.2 Definition Let X be a normal n-dimensional projective variety and L a line bundleon X.

(1) L is *nef* if $c_1(L) \cdot C \geq 0$ for all curves $C \subset X$.

(2) Assume L is nef. L is *big* if $c_1(f^*(L))^n > 0$.

Of course a divisor D is nef if $\mathcal{O}_X(D)$ is nef.

It is obvious that almost positive line bundles are big and nef. Note also that the definition (2.2) makes sense for all $\mathbb{Q}$-divisors! For Cartier divisors we have the following characterisation of "bigness" which explain also the name.

2.3 Proposition *Let X be a normal projective variety and D a nef divisor. Then D is big iff $\kappa(\mathcal{O}_X(D)) = n = \dim X$.*

Proof. By passing to a desingularisation, we may assume X smooth. Next note that

$$h^q(X, \mathcal{O}_X(mL)) \leq Cm^{n-1} \qquad (*)$$

for every nef line bundle. This is an easy exercise by induction on $\dim X$, taking hyperplane sections. Actually one has more: $h^q \leq Cm^{n-q}$. But now the claim follows immediately from Riemann-Roch:

$$\chi(X, \mathcal{O}_X(mD)) = \frac{D^n}{n!} m^n + o(m^n).$$

If D is not nef, then the property $D^n > 0$ is not very useful. In that case we **define** directly D to be **big** if $\kappa(\mathcal{O}_X(D)) = \dim X$.

Now we can state the *Kawamata-Viehweg vanishing theorem* [Ka82], [Vi82], we immediately choose the general form.

2.4 Theorem *Let X be a projective manifold and D a divisor on X. Assume that there is a numerical decomposition $D \equiv D_0 + \sum a_i A_i$ where D_0 is a $\mathbb{Q}$-divisor which is big and nef, where $0 \leq a_i < 1$ and where $\sum A_i$ has normal crossings. Then*

$$H^q(X, \mathcal{O}_X(D + K_X)) = 0, q \geq 1.$$

In the special case where all $a_i = 0$ we obtain the direct generalisation of the Grauert-Riemenschneider version: if D is big and nef, then

$$H^q(X, \mathcal{O}_X(D + K_X)) = 0$$

for $q \geq 1$. But the general version leaves much more flexibility: D itself does not have to be big or nef. We will see how to use this flexibility in the next section.

The original proof of Kawamata reduces (2.4) to (2.1) by covering techniques. We will however discuss here a completely different approach discovered by Nadel [Na89] and developped further by Demailly [De90, 93, 96] using analytic tools and the so-called multiplier ideal sheaves. The vanishing theorem of Nadel is an analytic analogue of the vanishing theorem of Kawamata-Viehweg, it actually contains it as special case. It also has the advantage that it can be proved directly, without a reduction to Kodaira vanishing.

First we recall the notion of plurisubharmonic functions.

2.5 Definition Let $G \subset \mathbb{C}^n$ be a domain. A function $\varphi : G \longrightarrow [-\infty, \infty)$ is called *plurisubharmonic* if and only if
(a) φ is upper semi-continous,
(b) for every complex line $L \subset \mathbb{C}^n$, the function $\varphi | L \cap G$ is subharmonic.

Recall that a function ψ on a domain in $\mathbb{C}$ is *subharmonic* if it satisfies the mean value inequality:

$$\psi(x_0) \leq \frac{1}{2\pi} \int_0^{2\pi} \psi(x_0 + e^{i\varphi}v) d\varphi.$$

For basic properties of plurisubharmonic functions and more references we refer to [De96] and [GPR94]. Just note here that for C^2-functions ψ, plurisubharmonicity means that the Levi form

$$\left(\frac{\partial^2 \psi}{\partial z_j \partial \bar{z}_k} \right)$$

is positive semidefinite at every point. Of course the definition (2.5) immediately can be extended to complex manifolds (even complex spaces) by using coordinate charts. The set of plurisubharmonic functions on the complex manifold X is usually denoted $\mathrm{PSH}(X)$.

2.6 Definition (*Nadel*) Let X be a complex manifold and $\varphi \in \mathrm{PSH}(X)$. The multiplier ideal sheaf $\mathcal{I}_\varphi \subset \mathcal{O}_X$ consists of all those germs $f \in \mathcal{O}_{X,x}, x \in X$, which are locally L^2 with respect to φ, i.e. $|f|^2 \epsilon^{-2\varphi}$ is locally integrable.

Multiplier ideal sheaves would be completely useless without

2.7 Proposition $\mathcal{I}_\varphi$ *is coherent.*

Proof. Let $\mathcal{I} = \mathcal{I}_\varphi$. The problem being local, it is sufficient to consider the case where X is a ball $U \subset \mathbb{C}^n$ centered around 0. Then we must prove that $H^0(U \; \mathcal{I})$ generates I over U to conclude coherence. Let $\mathcal{J} \subset \mathcal{O}_U$ be the ideal sheaf generated by $H^0(U, \mathcal{I})$. Then it comes down to show $\mathcal{I} = \mathcal{J}$, the inclusion $\mathcal{J} \subset \mathcal{I}$ being obvious. We claim that

$$\mathcal{I}_0 \subset \mathcal{J}_0 + m_0^k \qquad\qquad (*)$$

for all positive integers k; then the lemma of Krull gives $\mathcal{I}_0 = \mathcal{J}_0$. Here m_0 denotes of course the maximal ideal at 0.

To prove $(*)$ take a germ $g_0 \in \mathcal{I}_0$ represented by $g \in \mathcal{I}(V)$. We try to extend g to U; of course this is in general not possible. It is however possible in the C^∞ category: we take a C^∞-function ρ on U with compact support such that $\rho|V = 1$. Then we consider ρg. For reasons which become clear in a moment we slightly change φ by setting

$$\tilde{\varphi} = \varphi + (n + k) \log |z| + |z|^2, k \in \mathbb{N}.$$

Note that $d(\rho g) = (\bar{\partial}\rho) g$ is still L^2 with weight function $\tilde{\varphi}$. Since ρg is not holomorphic, we are going to change it by solving a $\bar{\partial}$-equation, however with estimates, since we do not want to loose the L^2-property. More precisely we obtain a function f on U such that

$$\bar{\partial} f = (\bar{\partial}\rho) g$$

and such that $|f|^2 e^{-\tilde{\varphi}}$ is integrable. This is possible by Hörmanders L^2-estimates with weight functions, see e.g. [De93]. Thus we obtain the following.

(a) $\rho g - f$ is holomorphic on U,

(b) $\rho g - f$ is L^2 w.r.t. $\tilde{\varphi}$, and finally

(c) $(\rho g - f - g)_0 \in m_0^k$.

(c) comes from the definition of $\tilde{\varphi}$, which forces f to vanish at 0 of order at least k, because otherwise f would not be locally L^2 w.r.t. $\tilde{\varphi}$.

From (a), (b) and (c) the claim $(*)$ is now obvious.

If we let

$$V = \{x \in X | \text{ there exists } f \in \mathcal{O}_{X,x} \text{ with } |f|^2 e^{-2\varphi} \text{ not integrable near } x\},$$

then (2.7) essentially says that V is an analytic set in X.

We next discuss an important functorial property of multiplier ideal sheaves. The proof is easy and straightforward and thus left to the reader.

2.8 Proposition *Let $\hat{X}$ and X be complex manifolds and $f : \hat{X} \longrightarrow X$ a bimeromorphic map, i.e. generically an isomorphism. Let $\varphi \in \mathrm{PSH}(X)$. Then $\varphi \circ f \in \mathrm{PSH}(X)$ and*

$$f_*(\omega_{\hat{X}} \otimes \mathcal{I}_{\varphi \circ f}) = \omega_X \otimes \mathcal{I}_{\varphi}.$$

In order to formulate Nadel's vanishing theorem we need the notion of a singular hermitian metric on a line bundle; we follow [De92].

2.9 Definition Let X be a compact manifold and L a line bundle on X. A *singular hermitian metric* on L is a "metric" which in a local trivialisation $\alpha : L|U \longrightarrow U \times \mathbb{C}$ is of the form

$$\|u\| = |\alpha(u)|e^{-\varphi(x)}$$

with a locally integrable function φ. The function φ is called (local) *weight function* of the metric.

If φ is C^0 or C^∞, we just obtain the usual hermitian metrics.

2.10 Remark Let L be a line bundle equipped with a singular metric with weight function φ. Then one can still define the curvature of L, it is however only a $(1,1)$-current T instead of a $(1,1)$-form. T is defined locally by $T = \frac{i}{\pi} \partial \bar{\partial} \varphi$, formally just as in the smooth case. Of course $\partial \bar{\partial} \varphi$ is understood in the sense of currents.

2.11 Examples (1) The most important example for us is the case of divisors. Let $D = \sum a_i D_i$ be a divisor with a_i integers. We put a singular metric on the line bundle L associated to D in the following way. Let s be a local section of L. This is nothing than a meromorphic function f with the pole order dictated by D in the usual way. Now let $\|s\| = |f|$. It is an obvious exercise to see that the local weight function associated with this metric is $\varphi = \sum a_i \log |g_i|$, where g_i locally defines D_i.

(2) Now assume in (1) that the a_i are **positive rational** numbers. Assume furthermore that the D_i are smooth and have normal crossings only. We attach to D the

plurisubharmonic functions $\varphi = \sum a_i \log |g_i|$. Here again the g_i are local equations for D_i. Then by definition

$$\mathcal{I}_\varphi(U) = \{ h \in \mathcal{O}_X(U) | \int_U |h|^2 \prod |g_i|^{-2a_i} d\mu < \infty \},$$

where μ is the Lebesgue measure. Now let $n_i = [a_i]$. Since by the normal crossing assumption we can choose coordinates z_i such that $z_i = g_i$, the integrability condition just means that h is divisible by $\prod g_i^{n_i}$. Hence

$$\mathcal{I}_\varphi = \mathcal{O}_X(-\sum [a_i] D_i) = \mathcal{O}_X(-[D]).$$

If in particular $0 < a_i < 1$, then $\mathcal{I}_\varphi = \mathcal{O}_X$. The ideal $\mathcal{I}_\varphi$ exists of course globally, see the remark after (2.12).

We rephrase the last remarks as follows. Let $D = \sum a_i D_i$ be an effective and let m be a positive integer. Let L be a line bundle with $L^m = \mathcal{O}_X(D)$ so that formally $L = \mathcal{O}_X(\frac{D}{m})$. Applying the above considerations to $\frac{D}{m}$ instead of D, we obtain a singular metric on L with local weight function φ. If we start with an "abstract" φ, then we still can state:

$$1 \in \mathcal{I}_\varphi \text{ iff } (X, \frac{D}{m}) \text{is klt.}$$

2.12 Nadel's Vanishing Theorem *Let X be a projective manifold and L a line bundle on X equipped with a singular metric with local weight function φ. Assume that $i\partial\bar{\partial}\varphi \geq \omega$, where ω is some positive $(1,1)$-form on X (i.e. the form associated to a smooth hermitian metric). Then*

$$H^q(X, \mathcal{I}_\varphi \otimes \mathcal{O}_X(K_X + L)) = 0, \ q \geq 1.$$

Some explanations have to be made. First, the inequality $i\partial\bar{\partial}\varphi \geq \omega$ means that

$$\int_X i\partial\bar{\partial}\varphi \wedge \eta \geq \int_X \omega \wedge \eta$$

for all positive $(n-1, n-1)$-forms on X, the positivity of such forms being defined in the same way as for $(1,1)$-forms by defining the local coefficient matrix to be positive definite. Second, we must explain the meaning of $\mathcal{I}_\varphi$ since φ exists only locally. If however ψ is another weight function, then it is immediately clear that $\mathcal{I}_\varphi = \mathcal{I}_\psi$ on the intersection of the domains of definition. Thus every singular metric on L defines a multiplier ideal sheaf.

2.13 Remarks (1) The vanishing theorem of Kodaira is just a special case of Nadel's vanishing theorem. In fact, if L is ample, then L admits a metric of positive curvature, i.e. we have local weight functions φ which are say C^∞ everywhere and which are strictly plurisubharmonic, i.e. the Levi form is positive definite at every point. Hence we can take $\omega = i\partial\bar{\partial}\varphi$ locally. Moreover we have $\mathcal{I}_\varphi = \mathcal{O}_X$ in that case, so Kodaira follows.

(2) Nadel's vanishing theorem is of course false, if we omit the multiplier ideal sheaf in the cohomology. In fact take any ample line bundle L on a projective manifold X and let E be an irreducible divisor. Then $L + E$ carries a singular metric such that the positivity condition in (2.12) holds. Namely, just take a metric on L with positive curvature and the metric on (the line bundle associated to) E constructed in (2.11). Then multiplying these metrics we get the desired metric on $L + E$. But $H^q(X, \mathcal{O}_X(K_X + L + E))$ is of course non-zero in general for $q \geq 1$. We leave it to the reader to figure out a concrete example.

Proof of (2.12). We follow [De93]. Let $\mathcal{A}^* = \mathcal{A}^{p,q}_{L^2,\varphi}$ be the complex of (p,q)-forms u with values in L which have measurable coefficients, which are locally L^2 w.r.t. φ and such that $\bar{\partial}u$ is also locally L^2 w.r.t. φ. Then $\mathcal{A}^*$ is a resolution of

$$\mathcal{I}_\varphi \otimes \mathcal{O}_X(K_X + L);$$

the local exactness being guaranteed by Hörmander's L^2-estimates with weight function on small balls. In other words, we have an exact sequence

$$0 \longrightarrow \mathcal{I}_\varphi \otimes \mathcal{O}_X(K_X + L) \longrightarrow \mathcal{A}^{n,0}_{L^2,\varphi} \longrightarrow \mathcal{A}^{n,1}_{L^2,\varphi} \longrightarrow \cdots$$

Hence

$$H^q(X, \mathcal{I}_\varphi \otimes \mathcal{O}_X(K_X + L)) \simeq H^q(\mathcal{A}^*).$$

The cohomology vanishing for $H^q(X, \mathcal{A}^*)$ is now provided by the global version of the L^2-estimates (which are usually done on Stein manifolds, for the projective analogue see [De93, 4.1]).

2.14 Proof of Kawamata-Viehweg vanishing. Of course we may assume $D = D_0 + \sum a_i A_i$ (rather than up to numerical equivalence). The strategy is to put a singular metric on the line bundle associated to D (we will not distinguish between both objects) whose multiplier ideal is trivial. Clearly we will use the decomposition of D to do this; the only slight difficulty is that D_0 (or a multiple) does not carry a natural metric. Instead we add to D_0 a "small ample divisor". To do this fix a hyperplane section H. Choose $m \gg 0$ such that mD_0 is Cartier and that

$$H^0(X, \mathcal{O}_X(mD_0 - H)) \neq 0.$$

The possibility of doing so is known as *Kodaira's lemma* and follows easily from the exact sequence

$$0 \longrightarrow \mathcal{O}_X(tmD_0 - H) \longrightarrow \mathcal{O}_X(tmD_0) \longrightarrow \mathcal{O}_H(tmD_0) \longrightarrow 0$$

and growth considerations of the attached spaces of global sections.

Fix $B \in |mD_0 - H|$. Let k be a positive integer. Then

$$D_0 = \frac{1}{mk}B + \frac{1}{k}L_k$$

with $L_k = (k-1)D_0 + \frac{1}{m}H$.

Since D_0 is nef and H ample, L_k is ample (see (3.12)). Take a metric h_k on L_k with positive curvature. Take the singular metric h_B on B and the singular metric h_A on $r \sum a_i A_i$ (r chosen such that ra_i are integers) as constructed in (2.11). Finally put

$$h := h_k^{\frac{1}{k}} h_B^{\frac{1}{mk}} h_A^r,$$

this is a singular metric on D. The positivity assumption in (2.12) is clearly fulfilled, so Nadel's vanishing theorem gives

$$H^q(X, \mathcal{I}_h \otimes \mathcal{O}_X(D + K_X)) = 0.$$

In order to compute $\mathcal{I}_h$, we can of course neglect h_k, since it defines a smooth curvature form. Therefore it follows from (2.11(2)) that $\mathcal{I}_h = \mathcal{O}_X$ for $k \gg 0$ which proves the Kawamata-Viehweg vanishing theorem.

We now discuss the relative version of the Kawamata-Viehweg vanishing theorem.

2.15 Theorem *Let X be a non-singular variety, $\pi : X \longrightarrow S$ a proper morphism to a normal variety S and D a divisor on X. Let $D = D_0 + \sum a_i A_i$ with a $\mathbb{Q}$-divisor D_0 and $0 \le a_i < 1$ and such that $\sum A_i$ has normal crossings. Assume*
(1) D_0 is π-nef, ie. $D_0 \cdot C \ge 0$ for all curves C with $\dim \pi(C) = 0$;
(2) D_0 is π-big, i.e. $D_0|F$ is big, F the general smooth fiber.
Then $R^q \pi_*(\mathcal{O}_X(K_X + D)) = 0$ for all $q \ge 1$.

Proof. We may assume that X and S are actually projective: first reduce to an affine S, the assertion being local in the base, then compactify suitably.

Choose H ample on S. Then $D_0 + \pi^*(mH))$ is ample for every positive integer m, hence Kawamata-Viehweg gives

$$H^q(X, \mathcal{O}_X(K_X + D_0 + \pi^*(H)) = 0, \; q \ge 1.$$

Then the claim follows by the same arguments as in the alternative proof of (1.7) for $m \gg 0$.

A special case is the (relative) Grauert-Riemenschneider vanishing theorem:

2.16 Corollary *Let $\pi : X \longrightarrow Y$ be a generically finite map from a non-singular variety X to a normal variety Y. Then $R^q \pi_*(\omega_X) = 0$ for $q \ge 1$.*

Proof. Let H be ample on Y. Then $\pi^*(H)$ is π-nef; it is also π-big, since π is generically finite. Therefore (2.15) together with the projection formula yield the claim.

Of course we have even $R^q\pi_*(\omega_X \otimes L) = 0$ for any nef line bundle L on X in the situation of (2.16). If $\dim Y < \dim X$, then the sheaf $R^q f_*(\omega_X)$ still has nice properties, e.g. it is torsion free. For this and much more see [Ko86].

3   The Ample Cone

In this section we discuss elementary properties of ample divisors and introduce the ample cone and the cone of curves. More subtle properties like the cone theorem will be treated in sect. 5. As reference we recommend [Ha70].

3.1 Definition Let X be a normal variety.

(1) Assume X compact. We let

$$N^1(X) = (\operatorname{Pic}(X)/\equiv) \otimes \mathbb{R},$$

$$N_1(X) = (C_1(X)/\equiv) \otimes \mathbb{R},$$

where $C_1(X)$ is the free abelian group generated by irreducible curves in X.

(2) If more generally $\pi : X \longrightarrow S$ is a proper morphism to a normal variety S, then we have a slightly different notion of numerical equivalence for divisors. For divisors D_i we let $D_1 \equiv_S D_2$ iff $D_1 \cdot C = D_2 \cdot C$ for all curves contracted by π. Then we define

$$N^1(X/S) = (\operatorname{Pic}X/\equiv_S) \otimes \mathbb{R},$$

$$N_1(X/S) = (C_1(X/S)/\equiv) \otimes \mathbb{R},$$

where $C_1(X/S)$ is the free abelian group generated by those curves which are contracted by π.

Often we identify elements in $N^1(X/S)$ or $N_1(X/S)$ with representatives.

3.2 Theorem and Definition *$N^1(X/S)$ and $N_1(X/S)$ are finite-dimensional and dual via the canonical pairing*

$$N^1(X/S) \times N_1(X/S) \longrightarrow \mathbb{R}, (D, C) \mapsto (D \cdot C).$$

We call $\rho(X) = \dim N^1(X)$ the Picard number of X. Moreover we call $\rho(X/S) = \dim N^1(X/S)$ is the relative Picard number of X over S.

Proof. The statement on the duality is obvious. We prove finite-dimensionality only in case S being a point. Using desingularisation we reduce to the case X smooth.

From Hodge decomposition and the Lefschetz theorem on $(n-1, n-1)$-classes (i.e. an element in $H^{n-1,n-1}(X) \cap H^{2n-2}(X, \mathbb{Q})$ is given by an algebraic cycle) we see that numerical equivalence is the same as homological equivalence. Thus a divisor D is numerically 0 iff and $c_1(\mathcal{O}_X(D)) = 0$ in $H^2(X; \mathbb{R})$. Since $H^2(X, \mathbb{R})$ is finite dimensional by a basic theorem in algebraic topology, we conclude for $N^1(X)$ and by duality also for $N_1(X)$. For the general case see [Kl66].

3.3 Definition (1) Let X be a normal projective variety. We let $Amp(X)$ denote the closed cone in $N^1(X)$ generated by (the classes of) the ample divisors. We define $\overline{NE}(X)$ to be the closed cone generated by the irreducible curves in X.

(2) Let $\pi : X \longrightarrow S$ be proper. Then $\overline{NE}(X/S)$ denotes the closed cone generated by the curves contracted by π.

In order to get more information about those cones, we need numerical characterisations for ampleness, which are more tractable than the definition by cohomology vanishing or embeddings into projective space.

3.4 Nakai-Moishezon criterion *Let X be a normal projective variety and D a Cartier divisor on X. Then D is ample if and only if $D^s \cdot Y > 0$ for all s and all irreducible reduced subspaces $Y \subset X$ of dimension s.*

Of course this condition is necessary for ampleness. For the (elementary) proof of the other direction we refer to [Ha70]. One should be aware that it is not sufficient to have $D \cdot C > 0$ for all curves $C \subset X$ to conclude D ample. The reason will become clear in a moment. For the next result due to Kleiman we again refer to [Ha70].

3.5 Theorem *Let X be normal projective and D a Cartier divisor on X. If D is nef, then*

$$D^s \cdot Y \geq 0$$

for all s and all s-dimensional irreducible reduced subspaces of X.

3.6 Corollary *If D is nef, then $[D] \in Amp(X)$.*

In fact, fix an ample divisor H. Let n be a positive integer. Then $nD + H$ is ample by virtue of (3.4) and (3.5). Hence $[D] + \frac{1}{n}[H] \in Amp(X)$, and we conclude, $Amp(X)$ being closed.

3.7 Corollary *Let $D \in N^1(X)$. Then $D \cdot C \geq 0$ for every curve if and only if $D \in Amp(X)$. In other words, $Amp(X)$ is the dual cone to $\overline{NE}(X)$.*

The last ampleness criterion we need is Seshadri's criterion. It explains the position of the ample divisors in the ample cone.

3.8 Theorem *Let X be normal projective and D Cartier on X. Then D is ample if and only if there exists $\epsilon > 0$ such that $D \cdot C \geq \epsilon m(C)$, where $m(C)$ is the maximal local multiplicity $m_p(C), p \in C$.*

One direction being clear, the key point to the other is the following way to compute $m_x(C)$: take the blow-up $f : \hat{X} \longrightarrow X$ in a smooth point $x \in X$, let E denote the exceptional divisor and $\hat{C}$ the strict transform of C in $\hat{X}$. Then $m_x(C) = E \cdot \hat{C}$. Having this in mind, the proof is a rather straightforward application of the Nakai-Moishezon criterion.

The following is an easy consequence:

3.9 Corollary *Let X be normal projective and D Cartier on X. Introduce a norm $\| \cdot \|$ on the finite dimensional vector space $N^1(X)$. Then D is ample if and only if there exists $\epsilon > 0$, such that $D \cdot C \geq \epsilon \| [C] \|$ for all curves $C \subset X$.*

We can reformulate (3.9) in the following way. D is ample if and only if $D \cdot C > 0$ for all $C \in \overline{NE}(X) \setminus \{0\}$. But note that on the boundary of $\overline{NE}(X)$ there might be points which are not represented by linear combinations of irreducible curves with positive integers, only by limits of those. This is the reason why it is not sufficient to have $D \cdot C > 0$ for all curves in order to conclude the ampleness of D. For an explicit example see [Ha70].

3.10 Corollary *Let D be Cartier on X. Then D is ample if and only if D belongs to the interior of $Amp(X)$.*

We will call every element $D \in N^1(X)$ nef, if $D \in Amp(X)$, and ample, if D is actually in the interior of $Amp(X)$.

Sometimes is the following Nakai-Moishezon criterion for an arbitrary $D \in N^1(X)$, i.e. for a $\mathbb{R}$-Cartier divisor, useful; it does not follow immediately from (3.4)! For the proof see [CP91].

3.11 Theorem *Let $D \in N^1(X)$. Then D is ample if and only if $D^s \cdot Y > 0$ for every s and every s-dimensional irreducible reduced subvariety $Y \subset X$.*

3.12 Here are some basic properties of nef divisors.
(1) If D_1 is ample, D_2 nef, then $D_1 + D_2$ is nef.
(2) If $f : Y \longrightarrow X$ is any map, D nef on X, then $f^*(D)$ is nef, the converse being true if f is surjective.
(3) If D_i are nef, then $D_1 + D_2$ is nef.
(4) If H is ample, D any divisor, then $mH - D$ is ample for $m \gg 0$.
(5) If D is big and nef, then there exists an effective $\mathbb{Q}$-divisor E such that $D - \epsilon E$ is ample for all $0 < \epsilon < 1$.
(6) If D is ample and $f : \hat{X} \longrightarrow X$ a blow-up with exceptional components E_i, then $f^*(D) - \sum e_i E_i$ is ample for suitable small positive e_i.

We give the proof of (5), the assertions (1)-(4) being clear from the previous theorems. Fix an ample divisor H. By Kodaira's lemma (see (2.14)) we find m_0 such that $m_0 D - H$ is effective, say F. Hence $m_0 D - F = H$ is ample. Now D being nef, $mD - F = (m - m_0)D + H$ is ample, hence, putting $E = \frac{1}{m_0}F$, our claim follows. Finally (6) is left to the reader.

We also note the relative version of Seshadri's criterion, for a proof see [Kl66]:

3.13 Theorem *Let $\pi : X \longrightarrow S$ be a proper morphism of normal varieties. Let D be Cartier on X. Then D is π-ample if and only if $D \cdot C > 0$ for every $C \in \overline{NE}(X/S) \setminus \{0\}$.*

3.14 Ampleness of a divisor on a projective manifold or rather its associated line bundle L can be characterised by differential geometry: L admits a metric of positive curvature. Therefore one might expect that nefness is the same as to say that L is semipositive, i.e. L carries a metric of semipositive curvature. Of course semipositive line bundles are nef, the converse however is not true. For an example take an elliptic curve C and a non split extension of the following form

$$0 \longrightarrow \mathcal{O}_C \longrightarrow E \longrightarrow \mathcal{O}_C \longrightarrow 0.$$

Let $X = \mathbb{P}(E)$ and $L = \mathcal{O}_{\mathbb{P}(E)}(1)$. Then L is nef but not semipositive. For details see [DPS94].

There is however a characterisation of nef line bundles. Fix a positive $(1,1)$-form on X (not necessarily closed). Then L is nef if and only if for every $\epsilon > 0$ there exists a metric h_ϵ on L such that the curvature $\Theta_{L,h_\epsilon} = \Theta$ fulfills the following inequality:

$$\Theta \geq -\epsilon\omega.$$

Since X is compact, this definition does not depend on the choice of ω.

4 The Base Point Free Theorem

We want to study projective manifolds X with K_X not nef. Having the basic facts on ample divisors from sect. 3 to our disposal, we are now in a position to formulate the general strategy. Fix an ample divisor H. Then there is a unique positive real number t_0 such that $K_X + t_0 H$ is nef but not ample. Of course $K_X + t_0 H$ does not have a priori a geometrical sense and is just understood as (a class of a) $\mathbb{R}$-divisor. The Rationality Theorem (to be treated in sect. 5) however says that t_0 is actually a rational number, therefore a multiple is Cartier and we can ask for sections. The Base Point Free Theorem, which is the main topic of this section, says that a multiple of $K_X + t_0 H$ is generated by global sections, hence it defines a morphism to a normal projective variety. These are the contractions we are looking for.

So we want to construct sections not in arbitrary line bundles, but in multiples of $K_X + tH$. By the way: line bundles of the form $K_X + mL$ with m a positive integer and L a mostly ample line bundle have been studied very intensively in the so-called adjunction theory; we will discuss this in sect.9. For technical reasons we will first discuss the Base Point Free Theorem and then, in the next section, the Rationality Theorem. The Base Point Free Theorem has many authors; for a "historical" note we refer to [KMM87].

4.1 Base Point Free Theorem *Let X be a projective variety with at most terminal singularities and D a Cartier divisor on X. Assume that D is nef and that $aD - K_X$ is big and nef for some positive integer a. Then mD is generated by global sections for $m \geq m_0$.*

There are more general versions in the log case and the relative case; these will be discussed at the end of the section. If D is merely big and nef (and not ample), then it is in general not true that mD is generated by global sections. Here we give the following example. Let C be a compact Riemann surfaces of genus $g \geq 2$. Let L be an ample line bundle with $H^1(L) \neq 0$. Therefore we have a non-split extension

$$0 \longrightarrow L \longrightarrow E \longrightarrow \mathcal{O}_C \longrightarrow 0.$$

Let $X = \mathbb{P}(E)$ and $\mathcal{O}_X(D) = \mathcal{O}_{\mathbb{P}(E)}(1)$. Then D is big and nef. Taking symmetric powers of the exact sequence we deduce that mD is never generated by global sections, since the map $S^m E \longrightarrow \mathcal{O}_C$ always induces the 0-map

$$H^0(X, S^m E) \longrightarrow H^0(X, \mathcal{O}_C).$$

A major ingredient in the proof of the Base Point Free Theorem is the Non-Vanishing Theorem of Shokurov. In the situation of (4.1) it says that mD has at least some section for $m \gg 0$, we need however a more general version.

4.2 Non-Vanishing Theorem *(Shokurov) Let X be a projective manifold, D a nef Cartier divisor on X and A a $\mathbb{Q}$-Cartier on X such that the fractional part of A has normal crossings and that the round up $\lceil A \rceil$ is effective. If D is nef and $aD + A - K_X$ is big and nef for some $a \in \mathbb{N}$, then*

$$H^0(X, \mathcal{O}_X(mD + \lceil A \rceil)) \neq 0$$

for $m \gg 0$.

The general principles behind the proof are very important also for other basic theorems mainly dealing with the existence of sections as the Base Point Free Theorem and the Fujita conjecture (see sect.9) as well as the Rationality Theorem (next section). Therefore we will explain the ideas very detailed.

Proof. We follow the main line of [KMM87].

(1) First we make two reductions. Namely we may assume that

(a) $aD + A - K_X$ is ample for large a,

(b) $D \not\equiv 0$.

(a) By (3.12(5)) there exists an effective $\mathbb{Q}$-divisor E such that

$$aD + A - K_X - \frac{1}{m}E$$

is ample for $m \gg 0$. Let $A' = A - \frac{1}{m}E$. We want to substitute A by A'. The only difficulty is that the fractional part $\{A'\}$ might not have normal crossings. Therefore we perform a suitable sequence of blow-ups $f : \hat{X} \longrightarrow X$ such that $\{f^*(A)\} + \sum E_i$ has normal crossings, where the E_i are the exceptional components of f. Choose $0 < e_i \ll 1$, such that

$$f^*(D) - \sum e_i E_i$$

is ample (3.12(6)). Write

$$K_{\hat{X}} = f^*(K_X) + \sum \lambda_i E_i,$$

with $\lambda_i \in \mathbb{N}$, X being smooth. Then $f^*(aD + A') - K_{\hat{X}} = f^*(aD + A' - K_X) - \sum \lambda_i E_i$. Now choose a_0 so large that

$$a_0 e_i \geq \lambda_i$$

for all i. Put

$$A'' = f^*(A') + \sum (a_0 e_i - \lambda_i) E_i.$$

Then $f^*(D) + A'' - K_{\hat{X}}$ is ample for $a \geq a_0$. It is now obvious to deduce the Non-Vanishing $D + \lceil A \rceil$ from that one for $f^*(D) + \lceil A'' \rceil$.

(b) Let $D \equiv 0$. Kawamata-Viehweg vanishing gives

$$H^q(X, \mathcal{O}_X(D + \lceil A \rceil)) = H^q(X, \mathcal{O}_X(\lceil A \rceil)) = 0, q \geq 1.$$

Hence

$$h^0(\mathcal{O}_X(D + \lceil A \rceil)) = \chi(X, \mathcal{O}_X(D + \lceil A \rceil)) = \chi(X, \mathcal{O}_X(\lceil A \rceil)) = h^0(\mathcal{O}_X(\lceil A \rceil)).$$

Since $\lceil A \rceil$ is effective, this last number is positive. The equality of the Euler characteristic comes from Riemann-Roch, since the involved line bundles are numerically equivalent and therefore have the same Chern class.

(2) We will proceed by induction on $n = \dim X$. For $n = 1$, the assertion is clear: by Riemann-Roch we have

$$\chi(X, \mathcal{O}_X(mD + \lceil A \rceil)) = 1 - g(X) + m \deg D + \deg \lceil A \rceil > 0$$

by our assumption on $aD + A - K_X$.

(3) We now assume $n \geq 2$. We shall formulate a general strategy and then reduce the proof step by step. Our aim is to construct a birational map, in fact a sequence of blow-ups, $f : Y \longrightarrow X$, and a bunch of divisors F_i which include the exceptional components of f with the following properties. There exists an effective divisor A', constructed from A, such that

$$H^0(Y, \mathcal{O}_Y(f^*(mD) + \lceil A' \rceil)) \neq 0. \tag{A1}$$

Of course this it is not sufficient, because we have to be able to come back to Y. This is guaranteed if

(A2) $f^*(\lceil A \rceil) - \lceil A' \rceil + \sum \alpha_i F_i$ is effective,

where $\alpha_i \neq 0$ at most if F_i is an exceptional component for f. From (A1) and (A2) the non-vanishing statement follows immediately.

We proceed by investigating (A1). Since we want to apply induction, we are looking for a smooth hypersurface $B \subset Y$ such that the canonical restriction map

$$H^0(Y, \mathcal{O}_Y(f^*(mD) + \lceil A' \rceil)) \longrightarrow H^0(B, \mathcal{O}_Y(f^*(mD) + \lceil A' \rceil))$$

is onto. This is guaranteed by

$$H^1(Y; \mathcal{O}_Y(f^*(mD) + \lceil A' \rceil - B)) = 0 \tag{A3}$$

and

$$H^0(B, \mathcal{O}_Y(f^*(mD) + \lceil A' \rceil | B)) \neq 0. \tag{A4}$$

For (A3) we need the ampleness of

$$f^*(mD) + A' - B - K_Y$$

in order to be able to apply Kawamata-Viehweg (actually big and nef would be sufficient). In order to satisfy (A4) we would like $f^*(mD) + A' - K_Y | B$ to be big and nef or ample, then (A4) follows by induction hypothesis. Introducing

$$N = f^*(mD) + A' - B - K_Y,$$

it therefore comes down for both (A3) and (A4) to prove that

(A5) N is ample.

The plan to check ampleness of N (once it is defined) is to decompose it into pieces we can control. We want

$$N \equiv s(f^*(mD)) + tf^*(qD + A - K_X)) - u\sum \delta_j F_j \tag{*}$$

(because these are the only divisors we can control) and then conclude ampleness. Therefore we define the $0 < \delta_j \ll 1$ such that for a fixed large q (to be specified in a moment) the divisor

$$f^*(qD + A - K_X) - \frac{1}{n+1}\sum \delta_j F_j$$

is ample; this is possible by (3.12(6)). The factor $\frac{1}{n+1}$ is just introduced to make some formula easier.

We certainly need to compare K_Y and K_X, so we put

$$K_Y = f^*(K_X) + \sum a_i F_i$$

with non-negative integers a_i (which are 0 if F_i is not f-exceptional). Furthermore we need to control $f^*(A)$, therefore we write

$$f^*(A) + \sum a_i F_i = \sum b_i F_i,$$

and we note that because of $\{A\} \geq 0$, we have $b_i > -1$.

(4) In order to adjust things such that ampleness holds in (*) and that (A1) is fulfilled, we need to investigate the linear system $|qD + A - K_X|$. Since we want to make explicit calculations, we need a specific element in the linear system (or rather a multiple of it). The idea is now not to take a smooth member but one which is as singular as possible at some point. We fix a positive integer a such that aA is Cartier. We moreover fix a point $x \notin \mathrm{Supp}A$. We want to construct a section in $ka(qD + A - K_X)$ which vanishes to high order at x; the precise statement is the following:

(**) there exists $s \in H^0(X, \mathcal{O}_X(ka(qD + A - K_X)))$ such that $M = (s)$ has

$$\mathrm{mult}_x M \geq ka(n+1)$$

for large k.

To verify (**) we first make a dimension count. If g is a local equation for M near x, then the inequality for the multiplicity just means that all terms of the Taylor expansion of g of order $< ka(n+1)$ vanish. These are

$$\binom{ka(n+1) - 1 + n}{n} = a^n(n+1)^n \frac{k^n}{n!} + o(k^n)$$

conditions. Therefore we need to see that for $q \gg 0$ we have

$$h^0(X, \mathcal{O}_X(ka(qD + A - K_X))) > a^n(n+1)^n \frac{k^n}{n!} + o(k^n). \qquad (+)$$

The proof of (+) is easy. First note that by Theorem B of Serre the left hand side of (+) is nothing than $\chi(X, \mathcal{O}_X(ka(qD + A - K_X))$. By Riemann-Roch

$$\chi(X, \mathcal{O}_X(ka(qD + A - K_X))) = a^n(qD + A - K_X)^n + o(k^n).$$

So we need
$$(qD + A - K_X)^n > (n+1)^n.$$

But this can certainly be achieved: write $qD + A - K_X = (pD + A - K_X) + (q-p)D$, take $q \gg p$, and apply Binomi, using $D \not\equiv 0$.

So (**) is proved.

(5) We enlarge the set of F_i's (up to now they are f-exceptional components and strict transforms of components in A) to have an equation

$$f^*(M) = \sum r_j F_j.$$

At the same time we should have said what finally f has to be: f is the composition $f_2 \circ f_1$ consisting of
(a) the blow-up of x in X,
(b) a sequence of blow-ups with smooth centers $f_2 : Y \longrightarrow X_1$ such that $\sum F_j$ is a divisor with normal crossings.

We adjust things such that F_1 is the strict transform of $f_1^{-1}(x)$.

Now the inequality $(**)$ can be reformulated in the following way

$$r_1 \geq ka(n+1). \tag{B}$$

Note moreover that

$$b_1 = a_1 = n - 1, \tag{C}$$

which holds since $x \notin \operatorname{Supp} A$.

(6) Now it remains to define A' and $B = F_0$ such that the assertions (A2) and (A5) (resp. $(*)$) hold. This is only a combinatorial problem. Examining $(*)$ it is pretty clear that A' should be of the form

$$A' = \sum_j (-\delta_j + \mu r_j + b_j) F_j.$$

To start with (A2) we make a naive guess $\alpha_i = a_i$ and see to what it leads. Namely we need in order to verify (A2) that

$$a_j - \delta_j + \mu r_j + b_j \geq 0.$$

It is sufficient to make $b_j - \mu r_j - \delta_j \leq 1$ (but positive); therefore we let

$$\mu = \min_j \frac{b_j + 1 - \delta_j}{r_j}.$$

Then (A2) holds.

It is convenient to perturbate the δ_j slightly so that the minimum is taken only for one j, which we call $j = 0$. (Note: we might have $F_0 = F_1$.) This is allowed since ampleness is an open condition. It follows that

$$\mu < \frac{b_1 + 1 - \delta_1}{r_1} < \frac{n}{ka(n+1)} \tag{D}$$

by (B) and (C). Putting everything into (∗), we see that in order to fulfill (∗) we slightly have to change the definition of A', namely

$$A' = \sum_{j \neq 0}(-\delta_j + b_j + \mu r_j)F_j.$$

Then (∗) holds with $s = m - q, t = 1 - \mu ka$ and $u = 1$. Thus we choose $m > q$. In view of the choice of the δ_j it is now sufficient to show that

$$\mu ka < n + 1;$$

this however follows from (D).

So (A5) holds and the proof of the Non-Vanishing Theorem is complete.

Why did we have to blow up X in the course of the proof? The reason is two-fold: first we needed to blow the point x in order to transform the statement on the multiplicity into a computable statement on our divisors. Second we needed to work with the divisor M which is far from being smooth; we need however normal crossings for the application of Kawamata-Viehweg.

Proof of 4.1. The proof of the Base Point Theorem uses essentially the Non-Vanishing Theorem and consists in a reduction step to make the potential base locus of mD smaller by increasing m. The proof being very similar to the one of the Non-Vanishing Theorem, we make the reduction step only in a very special case, to demonstrate the principle and then make some comments on the general case.

We will assume X smooth. This is not a serious restriction. Applying (4.2) with $A = 0$, we obtain

$$H^0(X, \mathcal{O}_X(mD)) \neq 0$$

for $m \geq m_0$. Now we assume that the base locus of $m_0 D$ is just rB with B a smooth hypersurface and that $m_0 D - rB$ is generated by global sections. We put

$$N = mD - B - K_X.$$

Then

$$N = \frac{1}{r}(m_0 D - rB) + (m - \frac{m_0}{r})D_0 - K_X,$$

hence N is big and nef if $m - \frac{m_0}{r} \geq a$; and therefore $H^1(X, \mathcal{O}_X(N)) = 0$. Hence the restriction map

$$H^0(X, \mathcal{O}_X(mD)) \longrightarrow H^0(B, \mathcal{O}_B(mD))$$

is onto. Since $bD - K_B = bD - K_X|B - B$ is big and nef for $b \geq a + m_0$, we have $H^0(B, \mathcal{O}_B(mD)) \neq 0$, and hence the base locus of mD is strictly contained in B for large m.

In general the situation is of course more complicated. One has to resolve the singularities of X and to blow-up such that $f^*(mD)$ has as base locus a normal crossing divisor E such that $f^*(mD) - E$ is generated by global sections. Then one applies to the resulting birational map $f : Y \longrightarrow X$ the same technique as in the proof of (4.2). The divisor M is somehow substituted by the base locus $E = \sum r_i F_i$. For details we refer to [KMM87].

We now state the most general version of (4.1); the proof is the same as in the absolute case with obvious modifications (see [KMM87]).

4.3 Base Point Free Theorem *Let (X, Δ) be klt. Let $\pi : X \longrightarrow S$ be a projective morphism to the normal variety S. Let D be a π-nef Cartier divisor on X. Assume that $aD - (K_X + \Delta)$ is π-ample for some positive integer a. Then mD is relatively generated by global sections for $m \geq m_0$, i.e. the canonical map*

$$\pi^* \pi_*(\mathcal{O}_X(mD)) \longrightarrow \mathcal{O}_X(mD)$$

is onto.

We now give the geometric version of the Base Point Free Theorem; again we leave the relative and the log case to the reader, but the transcription is obvious. First a notation: if H is a divisor, then we let

$$F_H = \{z \in \overline{NE}(X) | H \cdot z = 0\}$$

be the *face* associated with H.

4.4 Contraction Theorem *Let X be a projective variety with at most terminal singularities. Let H be a nef Cartier divisor such that*

$$F_H \subset \{z \in N_1(X) | K_X \cdot z < 0\}.$$

Then there exists a morphism $\varphi : X \longrightarrow Y$ to a normal projective variety with the following properties
(1) φ has connected fibers, i.e. $\varphi_*(\mathcal{O}_X) = \mathcal{O}_Y$;
(2) if $C \subset X$ is an irreducible curve, then $\dim \varphi(C) = 0$ if and only if $[C] \in F_H$, i.e. $H \cdot C = 0$;
(3) $H = \varphi^*(A)$ with some ample divisor A on Y
(4) $-K_X$ is φ-ample.
φ is called the extremal contraction associated with H.

Proof. To construct φ, note that there exists some $a \in \mathbb{N}$ such that $aH - K_X$ is ample. In fact, to see this we verify that $aH - K_X$ is positive on $\overline{NE}(X)$ which follows easily from the assumptions (H is 0 only on those elements in $\overline{NE}(X)$ on which $-K_X$ is positive !). Applying the Base Point Free Theorem, mH is generated by global sections for all large m and we define φ to be the Stein factorisation of the morphism given by $H^0(X, \mathcal{O}_X(mH))$. Then (1) and (2) hold automatically. Clearly φ is independent of the choice of m. To prove (3) consider the morphisms

$$\varphi = \varphi_{mH} : X \longrightarrow Y \subset \mathbb{P}_{n(m)}$$

and

$$\varphi = \varphi_{(m+1)H} : X \longrightarrow Y \subset \mathbb{P}_{n(m+1)}.$$

We have $\mathcal{O}_X(mH) = \varphi^*(\mathcal{O}_{\mathbb{P}_{n(m)}})$ and $\mathcal{O}_X((m+1)H) = \varphi^*(\mathcal{O}_{\mathbb{P}_{n(m+1)}})$, hence claim (3) follows. Claim (4) is a consequence of the fact that $-K_X$ is positive on $\overline{NE}(X/S)$ (3.13).

4.5 Definition Let $V \neq \{0\}$ be a linear subspace of $\overline{NE}(X)$ and $R = V \cap \overline{NE}(X)$ the associated half space. Then R is called an *extremal face* for $\overline{NE}(X)$, if

(a) $K_X \cdot Z < 0$ for all $Z \in R, Z \neq 0$ and if

(b) R is geometrically extremal, i.e. if $Z_1, Z_2 \in \overline{NE}(X)$ and if $Z_1 + Z_2 \in R$, then $Z_i \in \overline{NE}(X)$. If $\dim F = 1$, then F is called an *extremal ray*.

4.6 Corollary *Let H be nef. If $K_X \cdot F_H < 0$, then F_H is an extremal face.*

Proof. Let $Z_i \in \overline{NE}(X)$ with $Z_1 + Z_2 \in F_H$. Then $H \cdot Z_1 + Z_2 = 0$. Since H is nef, it follows $H \cdot Z_i = 0$.

We next give a converse to (4.4) (see [Mo82] and [KMM87]).

4.7 Proposition *Let X be a projective variety with at most terminal singularities. Let $\varphi : X \longrightarrow Y$ be a morphism to a normal projective variety with connected fibers. Assume that $-K_X$ is φ-ample. Then there exists a nef line bundle H on X such that φ is the contraction associated to the extremal face F_H. Moreover $\rho(X) = \rho(Y) + \dim F_H$.*

Proof. The first statement is clear: just choose an ample divisor A on Y and let $H = \varphi^*(A)$. For the second, we have to prove the following: let D be a line bundle on X. Then $D \equiv \varphi^*(M)$ if and only if $D \cdot C = 0$ for all curves C contracted by φ. One direction being clear, we assume that $D \cdot C = 0$ for all curves contracted by φ. Then by the relative version of the Base Point Free Theorem (having in mind that $-K_X$ is φ-ample) it follows that $\mathcal{O}_X(mD)$ is φ-generated, i.e. the canonical map

$$\varphi^* \varphi_*(\mathcal{O}_X(mD)) \longrightarrow \mathcal{O}_X(mD)$$

is surjective for large m. Let $g : X \longrightarrow \mathbb{P}(\varphi_*(\mathcal{O}_X(mD)))$ be the associated morphism. Then $\mathcal{O}_X(mD) = \varphi^*(\mathcal{O}_{\mathbb{P}}(1))$ and g contracts exactly the curves C with $D \cdot C = 0$. Therefore $g = \varphi$ and our claim follows.

We have seen that every extremal face F_H determines an extremal contraction. It is however not at all clear that there are any non-trivial extremal faces. The existence of those extremal faces (in case K_X is not nef, of course) will be the topic of the next section.

We close the section by having a first look at Fano manifolds and varieties of general type and see the implications of the Base Point Free Theorem.

4.8 Definition A projective manifold X is a *Fano manifold* if $-K_X$ is ample. More generally a normal projective $\mathbb{Q}$-Gorenstein variety with at most terminal singularities is called a $\mathbb{Q}$-*Fano variety* if $-K_X$ is ample. If X is Gorenstein, X is a *Fano variety*.

We will discuss Fano manifolds in detail in sect.6. Here we state as an application of the Base Point Free Theorem

4.9 Proposition *Let X be a $\mathbb{Q}$-Fano variety. Let D be a nef divisor on X. Then $\mathcal{O}_X(mD)$ is globally generated for $m \gg 0$.*

Concerning varieties of general type we have

4.10 Proposition *Let X be a projective variety with at most terminal singularities. Assume that K_X is nef and big. Then mK_X is generated by global sections for $m \gg 0$.*

4.11 Definition Assume the situation of (4.10). Let $f : X \longrightarrow Y$ be the Stein factorisation of the morphism determined by the sections of $mK_X, m \gg 0$. Then X is a normal projective variety with at most canonical singularities and it is called the *canonical model* of X.

The philosophy behind canonical models is to pass from varieties with K big and nef to varieties with K ample. The price one has to pay is that instead of terminal singularities we encounter canonical singularities. In case of surfaces, this procedure is classical: given a minimal surface of general type, we contract the (-2)-curves, i.e. smooth rational curves with normal bundle $\mathcal{O}(-2)$, to obtain the canonical model. This new surface has only rational double points as singularities.

Finally we are now able to explain the terminus "canonical singularities": they are just the singularities occuring on canonical models.

5 The Cone Theorem

In this section we discuss the Rationality Theorem and the Cone Theorem and derive geometric consequences for varieties with K_X not nef. The results in this section are due to Mori, Kawamata, Kollár, Shokurov and Reid; for more precise references we refer to [KMM87].

5.1 Rationality Theorem *Let X be a normal n-dimensional projective variety with at most terminal singularities. Let H be an ample divisor. Assume that K_X is not nef and let r be the unique number such that $H + rK_X$ is nef but not ample. Then r is rational.*

More precisely: if e is the index of X (i.e. the smallest positive number such that eK_X is Cartier) and if we write $\frac{r}{e} = \frac{p}{q}$ with p, q positive integers without common divisor, then there is the following estimate: $q \leq e(n+1)$.

The number r (or rather $\frac{1}{r}$) is often called the nef value of K_X with respect to H.

Comments on the Proof. We will deal only with the rationality and neglect the estimate for the denominator, since the proofs are very similar. Assume that r is irrational. Then we will be interested in divisors $D(x, y)$ of the form $D(x, y) = xH + yK_X$ with positive integers x, y such that $D(x, y)$ is not ample, i.e. $y - rx > 0$. The proof of (5.1) consists of two steps:

(1) to show that the base locus $B(x, y)$ of $D(x, y)$ is not empty for suitable x, y;

(2) to prove that $D(x, y)$ is actually generated by global sections, x, y suitable.

Then (2) is of course in contradiction to the definition of r. We will restrict ourselves to (1), since (2) again consists in a reduction of the base locus with the same methods as in the Non-Vanishing or Base Point Theorem (including an application of the latter one).

By passing to a desingularisation we may assume X smooth. We consider the polynomial

$$f(x, y) = \chi(X, \mathcal{O}_X(xH + yK_X)).$$

In order to compute sections, we want to see that

$$f(x, y) = h^0(X, \mathcal{O}_X(xH + yH)). \qquad (*)$$

Hence we want to apply the Kawamata-Viehweg vanishing theorem, for which application we have to know that $xH + yK_X - K_X$ is ample. This condition is equivalent to $y - rx < 1$. Therefore we consider the set

$$N = \left\{ (x, y) \in \mathbb{N} \;\middle|\; 0 < y - rx < \frac{1}{n+1} \right\}.$$

Note that the 1 in the inequality has been substituted by $\frac{1}{n+1}$ for a technical reason which becomes clear in a moment. Hence for $(tx, ty) \in N, 1 \leq t \leq n+1$, equation $(*)$ holds. Let furthermore $\tilde{N} = \{ \frac{y}{x} | (x, y) \in N \}$.

Now assume that r is not rational. Then N and $\tilde{N}$ are infinite (number theory!). Consider the function

$$g(z) = (H + zK_X)^n$$

for real z. Then g is a polynomial of degree at most n and $g \neq 0$, since $g(0) \neq 0$. Hence for all but finitely many

$$\frac{y}{x} \in \tilde{N}$$

we have $g(x,y) \neq 0$. Thus for all those (x,y) the polynomial

$$t \mapsto f(tx, ty)$$

is not 0 (Riemann-Roch). Hence for all those (x,y) it follows that there is some $1 \leq t \leq n+1$, such that $f(tx, ty) \neq 0$.

By $(*)$ and the fact that $ty - trx < 1$, we conclude the proof of (1).

5.2 Remarks (1) Theorem (5.1) holds of course more generally in the klt-category and also in the relative situation. The statement is exactly the same with one change in the estimate of the denominator of r. Namely, we have the estimate $v \leq e(d+1)$, where d is the maximal fiber dimension of the morphism $X \longrightarrow S$. We again refer to [KMM87] for details.

(2) The Rationality Theorem already implies the existence of extremal faces in $\overline{NE}(X)$ in case K_X is not nef: just take an ample line bundle H on X and choose r such that $H + rK_X$ is nef but not ample. Then r is rational and positive, so let $L = m(H + rK_X)$ with a suitable positive integer r. Hence F_L is an extremal face by the Base Point Free Theorem.

5.3 Remark One consequence of (5.1) is the following. If X is a projective manifold and H an ample divisor, then $K_X + tH$ is ample for all $t > n+1$. It is a classical result that if $K_X + (n+1)H$ is not ample, then $X \simeq \mathbb{P}_n$, see sect.9.

We are now approaching the Cone Theorem which describes the geometry of $\overline{NE}(X)$ in the half space $\{Z \in N_1(X) | K_X \cdot Z < 0\}$. We introduce the following notation

$$\overline{NE}_+(X) = \{Z \in \overline{NE}(X) | K_X \cdot Z \geq 0\}.$$

More generally, if (X, Δ) is klt and $\pi : X \longrightarrow S$ projective, we let

$$\overline{NE}^+_{K_X + \Delta}(X/S) = \{Z \in N_1(X/S) | K_X + \Delta \cdot Z \geq 0\}.$$

We state the cone theorem in the most general form, because it is convenient even for proving only the absolute case (with $\Delta = 0$) to apply relative techniques in the proof and therefore one should prove the relative case at once. It should be mentioned that the proof does not require any more geometry on X, it is just a deduction of the Rationality Theorem by doing some geometry in the cone of curves. Therefore an abstract lemma on certain cones in finite dimensional vector spaces could be formulated and then the Cone Theorem follows from the Rationality Theorem by applying this lemma; see [CKM88]. The Cone Theorem was first proved by Mori [Mo82] in the smooth case using (completely different) methods of characteristic p. We shall follow [KMM87].

5.4 Cone Theorem *Let (X, Δ) be klt. Let $\pi : X \longrightarrow S$ be a projective morphism. Then*

$$\overline{NE}(X/S) = \overline{NE}_{K_X + \Delta}^+(X/S) + \sum R_j,$$

where the R_j are the extremal rays in $\overline{NE}(X/S)$. Moreover the R_j are discrete in the half space $\{Z \in N_1(X) | K_X \cdot Z < 0\}$.

Proof. We only treat the case $\Delta = 0$ in order to keep notations easy. The proof consists of three steps.

(1) The crucial step is a "weak" decomposition

$$\overline{NE}(X/S) = \overline{NE}_+(X/S) + \overline{\sum F_H}, \qquad (*)$$

where the sum is taken over those $H \neq 0$ for which $K_X \cdot F_H < 0$. For the proof of $(*)$ assume that $K := \overline{NE}_+(X/S) + \sum F_H$ is strictly smaller than $\overline{NE}(X/S)$. Then we find a linear functional $f : N_1(X) \longrightarrow \mathbb{R}$ such that
(a) $f|K > 0$,
(b) $f(Z_0) < 0$ for some $Z_0 \in \overline{NE}(X/S)$.
Of course f is of the form $f(x) = x \cdot D$ with a real divisor D, but choosing a rational point in $N^1(X)$ sufficiently near to x we may take D to be Cartier. Furthermore we may assume that D is not a multiple of K_X. From (a) we derive easily (look at the dual cone to $\overline{NE}_+(X/S)$) a decomposition

$$D = H + pK_X$$

with an π-ample divisor H. Now let

$$L = H + rK_X$$

be π-nef but not π-ample; then r is rational by the relative version of (5.1). Note $L \not\equiv 0$. Since F_L is by definition contained in the cone K, we have $D \cdot F_L > 0$. Since $K_X \cdot F_L < 0$ and since $L \cdot F_L < 0$, we conclude $p < r$. Hence D is π-ample, since r is the "nef value", contradiction.

(2) The second step consists in showing that in $(*)$ it is sufficient to sum over those H such that $\dim F_H = 1$, i.e. over extremal *rays*. For this take some higher dimensional F_H. Let $\varphi : X \longrightarrow Z$ be the associated contraction (over S). Now

apply (1) to $F_H = \overline{NE}(X/Z)$. Since an extremal face for $\overline{NE}(X/Z)$ is clearly also one of $\overline{NE}(X/S)$, we have decomposed F_H into lower dimensional extremal faces. Continuing this process we finally reach at extremal rays.

(3) The last step consists in getting rid of the closure and proving discreteness; this follows easily from the estimate of the denominator in (5.1). We refer to [KMM87] for details.

5.5 Remark (1) In general it is not true that there are only finitely extremal rays on X. This is already shown by the classical examples of surfaces containing infinitely many (-1)-curves. It is however known that varieties of general type have only finitely many extremal rays.

(2) Kollár has shown in [Ko92b] that the cone theorem holds more generally for normal projective $\mathbb{Q}$-Gorenstein threefolds.

5.6 Corollary *Let X be a $\mathbb{Q}$-Fano variety. Then X has only finitely many extremal rays R_j and $\overline{NE}(X) = \sum R_j$. In particular $NE(X) = \overline{NE}(X)$.*

It follows that on a Fano variety a divisor D is ample if and only if $D \cdot C > 0$ for every irreducible curve C. A partial converse would be the following: say for simplicity X is smooth and $-K_X \cdot C > 0$ for every curve $C \subset X$. Is then X Fano? This is easy for surfaces and has been proved in dimension 3 by Serrano [Se95].

Given an extremal ray R on X one can find an irreducible curve C such that $R = \mathbb{R}_*[C]$. But much more is true:

5.7 Theorem *(Mori) Let X be a projective variety with at most terminal singularities and R an extremal ray on X. Then there exists a **rational** curve C such that $R = \mathbb{R}_+[C]$.*

Mori proved in [Mo82] actually the smooth case via reduction to characteristic p. The general case follows e.g. from the following result of Kawamata [Kw92, appendix] in case the contraction of R is birational; the non-birational case will be discussed in the next section.

5.8 Theorem *Let X be a normal projective variety with at most canonical singularities and $f : X \longrightarrow Y$ a birational map to a normal projective variety Y Assume that $-K_X$ is f-ample. Then every fiber of f is uniruled, i.e. has a covering family of rational curves.*

Of course (5.7) holds also in the relative case and also in the log case, see again [Kw92]. For the general theory of rational curves we refer to Miyaoka's article in this volume.

We conclude this section by showing how useful the log category is even for the investigation of manifolds. Our application is concerned with Calabi-Yau manifolds. These are projective manifolds X with $K_X \equiv 0$. Often one requires also X to be simply connected (or $\pi_1(X)$ to be finite) or that there are no q-forms ($q < \dim X$) on any finite étale cover of X. Here it does not matter what definition we take.

5.9 Theorem *Let X be a Calabi-Yau manifold and D an effective divisor on X which is not nef. Then X contains rational curves.*

Proof. Let $\Delta = \epsilon D$. If $\epsilon > 0$ is small enough, then (X, Δ) is klt. Now $K_X + \Delta = \Delta$ is not nef. Hence there exists a log contraction $f : X \longrightarrow Y$. In particular $-(K_X + \Delta)$ is f-ample. Therefore the result follows from [Kw92] (as mentioned above).

It is a very deep problem whether every Calabi-Yau manifold with $\pi_1(X)$ finite carries a rational curve. Of course (5.8) says nothing e.g. when $\rho(X) = 1$, which is the general case. Compare [Wi89], [Og93], [Pe91] and section 10.

6   Fano Manifolds and the Structure of Contractions of Extremal Rays

Let X be a projective variety such that K_X is not nef. Let $f : X \longrightarrow Y$ be the contraction of an extremal ray or "extremal contraction", for short. We want to investigate the structure of f, i.e. the structure of the fibers and the structure of Y. In order to keep notations easy, we will completely ignore the relative case and mention the log case only occasionally. We introduce the following notations.

6.1 Definition Let $f : X \longrightarrow Y$ be an extremal contraction. f is called
(1) of *fiber type*, if $\dim Y < \dim X$;
(2) *divisorial*, if f is birational and contracts some divisor;
(3) *small*, if f is birational and does not contract a divisor, i.e. f is an isomorphism in codimension 1.

6.2 Proposition *Let $f : X \longrightarrow Y$ be extremal. Then $R^q f_*(\mathcal{O}_X) = 0, q \geq 1$.*

Proof. If X is smooth, this follows directly from the relative Kawamata-Viehweg vanishing theorem (2.14). In the singular case we have to desingularise in order to be able to apply (2.14); the proof is as follows. Take a desingularisation $g : \check{X} \longrightarrow X$ such that the exceptional divisor of g has normal crossings. Let E_i denote the exceptional components of g. Choose small positive rational numbers δ_i such that

$$g^*(-K_X) - \sum \delta_i E_i$$

is $f \circ g$-ample, in particular g-ample. Write $K_Y = g^*(K_X) + \sum \lambda_i E_i$ with $\lambda_i > 0$. By (2.14) we obtain

$$R^q(f \circ g)_*(\mathcal{O}_{\hat{X}}(\sum \lceil \lambda_i \rceil E_i)) = 0, q \geq 1. \qquad (*)$$

In order to get back to X, we invoke the so-called Grothendieck spectral sequence (E_*^{pq}) with

$$E_2^{pq} = R^p g_*(R^q f_*(\mathcal{O}_{\hat{X}}(\sum \lceil \lambda_i \rceil E_i)))$$

converging to

$$R^{p+q}(f \circ g)_*(\mathcal{O}_{\hat{X}}(\sum \lceil \lambda_i \rceil E_i)).$$

Since

$$R^q g_*(\mathcal{O}_{\hat{X}}(\sum \lceil \lambda_i \rceil E_i)) = 0, q \geq 1,$$

again by (2.14), it follows from the spectral sequence that

$$R^q(f \circ g)_*(\mathcal{O}_{\hat{X}}(\sum \lceil \lambda_i \rceil E_i)) = R^q f_*(g_*(\mathcal{O}_{\hat{X}}(\sum \lceil \lambda_i \rceil E_i)).$$

Since $g_*(\mathcal{O}_{\hat{X}}(\lceil \lambda_i \rceil E_i)) = \mathcal{O}_X$, the claim follows.

The proof shows that it would have been sufficient to assume $(X, 0)$ to be log terminal and $-K_X$ to be f-ample. Of course (6.2) holds also in the log category. Kollár [Ko86, 7.4] proved the following:

6.3 Proposition *Let $f : X \longrightarrow Y$ be extremal. Then Y has only rational singularities.*

Proof. Consider a commutative "desingularisation" diagram

$$
\begin{array}{ccc}
\hat{X} & \xrightarrow{h} & X \\
\hat{f} \downarrow & & \downarrow f \\
\hat{Y} & \xrightarrow{g} & Y
\end{array}
$$

where $\hat{X}$ and $\hat{Y}$ are smooth and h, g are birational. We have

$$R^q h_*(\mathcal{O}_{\hat{X}}) = 0, q \geq 1.$$

since X has only rational singularities. We must prove

$$R^q g_*(\mathcal{O}_{\hat{Y}}) = 0. \qquad (*)$$

From the Grothendieck spectral sequence already used in (6.2) and from (6.2) itself it follows immediately that

$$R^q(f \circ h)_*(\mathcal{O}_{\hat{X}}) = 0, q \geq 1,$$

hence $R^q(g \circ \hat{f})_*(\mathcal{O}_{\hat{X}}) = 0$. Invoking the spectral sequence once more, we see that it suffices to prove

$$R^q \hat{f}_*(\mathcal{O}_{\hat{X}}) = 0 \qquad\qquad (**)$$

in order to deduce $(*)$. If f is birational, then $\hat{f}$ is birational and $(**)$ is clear since both $\hat{X}$ and $\hat{Y}$ are smooth. In case $\dim Y < \dim X$ one has to use deep results of Kollár. Namely, the main result of [Ko86] states that $R^q \hat{f}_*(\omega_{\hat{X}})$ is torsion free. Now this sheaf is in our situation also generically 0 since

$$H^q(\hat{F}, \omega_{\hat{X}}|\hat{F}) = H^q(\hat{F}, \omega_{\hat{F}}) = 0$$

for the general fiber $\hat{F}$ of $\hat{f}$. This last statement follows from the fact that $\hat{F}$ is a desingularisation of F, the general fiber of f, that F has only terminal singularities (see (6.4)) and that $H^q(F, \omega_F) = 0$ by (6.1) and duality.

We therefore conclude that $R^q \hat{f}_*(\omega_{\hat{X}}) = 0$ for $q < \dim - \dim Y = n - m$. In the top degree one has (see [Ko86, 7.6])

$$R^{n-m} \hat{f}_*(\omega_{\hat{X}}) \simeq \omega_{\hat{Y}}.$$

Now apply the Leray spectral sequence and Serre duality to deduce

$$H^q(\hat{X}, \mathcal{O}_{\hat{X}}) = H^q(\hat{Y}, \mathcal{O}_{\hat{Y}})$$

for all q. Again the Leray spectral sequence yields then $(*)$.

A slightly different way is to make $\hat{f}$ flat by Hironaka's flattening theorem [Hi75] and then apply relative duality [RRV71] directly to $(**)$ to get the vanishing. But still Kollár's torsion freeness result is the crucial point.

6.4 Proposition *Let $f : X \longrightarrow Y$ be extremal. Then the general fiber F of f is $\mathbb{Q}$-Fano with only terminal singularities.*

Proof. All statements except that $-K_F$ is ample (which is obvious once we know that F is $\mathbb{Q}$-Gorenstein) are local and can be reduced to the following. Let Z be affine with at most terminal singularities. Let L be a base point free linear system. Then for $W \in L$ general, W is normal, Cohen-Macaulay, K_W is $\mathbb{Q}$-Cartier and W has only terminal singularities. The details are left to the reader (consult e.g. [Re87]).

If in particular X is smooth in (6.4), F is a Fano manifold. So we want to have a closer look at this class of manifolds. From Kodaira vanishing we obtain

$$H^q(X, \mathcal{O}_X) = 0, q \geq 1.$$

We say that a manifold is *rationally connected* if for any two general points x, y there exists a chain of rational curves joining x and y.

6.5 Theorem *Every Fano manifold is rationally connected.*

This theorem is due to Campana [Ca92] and Kollár-Miyaoka-Mori [KoMiMo92a]. For this and much more about rational connectedness see Miyaoka's article in this volume. We just mention that without using methods of characteristic p, nobody knows how to construct a single rational curve on a general Fano manifold.

Here is an important consequence:

6.6 Corollary *Fano manifolds are simply connected.*

Proof. Actually one proves that rationally connected varieties are simply connected. Let $f : Y \longrightarrow X$ be the universal cover of X. All rational curves in X lift to Y since $\mathbb{P}_1$ is simply connected. Then it is quite plausible that Y should be compact; for a rigorous argument one needs the so-called "Shafarevitch" map, see [Ca92], [Ko92]. Hence $\pi_1(X)$ must be finite and Y compact.

Let us for simplicity again assume that X is Fano. So f is a finite unramified covering of degree, say, d. In particular Y is also Fano. We have

$$\chi(Y, \mathcal{O}_Y) = d\chi(X, \mathcal{O}_X),$$

on the other hand $\chi(Y, \mathcal{O}_Y) = \chi(X, \mathcal{O}_X) = 1$ by Kodaira vanishing, hence $d = 1$.

A completely different argument to prove the finiteness of the fundamental group is as follows (due to Campana [Ca91]). First one proves that if X is rationally connected, then one actually can find a family (C_t) of rational curves (see (6.19) joining every point with a fixed point x_0 in X. For this see [KoMiMo92, 2.1] and Miyaoka's article; the proof requires deformation techniques. Let T be the normalisation of the parameter space of (C_t), and $\mathcal{C}$ the desingularisation of the graph. Then one has maps $p : \mathcal{C} \longrightarrow X$ and $q : \mathcal{C} \longrightarrow Y$. Take an irreducible component Z of $p^{-1}(x_0)$ with $q(Z) = T$. This Z exists because every C_t goes through x_0. Now there is a general relation between $\pi_1(\mathcal{C}), \pi_1(Z)$, and π_1 of a general smooth fiber F of q. It says that the subgroup in $\pi_1(\mathcal{C})$ generated by $\pi_1(Z)$ and $\pi_1(F)$ has finite index (see [Ca91, 1.4]). Then $q_*(\pi_1(\mathcal{C}))$ has finite index in $\pi_1(X)$, moreover $q_*(\pi_1(Z)) = \{1\}$ and finally $\pi_1(F) = 0$. Hence our claim follows.

(6.7) Corollary (6.6) has been first proved in a weaker form by Kobayashi [Ko61], using differential geometric methods. Let us explain this. Kobayashi proved that if X is a compact Kähler manifold admitting a Kähler metric with positive Ricci curvature, then X is simply connected. A Kähler metric with positive curvature forces X to be Fano. Now assume conversely X to be Fano. Then $-K_X$ is ample and hence there exists a metric on the line bundle associated to $-K_X$ with positive curvature. But this metric might not be induced by a Kähler metric on X. This difficulty is very serious and was only overcome by Yau's solution of one of the Calabi conjectures [Ya77]. Yau's theorem is the following:

If X is compact Kähler and ω a closed form representing $c_1(X)$, then there exists a Kähler metric on X whose $(1,1)$-form is ω.

In our situation we find $\omega \in c_1(X)$ positive since $-K_X$ is ample and then apply Yau's theorem.

For a different proof of (6.6) and furthergoing results on the fundamental group of manifolds with $-K_X$ nef we refer to [DPS93].

A very important result due to Nadel [Na92], Kollár-Miyaoka-Mori [KoMiMo92a] and Campana[Ca92a] says that in principle Fano manifolds can be classified.

6.8 Theorem *There are finitely many families of Fano manifolds of a given dimension such that every Fano manifold appears in one of these families.*

A family of Fano manifolds is a proper surjective map $\pi : \mathcal{X} \longrightarrow T$, where every fiber of π is Fano and T is irreducible. The essential thing to be proved is a bound for $(-K_X)^n$ where $n = \dim X$. Again rational curves form the main technical tool to obtain the bound.

6.9 Remarks (1) Fano surfaces are classically called del Pezzo surfaces; they are completely understood. A del Pezzo surface is either $\mathbb{P}_2, \mathbb{P}_1 \times \mathbb{P}_1$ or $\mathbb{P}_2$ blown up in at most k points in general position, $k \leq 8$. "General position" means that no 3 points are on a line, no 6 points on a conic and and there is no cubic containing seven points and the last as a double point. We refer to [Dm80].

(2) Fano threefolds with $b_2 = 1$ were completely classified by Fano, Roth, Iskovskih, Shokurov and Mukai. We refer to [Is 77, 90], [Mu89]. If $b_2 \geq 2$, the methods are completely different: one has to use contractions of extremal rays and has to classify them (see also next section). This was done in [MM81, 81a].

(3) In higher dimensions a classification seems to be hopeless, at least from dimension 5 on. However one can say a lot under some restriction on the so-called index of X (see [Wi91a, 94]).

6.10 Definition Let X be a Fano manifold. The *index* $i(X)$ of X is the largest number r such that there exists a divisor D with $K_X = rD$. The *coindex* of X is defined by $c(X) = \dim X + 1 - i(X)$.

The most basic result on the index is the

6.11 Theorem of Kobayashi-Ochiai *Let X be a Fano manifold. Then*
(1) $i(X) \leq \dim X + 1$;
(2) if $i(X) = \dim X + 1 = n + 1$, then $X \simeq \mathbb{P}_n$;
(3) if $i(X) = n$, then X is the n-dimensional quadric Q_n.

Proof. (1) follows from the estimate of the denominator in (5.1) since $K_X + i(X)H$ is trivial but it can be proved much easier. In fact, if $i(X) > n + 1$, then from Kodaira vanishing and Serre duality we see at once that

$$H^q(X, \mathcal{O}_X(tH)) = 0$$

for all q and $0 > t \geq -(n+1)$, hence

$$\chi(X, \mathcal{O}_X(tH)) = 0$$

for those t. On the other hand $\chi(X, \mathcal{O}_X(tH))$ is (by Riemann-Roch) a polynomial of degree n, contradiction.

(2) Assume that $-K_X = (n+1)H$. Consider the following polynomials of degree n:

$$f(t) = \chi(X, \mathcal{O}_X(tH)), \quad g(t) = \chi(\mathbb{P}_n, \mathcal{O}_{\mathbb{P}_n}(t)).$$

Then $f(t) = g(t)$ for $0 \geq t \geq -n$, hence $f = g$. It follows $H^n = 1$ and

$$h^0(X; \mathcal{O}_X(H)) = \chi(X, \mathcal{O}_X(H)) = n + 1.$$

Let $h : X \dashrightarrow \mathbb{P}_n$ be the induced rational map. Using the ampleness and $H^n = 1$ it is easy to see that h is a morphism, i.e. H is base point free. Again by the ampleness of H, the map h has to be finite and $H^n = 1$ forces h to be bijective. Hence h is an isomorphism.

(3) The case of the quadric is somewhat more complicated, we refer to [KO73].

6.12 Remark (1) Fano manifolds of coindex $c(X) = 2$ have been classified by Fujita [Fu87] and Fano manifolds with $c(X) = 3$ are classified by Mukai [Mu89, 92] under the additional assumption that $|H|H|$ contains "enough" smooth members, so that one can cut down to a Fano threefold. It is expected that this additional assumption is always satisfied.

(2) One can roughly say that Fano manifolds get the more complicated the smaller the index is.

(3) For more informations on Fano manifolds with $b_2 \geq 2$, in particular under some restrictions on the index, we refer to [Ws91, 91a, 94].

Before we have a closer look to divisorial contraction, we consider the (rational) curves contracted by extremal contractions. By a deformation argument [Mo79, 82], Mori proved

6.13 Theorem *Let X be a n-dimensional projective manifold, $f : X \longrightarrow Y$ the contraction of the extremal ray R. Then there exists a rational curve C with $[C] \in R$ and $0 < -K_X \cdot C \leq n + 1$.*

This is based on the following basic fact: if X is a n-dimensional projective manifold and C a rational curve with $-K_X \cdot C \geq n+2$, then some deformation of C splits. If C is in the extremal ray R, then all the splitting components are also in R and one can repeat the process if necessary. For the deformation techniques necessary for the proof of the basic fact we refer to [Mo79] and Miyaoka's article.

In the singular (relative) log case see [Kw91a] for an estimate.

6.14 Definition (1) Let X be a projective manifold and R an extremal ray. A rational curve $C \subset X$ with $[C] \in R$ and $-K_X \cdot C \leq n+1$ is called an *extremal rational curve.*

(2) Let R be an extremal ray on X. We define the *length* $l(R)$ by

$$l(R) = \min\{-K_X \cdot C | C \in R\}.$$

By (6.13) it follows that $l(X) \leq \dim X + 1$.

Building up on previous work of Ionescu [Io88], Wisniewski proved in [Wi91] the following important inequality

6.15 Theorem *Let X be a projective manifold and R an extremal ray on X. Let A be the union of all curves $C, [C] \in R$. Let d be the minimal possible dimension of a non-trivial fiber of the contraction of R. Then A is an algebraic set in X and*

$$\dim A \geq \dim X + l(R) - d - 1.$$

The proof again relies on Mori's char p-technique. (6.15) is often useful in the investigation of extremal contractions (see sect. 9). Here we only give the following

6.16 Corollary *Let X be a n-dimensional projective manifold and R an extremal ray on X with contraction $f : X \longrightarrow Y$. Assume that f is birational with exceptional set A. Then $\dim A \geq \frac{n}{2}$.*

6.17 In particular we always have $\dim A \geq 2$, if $n \geq 3$. This can also be seen as follows. Assume $\dim A = 1$ and $n \geq 3$. Then every connected component of A consists of a tree of smooth rational curves. This follows from $R^1 f_*(\mathcal{O}_X) = 0$. Take one of the rational curves, say C with normalisation $g : \mathbb{P}_1 \to X$. Then $K_X \cdot C < 0$ and hence C deforms in an at least 1-dimensional family which is clearly a contradiction to $\dim A = 1$, since all deformations of C have to be contracted by f, too. In order to see that C moves let T denote a component of the Hilbert scheme of subspaces of X containing $[C]$. Then there is the following basic inequality (see Miyaoka's article):

$$\dim T \geq \chi(\mathbb{P}_1, g^*(T_X|C)) - 3.$$

Therefore $\dim T \geq 1$ by our assumption $K_X \cdot C < 0$ and Riemann-Roch.

Actually this last argument shows a little bit more than stated in (6.16), namely that A does not contain any 1-dimensional component while from (6.16) we can only say that A contains a component of dimension at least $\frac{n}{2}$.

It is very important that X is smooth in (6.16); we will see in sect.7 that (6.16) is false in the singular case.

In general it is rather difficult to say something on the structure of extremal contractions $f : X \longrightarrow Y$ in the fiber type case. In low dimensions there are much more precise results; see the next section. The only general result I know of is the following.

Let $f : X \longrightarrow Y$ be an extremal contraction from the projective manifold X to the normal projective variety Y with $\dim Y < \dim X$. If $\dim Y = \dim X - 1$ and if f is equidimensional, then f is a conic bundle [An85].

There are many papers dealing with the following situation. Let $f : X \longrightarrow Y$ be the contraction of an extremal face given by a multiple of $K_X + rL$ with L an ample line bundle. Then one can say a lot on the structure of f, if r is relatively large, i.e. say, $r \geq n - 3$. We refer e.g. to [ABW93], [AW93], [An85], [AW96], [Kc95].

6.18 Examples Here we just mention some obvious classes; more detailed examples will be given in the next sections.

(1) Let Y have at most terminal singularities and let X be a $\mathbb{P}_r$-bundle over Y in the usual topology. Then the projection is an extremal contraction.

(2) Let X and Y be projective manifolds. A conic bundle $f : X \longrightarrow Y$ is characterised by the fact that every fiber of f is a conic in $\mathbb{P}_2$. Actually X can be viewed as a hypersurface in a $\mathbb{P}_2$- bundle over Y. Now f is extremal if and only if $\rho(X) = \rho(Y) + 1$. See [Be77], [Mi83] for more informations on conic bundles.

(3) The standard birational (and divisorial) example is as follows: let Y be a projective manifold and $Z \subset Y$ a submanifold. Let $f : X \longrightarrow Y$ be the blow up of Y along Z. Then f is extremal.

We finish the discussion of varieties having an extremal contraction of fiber type by proving that they are uniruled.

6.19 Definition Let X be a normal projective variety.

(1) A *family of rational curves* is given by its graph, i.e there exists an irreducible projective scheme $\mathcal{C}$ and morphisms

$$p : \mathcal{C} \longrightarrow X, \ q : \mathcal{C} \longrightarrow T$$

onto an irreducible projective variety T such that the general fiber of q is an (irreducible) rational curve, that $p(q^{-1}(t_1)) \neq p(q^{-1}(t_2))$ and that p is generically an isomorphism on every fiber $q^{-1}(t)$. We write $(C_t)_{t \in T}$ for short. (C_t) is a *covering family*, if p is surjective.

(2) X is *uniruled* if X admits a covering family of rational curves.

It is easy to see that a projective manifold is uniruled if and only if there exists a generically finite dominant map $\mathbb{P}_1 \times Y \rightarrow X$ with a projective variety Y.

6.20 Proposition *Let X have an extremal contraction f of fiber type. Then X is uniruled.*

This follows from the following numerical characterisation of uniruled varieties of Miyaoka-Mori (see Miyaoka's article):

6.21 Theorem *Let X be a normal projective $\mathbb{Q}$-Gorenstein variety. Assume that there exists a Zariski open set $U \subset X$ such that through every point there exists an irreducible curve C such that $K_X \cdot C < 0$. Then X is uniruled.*

It is clear that a uniruled variety with at most terminal singularities has Kodaira dimension $-\infty$. The converse is one of the big conjectures in classification theory and holds true in dimension 3 (see sect. 8). It is clear that a uniruled variety does not necessarily carry an extremal contraction of fiber type, but some birational model should do.

We now turn to the birational divisorial case. A priori this only means that $f : X \longrightarrow Y$ contracts some irreducible divisor but there might be other components of the same dimension or less. This however fortunately does not happen as shown in the next proposition. However if X is singular, we need one more assumption on X :

6.22 Definition Let X be a normal variety. Then X is $\mathbb{Q}$-*factorial*, if X is Cohen-Macaulay and if every Weil divisor on X is $\mathbb{Q}$-Cartier.

The reason why we need $\mathbb{Q}$-factoriality is the following. Let $f : X \longrightarrow Y$ be a divisorial contraction contracting a 1-codimensional set A. Then certainly we want A to be $\mathbb{Q}$-Cartier, because otherwise Y would not even be $\mathbb{Q}$-Gorenstein.

6.23 Proposition *Let X be a normal projective $\mathbb{Q}$-factorial variety with at most terminal singularities and $f : X \longrightarrow Y$ be a divisorial contraction. Let A be the exceptional set for f. Then A is an irreducible divisor and Y is again $\mathbb{Q}$-factorial with at most terminal singularities.*

Proof. (1) Let E be a 1-codimensional component of A. By assumption E is $\mathbb{Q}$-Cartier. Let E' be another component of A. Let $C \subset E'$ be a general curve contracted by f. Then $C \not\subset E$ and hence $E \cdot C \geq 0$. Let R be the extremal ray defined by f. Then we obtain $E \cdot R \geq 0$. If however $B \subset E$ is a curve contracted by f, then we must have $E \cdot B < 0$ (otherwise E would be f-nef which is clearly absurd), contradiction.

(2) Since X has only rational singularities by (1.7), X is Cohen-Macaulay, see e.g. [Re87]. Now let D be a Weil divisor on Y, we may assume that D is effective and irreducible. Let $\hat{D}$ be the strict transform in X. Then $m\hat{D}$ is Cartier for some m, since X is $\mathbb{Q}$-factorial. Let $[l] \in R$. Take t such that $\hat{D} + tE \cdot l = 0$. Then $q(m\hat{D} + tE) = f^*(L)$ with a line bundle L on Y (4.7), hence $\mathcal{O}_Y(qmD) = L$ outside codimension at least 2, hence the equality holds everywhere.

(3) Finally it is easy to see that Y has only terminal singularities. Write $K_X = f^*(K_Y) + aA$. Since $-K_X|A$ is ample and since $-A|A$ is ample, it follows $a > 0$.

It remains to discuss small contractions. This will be done in the next two sections. Here we just want to see that small contractions behave badly.

6.24 Remark Let $f : X \longrightarrow Y$ be a small contraction. Then Y is never $\mathbb{Q}$-Gorenstein. In fact, assume that mK_Y is Cartier. Then mK_X and $f^*(mK_Y)$ coincide outside a set of codimension at least 2, hence $mK_X = f^*(mK_Y)$. But $K_X \cdot C < 0$ for all curves C contracted by f, contradiction.

6.25 Remark Is there a theory of extremal contractions etc., i.e. a theory of varieties with non-nef canonical bundle also on other manifolds? One should expect a reasonable theory on compact Kähler manifolds. Of course the term "nef" needs a redefinition in the non-algebraic case, since there are too few curves in general that the old definition makes really sense. The new definition is provided by the differential-geometric characterisation of nefness in (3.14). For first results on a non-algebraic "Mori theory" see [CP96] and [Pe96].

7   Surfaces and Threefolds

We are going to study extremal contractions on surfaces and threefolds. First we deal with surfaces and see that we can reconstruct a good part of the Enriques-Kodaira classification.

7.1 Let X be a smooth projective surface with K_X not nef. Let $f : X \longrightarrow Y$ be an extremal contraction. We treat the cases $\dim Y = 0, 1, 2$ separately.

7.1.1 $\dim Y = 0$.
Then $\rho(X) = 1$ and X is Fano. In particular
$H^q(X, \mathcal{O}_X) = 0, q \geq 1$ and $b_2(X) = \rho(X) = 1$.
We claim that $X \simeq \mathbb{P}_2$. For the proof let $\mathcal{O}_X(1)$ be an ample generator of $\text{Pic}X$ (modulo torsion). The intersection form on $H^2(X, \mathbb{Z})$ being unimodular, we have

$$c_1(\mathcal{O}_X(1))^2 = 1. \tag{$*$}$$

Write $-K_X \equiv \mathcal{O}_X(r)$ for some positive integer r. By the above cohomology vanishing we have $\chi(X, \mathcal{O}_X) = 1$, on the other hand we can compute $\chi(X, \mathcal{O}_X)$ by Riemann-Roch and obtain by comparing

$$r^2 + c_2(X) = 12.$$

We compute $c_2(X)$ by a theorem of Hopf (see e.g. [GH78]):

$$c_2(X) = \chi_{\text{top}}(X),$$

where $\chi_{\text{top}}(X)$ is the topological Euler characteristic. By Hodge decomposition we know $b_1(X) = 0$, hence we get $c_2(X) = 3$, hence $r = 3$. Now we compute from Riemann-Roch:

$$h^0(X, \mathcal{O}_X(1)) = \chi(X, \mathcal{O}_X(1)) = 3.$$

Let $g : X \dashrightarrow \mathbb{P}_2$ be the rational map given by $H^0(X, \mathcal{O}_X(1))$. We need to show that g is actually an isomorphism. First of all, it is clear that every $C \in |\mathcal{O}_X(1)|$ is irreducible and reduced by the choice of $\mathcal{O}_X(1)$. We claim that every C is smooth: in fact take a singular point $p \in C$ and choose another C' through p. Then $C \cdot C' \geq 2$, contradicting $(*)$. From the adjunction formula we then deduce $\deg K_C = -2$, hence $C \simeq \mathbb{P}_1$. Now look at the exact sequence

$$H^0(X, \mathcal{O}_X(1)) \longrightarrow H^0(C, \mathcal{O}_C(1)) \longrightarrow H^1(X, \mathcal{O}_X(1)).$$

Since the last group vanishes (Kodaira), every section on C lifts to X and from this it is almost obvious that f is an embedding.

7.1.2 $\dim Y = 1$.

In particular we have $\rho(X) = 2$. We claim that X is a ruled surface, i.e. a $\mathbb{P}_1$-bundle over a compact Riemann surface.

Let F_g be the general smooth fiber of f. By the adjunction formula, we have $F_g \simeq \mathbb{P}_1$. Now assume that there is a singular fiber F and write

$$F = \sum_{i=1}^{p} m_i F_i$$

with F_i the irreducible components. Note $K_X \cdot F = -2$. If R denotes the extremal ray of f, then $F_i \in R$ for all i. Hence $K_X \cdot F_i \leq -1$ and we are left with the following cases.

(a) $p = 1, m_1 = 1, K_X \cdot F_1 = -2$;
(b) $p = 1, m_1 = 2, K_X \cdot F_1 = -1$;
(c) $p = 2, m_1 = m_2 = 1, K_X \cdot F_i = -1$.

Now the cases (b) and (c) are immediately excluded, so that we are done.

7.1.3 $\dim Y = 2$.

Here we claim that f is the blow-up of a smooth point $p \in Y$.

Let E be the exceptional set of f. We already know that E is irreducible (6.20). Since $R^1 f_*(\mathcal{O}_X) = 0$, it is also clear that $E \simeq \mathbb{P}_1$. From the adjunction formula and $K_X \cdot C < 0$ it now follows that C is a (-1)-curve. Now the claim is just a semi-local argument (see e.g. [GH78]).

7.2 Conclusion Let X be a smooth projective surface such that K_X is not nef. Then either X is the projective plane or a ruled surface or X is not minimal. In the last case we blow down a (-1)-curve and ask the same question on Y. Hence finally we arrive – after blowing down finitely many (-1)-curves – at $\mathbb{P}_2$ a ruled surface or at a minimal surface with nef canonical bundle.

In surface classification it is then still necessary to investigate the structure of surfaces with K_X nef. The result here is that some multiple of $m K_X$ is generated by global sections which is the so-called Abundance Theorem for surfaces. But this is not our subject; we refer to standard books on surfaces and to Miyaoka's article.

We are now turning to smooth threefolds; all classification results here are due to Mori [Mo82]. We fix a smooth projective threefold and let $f : X \longrightarrow Y$ be an extremal contraction. If $\dim Y = 0$ then X is just a Fano threefold with $b_2 = 1$ and we refer to our previous remarks on Fano manifolds (sect.6). We discuss the cases $\dim Y > 0$ separately.

7.3 Theorem *Assume* $\dim Y = 1$. *Then every smooth fiber F is a del Pezzo surface and F is not the blow-up of $\mathbb{P}_2$ in two points. Every fiber is irreducible and reduced. Moreover:*
(a) if $F = \mathbb{P}_2$, then f is a $\mathbb{P}_2$-bundle;
(b) if $F = \mathbb{P}_1 \times \mathbb{P}_1$, then f is a quadric bundle, i.e. realised as a hypersurface in a $\mathbb{P}_3$-bundle over Y.

For the proof we refer to [Mo82].

7.4 Theorem *Assume* $\dim Y = 2$. *Then f is a conic bundle (possibly all fibers are smooth).*

Proof. (1) It is easy to see that Y must have terminal singularities, so that Y is smooth.

(2) Next we show that f cannot have a 2-dimensional fiber. So assume that F_0 is a 2-dimensional component of a fiber F. Here we consider F with the scheme-theoretic fiber structure, i.e. if $F = f^{-1}(p)$ set-theoretically, then the ideal of F is given by

$$I_F = \text{Im}(f^*(m_p) \longrightarrow \mathcal{O}_X).$$

Here of course m_p denotes the maximal ideal sheaf of p. Take a general fiber l of f. Then $F_0 \cdot l = 0$. So $F_0 \cdot R = 0$, where R is the defining extremal ray and we conclude

$$N_{F_0} = F_0|F_0 \equiv 0. \tag{$*$}$$

On the other hand write $F = \lambda F_0 \cup A$ scheme-theoretically, where A does not contain F_0. Then there is an exact sequence of conormal sheaves

$$N^*_{F|X} \longrightarrow N^*_{\lambda F_0|X} \longrightarrow N^*_{\lambda F_0|F} \longrightarrow 0.$$

Restrict this sequence to F_0. Since $N^*_{F|X}$ is generated by global sections and since the left hand arrow in the sequence is generically injective, we obtain global sections in $N^*_{\lambda F_0|X}|F_0 = N^{*\lambda}_{F_0|X}$. If $F \neq \lambda F_0$, the sequence shows us that we obtain sections with zeroes contradicting $(*)$. So $A = \emptyset$ and $F = \lambda F_0$; then we conclude that $N^{-\lambda}_{F_0|X}$ is generated by two sections, namely the differentials of the liftings of the coordinate functions at p. It is obvious that we cannot omit one of the two sections to generate $N^{-\lambda}_{F_0|X}$. This contradicts again $(*)$.

(3) So f is equidimensional. Let F be a singular fiber; we must show that F is a conic in $\mathbb{P}_2$, i.e. a line pair or a double line. As a 1-dimensional fiber, F is a complete intersection. Hence we can write $F = \sum_{i=1}^{p} m_i F_i$ where the F_i are the irreducible components and m_i the multiplicities. Since $K_X \cdot F_{gen} = -2$ and $F_i \in R$, we conclude that either

(a) $p = 2, m_1 = m_2 = 1$ or

(b) $p = 1, m_1 = 2$.

Now notice that $H^1(F, \mathcal{O}_F) = 0$, since $R^1 f_*(\mathcal{O}_X) = 0$. Hence F_0 is a tree of smooth rational curves. Hence in (a) we have a line pair and in (b) we have a double line.

7.5 Theorem *Let* $\dim Y = 3$. *Then* f *is divisorial, let* E *denote the irreducible exceptional divisor.*
(a) If $\dim f(E) = 0$, *then the pair* $(E, N_{E|X})$ *is one the following:*
   (a.1) $(\mathbb{P}_2, \mathcal{O}(-1))$,
   (a.2) $(\mathbb{P}_2, \mathcal{O}(-2))$,
   (a.3) $(\mathbb{P}_1 \times \mathbb{P}_1, \mathcal{O}(-1, -1))$,
   (a.4) $(Q_0, \mathcal{O}(-1))$, where Q_0 *is the quadric cone,*
(b) If $\dim f(E) = 1$, *then* Y *is smooth,* $f(E)$ *is smooth and* f *is the blow-up of* Y *along* $f(E)$.

Proof. We already know that f does not contract isolated curves (6.16) and that E is irreducible (6.20).

(a) Notice that by adjunction

$$K_E = K_X|E + N_E,$$

so that $-K_E$ is the sum of two ample line bundles. If E is smooth, then E is a del Pezzo surface and we see immediately that $E = \mathbb{P}_2$ or $E = \mathbb{P}_1 \times \mathbb{P}_1$ and moreover that the normal bundle N_E is as claimed. If E is singular but normal, one can use the classification of "normal Gorenstein del Pezzo surfaces" (see [Br80], [HW81]) to conclude that E is the quadric cone. The case "E not normal" is excluded by either the method used in [Pe89] to investigate non-normal Gorenstein surfaces (passing to the normalisation, the minimal desingularisation and then studying a minimal model) or by invoking Fujita's Δ-genus (see [Mo82], [Cu88]). We omit details.

(b) Let $C = f(E)$. It is a classical fact that away from the singularities of C and Y the map f is just the blow-up [Nk71]. Hence the general fiber F of $f|E$ is a $\mathbb{P}_1$ and we have

$$K_X \cdot F = E \cdot F = -1.$$

At least as a cycle every fiber of $f|E$ has therefore to be a smooth rational curve, $-K_X$ being f-ample. We conclude that, even if some fibers have embedded points, $-K_X$ is f-generated by global sections, i.e. the canonical map

$$f^* f_*(\mathcal{O}_X(-K_X)) \longrightarrow \mathcal{O}_X(-K_X)$$

is surjective. Now fix an ample line bundle L on Y. Then for $m \gg 0$, the bundle $f^*(mL) - K_X$ is generated by global sections. Let Z be a smooth member in its linear system. Then for a general fiber F of $f|E$ we have $Z \cdot F = 1$, hence $f|S$ is generically an isomorphism. In order to show that $f|S$ is actually an isomorphism, consider $Z' = f(Z)$. Note that Z' is $\mathbb{Q}$-Cartier by (6.23). Thus we have an equation of $\mathbb{Q}$-Cartier divisors

$$Z = f^*(Z') + \lambda E.$$

Since $Z \cdot F = -E \cdot F = 1$, it follows $\lambda = -1$. Hence Z' is even Cartier, in particular Z' is Cohen-Macaulay. Now Z' has only isolated singularities, hence it is normal. Therefore $f|Z$ has to be an isomorphism due to Zariski's main theorem. So Z' is smooth. From the exact sequence

$$0 \longrightarrow N^*_{Z'|Y} \longrightarrow \Omega^1_Y|Z' \longrightarrow \Omega^1_{Z'} \longrightarrow 0$$

it follows that Y is smooth along Z'. Since $C \subset Z'$, we conclude the smoothness of Y. Now again it is classical that C is smooth and f the blow-up (see [Nk71]).

7.6 Addendum *The analytic type of the singularity $p = f(E)$ of Y in (7.5.a) can also be determined. The result is the following [Mo82].*
(a.1) p is a smooth point;
(a.2) p is locally given as the quotient of $\mathbb{C}^3$ by the action $(x,y,z) \mapsto (-x,-y,-z)$;
(a.3) p is given by the equation $x^2 + y^2 + z^2 + w^2 = 0$;
(a.4) p is given by the equation $x^2 + y^2 + z^2 + w^3 = 0$.

For a proof see [Mo82].

7.7 Remark (1) The following observation is crucial. Most singularities occuring on Y are actually not only $\mathbb{Q}$-Gorenstein but Gorenstein. The one exception is the singularity (a.2); this singularity has index 2. In fact, we here have

$$K_X = j^*(K_Y) + \frac{1}{2}E :$$

the unique solution of $K_X + aE|E = 0$ is $a = \frac{1}{2}$. The consequences of this single exception are far-reaching, we will start to discuss this in (7.10) below.

(2) We remark that (7.5)/(7.6) has an analogue in the case where X is Gorenstein instead of being with almost the same result; see [Cu88].

In the following we show by examples that all types of contractions really occur (we do not mention projective bundles and blow-ups in manifolds any more explicitly).

7.8 Examples   (1) Let X be a smooth divisor $\mathbb{P}_2 \times \mathbb{P}_2$ of bidegree $(1, 2)$. Then X is Fano with $b_2(X) = 2$. By the cone theorem X has two contraction, given by the restrictions of the two projections. It is clear that, restricting to the first factor, we obtain a conic bundle $f : X \longrightarrow \mathbb{P}_2$. Since conics degenerate in codimension 1, it is clear that f has singular fibers.

Also all following examples are Fano with $b_2 = 2$.

(2) Here X is a two-sheeted covering of $\mathbb{P}_2 \times \mathbb{P}_1$ ramified over a smooth divisor of bidegree $(4, 2)$. Let $f : X \longrightarrow \mathbb{P}_1$ be the induced map. Then f has connected fibers, so it must be an extremal contraction. Let F be a general fiber of f. Then F is ramified over $\mathbb{P}_2$ along a smooth curve of degree 4, so that F is a del Pezzo surface with $K_F^2 = 2$.

(3) Same example as in (2), but now the ramification divisor is of degree $(2, 2)$. Here $f : X \longrightarrow \mathbb{P}_1$ is a quadric bundle.

(4) Put $X = \mathbb{P}(\mathcal{O}_{\mathbb{P}_2} \oplus \mathcal{O}_{\mathbb{P}_2}(2))$. Let $E = \mathbb{P}(\mathcal{O}) \subset X$. Then $E = \mathbb{P}_2$ with normal bundle $N = \mathcal{O}(-2)$. Let $f : X \longrightarrow Y$ be the contraction of E. In order to see that f is extremal, we need only to verify that Y is projective. In fact, let $L = \mathcal{O}_X(1)$. Then L is obviously generated by global sections and it is immediately verified that $L \cdot C = 0$ if and only if $C \subset E$. Therefore $L = f^*(L')$ with an ample line bundle L' on Y.

(5) Let X be the blow-up of $\mathbb{P}_3$ along a smooth curve C which is the complete intersection of a smooth quadric Q_2 and a smooth cubic. Let $E \subset X$ be the strict transform of Q_2 in X. Then $E \simeq Q_2 \simeq \mathbb{P}_1 \times \mathbb{P}_1$ and we compute easily that $N_{E|X} = \mathcal{O}(-1, -1)$. Let $f : X \longrightarrow Y$ be the blow-down of E. We leave it to the reader to verify that X is projective by exhibiting an ample line bundle on Y.

(6) If we take in (5) Q_2 to be a quadric cone rather than a smooth quadric, then we find a quadric cone $E \subset X$ with normal bundle $\mathcal{O}(-1)$ and again the corresponding blow-down gives the contraction we are looking for.

We now come to the case of small contractions. We already saw that small contractions do not exist on smooth threefolds. We have even more; Benveniste proved in [Be85] the following

7.9 Theorem   *Let X be a normal projective Gorenstein threefold with at most terminal singularities. Then there is no small contraction on X.*

The proof uses the techniques used in the proof e.g. of the Non-vanishing Theorem; we refer to [Be85]. What Benveniste actually shows is the following: if $C \subset X$ is an extremal rational curve on X which is only assumed to be $\mathbb{Q}$-Gorenstein, and if C cannot be deformed (i.e. defines a small contraction), then $0 < -K_X \cdot C < 1$.

7.10 Example Here we want to give a concrete example in dimension 3 that small contractions exist. We will construct a normal projective threefold having only one singularity of type (a.2) (7.6) and a smooth rational curve containing the singularity. This rational curve can be contracted and gives rise to an extremal contraction. This example is (at least locally) due to Francia [Fr80]. We show that this example is very natural if we just naively start with the situation we want to create and then study back.

(1) We first study the situation locally and call the construction the local Francia construction. Let U be a normal complex space with just one singularity x_0 of type (7.6, a.2). We assume that U contains a smooth rational curve through x_0 and moreover that C is contractible with $K_U \cdot C = -\frac{1}{2}$. Let $g : V \longrightarrow U$ be the blow-up of x_0. Then $E = g^{-1}(x_0) \simeq \mathbb{P}_2$ and $N_{E|V} = \mathcal{O}(-2)$. Let C_1 denote the strict transform of C in V. Note that C_1 meets E in one point transversally. It is easy to see that

$$N_{C_1|V} = \mathcal{O}(-1) \oplus \mathcal{O}(-1).$$

Next let $f : W \longrightarrow V$ be the blow-up of C_1. Let G be the exceptional divisor of f, then

$$N_{G|W} = \mathcal{O}(-1) \oplus \mathcal{O}(-1).$$

Hence W can be blown down along the other projection of G. We obtain a blow-down map $f' : W \longrightarrow V'$. Let $\hat{E}$ be the strict transform of E in W, then

$$\hat{E} \simeq \mathbb{F}_1 = \mathbb{P}(\mathcal{O} \oplus \mathcal{O}(-1)).$$

Let $E' = f'(\hat{E})$; then still $E' \simeq \mathbb{F}_1$, since $f'(G) \subset E'$. A short computation of intersection numbers shows

$$E' \cdot l = -1,$$

where l is a ruling line of E'. Therefore V' can be blown down along the projection of E' and we obtain a map $g' : V' \longrightarrow U'$ to a complex manifold U'. If $C' = g'(E')$, then C' is a smooth rational curve with normal bundle

$$N_{C'|U'} = \mathcal{O}(-1) \oplus \mathcal{O}(-2),$$

as another short calculation shows.

Now we can reverse the construction: start with U' and the curve C' and make all construction in reversed order. Then we finally arrive at U and C. The only thing which is not completely clear is the contractibility of C (in the analytic category). To see this we note that there is a birational map $h : U' \longrightarrow U$ which maps $U' \setminus C'$ biholomorphically to $U \setminus C$. Since $C' \subset U'$ has negative normal bundle, it has hence arbitrary small strongly pseudo-convex open neighborhoods [Gr62]. This pseudo-convexity is a property of the boundary of suitable neighborhoods in U'. Applying h it follows that also C has arbitrary small strongly pseudo-convex neighborhoods. Hence by [Gr62] again, C is contractible.

(2) We now construct a global example using the local Francia construction. Let

$$X' = \mathbb{P}(\mathcal{O} \oplus \mathcal{O}(1) \oplus \mathcal{O}(2)).$$

Let $C' = \mathbb{P}(\mathcal{O}) \subset X'$. Then $N_{C'|X'} = \mathcal{O}(-1) \oplus \mathcal{O}(-2)$. Now apply the construction to X' to obtain an algebraic variety X containing a smooth rational curve C and a singularity x_0 on C as we wanted. However it is not a priori clear that X is projective. Let $\varphi : X \longrightarrow Y$ be the blow-down of Y. First we check that Y is projective. For this consider the blow-down $\psi : X' \longrightarrow Y$ of C'. Now an ample line bundle L on Y is induced by $\mathcal{O}_{\mathbb{P}}(1)$. Since $-2K_X$ is φ-ample, the line bundle $L = \varphi^*(L^m) \otimes \mathcal{O}(-2K_X)$ is ample for large m, and X is projective. Finally φ is an extremal contraction, since Y is projective.

7.11 Example It is much easier to give an example of a small contraction in dimension 4, since then we can take X to be smooth. We consider the rank 3-bundle

$$E = \mathcal{O}(1) \oplus \mathcal{O}(1) \oplus \mathcal{O}$$

on $\mathbb{P}_2$ and let $X = \mathbb{P}(E)$. Then X is Fano, since

$$-K_X = \mathcal{O}_X(3) \otimes \pi^*(\mathcal{O}(1))$$

is ample. By the Cone Theorem we have a second extremal contraction on X. To see this contraction, let $D = \mathbb{P}(\mathcal{O}) \subset \mathbb{P}(E)$. Then $D \simeq \mathbb{P}_2$ and

$$N_{D|X} = \mathcal{O}(-1) \oplus \mathcal{O}(-1).$$

Let $f : X \longrightarrow Y$ be the blow-down of D. Then it only remains to show that Y is projective. But $\mathcal{O}_X(1)$ is generated by global sections, trivial on D and ample "outside", so that

$$\mathcal{O}_X(1) = f^*(L)$$

with an ample line bundle L on X (actually the sections of $\mathcal{O}_X(1)$ define f).

7.12 Remark It is not accidental that the exceptional locus of the small contraction in (7.11) is the projective plane. In fact Kawamata [Ka89] has shown that if $f :$

$X \longrightarrow Y$ is a small contraction from a smooth projective 4-fold, then the exceptional locus of f consists of disjoint projective planes with normal bundle $\mathcal{O}(-1) \oplus \mathcal{O}(-1)$. For some results in the higher-dimensional case, see Zhang [Zh91].

In higher dimensions it gets more and more difficult to obtain structure results for extremal contractions. In dimension 4 however quite a lot is known, see the recent papers [AW96], [Kc95]

8 Minimal Models

8.1 Let X be a projective manifold with K_X not nef. Then there exists an extremal contraction $f : X \longrightarrow Y$. So far we have studied the structure of f. If $\dim Y < \dim X$, then f is a $\mathbb{Q}$-Fano fibration over Y and X is uniruled. This is certainly all we can say in general. Of course, in special situations one can still ask for the structure of Y and f, e.g. if X is Fano. We now turn to the birational case. In that case we have not obtained any global information on X. Therefore – in analogy to the surface case – we want to repeat the procedure and ask whether K_Y is nef or not. So the question on the global structure of X is carried over to Y. In some sense Y is simpler than X since $\rho(Y) = \rho(X) - 1$. On the other hand it might happen that Y has singularities. For this reason we are forced to start with a singular X, even if we are only interested in the structure of projective manifolds. So we assume from the beginning that X is normal projective, $\mathbb{Q}$-factorial with only terminal singularities. If f is divisorial, then we have seen that Y has the same properties as X and we ask whether K_Y is nef or not. If K_Y is nef, then Y is a *minimal model* and we are happy from the point of view of these notes. Of course it is still necessary to study minimal models, but this is a different story, see e.g. Miyaoka's article. Basically one wants in that case a multiple mK_Y to be generated by global sections; this is known only in (at most) dimension 3 at the moment (solution of the "Abundance Conjecture" by Kawamata and Miyaoka). We observe that if K_Y is also big, then we already know that mK_Y is globally generated by the Base Point Free Theorem. The difficulty in the other cases is the lacking of an H^q-vanishing.

Now it might happen that f is small. Then we have seen that Y is no longer even $\mathbb{Q}$-Gorenstein and there is no hope to continue the programme. What one needs is to perform another birational process which improves the situation. These birational processes are called *flips*. It will be the task of this section to say something on these flips. Again we will mostly concentrate on the "absolute" case.

First we want to define minimal models and formulate (a weak form of) the Minimal Model Conjecture.

8.2 Definition Let X be a normal ($\mathbb{Q}$-factorial) projective variety with at most terminal singularities. X is called a *minimal variety* if K_X is nef.

Of course there are also notions of minimality in the relative and in the log case.

8.3 Minimal Model Conjecture *(weak form) Every normal projective variety X is birational to a minimal variety unless X is uniruled.*

We call (8.3) the Weak Minimal Model Conjecture (WMMC), since it does not make any predictions on the structure of the birational isomorphism resp. how to get to a minimal model. The strong version (SMMC) will make a precise assertion how to construct minimal models by contractions and flips. We will also discuss the WMMC in connection with the canonical ring and the Zariski decomposition.

8.4 Remark Before we start to discuss flips we want to convince ourselves that it is really necessary to consider a new birational process, different from extremal contractions. This is necessary because even in the case of a small contraction $f : X \longrightarrow Y$ one might argue that f just was the wrong choice and another choice of a contraction could be a fibration or divisorial. However the following example, due to Francia, shows that this is not the case; there are examples where there is only one contraction and this contraction is small.

Let X' be a smooth projective threefold with $K_{X'}$ ample and a smooth rational curve $C' \subset X'$ such that $N_{C'|X'} = \mathcal{O}(-1) \oplus \mathcal{O}(-2)$. Then we can perform on X' the local Francia construction and obtain an algebraic threefold X with one singularity of type (a.2) in the notation of (7.6); moreover X contains a smooth rational curve C which can be contracted. Let $\varphi : X \longrightarrow Y$ be the blow-down of C. Now K_X is not nef, therefore X carries an extremal contraction. We claim that this must be φ. In fact, we see easily from the ampleness of $K_{X'}$ that $K_X \cdot B > 0$ for every curve $B \neq C$. So C is the only curve in X with $K_X \cdot C < 0$, therefore φ must be extremal and is the only extremal contraction on X.

Let $f : X \longrightarrow Y$ be a small contraction. The Francia example suggests how to find a way out of the dilemma that Y has bad singularities. Namely one should take out the exceptional locus of f which is a tree of smooth rational curves and substitute it by another a tree of rational curves. One should do this in such a way that the situation improves, i.e. the new tree of rational curves should not make any trouble, which means that the canonical bundle should be ample on this tree, so that it cannot give rise to a small contraction.

8.5 Definition Let X be $\mathbb{Q}$-factorial with at most terminal singularities. Let $f : X \longrightarrow Y$ be a small contraction. A *flip* for f consists of a normal projective $\mathbb{Q}$-Gorenstein variety X^+ and a *small* birational morphism $f^+ : X^+ \longrightarrow Y$ such that K_{X^+} is f^+-ample. Also $\varphi = (f^+)^{-1} \circ f$ is called flip.

We shall call the minimal sets $A \subset X$ and $A^+ \subset X^+$ the *exceptional sets* of the flip.

Again one can define relative flips and log flips.

Of course it is not at all clear whether a flip exists. The first example of a flip is provided by (7.10). In (7.11) we had the example of a 4-dimensional small contraction. Let us construct the flip there.

8.6 Example Let $X = \mathbb{P}(\mathcal{O} \oplus \mathcal{O}(1)^{\oplus 2}) \longrightarrow \mathbb{P}_2$. Let $D = \mathbb{P}(\mathcal{O}) \subset X$ as in (7.11). Let $f : X \longrightarrow Y$ be the blow-down of D. To construct the flip for f, let $g : Z \longrightarrow X$ be the blow-up of D with exceptional divisor E. Then $E \simeq \mathbb{P}_1 \times \mathbb{P}_2$ with conormal bundle $N^*_{E|Z} = \mathcal{O}(1,1)$. Hence we can blow down Z along the projection to $\mathbb{P}_1$ and we obtain a morphism $h : Z \longrightarrow X^+$. A priori X^+ is only Moishezon. Let $C = h(E)$, then C is a smooth rational curve with normal bundle $\mathcal{O}(-1)^{\oplus 3}$. Therefore there exists a blow-down map $f^+ : X^+ \longrightarrow Y'$, but obviously $Y = Y'$. Since $-K_{X^+} \cdot C = 1, -K_{X^+}$ is f^+-ample, in particular X^+ is projective. Hence $f^+ : X^+ \longrightarrow Y$ is the flip of f. Note that here we have substituted the 2-dimensional D by the 1-dimensional C.

Before formulating the so-called Flip Conjectures we collect some easy properties of flips, see e.g. [KMM87].

8.7 Proposition Let X be $\mathbb{Q}$-factorial with at most terminal singularities. Let $f^+ : X^+ \longrightarrow Y$ be a flip of $f : X \longrightarrow Y$. Then the following assertions hold.

(1) X is again $\mathbb{Q}$-factorial with only terminal singularities.

(2) $\rho(X) = \rho(X^+)$.

(3) Let $\varphi : X \dashrightarrow X^+$ be the induced birational map. Let $g : Z \longrightarrow X$ be a finite sequence of blow-ups such that $h : Z \to X^+$ is a actually a morphism (elimination of indeterminacies). Let E be the exceptional set of f. Choose $F_i \subset Z$ such that we have simultaneously

$$K_Z = g^*(K_X) + \sum a_i F_i,$$

and

$$K_Z = h^*(K_{X^+}) + \sum a_i^+ F_i.$$

Then we have $a_i^+ \geq a_i$ with equality at the place i if and only if $g(F_i) \subset E$.

We should note that this last condition means that F_i comes from resolving the indeterminacies of φ and is not related to resolve singularities outside A. In particular, if X is smooth (thus necessarily of dimension at least 4), then we can choose g in such a way that we blow up only subvarieties "over" E, hence we have always have $a_i^+ > a_i$.

Proof. (see [KMM87]) (a) First we relate divisors on X and X^+. Let $W \subset X$ be an irreducible hypersurface (i.e. irreducible of codimension 1, but not necessarily

$\mathbb{Q}$-Cartier). Then its strict transform W^+ is a Weil divisor on X^+. Vice versa, W is the strict transform of W^+ in X. Therefore we obtain a canonical bijection

$$Z_1(X) \longrightarrow Z_1(X^+)$$

between the Weil divisors on X and on X^+. This yields an isomorphism

$$\alpha : Z_1(X) \otimes \mathbb{Q} \longrightarrow Z_1(X^+) \otimes \mathbb{Q}.$$

Since X is $\mathbb{Q}$-Cartier, we have $Z_1(X) \otimes \mathbb{Q} = \mathrm{Div} \otimes \mathbb{Q}$. So let $D \in \mathrm{Div} \otimes \mathbb{Q}$. Then we have

$$D = f^*(D') + r K_X,$$

since f is an extremal contraction. Let $D^+ = f^{+*}(D') + r K_{X^+}$. Then $D^+ = \alpha(\hat{D})$ outside a set of codimension at least 2, hence $D^+ = \alpha(D)$. Therefore

$$\alpha(\mathrm{Div}(X) \otimes \mathbb{Q}) = \mathrm{Div}(X^+) \otimes \mathbb{Q}.$$

In particular X^+ is $\mathbb{Q}$-factorial and (2) holds.

(b) Now we compare the canonical divisors. Let $g : Z \longrightarrow Y$ be a desingularisation such that $h = \varphi \circ g : Z \longrightarrow X^+$ is a morphism (elimination of indeterminacies). Write

$$K_Z = g^*(K_X) + \sum a_i F_i = h^*(K_{X^+}) + \sum a_i^+ F_i.$$

Since X has terminal singularities, we have $a_i > 0$. The only way to prove that X^+ has terminal singularities, too, is hence to prove the inequality

$$a_i^- \geq a_i. \tag{$*$}$$

But this follows easily from the fact that K_{X^+} is f^+-ample while K_X is f-negative (the shortest way to conclude is to look at relative global sections of $m K_{X^+}$ which give rise to relative global sections of $m K_X$ but with vanishing along A resp. the pull-back on Z).

(c) It remains finally to prove the second part of (3). But this follows at once from the last remark because the difference $a_i^+ - a_i$ is computed by the vanishing order of sections of $m K_X$ near A, which must be strictly positive because of the relative negativity of K_X. Outside A and A^+ however, everything coincides and the coefficients are the same.

In total we have improved the situation after performing the flip: the curves contracted by the small contraction have been turned into curves on which K_{X^+} is positive. However note that the inequality (8.7(3)) says that for every curve $C \not\subset A$ and its strict transform C^+ we have

$$K_{X^+} \cdot C^+ \leq K_X \cdot C.$$

Of course it might happen that all contractions on X^+ are again small so that we again have to make a flip. Therefore besides the question of the existence of a flip it is important to know whether there can be infinite sequences of flips. Before we address to the finiteness we shortly remark that flips, if all existent, are unique:

8.8 Remark Let $f : X \longrightarrow Y$ be small. If there exists a flip for f, then necessarily the "relative canonical ring" (which is a sheaf)

$$R(X|Y) = \bigoplus f_*(\mathcal{O}_X[mK_X])$$

is a finitely generated $\mathcal{O}_Y$-algebra and $X^+ = \operatorname{Proj} R(X|Y)$.

Conversely, if $R(X|Y)$ is finitely generated, then it is easily seen that

$$R(X|Y) \longrightarrow Y$$

is the flip for f. See [KMM87] for details. See also the last part of this section for more on the connection between finite generatedness and minimal models.

We are now ready to state the flip conjectures which are due to Reid.

8.9 Flip Conjectures *Let X be a normal projective $\mathbb{Q}$-factorial variety with at most terminal singularities.*

(I) For every small extremal contration $f : X \longrightarrow Y$ there exists a flip.

(II) There is no infinite sequence of flips.

Of course there are also relative and log versions.

The Strong Minimal Model Conjecture *(SMMC) can now be formulated as follows. Let X be a normal $\mathbb{Q}$-factorial projective variety with at most terminal singularities. Assume that X is not uniruled. Then X is via a finite sequence of divisorial contractions and flips birationally isomorphic to a minimal variety.*

We first address to (II). In order to prove finiteness one would ideally need an invariant (or a finite set of invariants) taking positive integers as values which drops at every flip. In dimension 3 such an invariant is known, the so-called difficulty, introduced by Shokurov [Sh88].

8.10 Definition Let X have at most terminal singularities. Let $f : \hat{X} \longrightarrow X$ be a desingularisation and write $K_{\hat{X}} = f^*(K_X) + \sum a_i E_i$. Then the *difficulty* $d(X)$ is defined by

$$d(X) = \operatorname{card}\{i | a_i < 1\}.$$

Of course one has to show that $d(X)$ does not depend on the choice of f which is easily done. From (8.7(3)) it is clear that $d(X) \geq d(X^+)$ in a flip $X \dashrightarrow X^+$. In order to make use of $d(X)$ one needs $d(X)$ to drop in a flip. Fortunately this is true in dimension 3, so that one has

8.11 Theorem *(Shokurov) Flip Conjecture II holds in dimension 3.*

Proof. Let A, A^+ be the exceptional sets of the flip $X \dashrightarrow X^+$. Let $f : Z \longrightarrow X$ and $g : Z \longrightarrow Y$ be the usual maps associated to the flip. Take an irreducible component $C \subset A^+$ and a component E_i of the exceptional divisor of f dominating C. Since the singularities of X^+ are isolated, h is generically a blow-up and we can choose E_i such that its coefficient $a_i^+ = 1$. By (8.7(3)) it follows that $a_i < 1$, hence $d(X^+) < d(X)$. So there cannot be an infinite sequence of flips.

In general it might happen that A^+ is contained in the singular locus of X^+ or has codimension larger than 2 and the argument breaks down. However by some additional argument, [KMM87] presents still a proof in dimension 4. It is clear that "in general" one cannot expect the difficulty to drop, so that there is a need for additional invariants. For example, if X is Gorenstein, then $d(X) = 0$, hence $d(X^+) = d(X)$. Note also that the difficulty seems to be the only possible invariant connected with inequalities of discrepancies. E.g. card$\{i|a_i < 2\}$ makes no sense since it depends on the choice of the resolution.

Turning to the existence of flips we have the deep result of Mori [Mo88]

8.12 Theorem *In dimension 3 flips always exist, hence SMMC holds.*

The proof of (8.12) is far beyond the scope of these notes. We will restrict ourselves to give some indications.

(1) (Kawamata [Kw88]) It is sufficient to prove the analytic version of the theorem: Let X be a threedimensional analytically $\mathbb{Q}$-factorial normal complex space X with at most terminal singularities. Let $A \subset X$ be a compact purely 1-dimensional analytic set such that X is a deformation retract of A. Assume that $f : X \longrightarrow Y$ is a bimeromorphic holomorphic map contracting exactly A such that $-K_X$ is f-ample. Then there exists a normal complex space X^+ with at most terminal singularities and a small contraction $f^+ : X^+ \longrightarrow Y$ with 1-dimensional exceptional set A^+ such that K_{X^+} is f^+-ample.

We also may assume that A is connected and of course we may make X smaller if we want. In other words we need only to consider the germ (X, A) of X along A. Such a germ (X, A) is called an *extremal* neighborhood.

(2) (Kawamata) One may assume that A is irreducible, i.e. $A \simeq \mathbb{P}_1$. The reason for this is that one can take an irreducible component of A, contract it, construct the flip, then take the next component etc. So from now on let A be a smooth rational curve.

(3) (Kawamata [Kw88, 8.5]) The next very important reduction step is the following. Let $D \in |-2K_X|$ be a general element (which is a *Weil* divisor) determining

a 2:1-cover $g : \tilde{X} \longrightarrow X$ branched along D. Now assume that D can be taken in such a way that $\tilde{X}$ has only **canonical** singularities. By adjunction

$$K_{\tilde{X}} = g^*(K_X) + \frac{1}{2}D = 0.$$

Instead of performing a *flip* on X we perform a *flop* on $\tilde{X}$ which is much easier, see [Kw88], [Ko89]. The definition of a flop is similar as that of a flip but the relative negativity resp. positivity of K_X is substituted by triviality, see (8.14) below. It is necessary to have canonical singularities on $\tilde{X}$ to perform the flop. Note that g induces a map $\tilde{Y} \longrightarrow Y$. Then, taking quotient, we obtain from the flop $g^+ : \tilde{X} \longrightarrow \tilde{Y}^+$ the flip $f^+ : X^+ \longrightarrow Y^+$. Note also that in [Kw88] flops are called "special log flips".

(4) Kawamata gives in [Kw88, 8.5] a "numerical" criterion when one can find an element D such that the induced cover has only canonical singularities. This in turn is used in (8.7) of that paper to prove the following. Consider the linear system $|-K_X|$. If one can find a member $D' \in |-K_X|$ with only rational double points, then the flip exists. The proof is by reducing to the situation (3).

(5) So what Mori proved is that one is able to find either a divisor D as in (3) or a divisor D' as in (4). This requires a detailed analysis of the singularities on (X, A) and is very complicated.

8.13 Remark After [Mo88] Kollár and Mori actually proved in [KM92] that one can always find an element $D \in |-K_X|$ having at most rational double points in the situation of (8.12). These elements D are also known as "general elephants" in the literature (Reid). [KM92] provides furthermore a classification of flips in dimension 3.

Now we define flops and state the main result on existence and finiteness.

8.14 Definition *Let X be a normal $\mathbb{Q}$-Gorenstein variety (or normal complex space) with at most canonical singularities. Let $f : X \longrightarrow Y$ be a small contraction to a normal $\mathbb{Q}$-Gorenstein variety Y with $\rho(X/Y) = 1$. Assume $K_X = f^*(K_Y)$. Assume furthermore that there is a Cartier divisor D on X such that $-D$ is f-ample. A D-flop for f is a small contraction $f^+ : X^+ \longrightarrow Y$ such $K_{X^+} = f^{+*}(K_Y)$ and such that the strict transform D^+ of D is f^+-ample.*

Buiding up on previous work of Reid and Kawamata, Kollár proved in [Ko89] the following result.

8.15 Theorem *Let X be a normal algebraic variety (or a normal complex space) with only canonical singularities. Let D be an **effective** Cartier divisor on X. Let $f : X \longrightarrow Y$ be as in (8.13). Then the D-flop of f exists.*
Moreover any sequence of D-flops is finite.

The notion of a sequence of D-flops is defined as follows. D is the divisor of the first flop; then the divisor of the second flop is just the strict transform of D and so on.

8.16 Example The easiest example of a flop is as follows. Take a projective manifold X containing a smooth rational curve C with normal bundle $N_{C|X} = \mathcal{O}(-1) \oplus \mathcal{O}(-1)$. Then C is contractible in the analytic category or Moishezon category. Clearly $K_X \cdot C = 0$. Let H be an ample line bundle on X and $D = -H$. The D-flop to f is constructed as follows.

Let $g : \hat{X} \longrightarrow X$ be the blow-up of C. Then the exceptional divisor $E \subset \hat{X}$ is nothing than $\mathbb{P}_1 \times \mathbb{P}_1$ with normal bundle $\mathcal{O}(-1) \oplus \mathcal{O}(-1)$, hence $\hat{X}$ can be blown down along the "other" projection.

We obtain a blow-down map $g^+ : \hat{X} \longrightarrow X^+$. Let $C^+ = g^+(E)$. Then C^+ is a smooth rational curve with normal bundle $\mathcal{O}(-1) \oplus \mathcal{O}(-1)$, hence there is a contraction $f^+ : X^+ \longrightarrow Y$ which is the flop we are looking for. In fact, let D^+ be the strict transform of D, then it is easily checked that $D^+ \cdot C^+ > 0$ so that D^+ is f^+ample. Note also that if D is any effective divisor with $-D$ being f-ample, i.e. $D \cdot C < 0$, then D^+ is f-ample.

8.17 Example Let us consider the problem of the existence of general elephants in the easiest possible situation. Namely, let (X, A) be an extremal neighborhood with A irreducible, i.e. a smooth rational curve and assume that X has only one singularity, namely the quotient singularity arising from the involution

$$(x, y, z) \longrightarrow (-x, -y, -z).$$

Notice that we must have a non-Gorenstein singularity on X by (7.9). We know already that X has a desingularisation $g : \hat{X} \longrightarrow X$ with exceptional set $E \simeq \mathbb{P}_2$ with normal bundle $\mathcal{O}(-2)$. We want to construct $D \in |-K_X|$ having only rational double points, i.e. canonical singularities. Let $\hat{C}$ be the strict transform of C in $\hat{X}$. Then we check easily $K_{\hat{X}} \cdot \hat{C} = 0$, so that $-K_{\hat{X}}$ is g-nef. Even more, $-K_{\hat{X}}$ is g-generated, hence generated by global sections. Let

$$\hat{D} \in |-K_{\hat{X}}|$$

be a general smooth member. Put $D = g(\hat{D})$. In order to see that D has only canonical singularities, we have to examine the exceptional locus of $g|\hat{D}$. Since

$$-K_{\hat{X}} \, E = \mathcal{O}(1),$$

the intersection $\hat{D} \cdot E$ is a line $l \subset E$. Since $E \cdot l = -2$ by our assumption on the normal bundle, we deduce that

$$N_{l|\hat{X}} = \mathcal{O}(1) \oplus \mathcal{O}(-2). \tag{$*$}$$

Now consider the exact sequence

$$0 \longrightarrow N_{l|\hat{D}} \longrightarrow N_{l|\hat{X}} \longrightarrow N_{\hat{D}|\hat{X}}|l \longrightarrow 0.$$

Since $\hat{D} \cdot l = 1$, we obtain from the last sequence and $(*)$ that

$$N_{l|\hat{D}} = \mathcal{O}(-2).$$

Hence D has only one rational double point, a singularity of type (A_1).

8.18 Remark In [Ko92] also the existence of log flips in dimension three is proved as well as finiteness of sequences of log flips. There are two proofs: one relying heavily on Mori's result on the existence of flips and the other one following a new approach of Shokurov [Sh91]. However also in the new approach the paper [Mo88] is still needed but there is a hope for an independent proof. We refer to [Ko92] for details.

In higher dimensions it seems very difficult or hopeless to get enough informations on terminal singularities in order to be able to construct flips. Therefore one is tempted to first look at WMMC rather than to SMMC and try to construct minimal models by other means. We start discussing alternative approaches by stating the following classical

8.19 Conjecture *Let X be a projective manifold of general type. Then its canonical ring*

$$R(X) = \bigoplus_{m \geq 0} H^0(X, \mathcal{O}_X(mK_X))$$

is a finitely generated $\mathbb{C}$-algebra.

The relation between this conjecture and minimal models comes from the following

8.20 Remark Let X be a normal projective variety and D a Cartier divisor on X. If some multiple $\mathcal{O}_X(kD)$ is generated by global sections, then

$$R(X, D) = \bigoplus_{m} H^0(X, \mathcal{O}_X(mD))$$

is finitely generated.

This follows easily from Serre's Theorem B on the projective space $\mathbb{P}_n$.

8.21 Remark Let X be a normal projective variety with at most terminal singularities such that K_X is big and nef. We have seen in (4.10) that mK_X is globally generated for suitable large m, hence $R(X)$ is finitely generated.

Therefore we can form $Proj(R(X))$, which is nothing than the canonical model X_{can} of X in the sense of (4.11). It is clear that X_{can} has only canonical singularities.

Reid [Re83] has shown more; namely if X is a projective manifold of general type and if $R(X)$ is finitely generated, then $Proj(R(X))$ has only canonical singularities. Of course this would be a consequence of the existence of minimal models.

The relation between the conjecture (8.19) and the existence of flips is given by

8.22 Proposition *(Moriwaki) If Conjecture (8.19) holds in dimension n, then Flip Conjecture I (existence of flips) holds in dimension n.*

As already noted (see (8.11)), Flip Conjecture I is equivalent to the following: if $f : X \longrightarrow Y$ is a small extremal contraction, then then the "relative canonical ring"

$$R(X/Y) = \bigoplus_m f_*(\mathcal{O}_X(mK_X))$$

is a finitely generated $\mathcal{O}_Y$-algebra. Now starting with such a contraction $f : X \longrightarrow Y$, we pass to a suitable covering $\tilde{X} \longrightarrow X$ such that $\tilde{X}$ is of general type and apply (8.19). For details we refer to [Mw86].

8.23 Remarks (1) (8.19) and Flip Conjecture II (finiteness) imply SMMC.

(2) WMMC for varieties of general type plus Flip Conjecture II imply SMMC. In fact, we already noted that WMMC implies (8.19).

To prove (8.19) means somehow to understand how far bigness for K_X is from being semi-ample (i.e. some multiple is globally generated). This is made precise by the notion of a "Zariski decomposition" introduced in [Kw87], [KMM87]

8.24 Definition Let X be a normal projective variety of general type with at most terminal singularities. Let $D \in \text{Div}(X) \otimes \mathbb{R}$. A *Zariski decomposition* of D is a decomposition

$$D = P + N,$$

where $P, N \in \text{Div}(X) \otimes \mathbb{R}$ such that P is nef, N is effective and moreover the canonical map

$$H^0(X, \mathcal{O}_X([mP])) \longrightarrow H^0(X, \mathcal{O}_X([mD]))$$

is bijective for all positive integers m.

P is usually called the *positive part* and N *the negative part*. The Zariski decomposition is unique if existent ([Kw87]). Historically this kind of decomposition was introduced by Zariski on surfaces only. Note also that there are different notions of Zariski decomposition in the literature (see [Fj86]).

Concerning existence one has the

8.25 Conjecture *Let X be a normal projective variety of general type with at most terminal singularities. Then there exists a birational morphism $f : \tilde{X} \longrightarrow X$ such that $f^*(K_X)$ has a Zariski decomposition.*

Without allowing the modification f the statement would be false. The importance of the conjecture is demonstrated by

8.26 Theorem *(Kawamata) Let X be a normal projective variety with only terminal singularities. Assume that K_X is big, so that X is of general type. Assume that the Zariski decomposition $K_X = P + N$ exists. Then mP is generated by global sections for large m.*

The proof uses essentially the technique from the Non-Vanishing and Base Point Free Theorem, for details we refer to [Kw87].

8.27 Corollary *Let X be a normal projective variety with only terminal singularities. Assume K_X big. If there exists a birational morphism $f : \hat{X} \longrightarrow X$ such that $f^*(K_X)$ has a Zariski decomposition, then $R(X)$ is finitely generated.*

Proof. Write $f^*(K_X) = P + N$. Introduce positive numbers λ_i such that

$$f^*(K_X) = K_{\hat{X}} - \sum \lambda_i E_i$$

where E_i are the exceptional divisors of f. Put $N' = N + \sum \lambda_i E_i$. Then it is obvious that $K_{\hat{X}} = P + N'$, hence by (8.27) mP is generated by global sections. Therefore $R(X, P) = R(X)$ is finitely generated. The last equality is just the last condition in the definition (8.24).

9   A View to Adjunction Theory

Let H be an ample divisor on a projective variety X. We have seen again and again that the adjoint bundles $K_X + tH$ play an important role for the geometry of X, in particular if $K_X + tH$ is nef but not ample. As a general reference for adjunction theory we recommend [Fj90] and in particular the new book of Beltrametti and Sommese [BS95] which gives the state of the art and hundreds of references. To make our life easy we will mostly refer to [BS95] instead of the original literature.

9.1 A pair (X, H) consisting of a normal projective variety X and an ample line bundle H will be called a *polarised variety*. We denote by $t = t(X, H)$ its nef value, i.e.

$$t = \sup\{s | K_X + sH \text{ is not nef }\}.$$

By the Rationality Theorem we have $t \in \mathbb{Q}$ and by the Base Point Free Theorem large multiples $m(K_X + tH)$ are generated by global sections and determine therefore a morphism

$$\varphi : X \longrightarrow Y$$

to a normal projective variety. φ is nothing than the contraction of an extremal face, if $t > 0$, and is often called the nef value morphism of (X, H). Therefore adjunction theory can be considered as a part of the theory of extremal rays etc., however we are now interested in the geometry of the pair (X, H) rather than of X alone.

We shall fix the notations of (9.1) for this section, however we always will assume X to be smooth for simplicity. For singular situations consult [BS95]. Let $n = \dim X$.

9.2 Theorem $K_X + nH$ *is nef unless* $(X, H) = (\mathbb{P}_n, \mathcal{O}(1))$. *In other words,* $t(X, H) \leq n$ *with the one above exception. In particular* $K_X + (n+1)H$ *is always nef.*

Proof. Assume $t > n$. Observe that φ factors over the contraction $\psi : X \longrightarrow Z$ of an extremal ray R. By our assumption, the length $l(R) = n + 1$. The dimension count (6.15) therefore yields $\dim Z = 0$. Hence X is Fano with $b_2(X) = 1$. Now apply the Kobayashi-Ochiai theorem to conclude.

Another proof proceeds by applying the estimate of the denominator in the Rationality Theorem and a Hilbert polynomial consideration. The details are merely an exercise.

9.3 Theorem *Assume that* $t(X, H) = n$, *i.e.* $K_X + nH$ *is nef but not ample. Then either* $K_X + nH$ *is big and nef or* X *is a quadric or a* $\mathbb{P}_{n-1}$*-bundle over a curve.*

Proof. If $K_X + nH$ is not big, then $\dim Y < n$. Let F be a general smooth fiber of φ. Then the adjunction formula gives $K_F = -nF$, so by Kobayashi-Ochiai we conclude that either X is a quadric or $\dim Y = 1$ and $F = \mathbb{P}_{n-1}$. Moreover we have in the last case that $H \cdot F = 1$. This implies that every fiber F is irreducible and reduced. Moreover a theorem of Fujita [Fj85] says that actually every fiber must be projective space (this can be viewed as a singular version of the Kobayashi-Ochiai theorem).

The case that $K_X + nH$ is big and nef does not occur in (9.3) due to a result of Ionescu:

9.4 Theorem *Assume that* $K_X + nH$ *is big and nef. Then* $K_X + nH$ *is ample.*

Proof. This follows from the dimension count (6.15).

9.5 Theorem *Assume that* $K_X + nH$ *is ample, i.e.* $t(X, H) < n$.
Then $K_X + (n-1)H$ *is nef, i.e.* $t(X, H) \leq n - 1$, *unless* $(X, H) = (\mathbb{P}_2, \mathcal{O}(2))$.

Proof. This follows again rather easily from the estimate of the denominator in the Rationality Theorem plus Kobayashi-Ochiai.

Of course the next task is to investigate the case where $K_X + (n-1)H$ is nef but not ample.

9.6 Theorem *Let $t(X, H) = n - 1$. Then either $K_X + (n-1)H$ is big or one of the following holds.*
(a) X is a del Pezzo manifold, i.e. Fano with index $n-1$;
(b) X is a quadric fibration over a smooth surface;
(c) X is a $\mathbb{P}_{n-2}$-bundle over a smooth curve, here $H \cdot F = 1$ for any fiber F (i.e. X is a "scroll").

Note that the projection morphisms in (b) and (c) are contractions of extremal rays. The proof of (9.6) is parallel to that one of (9.3). It remains to investigate the case where $K_X + (n-1)H$ is big and nef (but not ample).

9.7 Theorem *Assume that $K_X + (n-1)H$ is big and nef but not ample. Then the exceptional set of φ consists of a finite number of components $E_i \simeq \mathbb{P}_{n-1}$ with normal bundle $N_{E_i} = \mathcal{O}(-1)$ so that all E_i are contracted to smooth points of Y.*

Proof. Again φ factorises over the contraction of an extremal ray $\psi : X \longrightarrow Z$. The dimension count (6.15) shows that the exceptional set E' consists of divisors all contracted to points. Since ψ is an extremal contraction, we know that $E' = E_i$ is irreducible. By adjunction we have

$$K_{E_i} = K_X|E_i + N_{E_i} = -(n-1)H|E_i + N_{E_i}.$$

Since we must have $K_X \cdot l = -n + 1$ for an extremal rational curve l for ψ, we obtain that $N_{E_i} \equiv -H|E_i$, hence

$$K_{E_i} \equiv -nH|E_i.$$

Now we conclude by another singular version of the theorem of Kobayashi-Ochiai, see e.g. [BS95, 4.2.18] for details. Roughly speaking, one has enough vanishing to construct n sections in $H|E_i$ which give an isomorphism $E_i \longrightarrow \mathbb{P}_n$. The normal bundle formula is clear.

Now it is obvious that we also could have blown down first any other E_j since the curves in E_j must define an extremal ray (all the extremal rays span the extremal face which defines φ). Therefore we have $E_i \cap E_j = \emptyset$ which finishes the proof.

9.8 As in the theory of extremal rays we are not happy with the situation in (9.7) but want to investigate Y furthermore. However we need a polarisation on Y which comes "canonically" from X. This is done as follows. Define a divisor H' on Y by setting

$$\mathcal{O}_Y(H') = \varphi_*(\mathcal{O}_X(H)^{**}).$$

The right hand side is a priori only a reflexive sheaf but since Y is smooth it is actually locally free. Now note that

$$K_X + (n-1)H = \varphi^*(K_Y + (n-1)H'),$$

hence $K_Y + (n-1)H'$ is ample.

The pair (Y, H') is called the *first reduction* of (X, H) and we are now reduced to the study of polarised manifolds (X, H) such that $K_X + (n-1)H$ is ample.

9.9 Theorem *Assume that $n - 2 < t(X, H) < n - 1$. Then $n = 3$ or $n = 4$ and X is either projective space, a quadric or φ maps to a curve with general fiber $\mathbb{P}_2$.*

We leave it to the reader to work out H in all these cases. The proof of (9.9) is again essentially the Rationality Theorem. We omit the easy details.

We are now reduced to study polarised manifolds (X, H) with $t(X, H) \leq n - 2$.

9.10 Theorem *Assume $n \geq 3$. Assume that $K_X + (n-2)H$ is nef but not ample. Assume furthermore that K_X is not big. Then (X, H) is one of the following*
(a) X is Fano with $b_2 = 1$ and index $n - 2$;
(b) X is a del Pezzo fibration over a smooth curve;
(c) X is a quadric fibration ove a normal surface;
(d) X is a scroll over a normal threefold.

This is again clear from the dimension count and Kobayashi-Ochiai.

9.11 Of course it is much harder to describe the situation when $K_X + (n-2)H$ is big and nef but not ample, however the results are not surprising. First notice that we must have $\dim X \geq 4$ (by (6.15). Let E be the exceptional set of φ and let $A = \varphi(E)$. Then A is a disjoint union of smooth rational curves A_i and points p_j. The A_i are contained in the smooth part of Y and near A_i the morphism φ is just the blow-up of A_i. The p_j are singular points of Y and letting $E_j = \varphi^{-1}(p_j)$ we have

$$(E_j, N_{E_j}) = (\mathbb{P}_{n-1}, \mathcal{O}(-2)), (Q, \mathcal{O}(-1)),$$

where Q is a possibly singular quadric.

In other words, φ is the composition of "independent" contraction of extremal rays. Of course we have already met that list in dimension $n = 3$ (see sect. 7, however the nef value is different there).

We shall omit any details of the proof and refer again to [BS95].

9.12 We next define the so-called *second reduction*. Assume that $K_X + (n-2)H$ is big and nef but not ample and let $\varphi : X \longrightarrow Y$ be the associated contraction. Then $K_X + (n-1)H = \varphi^*(L)$ with L ample on Y. In order to get the polarisation on Y

we again define H' by $\mathcal{O}_Y(H') = \varphi_*(\mathcal{O}_X(H))^{**}$. Note however that sometimes Y is only 2–Gorenstein so that H' is merely 2–Cartier. Anyway we have

$$K_Y + (n-2)H' = L.$$

In the literature we still find the next step in adjunction theory, i.e. the investigation of the pair (Y, H'), however we will omit the sometimes slightly nasty details and instead refer to [BS95]. Also for singular versions (i.e. X is singular) we refer to [BS95].

9.13 Adjunction theory can still be seen in a more general context, the so-called vector bundle adjunction. We will briefly explain this. Let X be a projective manifold and E an ample vector bundle of rank r on X. We want to understand the adjoint bundle $K_X + \det E$. If $E = L^{\oplus r}$ with an ample line bundle L, then $K_X + \det E = K_X + rL$ and we are back in adjunction theory.

Here we state the main results on vector bundle adjunction; let $r = \mathrm{rk}E$. Assume that $K_X + \det E$ is not ample.

(a) We always have $r \le n+1$, where $n = \dim X$.

(b) If $r = n+1$ then $X = \mathbb{P}_n$ and $E = \mathcal{O}(1)^{\oplus(n+1)}$ or $E = T_{\mathbb{P}_n}$, the tangent bundle of projective space. In particular $K_X + \det E$ is always nef.

(c) Let $r = n$. Then (X, E) is one of the following:

   $(\mathbb{P}_n, \mathcal{O}(1)^{\oplus n})$,
   $(\mathbb{P}_n, \mathcal{O}(2) \oplus \mathcal{O}(1)^{\oplus(n-1)})$,
   $(Q_n, \mathcal{O}(1)^{\oplus n})$,

   or there is a vector bundle V over a curve C such that $X = \mathbb{P}(V)$, and $E|F \simeq \mathcal{O}(1)^{\oplus n}$ for every fiber of the projection $X \longrightarrow C$.

References: [Fj91], [YZ90], [Pe89, 90].

Also the case $r = n-1$ is completely settled, we refer to [ABW92], [PSW92]. For results in case $r = n-2$, see [AM95].

We should note also a very prominent special case of case (b): if X is a projective manifold with T_X ample, then X is projective space. This is the famous solution of Mori's of the Hartshorne-Frankel conjecture [Mo79]. It can be seen as the starting point of all the theory covered in these notes. For more information we also refer to [Pe96].

Our short discussion of adjunction theory would not be complete without mentioning the Fujita conjecture and the recent spectacular progress on it. The Fujita conjecture can on the hand be seen as the "very ample version" (of a part of) adjunction theory and on the other hand as an effective version of the Base Point Free Theorem.

9.14 Fujita Conjecture *Let X be a n-dimensional projective manifold with ample line bundle L. Then $K_X + (n+1)L$ is globally generated and $K_X + (n+2)L$ is very ample.*

9.15 The status of the Fujita conjecture is as follows; we restrict ourselves to the case of global generatedness.
(a) For curves the conjecture is classical.
(b) The case $n = 2$ has been done by Reider [Re88].
(c) The case $n = 3$ is settled by Ein-Lazarsfeld [EL93], the case $n = 4$ by Kawamata [Kw95].
(d) In higher dimension the conjecture is unsolved. However there are effective results concerning the spannedness of $K_X + mL$. Angehrn and Siu proved [AS95] that $K_X + mL$ is generated if $m \geq \frac{1}{2}(n^2 + n + 2)$. The methods are a mixture of techniques around multiplier ideal sheaves and the techniques of the Non-Vanishing Theorem.

9.16 (1) Concerning very ampleness there are results of Demailly [De96] and Siu [Si96] of the type: $2K_X + mL$ is globally generated for large m depending on $n = \dim X$. Demailly's bound is $2 + \binom{3n+1}{n}$. The first major result around the Fujita conjecture was actually [De93] and it very much influenced all other development.

(2) There also effective versions of the Base Point Free Theorem, i.e. when mL is generated for a nef line bundle L such that $aL - K_X$ is big and nef. See [Ko93], [Si93].

10 Calabi-Yau Threefolds

In this last short section we want to report on threefolds X with $K_X \equiv 0$ in connection with classification theory. At first glance this seems completely to fall out of the scope of these notes but a second look shows that there are actually close relations. We have met them already in (5.9). From an "absolute" point of view things of course have changed: K_X is no longer "not nef", so we might think of being in the limit case of Mori theory. On the other hand let D be an effective divisor. If we assume that X has only log terminal singularities, which we will always do of course, then the pair is $(X, \epsilon D)$ is klt for small ϵ, see (5.9). Then either $K_X + \epsilon D$ is not nef, hence we have a log contraction or $K_X + \epsilon D$ is nef and then the log generalisation of the Abundance Conjecture would predict that mD is generated by global sections for large m. Hence we can study the morphism defined by mD. Of course this is only interesting if D is not ample, so one has to look for effective non-ample divisors. These can clearly only exist if the Picard number $\rho(X) > 1$. The Calabi-Yau threefolds with $\rho = 1$ can be seen as analogues of the Fano manifolds with $\rho = 1$; they need special treatment.

We start by giving a rigorous definition.

10.1 Definition Let X be a normal ($\mathbb{Q}$-factorial) projective variety with at most terminal singularities. Then X is a *Calabi-Yau variety* if X is simply connected and if $K_X = \mathcal{O}_X$.

There is no unique definition of Calabi-Yau varieties; it depends on the purpose of the corresponding paper. Sometimes simply connectedness is relaxed to finiteness of $\pi_1(X)$, sometimes smoothness is required and often Calabi-Yau manifolds are just projective manifolds with trivial canonical bundles. In this last context the name is best explicable: a Calabi-Yau manifold carries a Kähler metric with vanishing Ricci curvature. Such a metric is called a Kähler-Einstein metric and its existence was conjectured by Calabi and proved by Yau [Ya77].

Examples of Calabi-Yau threefolds X can be obtained as follows. Let V be a Fano manifold and take X to be a smooth member of $|-K_V|$. In particular, the most prominent Calabi-Yau threefolds are the quintics in $\mathbb{P}_4$. Of course it is also easy to get examples by complete intersections. For many more examples, e.g. by the method of small resolutions, we refer to the references of [Wi89].

One of the most basic questions on Calabi-Yau varieties seems to be

10.2 Problem *Does every Calabi-Yau variety carry a rational curve ?*

Note that the Calabi-Yau surfaces are exactly the $K3$-surfaces (by definition), the existence of rational curves on those being proved in [MM82]. The proof is by degeneration to so-called Kummer surfaces so that it does not give any hint to higher dimensions. Not much seems to be known on the structure of Calabi-Yau varieties of dimension at least 4 so that from now on we restrict to threefolds. In dimension 3 Wilson's paper [Wi89] is the first systematic treatment on the existence of rational curve and the structure of Calabi-Yau threefolds.

If X carries an effective divisor which is not nef, then the existence of a rational curve follows from (5.9). In [Pe91a] it was shown more generally (and without log-theory) that X carries a rational curve if X admits a non-ample effective divisor. This was generalised and put in more conceptual context by Oguiso [Og93]; we will come to this in a moment.

10.3 Here we collect some problems on Calabi-Yau threefolds which arise from Mori theory.

(1) What is the structure of $\overline{NE}(X)$ resp. of $Amp(X)$?

(2) Let be given a rational point of the boundary of $Amp(X)$, i.e. a nef non-ample Cartier. Is mD generated by global sections? Since $D = K_X + D'$, this is a log abundance conjecture.

(3) Are there any rational points on $\partial Amp(X)$?

We will address to all three points in this section.

10.4 Definition (1) Let X be a Calabi-Yau threefold. A morphism $f : X \longrightarrow Y$ with connected fibers is called a Calabi-Yau contraction if Y is normal projective and

$\rho(Y) < \rho(X)$. If f cannot be factored further into Calabi-Yau contractions, it is called *primitive*.

(2) Assume f primitive. If f is birational, then
(a) f is of type I, if f is small;
(b) f is of type II, if f contracts exactly a surface to a point;
(c) f is of type III, if f contracts exactly a surface to a curve.

10.5 Remarks (1) Assume $1 \leq \dim Y \leq 2$. Then Oguiso has in [Og93] investigated the structure of f, here are some results.
(a) If Y is a curve, then $Y \simeq \mathbb{P}_1$ (because of simply connectedness) and the general fiber is a $K3$ surface or an abelian surface; both cases do appear;
(b) if $\dim Y = 2$, then the general fiber is (obviously) an elliptic curve, there exists an effective $\mathbb{Q}$-divisor Δ, possibly 0, on Y such that (Y, Δ) is log terminal and $K_Y + \Delta \equiv 0$. Moreover Y is rational with only quotient singularities.

(2) Assume that f is birational. Then Y has automatically only canonical singularities and $K_Y = \mathcal{O}_Y$. Those Y (arising from X) are called *Calabi-Yau models* by Wilson. Note that in contrary to contractions of extremal rays, the singularities on Y can be in codimension 2. Note also that if f is divisorial, then the exceptional locus of f is an irreducible surface. For more informations on the structure of E in that case we refer to [Wi95]. If f is small, then f contracts finitely many smooth rational curves (since $R^1 f_*(\mathcal{O}_X) = 0$, canonical singularities being rational).

Concerning the log abundance conjecture Oguiso proved in [Og93]

10.6 Theorem *Let X be a Calabi-Yau threefold. Let D be a nef and effective Cartier. Then mD is generated by global sections for large m.*

The proof uses [Kw92] and [Wi89] in an essential way.

10.7 Of course now the question arises whether a nef divisor is effective. If $D^3 > 0$. this is obvious, so assume $D^3 = 0$. Let us assume X smooth for simplicity. Then Riemann-Roch yields

$$\chi(X, \mathcal{O}_X(mD)) = \frac{m}{2} D \cdot c_2(X).$$

So let us assume $D \cdot c_2(X) \neq 0$. Then actually $D \cdot c_2(X) > 0$, because of the inequality
$$0 = c_1(X)^2 \cdot A \leq 3c_2(X) \cdot A$$

for every nef divisor A. This inequality comes from the existence of a Kähler-Einstein metric on X, see [Ya77] or, much easier, from the stability of the tangent bundle. If we have additionally

$$D^2 \cdot H \neq 0$$

for some ample divisor H, then we have for all positive m that

$$H^2(X, \mathcal{O}_X(mD)) = 0.$$

This comes from an easy generalisation of the Kawamata-Viehweg vanishing theorem to nef divisors which are not big:

let X be a projective manifold of dimension n and L a nef line bundle on X. Let k the maximal number such that $c_1(L)^k \neq 0$ in $H^{2k}(X, \mathbb{R})$, then

$$H^q(X, L \otimes K_X) = 0$$

for $q > n - k$.

For a proof see e.g. [SS85]. By the H^2-vanishing and Riemann-Roch it therefore follows that $h^0(X, \mathcal{O}_X(mD)) > 0$ for $m > 0$.

If $D^2 \cdot H = 0$ for all ample H, then $D^2 = 0$ and one needs additional arguments to proof that mD is effective; we refer to [Og93].

So it remains only to treat the case $D \cdot c_2(X) = 0$ which is unsolved at present. However Wilson proved in [Wi94] effectivity in case $D^2 \neq 0$ except in very special situations.

10.8 Independent of all existence questions, it is not difficult to see that if X admits a Calabi-Yau contraction, then X contains a rational curve [Og93].

Concerning existence there is the following theorem of Heath-Brown and Wilson [HW92] improving an earlier result of Wilson [Wi89]:

10.9 Theorem *Let X be a smooth Calabi-Yau threefolds with $\rho(X) > 13$. Then X admits a birational Calabi-Yau contraction and in particular X contains rational curves.*

The cubic cone $W = \{D^3 = 0\}$ in $\mathbb{P}(N^1(X))$ and its diophantine geometry play an important role in the proof.

10.10 It remains to say something on the structure of the ample cone $Amp(X)$ (here it is more convenient to look at the ample cone instead of $\overline{NE}(X)$.)

Away from the cubic cone the ample cone is locally finite rational polyhedral, see [Wi92], [Wi94a]. However the cone is not always finite rational polyhedral as there are examples of Calabi-Yau threefolds with infinite (but discrete) automorphism group, see [Bo91]. On the other hand it is suspected that if $c_2(X)$ is a positive function on $Amp(X) \setminus \{0\}$, then the ample cone is rational polyhedral. At least the automorphism group is finite in that case [Wi94a].

More generally Morrison conjectures in [Mr93] that inside the ample cone there exists a finite rational polyhedral cone K such that the translates $f(K), f \in Aut(X)$, cover the ample cone. A more geometric but weaker question is:

if $c_2(X) > 0$ (in the above sense), are there only finitely many Calabi-Yau contractions on the given Calabi-Yau threefold X?

In [OP95] this is proved for non-birational contractions.

10.11 All these considerations do not say anything about the case $\rho(X) = 1$. Rational curves in that case are in particular predicted by the so-called Mirror Symmetry, even with formulas how to calculate the numbers in a certain degree, if X is "general". We will completely ignore that. However we want to finish the discussion of Calabi-Yau threefolds with stating another analogy to Mori theory which makes the existence of rational curves very convincing. Since the dimension does not play any role here, we let X denote any Calabi-Yau manifold. The idea is to look at the cotangent bundle Ω_X^1 instead of K_X which is trivial. The basic fact is that Ω_X^1 is not nef. Here we say that a vector bundle E is nef if the line bundle $\mathcal{O}_{\mathbb{P}(E)}(1)$ is nef. In fact, if Ω_X^1 were nef, then we would have the Chern class inequality

$$c_1(\Omega_X^1)^2 \cdot H \geq c_2(\Omega_X^1) \cdot H,$$

for every ample H, see e.g. [FL83]. However we know already

$$c_1(\Omega_X^1)^2 \cdot H \leq 3c_2(\Omega_X^1) \cdot H,$$

and therefore we obtain $c_2(\Omega_X^1) = c_1(\Omega_X^1)^2 = 0$. But then there exists a finite unramified cover of X which is an abelian variety. This is a standard differential-geometric consequence of the existence of a Kähler-Einstein metric on X, see [Ya77]. Hence we obtain a contradiction to the simply connectedness of X.

So the problem on the existence of rational curves can be restated as follows:

let X be a projective manifold with $K_X = \mathcal{O}_X$. Assume that Ω_X^1 is not nef. Does X contain a rational curve?

References

[ABW93] Andreatta, M.; Ballico, E.; Wisniewski, J.: Two theorems on elementary contractions. Math. Ann. 297, 191–198 (1993)

[An85] Ando, T.: On extremal rays of the higher dimensional varieties. Inv. math. 81, 347–357 (1985)

[AS95] Angehrn, U.; Siu, Y.T.: Effective freeness and separation of points for adjoint bundles. Inv. Math. 122, 291–308 (1995

[Au76] Aubin, T.: Equations de type Monge-Ampère sur les variétés kählériennes compactes. Bull. Sci. Math. 102, 63–95 (1976)

[AW92] Andreatta, M.; Wisniewski, J.: Vector bundles and adjunction. Int. J. Math. 3, 331–340 (1992)

[AW93] Andreatta, M.; Wisniewski, J.: A note on non-vanishing and applications. Duke Math. J. 72, 739–755 (1993)

[AW96] Andreatta, M.; Wisniewski, J.: On good contractions of smooth varieties. Preprint 1996

[Br80] Brenton, L.: On singular complex surfaces with negative canonical bundle. Math. Ann. 248, 117–124 (1980)

[BS76] Banica, C.; Stanasila, O.: Algebraic methods in the global theory of complex spaces. Wiley 1976

[Be77] Beauville, A.: Variétés de Prym et jacobiennes intermédiares. Ann. Sci. Norm. Sup. 10, 309–391 (1977)

[Be78] Beauville, A.: Surfaces algébriques complexes. Astérisque 54, Soc. Math. de France, 1978

[Be83] Beauville, A.: Variétés kählériennes dont la première classe de Chern est nulle. J. Diff. Geom. 18, 755-782 (1983)

[Be85] Benveniste, X.: Sur le cone des 1-cycles effectifs en dimension 3. Math. Ann. 272, 257–265 (1985)

[Bl87] Beltrametti, M.: On d-folds whose canonical bundle is not numerically effective, according to Mori and Kawamata. Ann. Mat. Pura Appl. 147, 151–172 (1987)

[BPV84] Barth, W.; Peters, C; van den Ven, A.: Compact complex surfaces. Erg. d. Math., 3. Folge, Band 4. Springer 1984

[BS95] Beltrametti, M.; Sommese, A.J.: The adjunction theory of complex projective varieties. Expos. in Math. vol. 16, de Gruyter 1995

[Bo91] Borcea, C.: On desingularized Horrocks-Mumford quintics. Crelle's J. 421, 23–41 (1991)

[Ca85] Campana, F.: Réduction d'Albanese d'un morphisme propre et faiblement Kählérien. Comp. Math.54, 373–416 (1985)

[Ca91] Campana, F.: On twistor spaces of the class $\mathcal{C}$. J. Diff. Geom. 33, 541–549 (1991)

[Ca92] Campana, F.: Connexité rationnelle des variétés de Fano. Ann. scient. Ec. Norm. Sup. 25, 539–545 (1992)

[Ca92a] Campana, F.: Un théorème de finitude pour les variétés de Fano suffisamment reglées. Preprint 1992

[Ca93] Campana, F.: Fundamental group and positivity of cotangent bundles of compact Kähler manifolds. J. Alg. Geom. 4, 487–502 (1995)

[Ca94] Campana, F.: Remarques sur le revêtement universel des variétés Kähleriennes compactes. Bull. SMF 122, 255–284 (1994)

[CP91] Campana, F.; Peternell, T.: Algebraicity of the ample cone of projective varieties. J. reine u. angew. Math. 407, 160–166 (1990)

[CP96] Campana, F.; Peternell, T.: Towards a Mori theory on compact Kähler threefolds, I. To appear in Math. Nachrichten 1996

[CKM88] Clemens, H.; Kollár, J.; Mori, S.: Higherdimensional complex geometry. Astérisque 166. Soc. Math. de France, 1988

[De92] Demailly, J.P.: Singular hermitian metrics on positive line bundles. Proc. Alg. Geom. Bayreuth 1990, Lecture Notes in Math. 1507, 87–104 (1992)

[De93] Demailly, J.P.: A numerical criterion for very ample line bundles. J. Diff. Geom. 37, 323–374 (1993)

[De96] Demailly, J.P.: L^2 vanishing theorems for positive line bundles and adjunction theory. In: Transcendental methods in algebraic geometry. CIME session 1994, to appear in Lecture Notes in Math., Springer 1996

[De96a] Demailly, J.P.: Effective bounds for very ample line bundles. Inv. math. 124, 243–261 (1996)

[Dm80] Demazure, M.: Séminaire sur les singularités des surfaces. Lect. Notes in Math. 777, Springer 1980

[DPS93] Demailly, J.P.; Peternell, T.; Schneider, M.: Kähler manifolds with numerically effective Ricci class. Comp. math. 89, 217–240 (1993)

[DPS94a] Demailly, J.P.; Peternell, T.; Schneider, M.: Compact complex manifolds with numerically effective tangent bundles. J. Alg. Geom. 3, 295–345 (1994)

[DPS94b] Demailly, J.P.; Peternell, T.; Schneider, M.: Compact Kähler manifolds with hermitian semi-positive anticanonical bundle. Comp. math. 101 (1996)

[EL93] Ein, L.; Lazarsfeld, R.: Global generation of pluricanonical and adjoint linear series on projective smooth threefolds. J. Amer. Math. Soc. 6, 875–903 (1993)

[EV92] Esnault, H.; Viehweg, E.: Lectures on vanishing theorems. DMV Seminar, vol. 20. Birkhäuser 1992

[Fr80] Francia, P.: Some remarks on minimal models, I. Comp. math. 40, 301–313 (1980)

[Fj83] Fujita, T.: Semipositive line bundles. J. Fac. Sci. Univ. Tokyo 30, 353–378 (1983)

[Fj86] Fujita, T.: Zariski decomposition and canonical rings of elliptic threefolds. J. Math. Soc. Japan 38, 20–37 (1986)

[Fj87] Fujita, T.: On polarized manifolds whose adjoint bundles is not semipositive. Adv. Stud. Pure Math. 10, 167–178 (1987)

[Fj90] Fujita, T.: On adjoint bundles of ample vector bundles. Proc. Complex Alg. Var., Bayreuth 1990; Lecture Notes in Math. 105–112. Springer 1991

[Fj90a] Fujita, T.: Classification theories of polarized varieties. London Math. Soc. Lecture Note Ser. 155. Cambridge Univ. Press 1990

[Fj91] Fujita, T.: On Kodaira energy and adjoint reduction of polarized manifolds. manuscr. math. 76, 59–84 (1991)

[FL83] Fulton, W.; Lazarsfeld, R.: Positive polynomials for ample vector bundles. Ann. Math. 118, 35–60 (1983)

[GPR94] Grauert, H.; Peternell, Th.; Remmert, R.: Several complex variables VII. Enc. of Math. Sciences, vol. 74. Springer 1994

[Gr62] Grauert, H.: Über Modifikationen und exzeptionelle analytische Mengen. Math. Ann. 146, 331–368 (1962)

[GR70] Grauert, H.; Riemenschneider, O.: Verschwindungssätze für analytische Kohomologiegruppen auf komplexen Räumen. Inv. math. 11, 263–292 (1970)

[GH78] Griffiths, Ph.; Harris, J.: Principles of algebraic geometry. Wiley, 1978

[Gr66] Griffiths, Ph.: The extension problem in complex analysis 2. Amer. J. Math. 88, 366–446 (1966)

[Gr69] Griffiths, Ph.: Hermitian differential geometry, Chern classes and positive vector bundles. In: Global Analysis, Princeton Math. Series 29, 185–251 (1969)

[Ha66] Hartshorne, R.: Ample vector bundles. Publ. IHES 29, 319–394 (1966)

[Ha70] Hartshorne, R.: Ample subvarieties of algebraic varieties. Lecture Notes in Math. 156, Springer 1979

[Ha77] Hartshorne, R.: Algebraic geometry. Springer 1977

[Hi75] Hironaka, H.: Flattening theorem in complex analytic geometry. Am. J. Math. 97, 503–547 (1975)

[HW81] Hidaka, F.; Watanabe, K.: Normal Gorenstein surfaces with ample anticanonical divisor. Tokyo J. Math. 4, 319–330 (1981)

[HW92] Heath-Brown, D.R.; Wilson, P.M.H.: Calabi-Yau threefolds with $\rho > 13$. Math. Ann. 294, 49–57 (1992)

[Io88] Ionescu, P.: Generalized adjunction and applications. Math. Proc. Cambridge Phil. Soc. 99, 457–472 (1988)

[Is77] Iskovskih, V.: Fano 3-folds 1, 2. Russian Acad. Sci. Iszv., Math. (= Math. USSR Izv.) 11, 485–527 (1977); ibid. 12, 469–505 (1978)

[Is90] Iskovskih, V.: Double projection from a line on Fano threefolds of the first kind. Russian Acad. Sci., Sbornik, Math. (=USSR Sbornik) 66, 265–284 (1990)

[Kc95] Kachi, Y.: Extremal contractions from 4-dimensional manifolds to 3-folds. Preprint 1995

[Kl66] Kleiman, S.: Toward a numerical theory of ampleness. Ann. Math. 84, 293–344 (1966)

[KM92] Kollár, J.; Mori, S.: Classification of threedimensional flips. Journ. AMS 5, 533–703 (1992)

[KMM87] Kawamata, Y.; Matsuda, K.; Matsuki, K.: Introduction to the minimal model problem. Adv. Stud. Pure Mat. 10, 283–360 (1987)

[KO73] Kobayashi, S.; Ochiai, T.: Characterisations of complex projective spaces and hyperquadrics. J. Math. Kyoto Univ. 13, 31–47 (1973)

[Ko86] Kollár, J.: Higher direct images of dualising sheaves. Ann. Math. 123, 11–42 (1986)

[Ko87] Kollár, J.: Vanishing theorems. Proc. Symp. Pure Math. 46,

[Ko87a] Kollár, J.: The structure of algebraic threefolds – an introduction to Mori's program. Bull. Amer. Math. Soc. 17, 215–277 (1987)

[Ko89] Kollár, J.: Flops. Nagoya Math. J. 113, 14–36 (1989)

[Ko91] Kollár, J.: Flips, flops and minimal models. Surv. in Diff. Geom. 1, 113–199 (1991)

[Ko91a] Kollár, J.: Extremal rays on smooth threefolds. Ann. Sci. Ec. Norm. Sup. 24, 339–361 (1991)

[Ko92] Kollár, J.: Flips and abundance for algebraic threefolds. Astérisque 211. Soc. Math. France 1992

[Ko92a] Kollár, J.: Shafarevich maps and plurigenera of algebraic varieties. Inv. math. 113, 177–215 (1992)

[Ko92b] Kollár, J.: Cone theorems and bug-eyed covers. J. Alg. Geom. 1, 293–323 (1992)

[Ko93] Kollár, J.: Effective base point freeness. Math. Ann. 296, 595–605 (1993)

[KoMiMo92] Kollár, J.; Miyaoka, Y.; Mori, S.: Rationally connected varieties. J. Alg. Geom. 1, 429–448 (1992)

[KoMiMo92a] Kollár, J.; Miyaoka, Y.; Mori, S.: Rational connectedness and boundedness for Fano manifolds. J. Diff. Geom. 36, 765–769 (1992)

[Kw82] Kawamata, Y.: A generalisation of Kodaira-Ramanujan's vanishing theorem. Math. Ann. 261, 43–46 (1982)

[Kw84] Kawamata, Y.: The cone of curves of algebraic varieties. Ann. Math. 119, 603–633 (1984)

[Kw85] Kawamata, Y.: Pluricanonical system on minimal algebraic varieties. Inv. math. 79, 567–588 (1985)

[Kw87] Kawamata, Y.: The Zariski decomposition of log-canonical divisors. Symp. Pure Math. 46, 425–433

[Kw88] Kawamata, Y.: The crepant blowing-ups of 3-dimensional canonical singularities and its application to degenerations of surfaces. Ann. Math. 127, 93–163 (1988)

[Kw89] Kawamata, Y.: Small contractions of four dimensional manifolds. Math. Ann. 284, 595–600 (1989)

[Kw91a] Kawamata, Y.: On the length of an extremal rational curve. Inv. math. 105, 609–611 (1991)

[Kw92] Kawamata, Y.: Abundance theorem for minimal threefolds. Inv. math. 108, 229–246 (1992)

[Kw92a] Kawamata, Y.: Moderate degenerations of algebraic surfaces. Proc. Complex Alg. Var. Bayreuth 1990. Lecture Notes in Math. 1507, 113–132. Springer 1991

[La84] Lazarsfeld, R.: Some applications of the theory of ample vector bundles. Lecture Notes in Math. 1092, 29–61. Springer 1984

[Lm86] Lamotke, K.: Regular solids and isolated singularities. Vieweg 1986

[MM81] Mori, S; Mukai, S.: On Fano 3-folds with $B_2 \geq 2$. Adv. Stud. Pure Math. 1, 101–129 (1981)

[MM81a] Mori, S.; Mukai, S.: Classification of Fano 3-folds with $B_2 \geq 2$. Manuscript math. 36, 147–162 (1981)

[MM82] Mori, S.; Mukai, S.: The uniruledness of the moduli space of curves of genus 11. Lecture Notes in Math. 1016, 334–353 (1982)

[MM86] Miyaoka, Y.; Mori, S.: A numerical criterion of uniruledness. Ann. Math. 124, 65–69 (1986)

[Mi83] Miyanishi, M.: Algebraic threefolds. Adv. Stud. Pure Math. 1, 69–99 (1983)

[Mi88] Miyaoka, Y.: Abundance conjecture for threefolds: $\nu = 1$ case. Comp. math. 68, 203–220 (1988)

[Mo79] Mori, S.: Projective manifolds with ample tangent bundles. Ann. Math. 110, 593–606 (1979)

[Mo82] Mori, S.: Threefolds whose canonical bundles are not numerically effective. Ann. Math. 116, 133–176 (1982)

[Mo85] Mori, S.: Classification of higher dimensional varieties. Proc. Symp. Pure Math. 46, 269–331 (1987)

[Mo88] Mori, S.: Flip theorem and the existence of minimal models for 3-folds. J. Amer. Math. Soc. 1, 117–253 (1988)

[MS77] Mori, S.; Sumihiro, S.: On Hartshorne's conjecture. J. Math. Kyoto Univ. 18, 523–533 (1977)

[Mu81] Murre, J.: Classification of Fano threefolds according to Fano and Iskov-skih. Lecture Notes in Math. 947, 35–92 (1981)

[Mu89] Mukai, S.: New classification of Fano threefolds and Fano manifolds of coindex 3. Proc. Nat. Acad. Sci. USA 86, 3000–3002 (1989)

[Mu92] Mukai, S.: Fano 3-folds. In: London Math. Soc. Lect. Notes Ser. 179, 255–263. Cambridge Univ. Press. 1992

[Mw86] Moriwaki, A.: Semi-ampleness of the numerically effective part of Zariski decomposition. J. Math. Soc. Japan 26, 465–481 (1986)

[Na89] Nadel, A.: Multiplier ideal sheaves and Kähler-Einstein metrics of positive scalar curvature. Proc. Natl. Acad. Sci. USA 86, 7299–7300 and Ann. Math. 132, 549–596 (1989)

[Na91] Nadel, A.: The boundedness of degree of Fano varieties with Picard number 1. J. Amer. Math. Soc. 4, 681–692 (1991)

[Nk71] Nakano, S.: On the inverse of monoidal transformations. Publ. RIMS 6, 483–502 (1971)

[Og93] Oguiso, K.: On algebraic fiber space structures on a Calabi-Yau threefold. Intl. J. Math. 4, 439–465 (1993)

[Og96] Oguiso, K.: On the complete classification of Calabi-Yau threefolds of type III_0. Higher Dimensional Complex Varieties, Proceedings Trento 1994, ed. M. Andreatta, T. Peternell; 330–339 (1996)

[OP95] Oguiso, K.; Peternell, T.: On Calabi-Yau threefolds with positive second Chern class. preprint 1995

[Pe89] Peternell, T.: Singular surfaces and the rigidity of $\mathbb{P}_3$. manuscr. math. 63, 69–82 (1989)

[Pe90] Peternell, T.: A characterisation of $\mathbb{P}_n$ by vector bundles. Math. Z. 205, 487–490 (1990)

[Pe91] Peternell, T.: Ample vector bundles on Fano manifolds. Intl. J. Math. 2, 311–322 (1991)

[Pe91a] Peternell, T.: Calabi-Yau manifolds and a conjecture of Kobayashi Math. Z. 207, 305–318 (1991)

[Pe96] Peternell, T.: Towards a Mori theory on compact Kähler threefolds, II. Preprint 1996

[Pe96a] Peternell, T.: Kähler manifolds with semipositive curvature. In: Transcendental methods in algebraic geometry, CIME session 1994, to appear in Lecture Notes in Math., Springer 1996

[PSW92] Peternell, T.; Szurek, M.; Wisniewski, J.: Fano manifolds and vector bundles. Mathem. Ann. 294, 151–165 (1992)

[Re80] Reid, M.: Canonical 3-folds. Géometrie Algébrique Angers, ed. A. Beauville, 273–310, Sijthoff and Noordhoff 1980

[Re83] Reid, M.: Minimal models of canonical threefolds. Adv. Stud. Pure Math. 1, 131–180 (1983)

[Re87] Reid, M.: Young person's guide to canonical singularities. Proc. Symp. Pure Math. 46, 345–414 (1987)

[Ri88] Reider, I.: Vector bundles of rank 2 and linear systems on algebraic surfaces. Ann. Math. 127, 309–316 (1988)

[RRV71] Ramis, J.; Ruget, G.; Verdier, J.L.: Dualité rélative en géometrie analytique complexe. Inv. math. 13, 261–283 (1971)

[Se95] Serrano, F.: Strictly nef divisors and Fano threefolds. J. reine u. angew. Math. 464, 187–206 (1995)

[Sh86] Shokurov, V.V.: The non-vanishing theorem. Russian Math. Soc. Izv., Math. (= Math. USSR Izv.) 26, 591–604 (1986)

[Sh91] Shokurov, V.V.: 3-fold log flips. Izv. A. N. Ser. Mat. 56, 105–203 (1991), english transl. 1993

[Si93] Siu, Y.T.: An effective big Matsusaka theorem. Ann: Inst. Fourier 43, 1199–1209 (1993)

[Si96] Siu, Y.T.: Effective very ampleness. Inv. math. 124, 563–571 (1996)

[SS85] Shiffman, B.; Sommese, A.J.: Vanishing theorems on complex manifolds. Progr. in Math. vol. 56, Birkhäuser 1985

[SY80] Siu, Y.T.; Yau, S.T.: Compact Kähler manifolds with positive bisectional curvature. Inv. math. 59, 189–204 (1980)

[Ue75] Ueno, K.: Classification theory of algebraic varieties and compact complex spaces. Lecture Notes in Math. 439. Springer 1975

[Vi82] Viehweg, E.: Vanishing theorems. J. reine u. angew. Math. 335, 1–8 (1982)

[We80] Wells, R.O.: Differential analysis on complex manifolds. Graduate texts in Math. vol. 65, Springer 1980

[Wi87] Wilson, P.M.H.: Towards birational classification of algebraic varieties. Bull. London Math. Soc. 19, 1–48 (1987)

[Wi89] Wilson, P.M.H.: Calabi-Yau threefolds with large Picard number. Inv. math. 98, 139–155 (1989)

[Wi92] Wilson, P.M.H.: The Kähler cone on Calabi-Yau threefolds. Inv. math. 107, 561–583 (1992) and 114, 231–233 (1993)

[Wi94] Wilson, P.M.H.: The existence of elliptic fiber space structures on Calabi-Yau threefolds. Preprint 1994

[Wi94a] Wilson, P.M.H.: Minimal models of Calabi-Yau threefolds. Cont. Math. 162, 403–410 (1992)

[Wi95] Wilson, P.M.H.: Symplectic deformations of Calabi-Yau threefolds. Pre-
        print 1995

[Ws91] Wisniewski, J.: On contractions of extremal rays on Fano manifolds. J.
        reine u. angew. Math. 417, 141–157 (1991)

[Ws91a] Wisniewski, J.: On Fano manifolds of large index. manuscr. math. 70,
        145–152 (1991)

[Ws94] Wisniewski, J.: A report on Fano manifolds of middle index and $b_2 \geq 2$.
        In: Lecture Notes in pure and appl. Math. 166, 19–26. M. Dekker 1994

[Ya77] Yau, S.T.: Calabi's conjecture and some new results in algebraic geometry.
        Proc. Natl. Acad. Sci. USA 74, 1789–1799 (1977)

[YZ90] Ye, A.; Zhang, Q.: On ample vector bundles whose adjoint bundles are
        not numerically effective. Duke Math. J. 60, 671–687 (1990)

[Zh91] Zhang, Q.: Extremal rays in higher dimensional projective varieties. Math.
        Ann. 291, 497–504 (1991)

Index

DMV Seminar

Workshops, edited by the German Mathematics Society

The workshops organized by the Gesellschaft für mathematische Forschung in co-operation with the Deutsche Mathematiker-Vereinigung (German Mathematics Society) are primarily intended to introduce students and young mathematicians to current fields of research. By means of these well-organized seminars, scientists from other fields will also be introduced to new mathematical ideas. The publication of these workshops proceedings in the DMV-Seminar series will make the material available to an ever larger audience.